# Agricultural Statistics at a Glance 2008

सत्यमेव जयते

# Agricultural Statistics at a Glance 2008

Directorate of Economics & Statistics
Department of Agriculture & Cooperation
Ministry of Agriculture
Government of India

ACADEMIC FOUNDATION
NEW DELHI

*First Published in 2008 by*
**Academic Foundation**
4772-73 / 23 Bharat Ram Road, (23 Ansari Road), Darya Ganj, New Delhi - 110 002 (India).
Phones : 23245001 / 02 / 03 / 04. Fax : +91-11-23245005. E-mail : academic@vsnl.com
www : academicfoundation.com

*Published in association with :*
**Directorate of Economics & Statistics,**
**Department of Agriculture and Cooperation,**
**Ministry of Agriculture,**
**Government of India,**
**New Delhi.**



*Agricultural Statistics at a Glance 2008*
ISBN 10: 8171887171
ISBN 13: 9788171887170

Typeset by Italics India, New Delhi
Printed and bound in India.

# CONTENTS

## Appendix

**पी. के. मिश्र, पी एच डी,**
आई०ए०एस
सचिव, भारत सरकार
**P.K. Mishra,** PhD
Secretary IAS
Government of India

सत्यमेव जयते

**भारत सरकार**
कृषि मंत्रालय
कृषि एवं सहकारिता विभाग
Government of India
Ministry of Agriculture
Department of Agriculture & Cooperation

# Message

The significance of database for agriculture and allied activities lies in the value addition it makes to research and policy formulation. For ensuring public access to data generated through the untiring effort of various state and central agencies, an appropriate dissemination mechanism is of utmost importance. It is in this context that I felt it necessary to ensure wider circulation of the *Agricultural Statistics at a Glance* by getting it published as a priced document, through a reputed publisher.

I congratulate the team headed by Shri R.C. Ray, Economic and Statistical Adviser, for bringing out *Agricultural Statistics at a Glance 2008* as a priced publication in a record time. I am sure, this would be of enormous value to those engaged in academic research, policy formulation and implementation.

Kmishra
(P.K. Mishra)

Office: Krishi Bhawan, New Delhi - 110001, दूरभाष/Phone: 23382651, 23388444, फैक्स स०/Fax No.: 23386004
E-mail: secyagri.krishi@nic.in

सत्यमेव जयते

**Dr. S.M. Jharwal**
Principal Adviser

**कृषि एवं सहकारिता विभाग**
कृषि मंत्रालय
भारत सरकार
कृषि भवन
नई दिल्ली–110 001

**Department of Agriculture & Cooperation**
Ministry of Agriculture
Government of India
Krishi Bhavan
New Delhi-110 001
Telefax: 011-23382719
Email: *paa-moa@nic.in*

# Foreword

Agricultural sector in India, after having experienced an average growth rate of little more than 2 per cent for about 10 years, has clearly shown signs of revival. The average growth rate during the last three years has been 4.7 per cent. The production of foodgrains has also increased at an average growth rate of 5.2 per cent, during these three years. Such a revival is the result of a well worked out investment strategy and programmes, favourable weather conditions, close monitoring, and also quick response by the states and the farmers. However, to achieve an average growth rate of 4.1 per cent during the 11th Plan Five Year Plan period, this needs to be sustained.

Strengthening statistical inputs and disseminating them is of utmost importance for better planning and implementation. The "Agricultural Statistics at a Glance" is a much awaited publication by the persons who are active in the fields of policy, planning, research and academics, since this provides authentic information in a comprehensive manner and well in time.

Keeping in view the utility of vast pool of data, it was felt that the publication deserves professional and commercial handling to widen its reach through increased circulation. In fact, Shri P.K. Mishra, Secretary, Agriculture and Cooperation, stressed on the need for making it a priced publication. This was basically to enable everyone interested to buy this "Avatar" of *Agricultural Statistics at a Glance 2008* from the nearest store. In addition, the data would be placed on the official website. I am sure that this publication would be useful for policy formulation and further research in the agriculture sector by various stakeholders.

(Dr. S.M. Jharwal)

सत्यमेव जयते

**R.C. Ray**
Economic & Statistical Adviser

**आर्थिक एवं संख्यिकी निदेशालय**
**कृषि एवं सहकारिता विभाग**
कृषि मंत्रालय, भारत सरकार
कृषि भवन, नई दिल्ली–110 001

Directorate of Economics & Statistics
Department of Agriculture & Cooperation
Ministry of Agriculture, Government of India
Krishi Bhavan, New Delhi - 110 001
Tele: 011-23384036, 23382016
E-mail: *chandra.ray@nic.in*

# Preface

Agricultural sector has a vital place in the economic development of the country. Though the share of agriculture in both GDP and employment has declined over time, the pace of decline in its share in employment has been much slower than that in GDP. The share of agriculture, including the allied sectors of forestry and fishing, in GDP declined from 37 per cent in 1983 to 25 per cent in 1999-2000 and to around 18 per cent in 2007-08.

Agricultural Statistics at a Glance 2008 provides wide range of data on diverse aspects of Indian agriculture that are greatly useful for economists, policy makers, researchers, agricultural scientists, students and the public at large in the country. It will also meet the needs of different government and non-governmental organisations working in the agricultural sector. The book contains data on crop production and productivity across States/regions, markets and prices, terms of trade, price support and procurement, domestic and international trade, credit, insurance, etc.

It must be admitted that in the timely publication of this priced publication lies the keen interest evinced by the Secretary (Agriculture & Cooperation). I also express my gratitude to Dr. S.M. Jharwal, Principal Adviser, Department of Agriculture and Cooperation, without whose cooperation and guidance it would not have been possible to bring out this issue.

The Compilation Unit has endeavoured earnestly to prepare this issue under the guidance of Shri M.N. Lothe, Adviser, Shri P.S. Meena and Shri P.C. Bodh, Additional Economic Advisers. The timely and prompt cooperation extended by all the concerned departments and organisations is well acknowledged. I must also put on record the support given by IFFCO to have the publication in record time.

Any constructive suggestion(s) to improve the utility of this publication will be highly appreciated. Visit us at: *http://agricoop.nic.in*; *http://dacnet.nic.in/eands*

(R.C. Ray)

## Officials of the Directorate of Economics & Statistics, New Delhi associated in the preparation of this publication

**Compilation Section:**

| | |
|---|---|
| Shri I.K. Aggarwal | Economic Officer |
| Shri M.A. Raza | Statistical Officer |

**Computerised by:**

| | |
|---|---|
| Shri R.S. Pandey | DEO Grade–A |

**Under the Supervision of:**

| | |
|---|---|
| Shri M.N. Lothe | Adviser |
| Shri P.S. Meena | Additional Economic Adviser |
| Shri Nagender Jatav | Research Officer |

## Acronyms

| | | |
|---|---|---|
| AAY | – | Antyodaya Anna Yojana |
| AE | – | Advance Estimate |
| AGDP | – | Gross Domestic Product from Agriculture |
| Ant. | – | Anticipated |
| APL | – | Above Poverty Line |
| A & N | – | Andaman & Nicobar |
| BE | – | Budget Estimates |
| BPL | – | Below Poverty Line |
| CACP | – | Commission for Agricultural Costs and Prices. |
| CBB | – | Commercial Bank Branch |
| CCE | – | Crop Cutting Experiments |
| CCI | – | Cotton Corporation of India |
| CD | – | Community Development |
| CDSBO | – | Crude Degumed Soyabean Oil |
| CES | – | Crop Estimation Surveys |
| CFA | – | Central Financial Agency |
| CIF | – | Cost, Insurance & Freight |
| CIP | – | Central Issue Price |
| CPI-IW | – | Consumer Price Index for Industrial Workers |
| CWWG | – | Crop Weather Watch Group |
| DAC | – | Department of Agriculture & Cooperation |
| DAP | – | Di Ammonium Phosphate |
| DES | – | Directorate of Economics and Statistics |
| E | – | Estimated |

| | | |
|---|---|---|
| EARAS | – | Establishment of an Agency for Reporting of Agricultural Statistics |
| FAO | – | Food and Agriculture Organisation |
| FAQ | – | Fair Average Quality |
| F2 | – | Black Soil |
| F-414/H-777/J-34 | – | Cotton variety (of Medium Staple Length) |
| H-4/H-6 | – | Cotton variety (of Long Staple Length) |
| FCI | – | Food Corporation of India |
| FDI | – | Foreign Direct Investment |
| FFB | – | Fresh Fruit Bunch |
| FOD | – | Field Operations Division of NSSO |
| FRL | – | Full Reservoir Level |
| FYM | – | Farm Yard Manure |
| GCA | – | Gross Cropped Area |
| GCES | – | General Crop Estimation Survey |
| GCF | – | Gross Capital Formation |
| GDP | – | Gross Domestic Product |
| GNP | – | Gross National Product |
| GSDP | – | Gross State Domestic Product |
| Ha | – | Hectares |
| HYV | – | High Yielding Varieties |
| I | – | Irrigated |
| IAS | – | Improvement of Agricultural Statistics |
| ICOR | – | Incremental Capital Output Ratio |
| ICS | – | Improvement of Crop Statistics |
| K | – | Potash |
| Kgs | – | Kilograms |
| KMPH | – | Kilometers Per Hour |
| KWH | – | Kilo Watt Hour |
| LS | – | Light Soil |

| | | |
|---|---|---|
| LPCPD | – | Litres per Capita Per Day |
| MSCCGMF | – | Maharashtra State Co-operative Cotton Growers Marketing Federation |
| MIS | – | Market Intervention Scheme |
| MSP | – | Minimum Support Price |
| MSR | – | Marketed Surplus Ratio |
| N | – | Nitrogen |
| NA | – | Not Available/Not Announced |
| NC | – | Not Collected |
| NQ | – | Not Quoted |
| NAFED | – | National Agricultural Co-operative Marketing Federation of India Limited |
| NCAER | – | National Council of Applied Economic Research |
| NDDB | – | National Dairy Development Board |
| Neg. | – | Negligible |
| NNP | – | Net National Product |
| NR | – | Not Recommended/Not Reported |
| NSA | – | Net Sown Area |
| NSSO | – | National Sample Survey Organisation |
| OECD | – | Organisation of Economic Cooperation and Development |
| OPDP | – | Oil Palm Development Programme |
| P | – | Provisional/Phosphate |
| PACS | – | Primary Agricultural Credit Society |
| Pd. | – | Per day |
| PDS | – | Public Distribution System |
| PLDB | – | Primary Land Development Bank |
| PSE | – | Producers Support Estimate |
| PSEs | – | Public Sector Enterprises |
| PSS | – | Price Support Scheme |

| | | |
|---|---|---|
| QE | – | Quick Estimate |
| R | – | Revised |
| R&D | – | Research & Development |
| R&M | – | Rapeseed & Mustard |
| RMC | – | Regional Ministers Conference |
| RE | – | Revised Estimates |
| RRBB | – | Regional Rural Bank Branch |
| SCB | – | State Cooperative Bank |
| SDP | – | State Domestic Product |
| SASA | – | State Agricultural Statistical Authority |
| SLDB | – | State Land Development Bank |
| SRS | – | Sample Registration Scheme |
| SSP | – | Single Super Phosphate |
| T | – | Target |
| T E | – | Triennium Ending |
| TD-5 | – | A Variety of Jute |
| TPDS | – | Targeted Public Distribution System |
| TRQ | – | Tariff Rate Quota |
| TRRM | – | Tropical Rains Measuring Mission |
| TT | – | Tentative Targets |
| U.K. | – | United Kingdom |
| USA | – | United States of America |
| UI | – | Unirrigated |
| UT | – | Union Territory |
| WPI | – | Wholesale Price Index |
| WTO | – | World Trade Organisation |
| Y | – | Yield (Index Number of Yield) per hectare |

# Conversion Table

**Length**

| | | |
|---|---|---|
| 1 kilometre (km) | = | 1000 metres (m) |
| 1 km | = | 0.6214 miles |
| 1 m | = | 1.0936 yards |
| 1 m | = | 3.2808 feet |
| 1 mile | = | 1760 yards |
| 1 mile | = | 1.609 km |
| 1 yard | = | 0.9144 m |
| 1 foot | = | 0.3048 m |

**Area**

| | | |
|---|---|---|
| 1 $km^2$ | = | 100 hectares (ha) |
| 1 $km^2$ | = | 0.3861 square mile |
| 1 $km^2$ | = | 247.105 acres |
| 1$m^2$ | = | 10.7639 square feet |
| 1 ha | = | 10,000 $m^2$ |
| 1 ha | = | 2.4711 acres |
| 1 square mile | = | 2.591 $km^2$ |
| 1 acre | = | 0.4047 ha |
| 1 acre | = | 4046.86 $m^2$ |
| 1 acre | = | 4840 square yard |
| 1 square yard | = | 9 square feet |
| 1 square yard | = | 0.8361 $m^2$ |
| 1 square foot | = | 0.0929 $m^2$ |

**Weight**

| | | |
|---|---|---|
| 1 tonne | = | 1000 kg |
| 1 tonne | = | 1.1023 US ton |
| 1 US ton | = | 0.9072 tonnes |
| 1 kg | = | 2.2046 pounds (lb) |
| 1 kg | = | 35.274 ounce (oz) |
| 1 lb | = | 0.4536 kg |
| 1 oz | = | 28.3495 gram |

**Units**

| | | |
|---|---|---|
| 1 crore | = | 10 million |
| 1 million | = | 10 lakh |
| 1 lakh | = | 100,000 |
| 1 billion | = | 1000 million |

# 1

# Methodology of Crop Estimation

## 1.1 Introduction

The Directorate of Economics and Statistics (DES) releases estimates of area, production and yield in respect of principal crops of foodgrains, oilseeds, sugarcane, fibers and important commercial and horticulture crops. These crops together account for nearly 87% of agriculture output. The estimates of crop production are obtained by multiplication of area estimates by corresponding yield estimates. The estimates of area and yield rates assume prime importance in the entire gamut of agricultural statistics.

## 1.2 Area Statistics

From the point of view of collection of area statistics, the States in the country are divided into three broad categories:

i. The first category covers States and UTs which have been cadastrally surveyed and where area and land use statistics are built up as a part of the land records maintained by the revenue agencies (referred to as "Land Record States" or temporarily settled states). The system of land records is followed in 17 major states of Andhra Pradesh, Assam (excluding hilly districts), Bihar, Chhattisgarh, Gujarat, Haryana, Himachal Pradesh, Jammu & Kashmir, Jharkhand, Karnataka, Madhya Pradesh, Maharashtra, Punjab, Rajasthan, Tamil Nadu, Uttar Pradesh and Uttarakhand and 4 UTs of Chandigarh, Delhi, Dadra & Nagar Haveli and Pondicherry. These States/UTs account for about 86% of reporting area and are covered under Timely Reporting Scheme (TRS) under which 20% villages are selected at random for complete area remuneration.

ii. The second category covers States where area statistics are collected on the basis of sample surveys. A scheme for Establishment of an Agency for Reporting of Agricultural Statistics (EARAS) has been introduced in these states, *viz.*, Kerala, Orissa and West Bengal, and later extended to Arunachal Pradesh, Nagaland, Sikkim and Tripura. The scheme envisages, inter-alia, estimation of area through sample surveys in a sufficiently large sample of 20% villages/investigator zones. These states account for about 9% of reporting area.

iii. In the hilly districts of Assam, rest of the states in North-Eastern Region (Other than Arunachal Pradesh, Nagaland, Tripura and Sikkim), Goa, UTs of Andman & Nicobar Islands, Daman & Diu and Lakshwadeep where no reporting agency had been functioning, the work of collection of Agricultural Statistics is entrusted with the village headmen. The area statistics in these states are based on impressionistic approach. These areas/states account for 5% of the reporting area.

## 1.3 Yield Estimates

The second most important component of production statistics is yield rates. The yield estimates of major crops are obtained through analysis of Crop Cutting Experiments (CCE) conducted under scientifically designed General Crop Estimation Surveys (GCES). At present over 95% of the production of foodgrains is estimated on the basis of yield rates obtained from the CCEs. Field Operations Divisions (FOD) of the National Sample Survey Organization (NSSO) has been providing technical guidance to the States and Union territories for organising and conducting Crop Estimation Surveys for estimating yield rates of principal crops. In addition, NSSO in collaboration with States/Union Territories implements sample check programmes on area enumeration work, area aggregation and conduct of crop cutting experiments under the Scheme for Improvement of Crop Statistics (ICS). While executing the programme of sample checks on crop cutting experiments, the FOD associates itself with the operational aspects of the conduct of crop cutting experiments right from selection of sample villages, training of field staff to the supervision of field work, and in the process, gathers micro level information relating to conduct of crop cutting experiments for estimation of crop yield. The results of Crop Estimation Surveys are analysed and annual publication entitled "Consolidated Results of Crop Estimation Surveys on Principal Crops" is brought out by the NSSO regularly.

The primary objective of GCES is to obtain fairly reliable estimates of average yield of principal food and non-food crops for States and UTs which are important from the point of view of crop production. The estimates of yield rates thus arrived at are generally adopted for the purposes of planning, policy formulation and implementation. The CCEs consist of identification and marking of experimental plots of a specified size and shape in a selected field on the principle of random sampling, harvesting and threshing the produce and recording of the harvested produce for determining the percentage recovery of dry grains or the marketable form of the produce.

## 1.4 Coverage

The crop-wise details of number of experiments planned under GCES during 2004-05 are given in Table 1.1:

**Table 1.1**

*No. of Experiments Planned under GCES during 2004-05*

| *CropsType/Crop* | *No.of Experiments Planned* | | |
|---|---|---|---|
| *1. Food Crops* | *Kharif* | *Rabi* | *Total* |
| Paddy | 188271 | 33308 | 221579 |
| Wheat | 0 | 148501 | 148501 |
| *Jowar* | 23734 | 12472 | 36206 |
| *Bajra* | 29107 | 858 | 29965 |
| Maize | 38317 | 2360 | 40677 |
| Ragi | 7127 | 0 | 7127 |
| Barley | 0 | 3438 | 3438 |
| Gram | 0 | 27647 | 27647 |
| Sugarcane | 21333 | 0 | 21333 |
| **Total** | **307889** | **228584** | **536473** |

*contd. ...*

*...contd. ...*

| *CropsType/Crop* | *No.of Experiments Planned* | | |
|---|---|---|---|
| *2. Non Food Crops* | *Kharif* | *Rabi* | *Total* |
| Groundnut | 21340 | 5527 | 26867 |
| Sesamum | 11051 | 5061 | 16112 |
| Castor | 1940 | 0 | 1940 |
| Rapeseed & Mustard | 0 | 31230 | 31230 |
| Linseed | 0 | 5747 | 5747 |
| Cotton | 19486 | 0 | 19486 |
| Jute | 4313 | 0 | 4313 |
| Mesta | 761 | 0 | 761 |
| **Total** | **58891** | **47565** | **106456** |
| **Total (1+2)** | **366780** | **276149** | **642929** |
| Other Food Crops | 92852 | 57061 | 149913 |
| Other Non-Food Crops | 14895 | 8578 | 23473 |
| **Total All Crops** | **474527** | **341788** | **816315** |
| **Total Number of Experiments planned under GCES for 2004-05** | | | **816315** |

*Source*: FOD, NSSO.

## 1.5 Sampling Design

Stratified multi-stage random sampling design is generally adopted for carrying out GCES with *tehsils/taluks*/revenue inspector circles/CD blocks/*anchals* etc. as strata, revenue villages within a stratum as first stage unit of sampling, survey numbers/fields within each selected village as sampling unit at the second stage and experimental plot of a specified shape and size as the ultimate unit of sampling as depicted in Figure 1.1.

In each selected primary unit, generally 2 survey numbers/fields growing the experimental crop are selected for conducting CCE. Generally, 80-120 experiments are selected in a major crop growing district, where a district is considered as major for a given crop if the area under the crop in the district exceeds 80,000 hectares or lies between 40,000 and 80,000 hectares but exceeds the average area per district in the State. Otherwise, district is considered a minor for a given crop. Experiments in minor districts are so adjusted that the precision of the estimates is fairly high and the workload on the field staff is manageable. On an average, about 44 or 46 experiments are planned in a minor district. The number of experiments allotted to a district is distributed among the strata within the district roughly in proportion to the area under the crop in the stratum.

**Figure 1.1**

*Sampling Design for GCES*

*Tehsil/Taluk*

↓

Revenue Village

↓

Survey Number/Field

↓

Experimental Plot
(Specified size/shape)

## 1.6 Advance Estimates of Area and Production

The period of an agricultural crop year is from July to June, during which various farm operations from preparation of seed bed, nursery, sowing, transplanting various inter-culture operations, harvesting, threshing etc. are carried out. Different crops are grown during the agricultural seasons in the crop year.

Final estimates of production based on complete enumeration of area and yield through crop cutting experiments become available much after the crops are actually harvested. However, the Government requires advance estimates of production for taking various policy decisions relating to pricing, marketing, export/import, distribution, etc. Considering the genuine requirement of crop estimates much before the crops are harvested for various policy purposes, a time schedule of releasing the advance estimates has been evolved. These estimates of crops are prepared and released at four points of time during a year as enumerated below:

## 1.7 First Advance Estimates

The first official estimate of area and production of *kharif* crops is prepared in September every year, when south-west monsoon season is about to be over and *kharif* crops are at an advanced stage of maturity. This coincides with the holding of the National Conference of Agriculture for Rabi Campaign, where the State Governments give rough assessment of their respective *kharif* crops. The assessment, is made by the State Governments based on the reports from the field offices of the State Department of Agriculture. They are mainly guided by visual observations. These are validated on the basis of inputs from the proceedings of Crop Weather Watch Group (CWWG) meetings, and other feedback such as relevant availability of water in major reservoirs, availability/supply of important inputs including credit to farmers, rainfall, temperature, irrigation etc.

## 1.8 Second Advance Estimates

The second advance estimate is made in the month of January every year when the advance estimates of *kharif* crops prepared during the National Conference of Agriculture for *rabi* Campaign may undergo a revision in the light of flow of more precise information from the State Governments. Around this time, the first advance estimates of *rabi* crops are also prepared. The Second Advance Estimates then cover the second assessment in respect of *kharif* Crops and the first assessment in respect of *rabi* Crops.

## 1.9 Third Advance Estimates

The third advance estimates are prepared towards the end of March/beginning of April every year, when the National Conference on Agriculture for Kharif campaign is convened and the State Governments come up with their assessments for both *kharif* and *rabi* crops. The earlier advance estimates of both *kharif* and *rabi* seasons are firmed up/validated with the information available with State Agricultural Statistical Authorities (SASAs), remote sensing data, available with Space Application Centre, Ahmedabad as well as the proceedings of CWWG.

## 1.10 Fourth Advance Estimates

The fourth advance estimates are prepared in the month of June/July every year, when the National Workshop on Improvement of Agricultural Statistics is held. Since most of the *rabi* crops get harvested by the end of May, SASAs are in a position to supply the estimates of both *kharif* and *rabi* seasons as well as likely assessment of summer crops during the National Workshop. Like the third advance estimates, the fourth advance estimates are duly validated with the information available from other sources.

## 1.11 Final Estimates

Under the existing system of crop estimation, the fourth advance estimate is followed by final estimates in December/January of the following agricultural year. The main factors contributing to the relatively large number of crop estimates are the large variations in crop seasons across the country and the resulting delay in the compilation of yield estimates based on crop cutting experiments. As agriculture is a State subject, Central Government depends upon State Governments for accuracy of these estimates. For this purpose, State Governments have setup High Level Coordination Committees (HLCC) comprising, inter-alia, senior officers from their departments of Agriculture, Economics & Statistics, Land Records and NSSO (FOD), IASRI, DES from Central Government for sorting out problems in preparation of these estimates in a timely and orderly manner.

# Table 2.1

## *Key Economic Indicators of India*

| *Indicator* | *Unit* | *2003-04* | *2004-05* | *2005-06* | *2006-07* | *2007-08* |
|---|---|---|---|---|---|---|
| **GDP at Factor Cost** | | | | | | |
| at Current Prices | Rs. Billion | 25382 | 28777 | 32757 | 37901[1] | |
| at 1999-2000 prices | Rs. Billion | 22258 | 23884 | 26128 | 28643[1] | |
| **GDP in Agcultural & Allied Sectors** | | | | | | |
| at Current Prices | Rs. Billion | 5323 | 5524 | 6158 | 6954[1] | |
| at 1999-2000 prices | Rs. Billion | 4827 | 4824 | 5110 | 5302[1] | |
| **Population (as on 1st October)** | Million Number | 1072 | 1089 | 1106 | 1122 | 1138 |
| **Per Capita National Income (Net National Product at Factor Cost)** | | | | | | |
| at 1999-2000 prices | Rs. | 18317 | 19325 | 20358 | 22533[1] | |
| **Foodgrains Production** | Million Tonnes | 213.20 | 198.4 | 208.6 | 217.3 | 230.7[2] |
| **Inflation Rate** | | | | | | |
| WPI based (1993-94=100) | % | 5.5 | 6.5 | 4.4 | 5.4 | 4.7 |
| CPI-IW based (1982=100) | % | 3.9 | 3.8 | 4.4 | 6.7 | 6.2 |
| WPI for Agricultural commodities | | 182.9 | 186.7 | 190.7 | 204.1 | 218.1[3] |
| WPI for Non-Agricultural commodities | | 174.0 | 187.5 | 196.9 | 206.8 | 212.7[3] |
| **Foreign Trade** | | | | | | |
| Export | US$ billion | 63.8 | 83.5 | 103.1 | 126.4 | 111.0 |
| Import | US$ billion | 78.2 | 111.5 | 149.2 | 185.7 | 168.8 |
| Balance | US$ billion | -14.4 | -28.0 | -46.1 | -59.3 | -57.8 |
| Total Agricultural Imports | Rs. Crore | 21972.7 | 22811.8 | 21499.2 | 29637.9 | 29777.0 |
| % to National Imports | | 6.1 | 4.6 | 3.3 | 3.5 | 3.1 |
| Total Agricultural Exports | Rs. Crore | 37266.5 | 41602.7 | 49216.0 | 62411.4 | 77769.7 |
| % to National Exports | | 12.7 | 11.1 | 10.8 | 10.9 | 12.2 |
| **Total Long-Term Debt (End March)** | US$ billion | 107.21 | 115.25 | 118.56 | 143.39 | 159.73[5] |
| **Forex Reserves (End March)** | US$ billion | 107.4 | 135.60 | 145.1 | 191.9 | 281.2[6] |
| **Exchange Rate (at end of year)** | Rs./US $ | 45.95 | 44.93 | 44.27 | 42.25 | 39.44[3] |

*Notes*: 1 Quick Estimates 2 Fourth Advance Estimates 3 Average April-November 2007
4 End November 2007 5 Upto September, 2007

*Source*: (i) Central Statistical Organisation.
(ii) Directorate of Economics & Statistics, Ministry of Agriculture.
(iii) Economic Survey 2007-08.
(iv) Office of the Economic Adviser, Ministry of Commerce & Industry.

# Table 2.2

*Population and Agricultural Workers*

*(In Millions)*

| *Year* | *Total Population* | *Average Annual Exponential Growth Rate(%)* | *Rural Population* | *Agricultural Workers* | | |
|---|---|---|---|---|---|---|
| | | | | *Cultivators* | *Agricultural Labourers* | *Total* |
| (1) | (2) | (3) | (4) | (5) | (6) | (7) |
| 1951 | 361.1 | 1.25 | 298.6 (82.7) | 69.9 (71.9) | 27.3 (28.1) | 97.2 (100.0) |
| 1961 | 439.2 | 1.96 | 360.3 (82.0) | 99.6 (76.0) | 31.5 (24.0) | 131.1 (100.0) |
| 1971 | 548.2 | 2.22 | 439.0 (80.1) | 78.2 (62.2) | 47.5 (37.8) | 125.7 (100.0) |
| 1981 | 683.3 | 2.20 | 523.9 (76.7) | 92.5 (62.5) | 55.5 (37.5) | 148.0 (100.0) |
| 1991 | 846.4 | 2.14 | 628.9 (74.3) | 110.7 (59.7) | 74.6 (40.3) | 185.3 (100.0) |
| 2001 | 1028.7 | 1.95 | 742.6 (72.2) | 127.3 (54.4) | 106.8 (45.6) | 234.1 (100.0) |

*Notes*:

1. For 2001, figures include estimated figures for those of the three sub-divisions *viz.* Mao Maram, Paomata and Purul of Senapati district of Manipur as census results of 2001 Census in these three sub-divisions were cancelled due to technical and administrative reasons.
2. The 1991 Census could not be held owing to disturbed conditions prevailing in Jammu & Kashmir. Hence the population figures for 1991 of Jammu & Kashmir have been worked out by 'interpolation'. The data on workers in columns 5-7 exclude J&K.
3. The 1981 census could not be held in Assam. The figures for 1981 for Assam have been worked out by interpolation. The data on workers in col. 5-7 exclude Assam.
4. Figures within parentheses in column 4 are percentages to the total population.
5. Figures within parentheses in columns 5 and 6 are percentages to column 7.

*Source*: Registrar General of India, New Delhi.

# Table 2.3

## *Population, Growth Rate—2001*

| Sl. No. | India/States/ Union Territories | Population 2001 | | | Per cent Decadal Growth | |
|---|---|---|---|---|---|---|
| | | Persons | Males | Females | 1981-1991 | 1991-2001 |
| 1 | 2 | 3 | 4 | 5 | 6 | 7 |
| | **INDIA** | **1,028,737,436** | **532,223,090** | **496,514,346** | **23.87** | **21.54** |
| 1 | Jammu & Kashmir | 10,143,700 | 5,360,926 | 4,782,774 | 30.89 | 29.43 |
| 2 | Himachal Pradesh | 6,077,900 | 3,087,940 | 2,989,960 | 20.79 | 17.54 |
| 3 | Punjab | 24,358,999 | 12,985,045 | 11,373,954 | 20.81 | 20.10 |
| 4 | Chandigarh | 900,635 | 506,938 | 393,697 | 42.16 | 40.28 |
| 5 | Uttarakhand | 8,489,349 | 4,325,924 | 4,163,425 | 23.13 | 20.41 |
| 6 | Haryana | 21,144,564 | 11,363,953 | 9,780,611 | 27.41 | 28.43 |
| 7 | Delhi | 13,850,507 | 7,607,234 | 6,243,273 | 51.45 | 47.02 |
| 8 | Rajasthan | 56,507,188 | 29,420,011 | 27,087,177 | 28.44 | 28.41 |
| 9 | Uttar Pradesh | 166,197,921 | 87,565,369 | 78,632,552 | 25.61 | 25.85 |
| 10 | Bihar | 82,998,509 | 43,243,795 | 39,754,714 | 23.38 | 28.62 |
| 11 | Sikkim | 540,851 | 288,484 | 252,367 | 28.47 | 33.06 |
| 12 | Arunachal Pradesh | 1,097,968 | 579,941 | 518,027 | 36.83 | 27.00 |
| 13 | Nagaland | 1,990,036 | 1,047,141 | 942,895 | 56.08 | 64.53 |
| 14 | Manipur | 2,293,896 | 1,161,952 | 1,131,944 | 29.29 | 24.86 |
| 15 | Mizoram | 888,573 | 459,109 | 429,464 | 39.70 | 28.82 |
| 16 | Tripura | 3,199,203 | 1,642,225 | 1,556,978 | 34.30 | 16.03 |
| 17 | Meghalaya | 2,318,822 | 1,176,087 | 1,142,735 | 32.86 | 30.65 |
| 18 | Assam | 26,655,528 | 13,777,037 | 12,878,491 | 24.24 | 18.92 |
| 19 | West Bengal | 80,176,197 | 41,465,985 | 38,710,212 | 24.73 | 17.77 |
| 20 | Jharkhand | 26,945,829 | 13,885,037 | 13,060,792 | 24.03 | 23.36 |
| 21 | Orissa | 36,804,660 | 18,660,570 | 18,144,090 | 20.06 | 16.25 |
| 22 | Chhattisgarh | 20,833,803 | 10,474,218 | 10,359,585 | 25.73 | 18.27 |
| 23 | Madhya Pradesh | 60,348,023 | 31,443,652 | 28,904,371 | 27.24 | 24.26 |
| 24 | Gujarat | 50,671,017 | 26,385,577 | 24,285,440 | 21.19 | 22.66 |
| 25 | Daman & Diu | 158,204 | 92,512 | 65,692 | 28.62 | 55.73 |
| 26 | Dadra & Nagar Haveli | 220,490 | 121,666 | 98,824 | 33.57 | 59.22 |
| 27 | Maharashtra | 96,878,627 | 50,400,596 | 46,478,031 | 25.73 | 22.73 |
| 28 | Andhra Pradesh | 76,210,007 | 38,527,413 | 37,682,594 | 24.20 | 14.59 |
| 29 | Karnataka | 52,850,562 | 26,898,918 | 25,951,644 | 21.12 | 17.51 |
| 30 | Goa | 1,347,668 | 687,248 | 660,420 | 16.08 | 15.21 |
| 31 | Lakshadweep | 60,650 | 31,131 | 29,519 | 28.47 | 17.30 |
| 32 | Kerala | 31,841,374 | 15,468,614 | 16,372,760 | 14.32 | 9.43 |
| 33 | Tamil Nadu | 62,405,679 | 31,400,909 | 31,004,770 | 15.39 | 11.72 |
| 34 | Pondicherry | 974,345 | 486,961 | 487,384 | 33.64 | 20.62 |
| 35 | Andaman & Nicobar Islands | 356,152 | 192,972 | 163,180 | 48.70 | 26.90 |

*Notes*: 1. India and Manipur figures include estimated figures for those of the three sub-divisions *viz.* Mao Maram, Paomata and Purul of Senapati district of Manipur as census results of 2001 in these three sub-divisions were cancelled due to technical and administrative reasors.

2. The 1991 Census could not be held owing to disturbed conditions prevailing in Jammu & Kashmir. Hence the decadal growth rates for 1991-2001 are based on the interpolated population figures of 1991 for Jammu & Kashmir.

3. The 1981 census could not be held in Assam. Hence the decadal growth rates for 1981-1991 are based on the interpolated population figures of 1981 for Assam.

*Source*: Registrar General of India, New Delhi.

# Table 2.3(a)

## *State-wise Classification of Workers—2001*

*(in Numbers)*

| *Sl.* | *India/State/* | *Total Rural Urban* | *Persons Males Females* | *Total Population* | *Total Workers* | *Main Workers* | *Marginal Workers* | *Non-Workers* |
|---|---|---|---|---|---|---|---|---|
| *1* | *2* | *3* | *4* | *5* | *6* | *7* | *8* | *9* |
| | INDIA | Total | Persons | 1,025,251,059 | 402,512,190 | 313,173,394 | 89,338,796 | 622,738,869 |
| | | | Males | 530,422,415 | 275,463,736 | 240,520,672 | 34,943,064 | 254,958,679 |
| | | | Females | 494,828,644 | 127,048,454 | 72,652,722 | 54,395,732 | 367,780,190 |
| | | Rural | Persons | 740,255,371 | 310,655,339 | 229,672,348 | 80,982,991 | 429,600,032 |
| | | | Males | 380,438,194 | 199,199,602 | 169,333,233 | 29,866,369 | 181,238,592 |
| | | | Females | 359,817,177 | 111,455,737 | 60,339,115 | 51,116,622 | 248,361,440 |
| | | Urban | Persons | 284,995,688 | 91,856,851 | 83,501,046 | 8,355,805 | 193,138,837 |
| | | | Males | 149,984,221 | 76,264,134 | 71,187,439 | 5,076,695 | 73,720,087 |
| | | | Females | 135,011,467 | 15,592,717 | 12,313,607 | 3,279,110 | 119,418,750 |
| 1 | Jammu & Kashmir | Total | Persons | 10,069,917 | 3,688,875 | 2,536,509 | 1,152,366 | 6,381,042 |
| | | | Males | 5,300,574 | 2,641,348 | 2,189,225 | 452,123 | 2,659,226 |
| | | | Females | 4,769,343 | 1,047,527 | 347,284 | 700,243 | 3,721,816 |
| | | Rural | Persons | 7,564,608 | 2,869,319 | 1,798,722 | 1,070,597 | 4,695,289 |
| | | | Males | 3,925,846 | 1,929,238 | 1,520,448 | 408,790 | 1,996,608 |
| | | | Females | 3,638,762 | 940,081 | 278,274 | 661,807 | 2,698,681 |
| | | Urban | Persons | 2,505,309 | 819,556 | 737,787 | 81,769 | 1,685,753 |
| | | | Males | 1,374,728 | 712,110 | 668,777 | 43,333 | 662,618 |
| | | | Females | 1,130,581 | 107,446 | 69,010 | 38,436 | 1,023,135 |
| 2 | Himachal Pradesh | Total | Persons | 6,070,305 | 2,991,448 | 1,964,141 | 1,027,307 | 3,078,857 |
| | | | Males | 3,080,781 | 1,685,310 | 1,334,033 | 351,277 | 1,395,471 |
| | | | Females | 2,989,524 | 1,306,138 | 630,108 | 676,030 | 1,683,386 |
| | | Rural | Persons | 5,475,424 | 2,771,978 | 1,759,474 | 1,012,504 | 2,703,446 |
| | | | Males | 2,749,776 | 1,505,312 | 1,163,098 | 342,214 | 1,244,464 |
| | | | Females | 2,725,648 | 1,266,666 | 596,376 | 670,290 | 1,458,982 |
| | | Urban | Persons | 594,881 | 219,470 | 204,667 | 14,803 | 375,411 |
| | | | Males | 331,005 | 179,998 | 170,935 | 9,063 | 151,007 |
| | | | Females | 263,876 | 39,472 | 33,732 | 5,740 | 224,404 |
| 3 | Punjab | Total | Persons | 24,324,749 | 9,141,760 | 7,839,111 | 1,302,649 | 15,182,989 |
| | | | Males | 12,983,996 | 7,023,757 | 6,487,588 | 536,169 | 5,960,239 |
| | | | Females | 11,340,753 | 2,118,003 | 1,351,523 | 766,480 | 9,222,750 |
| | | Rural | Persons | 16,064,238 | 6,382,662 | 5,252,672 | 1,129,990 | 9,681,576 |
| | | | Males | 8,508,757 | 4,633,197 | 4,199,057 | 434,140 | 3,875,560 |
| | | | Females | 7,555,481 | 1,749,465 | 1,053,615 | 695,850 | 5,806,016 |
| | | Urban | Persons | 8,260,511 | 2,759,093 | 2,586,439 | 172,659 | 5,501,413 |
| | | | Males | 4,475,239 | 2,390,560 | 2,288,531 | 102,029 | 2,084,679 |
| | | | Females | 3,785,272 | 368,538 | 297,908 | 70,630 | 3,416,734 |
| 4 | Chandigarh* | Total | Persons | 900,914 | 339,021 | 328,906 | 10,115 | 561,893 |
| | | | Males | 508,224 | 285,136 | 278,377 | 6,759 | 223,088 |
| | | | Females | 392,690 | 53,885 | 50,529 | 3,356 | 338,805 |
| | | Rural | Persons | 92,118 | 39,993 | 38,202 | 1,791 | 52,125 |
| | | | Males | 56,837 | 36,352 | 35,109 | 1,243 | 20,485 |
| | | | Females | 35,281 | 3,641 | 3,093 | 548 | 31,640 |

*contd. ...*

*...contd. ...*

| 1 | 2 | 3 | 4 | 5 | 6 | 7 | 8 | 9 |
|---|---|---|---|---|---|---|---|---|
| | | Urban | Persons | 808,796 | 299,028 | 290,704 | 8,324 | 509,768 |
| | | | Males | 451,387 | 248,784 | 243,268 | 5,516 | 202,603 |
| | | | Females | 357,409 | 50,244 | 47,436 | 2,808 | 307,165 |
| 5 | Uttarakhand | Total | Persons | 8,483,355 | 3,133,281 | 2,323,518 | 809,763 | 5,350,074 |
| | | | Males | 4,321,041 | 2,005,858 | 1,647,759 | 358,099 | 2,315,183 |
| | | | Females | 4,162,314 | 1,127,423 | 675,759 | 451,664 | 3,034,891 |
| | | Rural | Persons | 6,312,415 | 2,501,791 | 1,752,827 | 748,964 | 3,810,624 |
| | | | Males | 3,146,031 | 1,446,825 | 1,135,357 | 311,468 | 1,699,206 |
| | | | Females | 3,166,384 | 1,054,966 | 617,470 | 437,496 | 2,111,418 |
| | | Urban | Persons | 2,170,940 | 631,490 | 570,691 | 60,799 | 1,539,450 |
| | | | Males | 1,175,010 | 559,033 | 512,402 | 46,631 | 615,977 |
| | | | Females | 995,930 | 72,457 | 58,289 | 14,168 | 923,473 |
| 6 | Haryana | Total | Persons | 21,082,989 | 8,382,890 | 6,245,402 | 2,137,488 | 12,700,099 |
| | | | Males | 11,327,658 | 5,718,804 | 4,941,571 | 777,233 | 5,608,854 |
| | | | Females | 9,755,331 | 2,664,086 | 1,303,831 | 1,360,255 | 7,091,245 |
| | | Rural | Persons | 14,968,850 | 6,455,946 | 4,521,184 | 1,934,762 | 8,512,904 |
| | | | Males | 8,017,622 | 4,079,735 | 3,424,251 | 655,484 | 3,937,887 |
| | | | Females | 6,951,228 | 2,376,211 | 1,096,933 | 1,279,278 | 4,575,017 |
| | | Urban | Persons | 6,114,139 | 1,926,944 | 1,724,218 | 202,726 | 4,187,195 |
| | | | Males | 3,310,036 | 1,639,069 | 1,517,320 | 121,749 | 1,670,967 |
| | | | Females | 2,804,103 | 287,875 | 206,898 | 80,977 | 2,516,228 |
| 7 | Delhi* | Total | Persons | 13,803,085 | 4,526,737 | 4,303,812 | 222,925 | 9,276,348 |
| | | | Males | 7,579,849 | 3,957,325 | 3,794,300 | 163,025 | 3,622,524 |
| | | | Females | 6,223,236 | 569,412 | 509,512 | 59,900 | 5,653,824 |
| | | Rural | Persons | 964,943 | 308,766 | 280,858 | 27,908 | 656,177 |
| | | | Males | 533,255 | 265,067 | 247,291 | 17,776 | 268,188 |
| | | | Females | 431,688 | 43,699 | 33,567 | 10,132 | 387,989 |
| | | Urban | Persons | 12,838,142 | 4,217,971 | 4,022,954 | 195,017 | 8,620,171 |
| | | | Males | 7,046,594 | 3,692,258 | 3,547,009 | 145,249 | 3,354,336 |
| | | | Females | 5,791,548 | 525,713 | 475,945 | 49,768 | 5,265,835 |
| 8 | Rajasthan | Total | Persons | 56,473,122 | 23,781,257 | 17,430,143 | 6,351,114 | 32,691,865 |
| | | | Males | 29,381,657 | 14,710,049 | 12,871,197 | 1,838,852 | 14,671,608 |
| | | | Females | 27,091,465 | 9,071,208 | 4,558,946 | 4,512,262 | 18,020,257 |
| | | Rural | Persons | 43,267,678 | 19,877,709 | 13,958,254 | 5,919,455 | 23,389,969 |
| | | | Males | 22,394,479 | 11,381,353 | 9,786,286 | 1,595,067 | 11,013,126 |
| | | | Females | 20,873,199 | 8,496,356 | 4,171,968 | 4,324,388 | 12,376,843 |
| | | Urban | Persons | 13,205,444 | 3,903,548 | 3,471,889 | 431,659 | 9,301,896 |
| | | | Males | 6,987,178 | 3,328,696 | 3,084,911 | 243,785 | 3,658,482 |
| | | | Females | 6,218,266 | 574,852 | 386,978 | 187,874 | 5,643,414 |
| 9 | Uttar Pradesh | Total | Persons | 166,174,612 | 54,180,232 | 39,452,928 | 14,727,304 | 111,994,380 |
| | | | Males | 87,560,236 | 41,380,607 | 34,655,654 | 6,724,953 | 46,179,629 |
| | | | Females | 78,614,376 | 12,799,625 | 4,797,274 | 8,002,351 | 65,814,751 |
| | | Rural | Persons | 131,570,779 | 44,865,277 | 31,331,667 | 13,533,610 | 86,705,502 |
| | | | Males | 69,117,312 | 33,065,880 | 27,181,863 | 5,884,017 | 36,051,432 |
| | | | Females | 62,453,467 | 11,799,397 | 4,149,804 | 7,649,593 | 50,654,070 |
| | | Urban | Persons | 34,603,833 | 9,314,955 | 8,121,261 | 1,193,694 | 25,288,878 |
| | | | Males | 18,442,924 | 8,314,727 | 7,473,791 | 840,936 | 10,128,197 |
| | | | Females | 16,160,909 | 1,000,228 | 647,470 | 352,758 | 15,160,681 |

*contd. ...*

...contd. ...

| 1 | 2 | 3 | 4 | 5 | 6 | 7 | 8 | 9 |
|---|---|---|---|---|---|---|---|---|
| 10 | Bihar | Total | Persons | 82,878,796 | 28,080,004 | 21,053,153 | 7,026,851 | 54,798,792 |
| | | | Males | 43,153,964 | 20,597,085 | 17,571,587 | 3,025,498 | 22,556,879 |
| | | | Females | 39,724,832 | 7,482,919 | 3,481,566 | 4,001,353 | 32,241,913 |
| | | Rural | Persons | 74,199,596 | 25,856,125 | 19,109,705 | 6,746,420 | 48,343,471 |
| | | | Males | 38,510,686 | 18,650,188 | 15,811,798 | 2,838,390 | 19,860,498 |
| | | | Females | 35,688,910 | 7,205,937 | 3,297,907 | 3,908,030 | 28,482,973 |
| | | Urban | Persons | 8,679,200 | 2,223,879 | 1,943,448 | 280,431 | 6,455,321 |
| | | | Males | 4,643,278 | 1,946,897 | 1,759,789 | 187,108 | 2,696,381 |
| | | | Females | 4,035,922 | 276,982 | 183,659 | 93,323 | 3,758,940 |
| 11 | Sikkim | Total | Persons | 540,493 | 263,320 | 212,478 | 50,842 | 277,173 |
| | | | Males | 288,217 | 165,957 | 146,666 | 19,291 | 122,260 |
| | | | Females | 252,276 | 97,363 | 65,812 | 31,551 | 154,913 |
| | | Rural | Persons | 480,488 | 239,025 | 190,152 | 48,873 | 241,463 |
| | | | Males | 255,386 | 147,482 | 129,620 | 17,862 | 107,904 |
| | | | Females | 225,102 | 91,543 | 60,532 | 31,011 | 133,559 |
| | | Urban | Persons | 60,005 | 24,295 | 22,326 | 1,969 | 35,710 |
| | | | Males | 32,831 | 18,475 | 17,046 | 1,429 | 14,356 |
| | | | Females | 27,174 | 5,820 | 5,280 | 540 | 21,354 |
| 12 | Arunachal Pradesh | Total | Persons | 1,096,702 | 482,206 | 413,521 | 68,685 | 614,496 |
| | | | Males | 579,158 | 293,549 | 266,922 | 26,627 | 285,609 |
| | | | Females | 517,544 | 188,657 | 146,599 | 42,058 | 328,887 |
| | | Rural | Persons | 873,633 | 405,998 | 343,181 | 62,817 | 467,635 |
| | | | Males | 458,482 | 234,433 | 210,871 | 23,562 | 224,049 |
| | | | Females | 415,151 | 171,565 | 132,310 | 39,255 | 243,586 |
| | | Urban | Persons | 223,069 | 76,208 | 70,340 | 5,868 | 146,861 |
| | | | Males | 120,676 | 59,116 | 56,051 | 3,065 | 61,560 |
| | | | Females | 102,393 | 17,092 | 14,289 | 2,803 | 85,301 |
| 13 | Nagaland | Total | Persons | 1,988,636 | 849,982 | 708,455 | 141,527 | 1,138,654 |
| | | | Males | 1,041,686 | 487,767 | 424,236 | 63,531 | 553,919 |
| | | | Females | 946,950 | 362,215 | 284,219 | 77,996 | 584,735 |
| | | Rural | Persons | 1,635,815 | 737,386 | 606,601 | 130,785 | 898,429 |
| | | | Males | 846,651 | 398,640 | 341,298 | 57,342 | 448,011 |
| | | | Females | 789,164 | 338,746 | 265,303 | 73,443 | 450,418 |
| | | Urban | Persons | 352,821 | 112,596 | 101,854 | 10,742 | 240,225 |
| | | | Males | 195,035 | 89,127 | 82,938 | 6,189 | 105,908 |
| | | | Females | 157,786 | 23,469 | 18,916 | 4,553 | 134,317 |
| 14 | Manipur | Total | Persons | 2,388,068 | 1,069,578 | 756,520 | 313,058 | 1,318,490 |
| | | | Males | 1,215,664 | 594,642 | 484,240 | 110,402 | 621,022 |
| | | | Females | 1,172,404 | 474,936 | 272,280 | 202,656 | 697,468 |
| | | Rural | Persons | 1,812,104 | 846,595 | 592,379 | 254,216 | 965,509 |
| | | | Males | 928,211 | 464,793 | 373,702 | 91,091 | 463,418 |
| | | | Females | 883,893 | 381,802 | 218,677 | 163,125 | 502,091 |
| | | Urban | Persons | 575,964 | 222,983 | 164,141 | 58,842 | 352,981 |
| | | | Males | 287,453 | 129,849 | 110,538 | 19,311 | 157,604 |
| | | | Females | 288,511 | 93,134 | 53,603 | 39,531 | 195,377 |
| 15 | Mizoram | Total | Persons | 891,058 | 469,597 | 363,858 | 105,739 | 421,461 |
| | | | Males | 459,783 | 264,162 | 226,637 | 37,525 | 195,621 |
| | | | Females | 431,275 | 205,435 | 137,221 | 68,214 | 225,840 |
| | | Rural | Persons | 450,018 | 257,488 | 202,257 | 55,231 | 192,530 |
| | | | Males | 233,718 | 139,111 | 120,915 | 18,196 | 94,607 |
| | | | Females | 216,300 | 118,377 | 81,342 | 37,035 | 97,923 |

contd. ...

...contd. ...

| 1 | 2 | 3 | 4 | 5 | 6 | 7 | 8 | 9 |
|---|---|---|---|---|---|---|---|---|
| | | Urban | Persons | 441,040 | 212,109 | 161,601 | 50,508 | 228,931 |
| | | | Males | 226,065 | 125,051 | 105,722 | 19,329 | 101,014 |
| | | | Females | 214,975 | 87,058 | 55,879 | 31,179 | 127,917 |
| 16 | Tripura | Total | Persons | 3,191,168 | 1,158,190 | 906,487 | 251,703 | 2,032,978 |
| | | | Males | 1,636,138 | 831,318 | 740,153 | 91,165 | 804,820 |
| | | | Females | 1,555,030 | 326,872 | 166,334 | 160,538 | 1,228,158 |
| | | Rural | Persons | 2,648,074 | 982,635 | 743,447 | 239,188 | 1,665,439 |
| | | | Males | 1,359,288 | 687,956 | 604,711 | 83,245 | 671,332 |
| | | | Females | 1,288,786 | 294,679 | 138,736 | 155,943 | 994,107 |
| | | Urban | Persons | 543,094 | 175,555 | 163,040 | 12,515 | 367,539 |
| | | | Males | 276,850 | 143,362 | 135,442 | 7,920 | 133,488 |
| | | | Females | 266,244 | 32,193 | 27,598 | 4,595 | 234,051 |
| 17 | Meghalaya | Total | Persons | 2,306,069 | 956,425 | 742,762 | 213,663 | 1,349,644 |
| | | | Males | 1,167,840 | 557,807 | 471,830 | 85,977 | 610,033 |
| | | | Females | 1,138,229 | 398,618 | 270,932 | 127,686 | 739,611 |
| | | Rural | Persons | 1,853,457 | 826,334 | 629,401 | 196,933 | 1,027,123 |
| | | | Males | 939,803 | 470,712 | 392,511 | 78,201 | 469,091 |
| | | | Females | 913,654 | 355,622 | 236,890 | 118,732 | 558,032 |
| | | Urban | Persons | 452,612 | 130,091 | 113,361 | 16,730 | 322,521 |
| | | | Males | 228,037 | 87,095 | 79,319 | 7,776 | 140,942 |
| | | | Females | 224,575 | 42,996 | 34,042 | 8,954 | 181,579 |
| 18 | Assam | Total | Persons | 26,638,407 | 9,557,064 | 7,083,046 | 2,474,018 | 17,081,343 |
| | | | Males | 13,787,799 | 6,884,451 | 5,839,050 | 1,045,401 | 6,903,348 |
| | | | Females | 12,850,608 | 2,672,613 | 1,243,996 | 1,428,617 | 10,177,995 |
| | | Rural | Persons | 23,248,994 | 8,473,127 | 6,078,938 | 2,394,189 | 14,775,867 |
| | | | Males | 11,983,157 | 5,963,601 | 4,968,101 | 995,500 | 6,019,556 |
| | | | Females | 11,265,837 | 2,509,526 | 1,110,837 | 1,398,689 | 8,756,311 |
| | | Urban | Persons | 3,389,413 | 1,083,937 | 1,004,108 | 79,829 | 2,305,476 |
| | | | Males | 1,804,642 | 920,850 | 870,949 | 49,901 | 883,792 |
| | | | Females | 1,584,771 | 163,087 | 133,159 | 29,928 | 1,421,684 |
| 19 | West Bengal | Total | Persons | 80,221,171 | 29,503,278 | 23,063,800 | 6,439,478 | 50,717,893 |
| | | | Males | 41,487,694 | 22,500,480 | 19,632,549 | 2,867,931 | 18,987,214 |
| | | | Females | 38,733,477 | 7,002,798 | 3,431,251 | 3,571,547 | 31,730,679 |
| | | Rural | Persons | 57,734,690 | 21,897,521 | 16,115,110 | 5,782,411 | 35,837,169 |
| | | | Males | 29,606,028 | 16,075,548 | 13,619,407 | 2,456,141 | 13,530,480 |
| | | | Females | 28,128,662 | 5,821,973 | 2,495,703 | 3,326,270 | 22,306,689 |
| | | Urban | Persons | 22,486,481 | 7,605,757 | 6,948,690 | 657,067 | 14,880,724 |
| | | | Males | 11,881,666 | 6,424,932 | 6,013,142 | 411,790 | 5,456,734 |
| | | | Females | 10,604,815 | 1,180,825 | 935,548 | 245,277 | 9,423,990 |
| 20 | Jharkhand | Total | Persons | 26,909,428 | 10,127,733 | 6,462,736 | 3,664,997 | 16,781,695 |
| | | | Males | 13,861,277 | 6,682,839 | 5,158,527 | 1,524,312 | 7,178,438 |
| | | | Females | 13,048,151 | 3,444,894 | 1,304,209 | 2,140,685 | 9,603,257 |
| | | Rural | Persons | 20,922,731 | 8,590,029 | 5,119,383 | 3,470,646 | 12,332,702 |
| | | | Males | 10,660,430 | 5,320,278 | 3,939,134 | 1,381,144 | 5,340,152 |
| | | | Females | 10,262,301 | 3,269,751 | 1,180,249 | 2,089,502 | 6,992,550 |
| | | Urban | Persons | 5,986,697 | 1,537,704 | 1,343,353 | 194,351 | 4,448,993 |
| | | | Males | 3,200,847 | 1,362,561 | 1,219,393 | 143,168 | 1,838,286 |
| | | | Females | 2,785,850 | 175,143 | 123,960 | 51,183 | 2,610,707 |
| 21 | Orissa | Total | Persons | 36,706,920 | 14,272,764 | 9,572,940 | 4,699,824 | 22,434,156 |
| | | | Males | 18,612,340 | 9,818,471 | 8,017,115 | 1,801,356 | 8,793,869 |
| | | | Females | 18,094,580 | 4,454,293 | 1,555,825 | 2,898,468 | 13,640,287 |

contd. ...

*...contd. ...*

| 1 | 2 | 3 | 4 | 5 | 6 | 7 | 8 | 9 |
|---|---|---|---|---|---|---|---|---|
| | | Rural | Persons | 31,210,602 | 12,587,554 | 8,058,257 | 4,529,297 | 18,623,048 |
| | | | Males | 15,711,853 | 8,386,624 | 6,684,924 | 1,701,700 | 7,325,229 |
| | | | Females | 15,498,749 | 4,200,930 | 1,373,333 | 2,827,597 | 11,297,819 |
| | | Urban | Persons | 5,496,318 | 1,685,210 | 1,514,683 | 170,527 | 3,811,108 |
| | | | Males | 2,900,487 | 1,431,847 | 1,332,191 | 99,656 | 1,468,640 |
| | | | Females | 2,595,831 | 253,363 | 182,492 | 70,871 | 2,342,468 |
| 22 | Chhattisgarh | Total | Persons | 20,810,983 | 9,685,260 | 7,059,731 | 2,625,529 | 11,125,723 |
| | | | Males | 10,462,337 | 5,542,010 | 4,753,649 | 788,361 | 4,920,327 |
| | | | Females | 10,348,646 | 4,143,250 | 2,306,082 | 1,837,168 | 6,205,396 |
| | | Rural | Persons | 16,626,637 | 8,384,439 | 5,888,778 | 2,495,661 | 8,242,198 |
| | | | Males | 8,294,598 | 4,502,529 | 3,783,290 | 719,239 | 3,792,069 |
| | | | Females | 8,332,039 | 3,881,910 | 2,105,488 | 1,776,422 | 4,450,129 |
| | | Urban | Persons | 4,184,346 | 1,300,821 | 1,170,953 | 129,868 | 2,883,525 |
| | | | Males | 2,167,739 | 1,039,481 | 970,359 | 69,122 | 1,128,258 |
| | | | Females | 2,016,607 | 261,340 | 200,594 | 60,746 | 1,755,267 |
| 23 | Madhya Pradesh | Total | Persons | 60,252,739 | 25,756,485 | 19,077,568 | 6,678,917 | 34,496,254 |
| | | | Males | 31,386,566 | 16,202,287 | 14,081,689 | 2,120,598 | 15,184,279 |
| | | | Females | 28,866,173 | 9,554,198 | 4,995,879 | 4,558,319 | 19,311,975 |
| | | Rural | Persons | 44,288,037 | 20,865,600 | 14,748,209 | 6,117,391 | 23,422,437 |
| | | | Males | 22,976,980 | 12,195,626 | 10,391,973 | 1,803,653 | 10,781,354 |
| | | | Females | 21,311,057 | 8,669,974 | 4,356,236 | 4,313,738 | 12,641,083 |
| | | Urban | Persons | 15,964,702 | 4,890,885 | 4,329,359 | 561,526 | 11,073,817 |
| | | | Males | 8,409,586 | 4,006,661 | 3,689,716 | 316,945 | 4,402,925 |
| | | | Females | 7,555,116 | 884,224 | 639,643 | 244,581 | 6,670,892 |
| 24 | Gujarat | Total | Persons | 48,387,270 | 20,368,797 | 16,288,598 | 4,080,199 | 28,018,473 |
| | | | Males | 25,208,865 | 13,870,929 | 12,920,179 | 950,750 | 11,337,936 |
| | | | Females | 23,178,405 | 6,497,868 | 3,368,419 | 3,129,449 | 16,680,537 |
| | | Rural | Persons | 30,160,219 | 14,341,413 | 10,590,157 | 3,751,256 | 15,818,806 |
| | | | Males | 15,506,359 | 8,620,383 | 7,816,150 | 804,233 | 6,885,976 |
| | | | Females | 14,653,860 | 5,721,030 | 2,774,007 | 2,947,023 | 8,932,830 |
| | | Urban | Persons | 18,227,051 | 6,027,384 | 5,698,441 | 328,943 | 12,199,667 |
| | | | Males | 9,702,506 | 5,250,546 | 5,104,029 | 146,517 | 4,451,960 |
| | | | Females | 8,524,545 | 776,838 | 594,412 | 182,426 | 7,747,707 |
| 25 | Daman & Diu* | Total | Persons | 158,059 | 72,654 | 67,603 | 5,051 | 85,405 |
| | | | Males | 92,478 | 60,627 | 59,034 | 1,593 | 31,851 |
| | | | Females | 65,581 | 12,027 | 8,569 | 3,458 | 53,554 |
| | | Rural | Persons | 100,740 | 52,413 | 48,521 | 3,892 | 48,327 |
| | | | Males | 63,576 | 45,069 | 43,857 | 1,212 | 18,507 |
| | | | Females | 37,164 | 7,344 | 4,664 | 2,680 | 29,820 |
| | | Urban | Persons | 57,319 | 20,241 | 19,082 | 1,159 | 37,078 |
| | | | Males | 28,902 | 15,558 | 15,177 | 381 | 13,344 |
| | | | Females | 28,417 | 4,683 | 3,905 | 778 | 23,734 |
| 26 | Dadra & Nagar Haveli* | Total | Persons | 220,451 | 114,121 | 96,333 | 17,788 | 106,330 |
| | | | Males | 121,731 | 75,939 | 71,355 | 4,584 | 45,792 |
| | | | Females | 98,720 | 38,182 | 24,978 | 13,204 | 60,538 |
| | | Rural | Persons | 169,995 | 91,545 | 75,136 | 16,409 | 78,450 |
| | | | Males | 91,887 | 56,320 | 52,161 | 4,159 | 35,567 |
| | | | Females | 78,108 | 35,225 | 22,975 | 12,250 | 42,883 |
| | | Urban | Persons | 50,456 | 22,576 | 21,197 | 1,379 | 27,880 |
| | | | Males | 29,844 | 19,619 | 19,194 | 425 | 10,225 |
| | | | Females | 20,612 | 2,957 | 2,003 | 954 | 17,655 |

*contd. ...*

...contd. ...

| 1 | 2 | 3 | 4 | 5 | 6 | 7 | 8 | 9 |
|---|---|---|---|---|---|---|---|---|
| 27 | Maharashtra | Total | Persons | 96,752,247 | 42,053,330 | 35,670,836 | 6,382,494 | 54,698,917 |
| | | | Males | 50,334,270 | 26,924,764 | 24,485,209 | 2,439,555 | 23,409,506 |
| | | | Females | 46,417,977 | 15,128,566 | 11,185,627 | 3,942,939 | 31,289,411 |
| | | Rural | Persons | 55,732,513 | 28,106,602 | 22,752,088 | 5,354,514 | 27,625,911 |
| | | | Males | 28,443,238 | 15,410,854 | 13,589,736 | 1,821,118 | 13,032,384 |
| | | | Females | 27,289,275 | 12,695,748 | 9,162,352 | 3,533,396 | 14,593,527 |
| | | Urban | Persons | 41,019,734 | 13,946,728 | 12,918,748 | 1,027,980 | 27,073,006 |
| | | | Males | 21,891,032 | 11,513,910 | 10,895,473 | 618,437 | 10,377,122 |
| | | | Females | 19,128,702 | 2,432,818 | 2,023,275 | 409,543 | 16,695,884 |
| 28 | Andhra Pradesh | Total | Persons | 76,111,243 | 34,865,117 | 29,001,326 | 5,863,791 | 41,246,126 |
| | | | Males | 38,490,506 | 21,725,810 | 19,517,524 | 2,208,286 | 16,764,696 |
| | | | Females | 37,620,737 | 13,139,307 | 9,483,802 | 3,655,505 | 24,481,430 |
| | | Rural | Persons | 55,296,875 | 28,159,160 | 22,943,260 | 5,215,900 | 27,137,715 |
| | | | Males | 27,890,340 | 16,309,047 | 14,484,219 | 1,824,828 | 11,581,293 |
| | | | Females | 27,406,535 | 11,850,113 | 8,459,041 | 3,391,072 | 15,556,422 |
| | | Urban | Persons | 20,814,368 | 6,705,957 | 6,058,066 | 647,891 | 14,108,411 |
| | | | Males | 10,600,166 | 5,416,763 | 5,033,305 | 383,458 | 5,183,403 |
| | | | Females | 10,214,202 | 1,289,194 | 1,024,761 | 264,433 | 8,925,008 |
| 29 | Karnataka | Total | Persons | 52,733,958 | 23,521,533 | 19,357,600 | 4,163,933 | 29,212,425 |
| | | | Males | 26,856,343 | 15,272,725 | 13,944,187 | 1,328,538 | 11,583,618 |
| | | | Females | 25,877,615 | 8,248,808 | 5,413,413 | 2,835,395 | 17,628,807 |
| | | Rural | Persons | 34,814,100 | 17,129,318 | 13,458,918 | 3,670,400 | 17,684,782 |
| | | | Males | 17,618,593 | 10,274,566 | 9,218,896 | 1,055,670 | 7,344,027 |
| | | | Females | 17,195,507 | 6,854,752 | 4,240,022 | 2,614,730 | 10,340,755 |
| | | Urban | Persons | 17,919,858 | 6,392,215 | 5,898,682 | 493,533 | 11,527,643 |
| | | | Males | 9,237,750 | 4,998,159 | 4,725,291 | 272,868 | 4,239,591 |
| | | | Females | 8,682,108 | 1,394,056 | 1,173,391 | 220,665 | 7,288,052 |
| 30 | Goa | Total | Persons | 1,343,998 | 522,565 | 425,702 | 96,863 | 821,433 |
| | | | Males | 685,617 | 376,113 | 328,349 | 47,764 | 309,504 |
| | | | Females | 658,381 | 146,452 | 97,353 | 49,099 | 511,929 |
| | | Rural | Persons | 675,129 | 274,744 | 205,248 | 69,496 | 400,385 |
| | | | Males | 339,626 | 186,286 | 154,753 | 31,533 | 153,340 |
| | | | Females | 335,503 | 88,458 | 50,495 | 37,963 | 247,045 |
| | | Urban | Persons | 668,869 | 247,821 | 220,454 | 27,367 | 421,048 |
| | | | Males | 345,991 | 189,827 | 173,596 | 16,231 | 156,164 |
| | | | Females | 322,878 | 57,994 | 46,858 | 11,136 | 264,884 |
| 31 | Lakshadweep * | Total | Persons | 60,595 | 15,349 | 11,739 | 3,610 | 45,246 |
| | | | Males | 31,118 | 13,229 | 10,332 | 2,897 | 17,889 |
| | | | Females | 29,477 | 2,120 | 1,407 | 713 | 27,357 |
| | | Rural | Persons | 33,647 | 7,998 | 5,910 | 2,088 | 25,649 |
| | | | Males | 17,196 | 6,999 | 5,240 | 1,759 | 10,197 |
| | | | Females | 16,451 | 999 | 670 | 329 | 15,452 |
| | | Urban | Persons | 26,948 | 7,351 | 5,829 | 1,522 | 19,597 |
| | | | Males | 13,922 | 6,230 | 5,092 | 1,138 | 7,692 |
| | | | Females | 13,026 | 1,121 | 737 | 384 | 11,905 |
| 32 | Kerala | Total | Persons | 31,838,619 | 10,291,258 | 8,236,741 | 2,054,517 | 21,547,361 |
| | | | Males | 15,468,664 | 7,790,522 | 6,479,350 | 1,311,172 | 7,678,142 |
| | | | Females | 16,369,955 | 2,500,736 | 1,757,391 | 743,345 | 13,869,219 |
| | | Rural | Persons | 23,571,484 | 7,675,096 | 5,996,288 | 1,678,808 | 15,896,388 |
| | | | Males | 11,450,785 | 5,750,087 | 4,689,974 | 1,060,113 | 5,700,698 |
| | | | Females | 12,120,699 | 1,925,009 | 1,306,314 | 618,695 | 10,195,690 |

contd. ...

*...contd. ...*

| 1 | 2 | 3 | 4 | 5 | 6 | 7 | 8 | 9 |
|---|---|---|---|---|---|---|---|---|
| | | Urban | Persons | 8,267,135 | 2,616,162 | 2,240,453 | 375,709 | 5,650,973 |
| | | | Males | 4,017,879 | 2,040,435 | 1,789,376 | 251,059 | 1,977,444 |
| | | | Females | 4,249,256 | 575,727 | 451,077 | 124,650 | 3,673,529 |
| 33 | Tamil Nadu | Total | Persons | 62,110,839 | 27,811,647 | 23,684,611 | 4,127,036 | 34,299,192 |
| | | | Males | 31,268,654 | 18,153,275 | 16,346,879 | 1,806,396 | 13,115,379 |
| | | | Females | 30,842,185 | 9,658,372 | 7,337,732 | 2,320,640 | 21,183,813 |
| | | Rural | Persons | 34,869,286 | 17,572,083 | 14,290,211 | 3,281,872 | 17,297,203 |
| | | | Males | 17,508,985 | 10,396,912 | 9,067,457 | 1,329,455 | 7,112,073 |
| | | | Females | 17,360,301 | 7,175,171 | 5,222,754 | 1,952,417 | 10,185,130 |
| | | Urban | Persons | 27,241,553 | 10,239,564 | 9,394,400 | 845,164 | 17,001,989 |
| | | | Males | 13,759,669 | 7,756,363 | 7,279,422 | 476,941 | 6,003,306 |
| | | | Females | 13,481,884 | 2,483,201 | 2,114,978 | 368,223 | 10,998,683 |
| 34 | Pondicherry * | Total | Persons | 973,829 | 342,101 | 316,851 | 25,250 | 631,728 |
| | | | Males | 486,705 | 259,309 | 245,878 | 13,431 | 227,396 |
| | | | Females | 487,124 | 82,792 | 70,973 | 11,819 | 404,332 |
| | | Rural | Persons | 325,596 | 127,635 | 113,318 | 14,317 | 197,961 |
| | | | Males | 163,586 | 89,102 | 82,273 | 6,829 | 74,484 |
| | | | Females | 162,010 | 38,533 | 31,045 | 7,488 | 123,477 |
| | | Urban | Persons | 648,233 | 214,466 | 203,533 | 10,933 | 433,767 |
| | | | Males | 323,119 | 170,207 | 163,605 | 6,602 | 152,912 |
| | | | Females | 325,114 | 44,259 | 39,928 | 4,331 | 280,855 |
| 35 | Andaman & Nicobar Islands * | Total | Persons | 356,265 | 136,331 | 113,929 | 22,402 | 219,934 |
| | | | Males | 192,985 | 109,475 | 97,842 | 11,633 | 83,510 |
| | | | Females | 163,280 | 26,856 | 16,087 | 10,769 | 136,424 |
| | | Rural | Persons | 239,858 | 94,035 | 73,635 | 20,400 | 145,823 |
| | | | Males | 128,837 | 73,497 | 63,502 | 9,995 | 55,340 |
| | | | Females | 111,021 | 20,538 | 10,133 | 10,405 | 90,483 |
| | | Urban | Persons | 116,407 | 42,296 | 40,294 | 2,002 | 74,111 |
| | | | Males | 64,148 | 35,978 | 34,340 | 1,638 | 28,170 |
| | | | Females | 52,259 | 6,318 | 5,954 | 364 | 45,941 |

*Notes*:

1. The figures for India and Gujarat exclude the data for the entire Kachchh district, Morvi, Maliya-Miyana and Wankaner talukas of Rajkot district Jodiya taluka of Jamnagar district of Gujarat State where population enumeration of Census of India, 2001 could not be conducted due to earthquake.
2. The population figures of India and certain States presented here are different from that published earlier in Provisional Paper 1 because the basic compilation sources are different.
3. Agricultural Labourers: A Person Who works on another person's land for wages in money or kind or share is regarded as an Agricultural labourer.
4. Household Industry Workers: Household industry is defined as an industry conducted by one or more memberd of the houschold at home or within the village in rural areas and only within the precincts of the house where the household lives in urban areas.
5. Other Workers: All workers, i.e., those who have been engaged in some economic activity during the last one year, but are not Cultivators or agricultural labourers or in Household Industry, are 'Other workers (OW)'.

*Source*: Registrar General of India, New Delhi.

# Table 2.3(b)

*State-wise Classification of Workers—2001*

(in Numbers)

| Sl. No. | India/State/ Union Territory* | Total Rural Urban | Persons Males Females | Workers (Main+ Marginal) | Cultivators | Agricultural Labourers | Household Industry Workers | Other Workers |
|---|---|---|---|---|---|---|---|---|
| 1 | 2 | 3 | 4 | 5 | 6 | 7 | 8 | 9 |
| | INDIA | Total | Persons | 402,512,190 | 127,628,287 | 107,447,725 | 16,395,870 | 151,040,308 |
| | | | Males | 275,463,736 | 86,328,447 | 57,354,281 | 8,312,191 | 123,468,817 |
| | | | Females | 127,048,454 | 41,299,840 | 50,093,444 | 8,083,679 | 27,571,491 |
| | | Rural | Persons | 310,655,339 | 124,682,055 | 103,122,189 | 11,709,533 | 71,141,562 |
| | | | Males | 199,199,602 | 84,046,644 | 54,749,291 | 5,642,112 | 54,761,555 |
| | | | Females | 111,455,737 | 40,635,411 | 48,372,898 | 6,067,421 | 16,380,007 |
| | | Urban | Persons | 91,856,851 | 2,946,232 | 4,325,536 | 4,686,337 | 79,898,746 |
| | | | Males | 76,264,134 | 2,281,803 | 2,604,990 | 2,670,079 | 68,707,262 |
| | | | Females | 15,592,717 | 664,429 | 1,720,546 | 2,016,258 | 11,191,484 |
| 1 | Jammu & Kashmir | Total | Persons | 3,688,875 | 1,599,656 | 248,577 | 229,610 | 1,611,032 |
| | | | Males | 2,641,348 | 1,023,149 | 194,887 | 118,995 | 1,304,317 |
| | | | Females | 1,047,527 | 576,507 | 53,690 | 110,615 | 306,715 |
| | | Rural | Persons | 2,869,319 | 1,561,798 | 229,740 | 174,675 | 903,106 |
| | | | Males | 1,929,238 | 994,104 | 178,980 | 86,454 | 669,700 |
| | | | Females | 940,081 | 567,694 | 50,760 | 88,221 | 233,406 |
| | | Urban | Persons | 819,556 | 37,858 | 18,837 | 54,935 | 707,926 |
| | | | Males | 712,110 | 29,045 | 15,907 | 32,541 | 634,617 |
| | | | Females | 107,446 | 8,813 | 2,930 | 22,394 | 73,309 |
| 2 | Himachal Pradesh | Total | Persons | 2,991,448 | 1,960,840 | 92,761 | 50,380 | 887,467 |
| | | | Males | 1,685,310 | 835,574 | 55,116 | 32,342 | 762,278 |
| | | | Females | 1,306,138 | 1,125,266 | 37,645 | 18,038 | 125,189 |
| | | Rural | Persons | 2,771,978 | 1,952,202 | 91,333 | 47,273 | 681,170 |
| | | | Males | 1,505,312 | 831,519 | 54,045 | 30,069 | 589,679 |
| | | | Females | 1,266,666 | 1,120,683 | 37,288 | 17,204 | 91,491 |
| | | Urban | Persons | 219,470 | 8,638 | 1,428 | 3,107 | 206,297 |
| | | | Males | 179,998 | 4,055 | 1,071 | 2,273 | 172,599 |
| | | | Females | 39,472 | 4,583 | 357 | 834 | 33,698 |
| 3 | Punjab | Total | Persons | 9,141,760 | 2,099,330 | 1,498,976 | 306,980 | 5,236,474 |
| | | | Males | 7,023,757 | 1,823,594 | 1,119,558 | 158,349 | 3,922,256 |
| | | | Females | 2,118,003 | 275,736 | 379,418 | 148,631 | 1,314,218 |
| | | Rural | Persons | 6,382,662 | 2,011,368 | 1,404,035 | 199,909 | 2,767,350 |
| | | | Males | 4,633,197 | 1,742,355 | 1,040,346 | 82,775 | 1,767,721 |
| | | | Females | 1,749,465 | 269,013 | 363,689 | 117,134 | 999,629 |
| | | Urban | Persons | 2,759,098 | 87,962 | 94,941 | 107,071 | 2,469,124 |
| | | | Males | 2,390,560 | 81,239 | 79,212 | 75,574 | 2,154,535 |
| | | | Females | 368,538 | 6,723 | 15,729 | 31,497 | 314,589 |
| 4 | Chandigarh* | Total | Persons | 339,021 | 1,585 | 387 | 3,538 | 333,511 |
| | | | Males | 285,136 | 1,472 | 333 | 2,248 | 281,083 |
| | | | Females | 53,885 | 113 | 54 | 1,290 | 52,428 |
| | | Rural | Persons | 39,993 | 969 | 258 | 704 | 38,062 |
| | | | Males | 36,352 | 905 | 237 | 434 | 34,776 |
| | | | Females | 3,641 | 64 | 21 | 270 | 3,286 |

contd. ...

*...contd. ...*

| 1 | 2 | 3 | 4 | 5 | 6 | 7 | 8 | 9 |
|---|---|---|---|---|---|---|---|---|
| | | Urban | Persons | 299,028 | 616 | 129 | 2,834 | 295,449 |
| | | | Males | 248,784 | 567 | 96 | 1,814 | 246,307 |
| | | | Females | 50,244 | 49 | 33 | 1,020 | 49,142 |
| 5 | Uttarakhand | Total | Persons | 3,133,281 | 1,559,415 | 258,752 | 69,996 | 1,245,118 |
| | | | Males | 2,005,858 | 685,921 | 190,836 | 42,964 | 1,086,137 |
| | | | Females | 1,127,423 | 873,494 | 67,916 | 27,032 | 158,981 |
| | | Rural | Persons | 2,501,791 | 1,544,824 | 243,877 | 51,808 | 661,282 |
| | | | Males | 1,446,825 | 676,240 | 178,618 | 30,016 | 561,951 |
| | | | Females | 1,054,966 | 868,584 | 65,259 | 21,792 | 99,331 |
| | | Urban | Persons | 631,490 | 14,591 | 14,875 | 18,188 | 583,836 |
| | | | Males | 559,033 | 9,681 | 12,218 | 12,948 | 524,186 |
| | | | Females | 72,457 | 4,910 | 2,657 | 5,240 | 59,650 |
| 6 | Haryana | Total | Persons | 8,382,890 | 3,046,091 | 1,276,143 | 207,135 | 3,853,521 |
| | | | Males | 5,718,804 | 1,873,479 | 712,404 | 125,524 | 3,007,397 |
| | | | Females | 2,664,086 | 1,172,612 | 563,739 | 81,611 | 846,124 |
| | | Rural | Persons | 6,455,946 | 2,975,648 | 1,222,279 | 134,639 | 2,123,380 |
| | | | Males | 4,079,735 | 1,820,096 | 679,191 | 74,640 | 1,505,808 |
| | | | Females | 2,376,211 | 1,155,552 | 543,088 | 59,999 | 617,572 |
| | | Urban | Persons | 1,926,944 | 70,443 | 53,864 | 72,496 | 1,730,141 |
| | | | Males | 1,639,069 | 53,383 | 33,213 | 50,884 | 1,501,589 |
| | | | Females | 287,875 | 17,060 | 20,651 | 21,612 | 228,552 |
| 7 | Delhi * | Total | Persons | 4,526,737 | 36,588 | 13,559 | 133,720 | 4,342,870 |
| | | | Males | 3,957,325 | 26,584 | 9,613 | 107,401 | 3,813,727 |
| | | | Females | 569,412 | 10,004 | 3,946 | 26,319 | 529,143 |
| | | Rural | Persons | 308,766 | 25,969 | 8,510 | 6,989 | 267,298 |
| | | | Males | 265,067 | 18,317 | 6,027 | 4,936 | 235,787 |
| | | | Females | 43,699 | 7,652 | 2,483 | 2,053 | 31,511 |
| | | Urban | Persons | 4,217,971 | 10,619 | 5,049 | 126,731 | 4,075,572 |
| | | | Males | 3,692,258 | 8,267 | 3,586 | 102,465 | 3,577,940 |
| | | | Females | 525,713 | 2,352 | 1,463 | 24,266 | 497,632 |
| 8 | Rajasthan | Total | Persons | 23,781,257 | 13,166,777 | 2,529,225 | 651,421 | 7,433,834 |
| | | | Males | 14,710,049 | 7,086,876 | 1,038,287 | 401,191 | 6,183,695 |
| | | | Females | 9,071,208 | 6,079,901 | 1,490,938 | 250,230 | 1,250,139 |
| | | Rural | Persons | 19,877,709 | 12,923,353 | 2,442,347 | 424,927 | 4,087,082 |
| | | | Males | 11,381,353 | 6,936,535 | 993,343 | 254,954 | 3,196,521 |
| | | | Females | 8,496,356 | 5,986,818 | 1,449,004 | 169,973 | 890,561 |
| | | Urban | Persons | 3,903,548 | 243,424 | 86,878 | 226,494 | 3,346,752 |
| | | | Males | 3,328,696 | 150,341 | 44,944 | 146,237 | 2,987,174 |
| | | | Females | 574,852 | 93,083 | 41,934 | 80,257 | 359,578 |
| 9 | Uttar Pradesh | Total | Persons | 54,180,232 | 22,172,563 | 13,604,812 | 2,885,596 | 15,517,261 |
| | | | Males | 41,380,607 | 17,779,985 | 8,329,123 | 1,818,419 | 13,453,080 |
| | | | Females | 12,799,625 | 4,392,578 | 5,275,689 | 1,067,177 | 2,064,181 |
| | | Rural | Persons | 44,865,277 | 21,724,771 | 13,141,632 | 2,036,660 | 7,962,214 |
| | | | Males | 33,065,880 | 17,379,801 | 7,957,247 | 1,225,312 | 6,503,520 |
| | | | Females | 11,799,397 | 4,344,970 | 5,184,385 | 811,348 | 1,458,694 |
| | | Urban | Persons | 9,314,955 | 447,792 | 463,180 | 848,936 | 7,555,047 |
| | | | Males | 8,314,727 | 400,184 | 371,876 | 593,107 | 6,949,560 |
| | | | Females | 1,000,228 | 47,608 | 91,304 | 255,829 | 605,487 |
| 10 | Bihar | Total | Persons | 28,080,004 | 8,192,108 | 13,527,884 | 1,086,638 | 5,273,374 |
| | | | Males | 20,597,085 | 6,514,731 | 8,798,532 | 639,347 | 4,644,475 |
| | | | Females | 7,482,919 | 1,677,377 | 4,729,352 | 447,291 | 628,899 |

*contd. ...*

*...contd. ...*

| 1 | 2 | 3 | 4 | 5 | 6 | 7 | 8 | 9 |
|---|---|---|---|---|---|---|---|---|
| | | Rural | Persons | 25,856,125 | 8,062,095 | 13,253,215 | 944,169 | 3,596,646 |
| | | | Males | 18,650,188 | 6,401,717 | 8,600,543 | 540,240 | 3,107,688 |
| | | | Females | 7,205,937 | 1,660,378 | 4,652,672 | 403,929 | 488,958 |
| | | Urban | Persons | 2,223,879 | 130,013 | 274,669 | 142,469 | 1,676,728 |
| | | | Males | 1,946,897 | 113,014 | 197,989 | 99,107 | 1,536,787 |
| | | | Females | 276,982 | 16,999 | 76,680 | 43,362 | 139,941 |
| 11 | Sikkim | Total | Persons | 263,320 | 131,422 | 16,939 | 3,250 | 111,709 |
| | | | Males | 165,957 | 70,154 | 8,614 | 2,287 | 84,902 |
| | | | Females | 97,363 | 61,268 | 8,325 | 963 | 26,807 |
| | | Rural | Persons | 239,025 | 131,400 | 16,914 | 3,057 | 87,654 |
| | | | Males | 147,482 | 70,137 | 8,592 | 2,141 | 66,612 |
| | | | Females | 91,543 | 61,263 | 8,322 | 916 | 21,042 |
| | | Urban | Persons | 24,295 | 22 | 25 | 193 | 24,055 |
| | | | Males | 18,475 | 17 | 22 | 146 | 18,290 |
| | | | Females | 5,820 | 5 | 3 | 47 | 5,765 |
| 12 | Arunachal Pradesh | Total | Persons | 482,206 | 281,822 | 18,569 | 4,130 | 177,685 |
| | | | Males | 293,549 | 137,292 | 10,097 | 2,146 | 144,014 |
| | | | Females | 188,657 | 144,530 | 8,472 | 1,984 | 33,671 |
| | | Rural | Persons | 405,998 | 277,148 | 17,334 | 2,922 | 108,594 |
| | | | Males | 234,433 | 135,257 | 9,362 | 1,325 | 88,489 |
| | | | Females | 171,565 | 141,891 | 7,972 | 1,597 | 20,105 |
| | | Urban | Persons | 76,208 | 4,674 | 1,235 | 1,208 | 69,091 |
| | | | Males | 59,116 | 2,035 | 735 | 821 | 55,525 |
| | | | Females | 17,092 | 2,639 | 500 | 387 | 13,566 |
| 13 | Nagaland | Total | Persons | 849,982 | 544,433 | 33,852 | 18,072 | 253,625 |
| | | | Males | 487,767 | 271,608 | 18,141 | 6,529 | 191,489 |
| | | | Females | 362,215 | 272,825 | 15,711 | 11,543 | 62,136 |
| | | Rural | Persons | 737,386 | 537,861 | 32,842 | 14,702 | 151,981 |
| | | | Males | 398,640 | 268,540 | 17,673 | 4,923 | 107,504 |
| | | | Females | 338,746 | 269,321 | 15,169 | 9,779 | 44,477 |
| | | Urban | Persons | 112,596 | 6,572 | 1,010 | 3,370 | 101,644 |
| | | | Males | 89,127 | 3,068 | 468 | 1,606 | 83,985 |
| | | | Females | 23,469 | 3,504 | 542 | 1,764 | 17,659 |
| 14 | Manipur | Total | Persons | 1,069,578 | 492,696 | 120,991 | 97,955 | 357,936 |
| | | | Males | 594,642 | 277,583 | 53,227 | 19,603 | 244,229 |
| | | | Females | 474,936 | 215,113 | 67,764 | 78,352 | 113,707 |
| | | Rural | Persons | 846,595 | 452,841 | 99,550 | 69,187 | 225,017 |
| | | | Males | 464,793 | 250,413 | 43,655 | 13,308 | 157,417 |
| | | | Females | 381,802 | 202,428 | 55,895 | 55,879 | 67,600 |
| | | Urban | Persons | 222,983 | 39,855 | 21,441 | 28,768 | 132,919 |
| | | | Males | 129,849 | 27,170 | 9,572 | 6,295 | 86,812 |
| | | | Females | 93,134 | 12,685 | 11,869 | 22,473 | 46,107 |
| 15 | Mizoram | Total | Persons | 469,597 | 253,167 | 27,494 | 6,572 | 182,364 |
| | | | Males | 264,162 | 128,836 | 13,010 | 3,164 | 119,152 |
| | | | Females | 205,435 | 124,331 | 14,484 | 3,408 | 63,212 |
| | | Rural | Persons | 257,488 | 200,316 | 10,086 | 2,166 | 44,920 |
| | | | Males | 139,111 | 101,594 | 4,793 | 1,093 | 31,631 |
| | | | Females | 118,377 | 98,722 | 5,293 | 1,073 | 13,289 |
| | | Urban | Persons | 212,109 | 52,851 | 17,408 | 4,406 | 137,444 |
| | | | Males | 125,051 | 27,242 | 8,217 | 2,071 | 87,521 |
| | | | Females | 87,058 | 25,609 | 9,191 | 2,335 | 49,923 |

*contd. ...*

*...contd. ...*

| 1 | 2 | 3 | 4 | 5 | 6 | 7 | 8 | 9 |
|---|---|---|---|---|---|---|---|---|
| 16 | Tripura | Total | Persons | 1,158,190 | 311,378 | 278,334 | 33,535 | 534,943 |
| | | | Males | 831,318 | 221,211 | 163,926 | 13,568 | 432,613 |
| | | | Females | 326,872 | 90,167 | 114,408 | 19,967 | 102,330 |
| | | Rural | Persons | 982,635 | 308,613 | 274,869 | 30,428 | 368,725 |
| | | | Males | 687,956 | 218,602 | 160,996 | 11,986 | 296,372 |
| | | | Females | 294,679 | 90,011 | 113,873 | 18,442 | 72,353 |
| | | Urban | Persons | 175,555 | 2,765 | 3,465 | 3,107 | 166,218 |
| | | | Males | 143,362 | 2,609 | 2,930 | 1,582 | 136,241 |
| | | | Females | 32,193 | 156 | 535 | 1,525 | 29,977 |
| 17 | Meghalaya | Total | Persons | 956,425 | 457,195 | 172,975 | 17,956 | 308,299 |
| | | | Males | 557,807 | 250,376 | 90,624 | 7,985 | 208,822 |
| | | | Females | 398,618 | 206,819 | 82,351 | 9,971 | 99,477 |
| | | Rural | Persons | 826,334 | 450,097 | 166,572 | 16,129 | 193,536 |
| | | | Males | 470,712 | 246,680 | 87,530 | 6,884 | 129,618 |
| | | | Females | 355,622 | 203,417 | 79,042 | 9,245 | 63,918 |
| | | Urban | Persons | 130,091 | 7,098 | 6,403 | 1,827 | 114,763 |
| | | | Males | 87,095 | 3,696 | 3,094 | 1,101 | 79,204 |
| | | | Females | 42,996 | 3,402 | 3,309 | 726 | 35,559 |
| 18 | Assam | Total | Persons | 9,557,064 | 3,741,912 | 1,289,902 | 328,541 | 4,196,709 |
| | | | Males | 6,884,451 | 2,661,619 | 849,434 | 117,647 | 3,255,751 |
| | | | Females | 2,672,613 | 1,080,293 | 440,468 | 210,894 | 940,958 |
| | | Rural | Persons | 8,473,127 | 3,722,180 | 1,280,776 | 300,014 | 3,170,157 |
| | | | Males | 5,963,601 | 2,645,554 | 843,146 | 100,794 | 2,374,107 |
| | | | Females | 2,509,526 | 1,076,626 | 437,630 | 199,220 | 796,050 |
| | | Urban | Persons | 1,083,937 | 19,732 | 9,126 | 28,527 | 1,026,552 |
| | | | Males | 920,850 | 16,065 | 6,288 | 16,853 | 881,644 |
| | | | Females | 163,087 | 3,667 | 2,838 | 11,674 | 144,908 |
| 19 | West Bengal | Total | Persons | 29,503,278 | 5,613,113 | 7,350,988 | 2,153,359 | 14,385,818 |
| | | | Males | 22,500,480 | 4,672,284 | 5,081,540 | 894,970 | 11,851,686 |
| | | | Females | 7,002,798 | 940,829 | 2,269,448 | 1,258,389 | 2,534,132 |
| | | Rural | Persons | 21,897,521 | 5,552,488 | 7,235,021 | 1,708,040 | 7,401,972 |
| | | | Males | 16,075,548 | 4,616,800 | 4,990,218 | 648,154 | 5,820,376 |
| | | | Females | 5,821,973 | 935,688 | 2,244,803 | 1,059,886 | 1,581,596 |
| | | Urban | Persons | 7,605,757 | 60,625 | 115,967 | 445,319 | 6,983,846 |
| | | | Males | 6,424,932 | 55,484 | 91,322 | 246,816 | 6,031,310 |
| | | | Females | 1,180,825 | 5,141 | 24,645 | 198,503 | 952,536 |
| 20 | Jharkhand | Total | Persons | 10,127,733 | 3,908,235 | 2,861,939 | 420,118 | 2,937,441 |
| | | | Males | 6,682,839 | 2,430,303 | 1,491,828 | 229,235 | 2,531,473 |
| | | | Females | 3,444,894 | 1,477,932 | 1,370,111 | 190,883 | 405,968 |
| | | Rural | Persons | 8,590,029 | 3,875,645 | 2,822,343 | 374,301 | 1,517,740 |
| | | | Males | 5,320,278 | 2,405,683 | 1,465,957 | 196,252 | 1,252,386 |
| | | | Females | 3,269,751 | 1,469,962 | 1,356,386 | 178,049 | 265,354 |
| | | Urban | Persons | 1,537,704 | 32,590 | 39,596 | 45,817 | 1,419,701 |
| | | | Males | 1,362,561 | 24,620 | 25,871 | 32,983 | 1,279,087 |
| | | | Females | 175,143 | 7,970 | 13,725 | 12,834 | 140,614 |
| 21 | Orissa | Total | Persons | 14,272,764 | 4,238,347 | 5,001,075 | 689,173 | 4,344,169 |
| | | | Males | 9,818,471 | 3,371,717 | 2,580,522 | 310,153 | 3,556,079 |
| | | | Females | 4,454,293 | 866,630 | 2,420,553 | 379,020 | 788,090 |
| | | Rural | Persons | 12,587,554 | 4,182,136 | 4,923,637 | 631,779 | 2,850,002 |
| | | | Males | 8,386,624 | 3,321,937 | 2,532,952 | 272,646 | 2,259,089 |
| | | | Females | 4,200,930 | 860,199 | 2,390,685 | 359,133 | 590,913 |

*contd. ...*

*...contd. ...*

| 1 | 2 | 3 | 4 | 5 | 6 | 7 | 8 | 9 |
|---|---|---|---|---|---|---|---|---|
| | | Urban | Persons | 1,685,210 | 56,211 | 77,438 | 57,394 | 1,494,167 |
| | | | Males | 1,431,847 | 49,780 | 47,570 | 37,507 | 1,296,990 |
| | | | Females | 253,363 | 6,431 | 29,868 | 19,887 | 197,177 |
| 22 | Chhattisgarh | Total | Persons | 9,685,260 | 4,316,981 | 3,088,216 | 201,591 | 2,078,472 |
| | | | Males | 5,542,010 | 2,482,223 | 1,249,632 | 120,116 | 1,690,039 |
| | | | Females | 4,143,250 | 1,834,758 | 1,838,584 | 81,475 | 388,433 |
| | | Rural | Persons | 8,384,439 | 4,259,413 | 3,018,851 | 154,315 | 951,860 |
| | | | Males | 4,502,529 | 2,443,874 | 1,219,636 | 92,549 | 746,470 |
| | | | Females | 3,881,910 | 1,815,539 | 1,799,215 | 61,766 | 205,390 |
| | | Urban | Persons | 1,300,821 | 57,568 | 69,365 | 47,276 | 1,126,612 |
| | | | Males | 1,039,481 | 38,349 | 29,996 | 27,567 | 943,569 |
| | | | Females | 261,340 | 19,219 | 39,369 | 19,709 | 183,043 |
| 23 | Madhya Pradesh | Total | Persons | 25,756,485 | 11,058,500 | 7,380,878 | 1,010,067 | 6,307,040 |
| | | | Males | 16,202,287 | 6,935,121 | 3,485,987 | 501,369 | 5,279,810 |
| | | | Females | 9,554,198 | 4,123,379 | 3,894,891 | 508,698 | 1,027,230 |
| | | Rural | Persons | 20,865,600 | 10,724,404 | 7,116,698 | 693,790 | 2,330,708 |
| | | | Males | 12,195,626 | 6,685,217 | 3,347,264 | 334,286 | 1,828,859 |
| | | | Females | 8,669,974 | 4,039,187 | 3,769,434 | 359,504 | 501,849 |
| | | Urban | Persons | 4,890,885 | 334,096 | 264,180 | 316,277 | 3,976,332 |
| | | | Males | 4,006,661 | 249,904 | 138,723 | 167,083 | 3,450,951 |
| | | | Females | 884,224 | 84,192 | 125,457 | 149,194 | 525,381 |
| 24 | Gujarat | Total | Persons | 20,368,797 | 5,613,185 | 4,987,657 | 381,601 | 9,386,354 |
| | | | Males | 13,870,929 | 3,804,880 | 2,404,180 | 222,579 | 7,439,290 |
| | | | Females | 6,497,868 | 1,808,305 | 2,583,477 | 159,022 | 1,947,064 |
| | | Rural | Persons | 14,341,413 | 5,491,522 | 4,816,260 | 224,102 | 3,809,529 |
| | | | Males | 8,620,383 | 3,707,298 | 2,301,215 | 141,209 | 2,470,661 |
| | | | Females | 5,721,030 | 1,784,224 | 2,515,045 | 82,893 | 1,338,868 |
| | | Urban | Persons | 6,027,384 | 121,663 | 171,397 | 157,499 | 5,576,825 |
| | | | Males | 5,250,546 | 97,582 | 102,965 | 81,370 | 4,968,629 |
| | | | Females | 776,838 | 24,081 | 68,432 | 76,129 | 608,196 |
| 25 | Daman & Diu* | Total | Persons | 72,654 | 3,896 | 1,287 | 1,148 | 66,323 |
| | | | Males | 60,627 | 1,936 | 321 | 362 | 58,008 |
| | | | Females | 12,027 | 1,960 | 966 | 786 | 8,315 |
| | | Rural | Persons | 52,413 | 3,767 | 1,241 | 525 | 46,880 |
| | | | Males | 45,069 | 1,861 | 303 | 167 | 42,738 |
| | | | Females | 7,344 | 1,906 | 938 | 358 | 4,142 |
| | | Urban | Persons | 20,241 | 129 | 46 | 623 | 19,443 |
| | | | Males | 15,558 | 75 | 18 | 195 | 15,270 |
| | | | Females | 4,683 | 54 | 28 | 428 | 4,173 |
| 26 | Dadra & Nagar Haveli* | Total | Persons | 114,121 | 39,523 | 14,743 | 828 | 59,027 |
| | | | Males | 75,939 | 18,129 | 5,374 | 503 | 51,933 |
| | | | Females | 38,182 | 21,394 | 9,369 | 325 | 7,094 |
| | | Rural | Persons | 91,545 | 37,925 | 14,226 | 676 | 38,718 |
| | | | Males | 56,320 | 17,274 | 5,180 | 395 | 33,471 |
| | | | Females | 35,225 | 20,651 | 9,046 | 281 | 5,247 |
| | | Urban | Persons | 22,576 | 1,598 | 517 | 152 | 20,309 |
| | | | Males | 19,619 | 855 | 194 | 108 | 18,462 |
| | | | Females | 2,957 | 743 | 323 | 44 | 1,847 |
| 27 | Maharashtra | Total | Persons | 42,053,330 | 12,009,903 | 11,290,945 | 1,046,149 | 17,706,333 |
| | | | Males | 26,924,764 | 6,765,759 | 4,928,793 | 540,611 | 14,689,601 |
| | | | Females | 15,128,566 | 5,244,144 | 6,362,152 | 505,538 | 3,016,732 |

*contd. ...*

...contd. ...

| 1 | 2 | 3 | 4 | 5 | 6 | 7 | 8 | 9 |
|---|---|---|---|---|---|---|---|---|
| | | Rural | Persons | 28,106,602 | 11,716,298 | 10,790,748 | 586,790 | 5,012,766 |
| | | | Males | 15,410,854 | 6,547,568 | 4,669,390 | 304,430 | 3,889,466 |
| | | | Females | 12,695,748 | 5,168,730 | 6,121,358 | 282,360 | 1,123,300 |
| | | Urban | Persons | 13,946,728 | 293,605 | 500,197 | 459,359 | 12,693,567 |
| | | | Males | 11,513,910 | 218,191 | 259,403 | 236,181 | 10,800,135 |
| | | | Females | 2,432,818 | 75,414 | 240,794 | 223,178 | 1,893,432 |
| 28 | Andhra Pradesh | Total | Persons | 34,865,117 | 7,903,635 | 13,818,754 | 1,569,699 | 11,573,029 |
| | | | Males | 21,725,810 | 5,287,272 | 6,431,834 | 666,358 | 9,340,346 |
| | | | Females | 13,139,307 | 2,616,363 | 7,386,920 | 903,341 | 2,232,683 |
| | | Rural | Persons | 28,159,160 | 7,784,145 | 13,379,667 | 1,198,770 | 5,796,578 |
| | | | Males | 16,309,047 | 5,188,706 | 6,188,479 | 481,882 | 4,449,980 |
| | | | Females | 11,850,113 | 2,595,439 | 7,191,188 | 716,888 | 1,346,598 |
| | | Urban | Persons | 6,705,957 | 119,490 | 439,087 | 370,929 | 5,776,451 |
| | | | Males | 5,416,763 | 98,566 | 243,355 | 184,476 | 4,890,366 |
| | | | Females | 1,289,194 | 20,924 | 195,732 | 186,453 | 886,085 |
| 29 | Karnataka | Total | Persons | 23,521,533 | 6,936,121 | 6,209,153 | 935,712 | 9,440,547 |
| | | | Males | 15,272,725 | 4,909,653 | 2,595,871 | 388,441 | 7,378,760 |
| | | | Females | 8,248,808 | 2,026,468 | 3,613,282 | 547,271 | 2,061,787 |
| | | Rural | Persons | 17,129,318 | 6,704,656 | 5,890,452 | 581,561 | 3,952,649 |
| | | | Males | 10,274,566 | 4,719,336 | 2,434,992 | 237,772 | 2,882,466 |
| | | | Females | 6,854,752 | 1,985,320 | 3,455,460 | 343,789 | 1,070,183 |
| | | Urban | Persons | 6,392,215 | 231,465 | 318,701 | 354,151 | 5,487,898 |
| | | | Males | 4,998,159 | 190,317 | 160,879 | 150,669 | 4,496,294 |
| | | | Females | 1,394,056 | 41,148 | 157,822 | 203,482 | 991,604 |
| 30 | Goa | Total | Persons | 522,565 | 50,663 | 36,150 | 14,094 | 421,658 |
| | | | Males | 376,113 | 26,008 | 16,211 | 8,551 | 325,343 |
| | | | Females | 146,452 | 24,655 | 19,939 | 5,543 | 96,315 |
| | | Rural | Persons | 274,744 | 45,258 | 31,510 | 8,317 | 189,659 |
| | | | Males | 186,286 | 23,340 | 13,867 | 4,869 | 144,210 |
| | | | Females | 88,458 | 21,918 | 17,643 | 3,448 | 45,449 |
| | | Urban | Persons | 247,821 | 5,405 | 4,640 | 5,777 | 231,999 |
| | | | Males | 189,827 | 2,668 | 2,344 | 3,682 | 181,133 |
| | | | Females | 57,994 | 2,737 | 2,296 | 2,095 | 50,866 |
| 31 | Lakshadweep* | Total | Persons | 15,349 | 0 | 0 | 910 | 14,439 |
| | | | Males | 13,229 | 0 | 0 | 542 | 12,687 |
| | | | Females | 2,120 | 0 | 0 | 368 | 1,752 |
| | | Rural | Persons | 7,998 | 0 | 0 | 441 | 7,557 |
| | | | Males | 6,999 | 0 | 0 | 283 | 6,716 |
| | | | Females | 999 | 0 | 0 | 158 | 841 |
| | | Urban | Persons | 7,351 | 0 | 0 | 469 | 6,882 |
| | | | Males | 6,230 | 0 | 0 | 259 | 5,971 |
| | | | Females | 1,121 | 0 | 0 | 210 | 911 |
| 32 | Kerala | Total | Persons | 10,291,258 | 740,403 | 1,653,601 | 364,770 | 7,532,484 |
| | | | Males | 7,790,522 | 622,724 | 1,103,317 | 181,935 | 5,882,546 |
| | | | Females | 2,500,736 | 117,679 | 550,284 | 182,835 | 1,649,938 |
| | | Rural | Persons | 7,675,096 | 685,163 | 1,514,706 | 262,066 | 5,213,161 |
| | | | Males | 5,750,087 | 578,665 | 1,009,620 | 125,224 | 4,036,578 |
| | | | Females | 1,925,009 | 106,498 | 505,086 | 136,842 | 1,176,583 |
| | | Urban | Persons | 2,616,162 | 55,240 | 138,895 | 102,704 | 2,319,323 |
| | | | Males | 2,040,435 | 44,059 | 93,697 | 56,711 | 1,845,968 |
| | | | Females | 575,727 | 11,181 | 45,198 | 45,993 | 473,355 |

contd. ...

...contd. ...

| 1 | 2 | 3 | 4 | 5 | 6 | 7 | 8 | 9 |
|---|---|---|---|---|---|---|---|---|
| 33 | Tamil Nadu | Total | Persons | 27,811,647 | 5,114,384 | 8,665,020 | 1,458,546 | 12,573,697 |
| | | | Males | 18,153,275 | 3,305,413 | 4,277,140 | 619,096 | 9,951,626 |
| | | | Females | 9,658,372 | 1,808,971 | 4,387,880 | 839,450 | 2,622,071 |
| | | Rural | Persons | 17,572,083 | 4,725,890 | 7,565,439 | 815,009 | 4,465,745 |
| | | | Males | 10,396,912 | 3,028,113 | 3,667,853 | 324,381 | 3,376,565 |
| | | | Females | 7,175,171 | 1,697,777 | 3,897,586 | 490,628 | 1,089,180 |
| | | Urban | Persons | 10,239,564 | 388,494 | 1,099,581 | 643,537 | 8,107,952 |
| | | | Males | 7,756,363 | 277,300 | 609,287 | 294,715 | 6,575,061 |
| | | | Females | 2,483,201 | 111,194 | 490,294 | 348,822 | 1,532,891 |
| 34 | Pondicherry* | Total | Persons | 342,101 | 11,284 | 72,095 | 6,096 | 252,626 |
| | | | Males | 259,309 | 10,186 | 41,990 | 3,097 | 204,036 |
| | | | Females | 82,792 | 1,098 | 30,105 | 2,999 | 48,590 |
| | | Rural | Persons | 127,635 | 8,774 | 60,251 | 2,103 | 56,507 |
| | | | Males | 89,102 | 7,826 | 34,174 | 1,051 | 46,051 |
| | | | Females | 38,533 | 948 | 26,077 | 1,052 | 10,456 |
| | | Urban | Persons | 214,466 | 2,510 | 11,844 | 3,993 | 196,119 |
| | | | Males | 170,207 | 2,360 | 7,816 | 2,046 | 157,985 |
| | | | Females | 44,259 | 150 | 4,028 | 1,947 | 38,134 |
| 35 | A & N Islands* | Total | Persons | 136,331 | 21,136 | 5,092 | 6,984 | 103,119 |
| | | | Males | 109,475 | 14,795 | 3,979 | 4,564 | 86,137 |
| | | | Females | 26,856 | 6,341 | 1,113 | 2,420 | 16,982 |
| | | Rural | Persons | 94,035 | 21,118 | 4,970 | 6,590 | 61,357 |
| | | | Males | 73,497 | 14,780 | 3,867 | 4,278 | 50,572 |
| | | | Females | 20,538 | 6,338 | 1,103 | 2,312 | 10,785 |
| | | Urban | Persons | 42,296 | 18 | 122 | 394 | 41,762 |
| | | | Males | 35,978 | 15 | 112 | 286 | 35,565 |
| | | | Females | 6,318 | 3 | 10 | 108 | 6,197 |

*Notes*:

1. The figures for India and Gujarat exclude the data for the entire Kachchh district, Morvi, Maliya-Miyana and Wankaner talukas of Rajkot District and Jodiya taluka of Jamnager District of Gujrat State where Population enumeration of Census of India 2001.
2. Agricultural Labourers: A person Who works on another person's land for wages in money or kind or share is regarded as an Agricultural labourer.
3. Household Industry: Household industry is defined as an industry conducted by one or more members of the household at home or within the village in rural areas and only within the precincts of the house where the household lives in urban areas.
4. Other Workers: All Workers, i.e., those who have been engaged in some economic activity during the last one year, but are not Cultivators or agricultural labourers or in Household Industry, are 'Other Workers(OW)'.

*Source*: Registrar General of India, New Delhi.

## Table 2.4 (a)

*State-wise Poverty Line—2004-05*

*(Rs. per capita per month)*

| *Sl. No.* | *States/Uts* | *Rural* | *Urban* |
|---|---|---|---|
| *1* | *2* | *3* | *4* |
| 1 | Andhra Pradesh | 292.95 | 542.89 |
| 2 | Assam | 387.64 | 378.84 |
| 3 | Bihar | 354.36 | 435.00 |
| 4 | Chhattisgarh | 322.41 | 560.00 |
| 5 | Delhi | 410.38 | 612.91 |
| 6 | Goa | 362.25 | 665.90 |
| 7 | Gujarat | 353.93 | 541.16 |
| 8 | Haryana | 414.76 | 504.49 |
| 9 | Himachal Pradesh | 394.28 | 504.49 |
| 10 | Jammu & Kashmir | 391.26 | 553.77 |
| 11 | Jharkhand | 366.56 | 451.24 |
| 12 | Karnataka | 324.17 | 599.66 |
| 13 | Kerala | 430.12 | 559.39 |
| 14 | Madhya Pradesh | 327.78 | 570.15 |
| 15 | Maharashtra | 362.25 | 665.90 |
| 16 | Orissa | 325.79 | 528.49 |
| 17 | Punjab | 410.38 | 466.16 |
| 18 | Rajasthan | 374.57 | 559.63 |
| 19 | Tamil Nadu | 351.86 | 547.42 |
| 20 | Uttar Pradesh | 365.84 | 483.26 |
| 21 | Uttarakhand | 478.02 | 637.67 |
| 22 | West Bengal | 382.82 | 449.32 |
| 23 | Dadra & Nagar Haveli | 362.25 | 665.90 |
| | All-India* | 356.30 | 538.60 |

* The poverty line (implicit) at all-India level is worked out from the expenditure class-wise distribution of persons (based on Uniform Recall Period-consumption, i.e., consumption data collected from 30-day recall period for all items) and the poverty ratio at all-India level. The povery ratio at all-India is obtained as the weighted average of the state-wise poverty ratio.

*Note*: State specific poverty lines identified by the Expert Group and Updating them to 2004-05 prices using the Consumer Price Index of Agricultural Labourers (CPIAL) for rural poverty lines and Consumer Price Index for Industrial Workers (CPIIW) for urban poverty lines.

*Source*: Planning Commission.

## Table 2.4 (b)

*State-wise Population Below Poverty Line—2004-05*

*(Based on URP-Consumption)*

| Sl. No. | States/UTs | Rural | | Urban | | Combined | |
|---|---|---|---|---|---|---|---|
| | | No. of Persons (Lakhs) | % Share | No. of Persons (Lakhs) | % Share | No. of Persons (Lakhs) | % Share |
| 1 | 2 | 3 | 4 | 5 | 6 | 7 | 8 |
| 1 | Andhra Pradesh | 64.70 | 11.20 | 61.40 | 28.00 | 126.10 | 15.80 |
| 2 | Arunachal Pradesh | 1.94 | 22.30 | 0.09 | 3.30 | 2.03 | 17.60 |
| 3 | Assam | 54.50 | 22.30 | 1.28 | 3.30 | 55.77 | 19.70 |
| 4 | Bihar | 336.72 | 42.10 | 32.42 | 34.60 | 369.15 | 41.40 |
| 5 | Chhattisgarh | 71.50 | 40.80 | 19.47 | 41.20 | 90.96 | 40.90 |
| 6 | Delhi | 0.63 | 6.90 | 22.30 | 15.20 | 22.93 | 14.70 |
| 7 | Goa | 0.36 | 5.40 | 1.64 | 21.30 | 2.01 | 13.80 |
| 8 | Gujarat | 63.49 | 19.10 | 27.19 | 13.00 | 90.69 | 16.80 |
| 9 | Haryana | 21.49 | 13.60 | 10.60 | 15.10 | 32.10 | 14.00 |
| 10 | Himachal Pradesh | 6.14 | 10.70 | 0.22 | 3.40 | 6.36 | 10.00 |
| 11 | Jammu & Kashmir | 3.66 | 4.60 | 2.19 | 7.90 | 5.85 | 5.40 |
| 12 | Jharkhand | 103.19 | 46.30 | 13.20 | 20.20 | 116.39 | 40.30 |
| 13 | Karnataka | 75.05 | 20.80 | 63.83 | 32.60 | 138.89 | 25.00 |
| 14 | Kerala | 32.43 | 13.20 | 17.17 | 20.20 | 49.60 | 15.00 |
| 15 | Madhya Pradesh | 175.65 | 36.90 | 74.03 | 42.10 | 249.68 | 38.30 |
| 16 | Maharashtra | 171.13 | 29.60 | 146.25 | 32.20 | 317.38 | 30.70 |
| 17 | Manipur | 3.76 | 22.30 | 0.20 | 3.30 | 3.95 | 17.30 |
| 18 | Meghalaya | 4.36 | 22.30 | 0.16 | 3.30 | 4.52 | 18.50 |
| 19 | Mizoram | 1.02 | 22.30 | 0.16 | 3.30 | 1.18 | 12.60 |

contd. ...

*...contd. ...*

| 1 | 2 | 3 | 4 | 5 | 6 | 7 | 8 |
|---|---|---|---|---|---|---|---|
| 20 | Nagaland | 3.87 | 22.30 | 0.12 | 3.30 | 3.99 | 19.00 |
| 21 | Orissa | 151.75 | 46.80 | 26.74 | 44.30 | 178.49 | 46.40 |
| 22 | Punjab | 15.12 | 9.10 | 6.50 | 7.10 | 21.63 | 8.40 |
| 23 | Rajasthan | 87.38 | 18.70 | 47.51 | 32.90 | 134.89 | 22.10 |
| 24 | Sikkim | 1.12 | 22.30 | 0.02 | 3.30 | 1.14 | 20.10 |
| 25 | Tamil Nadu | 76.50 | 22.80 | 69.13 | 22.20 | 145.62 | 22.50 |
| 26 | Tripura | 6.18 | 22.30 | 0.20 | 3.30 | 6.38 | 18.90 |
| 27 | Uttar Pradesh | 473.00 | 33.40 | 117.03 | 30.60 | 590.03 | 32.80 |
| 28 | Uttarakhand | 27.11 | 40.80 | 8.85 | 36.50 | 35.96 | 39.60 |
| 29 | West Bengal | 173.22 | 28.60 | 35.14 | 14.80 | 208.36 | 24.70 |
| 30 | A&N Islands | 0.60 | 22.90 | 0.32 | 22.20 | 0.92 | 22.60 |
| 31 | Chandigarh | 0.08 | 7.10 | 0.67 | 7.10 | 0.74 | 7.10 |
| 32 | Dadra & Nagar Haveli | 0.68 | 39.80 | 0.15 | 19.10 | 0.84 | 33.20 |
| 33 | Damn & Diu | 0.07 | 5.40 | 0.14 | 21.20 | 0.21 | 10.50 |
| 34 | Lakshadweep | 0.06 | 13.30 | 0.06 | 20.20 | 0.11 | 16.00 |
| 35 | Pondicherry | 0.78 | 22.90 | 1.59 | 22.20 | 2.37 | 22.40 |
| | All-India | 2209.24 | 28.30 | 807.96 | 25.70 | 3017.20 | 27.50 |

URP consumption = Uniform Recall Period consumption in which the consumer expenditure data for all the items are collected from 30-day recall period.

*Notes:*

1. Poverty Ratio of Assam is used for Sikkim, Arunachal Pradesh, Meghalaya, Mizoram, Manipur, Nagaland and Tripura.
2. Poverty line of Maharashtra and expenditure distribution of Goa is used to estimate poverty ratio of Goa.
3. Poverty ratio of Tamil Nadu is used for Pondicherry and A&N Islands.
4. Urban Poverty ratio of Punjab used for both rural and urban poverty of Chandigarh.
5. Poverty line of Maharashtra and expenditure distribution of Dadra & Nagar Haveli is used to estimate poverty ratio of Dadra & Nagar Haveli.
6. Poverty ratio of Goa is used for Daman & Diu.
7. Poverty ratio of Kerala is used for Lakshadweep.

*Source:* Planning Commission.

# Table 2.4 (c)

*State-wise Population Below Poverty Line—2004-05*

*(Based on MRP-Consumption)*

| Sl. No. | States/UTs | Rural | | Urban | | Combined | |
|---|---|---|---|---|---|---|---|
| | | No. of Persons (Lakhs) | % Share | No. of Persons (Lakhs) | % Share | No. of Persons (Lakhs) | % Share |
| 1 | 2 | 3 | 4 | 5 | 6 | 7 | 8 |
| 1 | Andhra Pradesh | 43.21 | 7.50 | 45.50 | 20.70 | 88.71 | 11.10 |
| 2 | Arunachal Pradesh | 1.47 | 17.00 | 0.07 | 2.40 | 1.54 | 13.40 |
| 3 | Assam | 41.46 | 17.00 | 0.93 | 2.40 | 42.39 | 15.00 |
| 4 | Bihar | 262.92 | 32.90 | 27.09 | 28.90 | 290.01 | 32.50 |
| 5 | Chhattisgarh | 54.72 | 31.20 | 16.39 | 34.70 | 71.11 | 32.00 |
| 6 | Delhi | 0.01 | 0.10 | 15.83 | 10.80 | 15.83 | 10.20 |
| 7 | Goa | 0.13 | 1.90 | 1.62 | 20.90 | 1.74 | 12.00 |
| 8 | Gujarat | 46.25 | 13.90 | 21.18 | 10.10 | 67.43 | 12.50 |
| 9 | Haryana | 14.57 | 9.20 | 7.99 | 11.30 | 22.56 | 9.90 |
| 10 | Himachal Pradesh | 4.10 | 7.20 | 0.17 | 2.60 | 4.27 | 6.70 |
| 11 | Jammu & Kashmir | 2.20 | 2.70 | 2.34 | 8.50 | 4.54 | 4.20 |
| 12 | Jharkhand | 89.76 | 40.20 | 10.63 | 16.30 | 100.39 | 34.80 |
| 13 | Karnataka | 43.33 | 12.00 | 53.28 | 27.20 | 96.60 | 17.40 |
| 14 | Kerala | 23.59 | 9.60 | 13.92 | 16.40 | 37.51 | 11.40 |
| 15 | Madhya Pradesh | 141.99 | 29.80 | 68.97 | 39.30 | 210.97 | 32.40 |
| 16 | Maharashtra | 128.43 | 22.20 | 131.40 | 29.00 | 259.83 | 25.20 |
| 17 | Manipur | 2.86 | 17.00 | 0.14 | 2.40 | 3.00 | 13.20 |
| 18 | Meghalaya | 3.32 | 17.00 | 0.12 | 2.40 | 3.43 | 14.10 |
| 19 | Mizoram | 0.78 | 17.00 | 0.11 | 2.40 | 0.89 | 9.50 |

contd. ...

*...contd. ...*

| 1 | 2 | 3 | 4 | 5 | 6 | 7 | 8 |
|---|---|---|---|---|---|---|---|
| 20 | Nagaland | 2.94 | 17.00 | 0.09 | 2.40 | 3.03 | 14.50 |
| 21 | Orissa | 129.29 | 39.80 | 24.30 | 40.30 | 153.59 | 39.90 |
| 22 | Punjab | 9.78 | 5.90 | 3.52 | 3.80 | 13.30 | 5.20 |
| 23 | Rajasthan | 66.69 | 14.30 | 40.50 | 17.50 | 107.18 | 17.50 |
| 24 | Sikkim | 0.85 | 17.00 | 0.02 | 15.20 | 0.87 | 15.20 |
| 25 | Tamil Nadu | 56.51 | 16.90 | 58.59 | 17.80 | 115.10 | 17.80 |
| 26 | Tripura | 4.70 | 17.00 | 0.14 | 14.40 | 4.85 | 14.40 |
| 27 | Uttar Pradesh | 357.68 | 25.30 | 100.47 | 25.50 | 458.15 | 25.50 |
| 28 | Uttarakhand | 21.11 | 31.70 | 7.75 | 31.80 | 28.86 | 31.80 |
| 29 | West Bengal | 146.59 | 24.20 | 26.64 | 20.60 | 173.23 | 20.60 |
| 30 | A&N Islands | 0.44 | 16.90 | 0.27 | 17.60 | 0.71 | 17.60 |
| 31 | Chandigarh | 0.04 | 3.80 | 0.36 | 3.80 | 0.40 | 3.80 |
| 32 | Dadra & Nagar Haveli | 0.62 | 36.00 | 0.16 | 30.60 | 0.77 | 30.80 |
| 33 | Damn & Diu | 0.03 | 1.90 | 0.14 | 8.00 | 0.16 | 8.00 |
| 34 | Lakshadweep | 0.04 | 9.60 | 0.05 | 12.30 | 0.09 | 12.30 |
| 35 | Pondicherry | 0.58 | 16.90 | 1.34 | 18.20 | 1.92 | 18.20 |
| | All-India | 1702.99 | 21.80 | 582.00 | 21.80 | 2384.99 | 21.80 |

MRP consumption = Mixed Recall Period consumption in which the consumer expenditure data for five non-food items, namely, clothing, footwear, durable goods, education and institutional medical expenses are collected from 365-day recall period and the consumption data for the remaining items are collected from 30-day recall period.

*Notes*:

1. Poverty Ratio of Assam is used for Sikkim, Arunachal Pradesh, Meghalaya, Mizoram, Manipur, Nagaland and Tripura.
2. Poverty line of Maharashtra and expenditure distribution of Goa is used to estimate poverty ratio of Goa.
3. Poverty ratio of Tamil Nadu is used for Pondicherry and A&N Islands.
4. Urban Poverty ratio of Punjab used for both rural and urban poverty of Chandigarh.
5. Poverty line of Maharashtra and expenditure distribution of Dadra & Nagar Haveli is used to estimate poverty ratio of Dadra & Nagar Haveli.
6. Poverty ratio of Goa is used for Daman & Diu.
7. Poverty ratio of Kerala is used for Lakshadweep.

*Source*: Planning Commission.

# Table 2.5

*Some Important Indicators of Social Development at the All-India Level*

| *Year* | *Indicators (per thousand)* | | |
|---|---|---|---|
| | *Birth Rate* | *Death Rate* | *Infant Mortality Rate* |
| *1* | *2* | *3* | *4* |
| 1991* | 29.5 | 9.8 | 80 |
| 1992* | 29.2 | 10.1 | 79 |
| 1993* | 28.7 | 9.3 | 74 |
| 1994* | 28.7 | 9.3 | 74 |
| 1995*^ | 28.3 | 9.0 | 74 |
| 1996*^ | 27.5 | 9.0 | 72 |
| 1997*^ | 27.2 | 8.9 | 71 |
| 1998^ | 26.5 | 9.0 | 72 |
| 1999^ | 26.1 | 8.7 | 70 |
| 2000^ | 25.8 | 8.5 | 68 |
| 2001^ | 25.4 | 8.4 | 66 |
| 2002^ | 25.0 | 8.1 | 63 |
| 2003^ | 24.8 | 8.0 | 60 |
| 2004^ | 24.1 | 7.5 | 58 |
| 2005^ | 23.8 | 7.6 | 58 |
| 2006^ | 23.5 | 7.5 | 57 |

| | *1991** | *2001* |
|---|---|---|
| Literacy Rate (%) | | |
| 15-34 years | 56.86 | 71.03 |
| 7 Years & above | 52.22 | 64.84 |
| Electricity as a source of Lighting (Per cent of Households) | | |
| Rural | 30.54 | 43.52 |
| Urban | 75.78 | 87.59 |

*Note*: Estimates of vital rates at the national level upto 1995 do not include Mizoram as the SRS was not operational in Mizoram till 1995.

* Excludes Jammu & Kashmir due to non-receipt of returns.

^ Excludes Nagaland rural due to part receipt of returns.

*Source*: Registrar General of India, New Delhi.

## Table 2.6

*Projected Growth in Population and Labour Force*

*(per cent per annum)*

| | 2002-2007 | 2007-2012 | 2012-2017 |
|---|---|---|---|
| 1 | 2 | 3 | 4 |
| Population | | | |
| All age groups | 1.54 | 1.37 | 1.22 |
| 15-59 years | 2.43 | 2.04 | 1.54 |
| 15+ | 2.54 | 2.20 | 1.82 |
| Labour Force* | 2.31 | 1.96 | 1.62 |
| 15-59 years | 2.31 | 1.95 | 1.53 |
| 15+ | 2.37 | 2.04 | 1.68 |

* Labour force projections are on the basis of labour force participation rate for each quinquennial age group remaining unchanged, i.e. the changes in labour force growth in relation to population are due to changes in the age composition of the population.

*Source*: LEM Division, Planning Commission, New Delhi.

# Table 2.7 (a)

*Gross Domestic Product by Economic Activity (At Current Prices)*

(Rs. Crore)

| Sector/Sub-Sector | 1999-2000 | 2000-01 | 2001-02 | 2002-03 | 2003-04 | 2004-05 | 2005-06 | 2006-07(Q) | 2006-07(A) |
|---|---|---|---|---|---|---|---|---|---|
| 1 | 2 | 3 | 4 | 5 | 6 | 7 | 8 | 9 | 10 |
| 1. Agriculture, Forestry & Fishing | 446515 | 449565 | 486617 | 472060 | 532342 | 552422 | 615845 | 695424 | 754561 |
| 1.1 Agriculture | 409660 | 408932 | 442464 | 425521 | 483030 | 501415 | 557118 | 634519 | |
| 1.2 Forestry & Logging | 17915 | 19298 | 20913 | 21048 | 22374 | 23351 | 25839 | 26855 | |
| 1.3 Fishing | 18939 | 21336 | 23240 | 25491 | 26938 | 27656 | 32888 | 34050 | |
| 2. Mining & Quarrying | 41594 | 45706 | 47871 | 62742 | 63882 | 84776 | 94153 | 101816 | 110757 |
| 3. Manufacturing | 264113 | 300392 | 315314 | 346029 | 388549 | 453603 | 519387 | 617954 | 708748 |
| 3.1 Registered | 173003 | 199262 | 213677 | 235734 | 265303 | 312622 | 360053 | 428909 | |
| 3.2 Unregistered | 91110 | 101130 | 101637 | 110295 | 123246 | 140981 | 159334 | 189045 | |
| 4. Electricity, Gas & Water Supply | 44526 | 46040 | 47482 | 54531 | 56675 | 59892 | 64406 | 70563 | 76505 |
| 5. Construction | 102007 | 111999 | 120865 | 135172 | 156806 | 212812 | 264616 | 319497 | 365140 |
| 6. Trade, Hotels & Restaurants | 254143 | 280934 | 313105 | 347255 | 395310 | 461287 | 538542 | 630364 | 1084315# |
| 6.1 Trade | 231878 | 256100 | 285731 | 317021 | 361446 | 420529 | 489535 | 570888 | |
| 6.2 Hotels & Restaurants | 22265 | 24834 | 27374 | 30233 | 33864 | 40758 | 49007 | 59476 | |
| 7. Transport, Storage & Communication | 133371 | 147921 | 163125 | 178713 | 209373 | 244786 | 277156 | 319104 | |
| 7.1 Railways | 21122 | 21443 | 22408 | 24678 | 26919 | 29645 | 33622 | 39308 | |
| 7.2 Transport by Other Means | 82376 | 93123 | 100979 | 114732 | 135662 | 159385 | 178219 | 200722 | |
| 7.3 Storage | 1427 | 1577 | 1651 | 1697 | 1825 | 2161 | 2272 | 2535 | |
| 7.4 Communication | 28446 | 31778 | 38087 | 37606 | 44967 | 53595 | 63043 | 76539 | |

contd. ...

...contd. ...

| 1 | 2 | 3 | 4 | 5 | 6 | 7 | 8 | 9 | 10 |
|---|---|---|---|---|---|---|---|---|---|
| 8. Financing, Insurance, Real Estate & Business Services | 233550 | 254772 | 282862 | 330685 | 371452 | 405081 | 452593 | 576755 | 611755 |
| 8.1 Banking & Insurance | 105662 | 105715 | 125013 | 144801 | 161985 | 167831 | 180519 | 213381 | |
| 8.2 Real Estate, Ownership of Dwellings, Business & Legal services | 127888 | 149057 | 167849 | 185884 | 209467 | 237250 | 272074 | 313374 | |
| 9. Community, Social & Personal Services | 266707 | 287688 | 310485 | 334228 | 363781 | 403047 | 448972 | 508586 | 571259 |
| 9.1 Public Administration & Defence | 122384 | 129145 | 138728 | 146629 | 255925 | 173491 | 194800 | 218883 | |
| 9.2 Other Services | 144333 | 158543 | 171757 | 187599 | 207856 | 229556 | 254172 | 289703 | |
| 10. Gross Domestic Product at Factor Cost (1 to 9) | 1786525 | 1975917 | 2097728 | 2261415 | 2528171 | 2877706 | 3275670 | 3790063 | 4283040 |
| 11. Total Net Domestic Product at Factor Cost | 1605103 | 1723200 | 1869428 | 2010938 | 2258189 | 2548783 | 2896866 | 3355595 | 3790727 |
| 12. Net Factor Income from Abroad | -15431 | -22733 | -20068 | -16690 | -8250 | -22375 | -26116 | -29778 | -20425 |
| 13. NNP at Factor Cost | 1589672 | 1700467 | 1849360 | 1994244 | 2239939 | 2526408 | 2870750 | 3325817 | 3770302 |
| 14. Per Capita Income (in Rupees) | 15881 | 16688 | 17782 | 18885 | 20895 | 23199 | 25956 | 29642 | 33131 |

(Q) Quick Estimate. # Includes Transport and Communication.

(A) Advance Estimates

*Note*: Blanks in column 9 indiates that these estimates are not available

*Source*: Central Statistical Organisation, New Delhi.

# Table 2.7(b)

*Gross Domestic Product by Economic Acitivity (At 1999-2000 Prices)*

Rs. Crores

| Sector/Sub-Sector | 1999-2000 | 2000-01 | 2001-02 | 2002-03 | 2003-04 | 2004-05 | 2005-06 | 2006-07(Q) | 2006-07(A) |
|---|---|---|---|---|---|---|---|---|---|
| 1 | 2 | 3 | 4 | 5 | 6 | 7 | 8 | 9 | 10 |
| 1. Agriculture, Forestry & Fishing | 446515 | 445403 | 473249 | 438966 | 482676 | 482446 | 511013 | 530236 | 543984 |
| 1.1 Agriculture | 409660 | 407176 | 433475 | 398206 | 441360 | 441183 | 468013 | 485937 | |
| 1.2 Forestry & Logging | 17916 | 18399 | 18964 | 19090 | 18872 | 19276 | 19636 | 19991 | |
| 1.3 Fishing | 18939 | 19828 | 20810 | 21671 | 22444 | 21987 | 23364 | 24308 | |
| 2. Mining & Quarrying | 41594 | 42589 | 43335 | 47168 | 48626 | 52591 | 55150 | 58294 | 60267 |
| 3. Manufacturing | 264113 | 284571 | 291803 | 311685 | 332363 | 361115 | 393557 | 440770 | 482358 |
| 3.1 Registered | 173003 | 186570 | 195107 | 209909 | 224919 | 245343 | 267815 | 299931 | |
| 3.2 Unregistered | 91110 | 98001 | 96696 | 1011776 | 107444 | 115772 | 125742 | 140839 | |
| 4. Electricity Gas & Water Supply | 44526 | 45439 | 46228 | 48423 | 50735 | 54745 | 57309 | 60737 | 65492 |
| 5. Construction | 102007 | 108362 | 112692 | 121650 | 136224 | 158217 | 184255 | 206338 | 226200 |
| 6. Trade, Hotels & Restaurants | 254143 | 267326 | 293075 | 313221 | 344743 | 371155 | 406203 | 440677 | 860854# |
| 6.1 Trade | 231878 | 243505 | 267382 | 286060 | 315240 | 338283 | 368918 | 398367 | |
| 6.2 Hotels & Restaurants | 22265 | 23821 | 25693 | 27161 | 29503 | 32872 | 37285 | 42310 | |
| 7. Transport, Storage & Communication | 133371 | 148324 | 160772 | 183471 | 211627 | 244693 | 280535 | 327207 | |
| 7.1 Railways | 21122 | 21996 | 23535 | 24864 | 26336 | 28257 | 30885 | 33668 | |
| 7.2 Transport by Other Means | 82376 | 88735 | 92311 | 101779 | 113907 | 127742 | 138076 | 151042 | |
| 7.3 Storage | 1427 | 1514 | 1523 | 1421 | 1493 | 1695 | 1718 | 1768 | |
| 7.4 Communication | 28446 | 36079 | 43403 | 55407 | 69891 | 86999 | 109856 | 140729 | |

contd. ...

...contd. ...

| 1 | 2 | 3 | 4 | 5 | 6 | 7 | 8 | 9 | 10 |
|---|---|---|---|---|---|---|---|---|---|
| 8. Financing, Insurance, Real Estate & Business Services | 233550 | 243048 | 260737 | 281550 | 297250 | 323080 | 359942 | 410030 | 458097 |
| 8.1 Banking & Insurance | 105662 | 103571 | 113036 | 125845 | 128659 | 139940 | 159881 | 191387 | |
| 8.2 Real Estate, Ownership of Dwellings, Business & Legal services | 127888 | 139477 | 147701 | 155705 | 168591 | 183140 | 200061 | 218643 | |
| 9. Community, Social & Personal Services | 266707 | 279239 | 290715 | 302153 | 318514 | 340342 | 364883 | 390020 | 417201 |
| 9.1 Public Administration & Defence | 122384 | 124700 | 128351 | 130409 | 133789 | 142522 | 153061 | 161547 | |
| 9.2 Other Services | 144323 | 154539 | 162364 | 171744 | 184725 | 197820 | 211822 | 228473 | |
| 10. Gross Domestic Product at Factor Cost (1 to 9) | 1786525 | 1864300 | 1972606 | 2048287 | 2222758 | 2388384 | 2612847 | 2864309 | 3114452 |
| 11. Total Net Domestic Product at Factor Cost | 1605103 | 1670448 | 1764137 | 1824635 | 1981389 | 2126018 | 2326581 | 2549648 | 2772772 |
| 12. Net Factor Income from Abroad | -15431 | -22545 | -20671 | -18805 | -17845 | -21498 | -19687 | -19154 | -12447 |
| 13. NNP at Factor Cost | 1589672 | 1647903 | 1743466 | 1805830 | 1963544 | 2104520 | 2306894 | 2530494 | 2760325 |
| 14. Per Capita income (in Rupees) | 15881 | 16172 | 16764 | 17101 | 18317 | 19325 | 20858 | 22553 | 24256 |

(Q) Quick Estimates # Includes Transport and Communication. (A) Advance Estimates

*Source*: Central Statistical Organisation, New Delhi.

## Table 2.8

*Gross Domestic Product (G.D.P.) at Factor Cost from Agriculture (Excluding Forestry & Logging and Fishing)*

(Rs. Crore)

| Year | G.D.P. (Total) | G.D.P. (Agriculture) | | Percentage Share of Agriculture | | |
|---|---|---|---|---|---|---|
| | At Current Prices | At 1999-00 Prices | At Current Prices | At 1999-00 Prices | At Current Prices | At 1999-00 Prices |
| 1 | 2 | 3 | 4 | 5 | 6 | 7 |
| 1999-2000 | 1786525 | 1786525 | 409660 | 409660 | 22.9 | 22.9 |
| 2000-01 | 1925017 | 1864300 | 408932 | 407176 | 21.2 | 21.8 |
| 2001-02 | 2097726 | 1972606 | 442464 | 433475 | 21.1 | 22.0 |
| 2002-03 | 2261415 | 2048287 | 425521 | 398206 | 18.8 | 19.4 |
| 2003-04 | 2538171 | 2222758 | 483030 | 441360 | 19.0 | 19.9 |
| 2004-05 | 2877706 | 2388384 | 501415 | 441183 | 17.4 | 18.5 |
| 2005-06 | 3275670 | 2612847 | 557118 | 468013 | 17.0 | 17.9 |
| 2006-07(Q) | 3790063 | 2864309 | 634519 | 485937 | 16.7 | 17.0 |

(Q) : Quick Estimates

*Source*: Central Statistical Organisation, New Delhi.

## Table 2.9

*Gross Domestic Product (G.D.P.) at Factor Cost from Agriculture and Allied Sectors (Including Forestry & Logging and Fishing)*

(Rs. Crore)

| Year | G.D.P. (Total) | G.D.P. (Agriculture) | | Percentage Share of Agriculture | | |
|---|---|---|---|---|---|---|
| | At Current Prices | At 1999-00 Prices | At Current Prices | At 1999-00 Prices | At Current Prices | At 1999-00 Prices |
| 1 | 2 | 3 | 4 | 5 | 6 | 7 |
| 1999-2000 | 1786525 | 1786525 | 446515 | 446515 | 25.0 | 25.0 |
| 2000-01 | 1925017 | 1864300 | 449565 | 445403 | 23.4 | 23.9 |
| 2001-02 | 2097726 | 1972606 | 486617 | 473249 | 23.2 | 24.0 |
| 2002-03 | 2261415 | 2048287 | 472060 | 438966 | 20.9 | 21.4 |
| 2003-04 | 2538171 | 2222758 | 532342 | 482676 | 21.0 | 21.7 |
| 2004-05 | 2877706 | 2388384 | 552422 | 482446 | 19.2 | 20.2 |
| 2005-06 | 3275670 | 2612847 | 615845 | 511013 | 18.8 | 19.6 |
| 2006-07(Q) | 3790063 | 2864309 | 695424 | 530236 | 18.3 | 18.5 |

(Q) : Quick Estimates

*Source*: Central Statistical Organisation, New Delhi.

## Table 2.10

*Gross State Domestic Product from Agriculture and Allied Sector at Constant (1999-2000 Prices)**

(Rs Lakh)

| Sl.No. | State/UT | 1999-2000 | 2000-2001 | 2001-2002 | 2002-03 | 2003-04 | 2004-05 | 2005-06 | 2006-07 |
|---|---|---|---|---|---|---|---|---|---|
| 1 | 2 | 3 | 4 | 5 | 6 | 7 | 8 | 9 | 10 |
| 1 | Andhra Pradesh | 3712588 | 4201200 | 4136852 | 3815800 | 4393631 | 4588908 | 4952950 | 5147746 |
| 2 | Arunachal Pradesh | 52476 | 59344 | 56236 | 55862 | 57919 | 56395 | 57631 | NA |
| 3 | Assam | 1226553 | 1210071 | 1202374 | 1217058 | 1235493 | 1301620 | 1344444 | 1386632 |
| 4 | Bihar | 1680972 | 2259353 | 1825980 | 2275813 | 1905670 | 2162930 | 1937959 | 2527972 |
| 5 | Jharkhand | 578766 | 543072 | 728939 | 600379 | 619362 | 723839 | 759400 | 793422 |
| 6 | Goa | 77730 | 67067 | 63934 | 69831 | 78187 | 77572 | 81908 | NA |
| 7 | Gujarat | 1770000 | 1566100 | 2045900 | 1906100 | 2666400 | 2443800 | 3093900 | NA |
| 8 | Haryana | 1630178 | 1687288 | 1685227 | 1666568 | 1795743 | 1856343 | 1824991 | 2001282 |
| 9 | Himachal Pradesh | 319849 | 371990 | 403163 | 412882 | 460381 | 487325 | 500073 | 518738 |
| 10 | Jammu & Kashmir | 490280 | 487991 | 516930 | 529813 | 576838 | 583974 | NA | NA |
| 11 | Karnataka | 2670413 | 2833761 | 2509134 | 2330345 | 2021753 | 2463175 | 1632569 | NA |
| 12 | Kerala | 1517257 | 1552981 | 1574983 | 1600121 | 1583167 | 1699081 | 1783543 | 1891798 |
| 13 | Madhya Pradesh | 2383910 | 1757729 | 271516 | 1767789 | 2417398 | 2313266 | 2405050 | 2462285 |
| 14 | Chhattisgarh | 705169 | 546797 | 778141 | 576458 | 808515 | 767437 | 956459 | NA |
| 15 | Maharashtra | 3885998 | 3726282 | 3995007 | 4043798 | 4277052 | 4047959 | 4314998 | NA |
| 16 | Manipur | 87726 | 91421 | 94669 | 91435 | 105116 | 115062 | 117991 | 122577 |

contd. ...

...*contd.* ...

| 1 | 2 | 3 | 4 | 5 | 6 | 7 | 8 | 9 | 10 |
|---|---|---|---|---|---|---|---|---|---|
| 17 | Meghalaya | 82056 | 86912 | 89659 | 95380 | 97425 | 99579 | 107247 | 112656 |
| 18 | Mizoram | 34910 | 32368 | 32607 | 33923 | 33470 | 34927 | 36189 | 37327 |
| 19 | Nagaland | 78707 | 102662 | 115073 | 129461 | 141748 | 153483 | NA | NA |
| 20 | Orissa | 1314535 | 1217387 | 1413458 | 1171655 | 1438858 | 1486254 | 1533530 | 1562229 |
| 21 | Punjab | 2520771 | 2556676 | 2577276 | 2545812 | 2692576 | 2750829 | 2796906 | 2910231 |
| 22 | Rajasthan | 2454366 | 2303891 | 2909013 | 1934149 | 3507891 | 3028734 | 2953383 | 3243363 |
| 23 | Sikkim | 19578 | 20857 | 22323 | 23698 | 25538 | 26864 | 28196 | 29700 |
| 24 | Tamil Nadu | 2272387 | 2389881 | 2334151 | 1883883 | 1848892 | 2148246 | 2387589 | 2497176 |
| 25 | Tripura | 145327 | 131562 | 159962 | 158415 | 166674 | 172367 | 177883 | NA |
| 26 | Uttar Pradesh | 6025932 | 6066126 | 6164923 | 6173077 | 6412572 | 6343831 | 6382017 | 6734537 |
| 27 | Uttarakhand | 381794 | 403044 | 381625 | 393088 | 415238 | 434486 | 441213 | 453283 |
| 28 | West Bengal | 4089451 | 4075177 | 4427790 | 4349880 | 4503657 | 4600965 | 4703214 | NA |
| 29 | A&N Islands | 30650 | 30545 | 29409 | 30645 | 30868 | 23972 | 24637 | NA |
| 30 | Chandigarh | 5225 | 5032 | 5071 | 5191 | 5491 | 5199 | 5525 | NA |
| 31 | Delhi | 86913 | 78537 | 77360 | 75883 | 77195 | 77687 | 69978 | NA |
| 32 | Pondicherry | 24122 | 24370 | 23706 | 22506 | 23470 | 21752 | 21844 | 22020 |

* As on 28.02.2008

NA-Not Available

*Source*: Central Statistical Organisational.

## Table 2.11

*Share of Agriculture and Allied in Gross State Domestic Product during 2005-06 and 2006-07 (At 1999-2000 Prices)**

*Rs. Lakh*

| Sl. No. | State/UT | 2005-06 Agriculture & Allied SDP | 2005-06 % Share of Agriculture & Allied SDP to GSDP | 2006-07 Agriculture & Allied SDP | 2006-07 % Share of Agriculture & Allied SDP to GSDP |
|---|---|---|---|---|---|
| 1 | 2 | 3 | 4 | 5 | 6 |
| 1 | Andhra Pradesh | 4952950 | 25.9 | 5147746 | 24.8 |
| 2 | Arunachal Pradesh | 57631 | 23.0 | NA | NA |
| 3 | Assam | 1344444 | 29.0 | 1386632 | 27.8 |
| 4 | Bihar | 1937959 | 29.4 | 2527972 | 33.0 |
| 5 | Jharkhand | 759400 | 15.3 | 793422 | 15.0 |
| 6 | Goa | 81908 | 9.5 | NA | NA |
| 7 | Gujarat | 3093900 | 18.3 | NA | NA |
| 8 | Haryana | 1824991 | 22.1 | 2001282 | 21.7 |
| 9 | Himachal Pradesh | 500073 | 23.9 | 518738 | 22.7 |
| 10 | Jammu & Kashmir | NA | NA | NA | NA |
| 11 | Karnataka | 2632569 | 19.4 | NA | NA |
| 12 | Kerala | 1783543 | 18.0 | 1891798 | 17.7 |
| 13 | Madhya Pradesh | 2405050 | 26.0 | 2462285 | 25.6 |
| 14 | Chhattisgarh | 956459 | 22.7 | NA | NA |
| 15 | Maharashtra | 4314998 | 12.7 | NA | NA |
| 16 | Manipur | 11799 | 23.8 | 122577 | 22.2 |
| 17 | Meghalaya | 107247 | 21.0 | 112656 | 20.9 |
| 18 | Mizoram | 36189 | 16.4 | 37327 | 15.8 |
| 19 | Nagaland | NA | NA | NA | NA |
| 20 | Orissa | 1533530 | 24.8 | 1562229 | 23.1 |
| 21 | Punjab | 2796906 | 32.6 | 2910231 | 31.9 |
| 22 | Rajasthan | 2953383 | 28.5 | 3243363 | 29.3 |
| 23 | Sikkim | 28196 | 20.1 | 29700 | 19.6 |
| 24 | Tamil Nadu | 2387589 | 13.3 | 2497176 | 12.9 |
| 25 | Tripura | 177883 | 22.9 | NA | NA |
| 26 | Uttar Pradesh | 6382017 | 29.1 | 6734537 | 28.5 |
| 27 | Uttarakhand | 441213 | 20.9 | 453283 | 19.5 |
| 28 | West Bengal | 4703214 | 24.7 | NA | NA |
| 29 | A&N Islands | 24637 | 20.1 | NA | NA |
| 30 | Chandigarh | 5525 | 0.7 | NA | NA |
| 31 | Delhi | 69978 | 0.8 | NA | NA |
| 32 | Pondicherry | 21844 | 5.1 | 22020 | 4.8 |

NA: Not Available. * As on 28.02.2008

*Note:* Data for the remaining States/Uts is not available.

*Source:* Central Statistical Organisation, New Delhi.

# Table 2.12

*State-Wise Gross State Domestic Product from Agriculture & Allied during 2005-06 and 2006-07 (At Current Prices)**

(Rs. In Lakhs)

| Sl. No. | States/Uts | | | % Growth Over Previous Year | |
|---|---|---|---|---|---|
| | | 2005-06 | 2006-07 | 2005-06 | 2006-07 |
| 1 | 2 | 3 | 4 | 5 | 6 |
| 1 | Andhra Pradesh | 6107329 | 6609890 | 11.6 | 8.2 |
| 2 | Arunachal Pradesh | 72798 | NA | 10.0 | -4.0 |
| 3 | Assam | 1641995 | 1855785 | 13.0 | 13.0 |
| 4 | Bihar | 2231081 | 2859662 | -11.5 | 28.2 |
| 5 | Jharkhand | 836975 | 865240 | -11.5 | 3.4 |
| 6 | Goa | 105667 | NA | 8.7 | NA |
| 7 | Gujarat | 3642300 | NA | 61.0 | NA |
| 8 | Haryana | 2208070 | 2709621 | 11.3 | 22.7 |
| 9 | Himachal Pradesh | 605188 | 638341 | 14.0 | 5.5 |
| 10 | Jammu & Kashmir | NA | NA | 9.1 | NA |
| 11 | Karnataka | 3229037 | NA | -12.3 | NA |
| 12 | Kerala | 2001466 | 2177323 | 2.7 | 8.8 |
| 13 | Madhya Pradesh | 3190202 | 3604691 | 44.4 | 13.0 |
| 14 | Chhattisgarh | 1076217 | NA | 58.1 | NA |
| 15 | Maharashtra | 5361981 | NA | -3.0 | NA |
| 16 | Manipur | 137020 | 144556 | 4.9 | 5.5 |
| 17 | Meghalaya | 127159 | 135137 | 5.0 | 6.3 |
| 18 | Mizoram | 46563 | 47971 | NA | 3.0 |
| 19 | Nagaland | NA | NA | NA | 0.0 |
| 20 | Orissa | 1934073 | 2193160 | 33.3 | 13.4 |
| 21 | Punjab | 3367041 | 3851501 | 10.0 | 14.4 |
| 22 | Rajasthan | 3271115 | 3836794 | 78.7 | 17.3 |
| 23 | Sikkim | 35308 | 38234 | 10.0 | 8.3 |
| 24 | Tamil Nadu | 3012174 | 3109976 | 3.3 | 3.2 |
| 25 | Tripura | 210262 | NA | 4.8 | 0.0 |
| 26 | Uttar Pradesh | 8603215 | 9460843 | 9.5 | 10.0 |
| 27 | Uttarakhand | 568176 | 606922 | 7.2 | 6.8 |
| 28 | West Bengal | 5491787 | NA | 10.4 | 0.0 |
| 29 | A&N Islands | 33021 | NA | NA | 0.0 |
| 30 | Chandigarh | 6401 | NA | 4.9 | 0.0 |
| 31 | Delhi | 87674 | NA | 1.1 | 0.0 |
| 32 | Pondicherry | 29055 | 30860 | 0.5 | 6.2 |

NA: Not Available. * As on 28.02.2008

*Note*: Data for the remaining States/Uts is not available.

*Source*: Central Statistical Organisation, New Delhi.

## Table 2.13

*Share of Agriculture & Allied in Gross State Domestic Product during 2005-06 and 2006-07 (At Current Prices)**

*(Rs. Lakh)*

| Sl. No. | State/UT | 2005-06 Agriculture & Allied SDP | 2005-06 % Share of Agriculture & Allied SDP to GSDP | 2006-07 Agriculture & Allied SDP | 2006-07 % Share of Agriculture & Allied SDP to GSDP |
|---|---|---|---|---|---|
| 1 | 2 | 3 | 4 | 5 | 6 |
| 1 | Andhra Pradesh | 6107329 | 25.9 | 6609890 | 24.6 |
| 2 | Arunachal Pradesh | 72798 | 24.4 | NA | NA |
| 3 | Assam | 1641995 | 28.5 | 1855785 | 28.5 |
| 4 | Bihar | 2231081 | 28.0 | 2859662 | 30.3 |
| 5 | Jharkhand | 836975 | 13.3 | 865240 | 12.4 |
| 6 | Goa | 105667 | 8.5 | NA | NA |
| 7 | Gujarat | 3642300 | 16.8 | NA | NA |
| 8 | Haryana | 2208070 | 20.8 | 2709621 | 21.4 |
| 9 | Himachal Pradesh | 605188 | 23.8 | 638341 | 22.6 |
| 10 | Jammu & Kashmir | NA | NA | NA | NA |
| 11 | Karnataka | 3229037 | 18.9 | NA | NA |
| 12 | Kerala | 2001466 | 16.8 | 2177323 | 16.4 |
| 13 | Madhya Pradesh | 3190202 | 27.4 | 3604691 | 28.1 |
| 14 | Chhattisgarh | 1076217 | 20.7 | NA | NA |
| 15 | Maharashtra | 5361981 | 12.4 | NA | NA |
| 16 | Manipur | 137020 | 24.0 | 144556 | 22.0 |
| 17 | Meghalaya | 127159 | 19.7 | 135137 | 19.2 |
| 18 | Mizoram | 46563.0 | 17.3 | 47971.0 | 16.1 |
| 19 | Nagaland | NA | NA | NA | NA |
| 20 | Orissa | 1934073 | 24.6 | 2193160 | 24.1 |
| 21 | Punjab | 3367041 | 30.7 | 3851501 | 31.2 |
| 22 | Rajasthan | 3271115 | 26.3 | 3836794 | 27.0 |
| 23 | Sikkim | 35308 | 19.6 | 38234 | 18.7 |
| 24 | Tamil Nadu | 3012174 | 13.5 | 3109976 | 12.6 |
| 25 | Tripura | 210262.0 | 23.0 | NA | NA |
| 26 | Uttar Pradesh | 8603215.0 | 30.8 | 9460843 | 30.2 |
| 27 | Uttarakhand | 568176 | 22.0 | 606922 | 20.3 |
| 28 | West Bengal | 5491787 | 23.3 | NA | NA |
| 29 | A&N Islands | 33021 | 21.1 | NA | NA |
| 30 | Chandigarh | 6401 | 0.6 | NA | NA |
| 31 | Delhi | 87674 | 0.8 | NA | NA |
| 32 | Pondicherry | 29055 | 5.1 | 30860 | 4.9 |

* As on 28.02.2008 NA: Not Available

*Note*: Data for the remaining States

GSDP = Gross State Domestic Product

*Source*: Central Statistical Organisation, New Delhi.

# Table 2.14

## *Annual Growth Rate of GDP at Factor Cost in Agriculture during 2000-01 to 2007-08*

*(Per cent)*

| Year | At Current Prices | | At 1999-2000 Prices | |
|---|---|---|---|---|
| | *Agriculture* | *Agriculture, Forestry & Fishing* | *Agriculture* | *Agriculture, Forestry & Fishing* |
| *1* | 2 | 3 | 4 | 5 |
| 2000-01 | 0.3 | 0.7 | -0.8 | -0.2 |
| 2001-02 | 8.2 | 8.2 | 6.5 | 6.3 |
| 2002-03 | -3.8 | -3.0 | -8.1 | -7.2 |
| 2003-04 | 8.2 | 12.8 | 10.8 | 10.0 |
| 2004-05 | 3.8 | 3.8 | 0.0 | 0.0 |
| 2005-06 | 11.1 | 11.5 | 6.1 | 5.9 |
| 2006-07(Q) | 13.9 | 12.9 | 3.8 | 3.8 |
| 2007-08(A) | NA | 8.5 | NA | 2.6 |

(Q) Quick Estimates (A) Advance Estimates

NA - Not Available.

*Source*: Central Statistical Organisation, New Delhi.

## Table 2.15

*Plan-wise Trend of Growth of Total GDP and Agriculture GDP (including Allied Sectors) (at 1993-94 prices)*

| Plan | Average Growth Rate (% Per Annum) | | Compound Growth Rate (% Per Annum) | | Trend Growth Rate (% Per Annum) | | Targeted Average Annual Growth Rate % |
|---|---|---|---|---|---|---|---|
| | GDP | Agri-GDP | GDP | Agri-GDP | GDP | Agri-GDP | Agri-GDP |
| 1 | 2 | 3 | 4 | 5 | 6 | 7 | 8 |
| Fifth Plan (1974-1979) | 4.9 | 3.6 | 4.8 | 3.4 | 5.3 | 4.0 | 3.3 |
| Sixth Plan (1980-1985) | 5.7 | 5.7 | 5.6 | 5.6 | 5.1 | 3.8 | 9.0 |
| Seventh Plan (1985-1990) | 6.0 | 3.1 | 5.9 | 3.0 | 6.3 | 4.0 | 2.5 |
| Two Annual Plans (1990-1992) | 3.5 | 1.3 | 3.4 | 1.2 | - | - | - |
| Eighth Plan (1992-1997) | 6.7 | 4.7 | 7.0 | 4.7 | 6.9 | 3.8 | 3.1 |
| Ninth Plan (1997-2002) | 5.5 | 2.1 | 5.5 | 2.0 | 5.4 | 2.5 | 3.9 |
| Tenth Plan (2002-2007)* | 7.6 | 2.3 | 7.6 | 2.1 | 8.2 | 4.2 | 4.0 |

*At 1999-2000 Prices

*Source:* Original data from CSO's National Accounts Statistics,

# Table 2.15.1

*Quarterly Estimate of GDP at Factor Cost in 2006-07 and 2007-08 (at 1999-2000 Prices)*

| *Industry* | *Gross Domestic Product* | | | | *(Rs. in Crore)* | | | | *Percentage Change Over Previous Year* | | | | | | | |
|---|---|---|---|---|---|---|---|---|---|---|---|---|---|---|---|---|
| | 2006-07 | | | | 2007-08 | | | | 2006-07 | | | | 2007-08 | | | |
| | *Q1* | *Q2* | *Q3* | *Q4* | *Q1* | *Q2* | *Q3* | *Q4* | *Q1* | *Q2* | *Q3* | *Q4* | *Q1* | *Q2* | *Q3* | *Q4* |
| *1* | 2 | 3 | 4 | 5 | 6 | 7 | 8 | 9 | *10* | *11* | *12* | *13* | *14* | *15* | *16* | *17* |
| 1. Agriculture, Forestry & Fishing | 122660 | 98606 | 168666 | 140305 | 128042 | 103199 | 178741 | 144354 | 2.7 | 3.2 | 4.0 | 4.9 | 4.4 | 4.7 | 6.0 | 2.9 |
| 2. Miming & Quarrying | 13852 | 12916 | 14863 | 16663 | 14083 | 13621 | 15706 | 17651 | 4.1 | 3.9 | 6.0 | 8.2 | 1.7 | 5.5 | 5.7 | 5.9 |
| 3. Manufactruring | 103009 | 106853 | 110948 | 119960 | 114270 | 116685 | 121574 | 126928 | 11.7 | 12.2 | 11.3 | 12.8 | 10.9 | 9.2 | 9.6 | 5.8 |
| 4. Electricity, Gas & Water Supply | 14942 | 14961 | 15372 | 15462 | 16121 | 15990 | 16109 | 16324 | 4.3 | 6.6 | 7.6 | 5.4 | 7.9 | 6.9 | 4.8 | 5.6 |
| 5. Construction | 48967 | 48745 | 52745 | 55880 | 52720 | 54478 | 56474 | 62907 | 13.1 | 12.0 | 10.8 | 12.2 | 7.7 | 11.8 | 7.1 | 12.6 |
| 6. Trade, Hotels, Transport & Communication | 172481 | 178064 | 200900 | 216439 | 195068 | 197728 | 224091 | 243325 | 10.9 | 12.7 | 12.1 | 11.6 | 13.1 | 11.0 | 11.5 | 12.4 |
| 7. Financing, Ins., Real Est. & Buisness | 97331 | 99526 | 103342 | 109831 | 109559 | 111833 | 115639 | 121334 | 13.6 | 13.9 | 14.7 | 13.4 | 12.6 | 12.4 | 11.9 | 10.5 |
| 8. Community, Social & Personal Services | 90404 | 96043 | 94495 | 109078 | 95086 | 103451 | 100325 | 119446 | 10.3 | 7.2 | 5.6 | 5.1 | 5.2 | 7.7 | 6.2 | 9.5 |
| GDP at factor cost | 663645 | 655713 | 761333 | 783618 | 724949 | 716984 | 828659 | 852270 | 9.6 | 10.1 | 9.3 | 9.7 | 9.2 | 9.3 | 8.8 | 8.8 |

*Source:* Central Statistical Organisation, New Delhi.

# Table 3.1

*Share of Public Sector Outlays and Expenditure under Agriculture and Allied Activities during Ninth Plan (1997-98 to 2001-02), Tenth Plan (2002-03 to 2006-07) and Annual Plan (2007-08)and Eleventh Plan (2007-12) ^*

(Rs. Crore)

| Five Year Plan/Annual Plan | Agriculture and Allied Activities* | | Total | | Per cent Share of Agriculture and Allied Activities to Total | |
|---|---|---|---|---|---|---|
| | Plan Outlays | Actual Expenditure | Plan Outlays | Actual Expenditure | Plan Outlays | Actual Expenditur |
| 1 | 2 | 3 | 4 | 5 | 6 | 7 |
| Ninth Plan (1997-02) | 42462 | 37239$ | 859200 | 941041$ | 4.9 | 4.0 |
| Annual Plan (1997-98) | 6974 | 5929 | 155905 | 129757 | 4.5 | 4.6 |
| Annual Plan (1998-99) | 8687 | 7698 | 185907 | 151581 | 4.7 | 5.1 |
| Annual Plan (1999-00)** | 8796 | 7365 | 192263 | 160608 | 4.6 | 4.6 |
| Annual Plan (2000-01)** | 8281 | 7577 | 203359 | 164479 | 4.1 | 4.6 |
| Annual Plan (2001-02)** | 9097 | 8248 | 228893 | 186315 | 4.0 | 4.4 |
| Tenth Plan (2002-07) | 58933 | 56997 | 1525639 | 1489395 | 3.9 | 3.8 |
| Annual Plan (2002-03)** | 9977 | 7655 | 247897 | 210203 | 4.0 | 3.8 |
| Annual Plan (2003-04)** | 9948 | 8818 | 255883 | 236664 | 3.9 | 3.7 |
| Annual Plan (2004-05)** | 11109 | 10778 | 287843 | 263434 | 3.9 | 4.1 |
| Annual Plan (2005-06)** | 13840 | 12760 | 361239 | 347857 | 3.8 | 3.7 |
| Annual Plan (2006-07)*** | 16163 | 16987 | 441285 | 431237 | 3.7 | 3.9 |
| Eleventh Plan (2007-12) | 136381 | NA | 3644718 | NA | 3.7 | NA |
| Annual Plan (2007-08)# | 8558 | 8541 | 319992 | 292337 | 2.7 | 2.9 |
| Annual Plan (2008-09)## | 10075 | NA | 375485 | NA | 2.7 | NA |

* Includes Crop Husbandry, Animal Husbandry, Dairy Development, Agricultural Research & Education, Forestry & Wild life, Plantation, Agicultural Marketing, Food Storage & Warehousing, Soil and Water Conservation, Agricutlural Financial Institutions, other Agricultural Programmes and Cooperation.

** For the Centre, some of the revised estimate figures have been taken into account for actual expenditure figures, as the figures for some of the sub-heads are not available.

*** Revised Expenditure has been taken into account for some Head of Development for Centre in place of actual Expenditure.

# For Centre only, as figures for States and UTs are not available yet. Only Revised figures for Centre has been givien.

## For Centre only, as figures for States and Uts are not available yet.

^ Data pertain to eleventh Five Year Plan are as approved by the National Development Council (NDC) in its meeting held on 19th December, 2007.

NA: Not Available

*Source*: Planning Commission, New Delhi.

# Table 3.2

*Public Sector Outlays and Expenditure during Ninth Five Year Plan (1997-98 to 2001-02), Tenth Plan (2002-03 to 2006-07) and Eleven Five Year Plan (2007-2012) under various Heads of Development*

(Rs. Crore)

| Sl. No. | Sector | 9th Plan Outlays$ | 1997-98 | | 1998-99 | | 1999-2000 | | 2000-2001 | | 2001-02 | | 10th Plan Outlays | 2002-03 | |
|---|---|---|---|---|---|---|---|---|---|---|---|---|---|---|---|
| | | | BE | AE** | BE | AE** | BE | AE | BE | AE** | BE | AE** | | BE | AE** |
| 1 | 2 | 3 | 4 | 5 | 6 | 7 | 8 | 9 | 10 | 11 | 12 | 13 | 14 | 15 | 16 |
| 1 | Agriculture and Allied Activities# | 42462 | 6974 | 5929 | 8687 | 7698 | 8796 | 7365 | 8281 | 7577 | 9097 | 8248 | 58933 | 9977 | 7655 |
| 2 | Rural Development | 74686 | 11156 | 10074 | 12203 | 10985 | 12252 | 11281 | 11349 | 9852 | 13444 | 14235 | 121928 | 15778 | 19753 |
| 3 | Special Area Programmes | 3649 | 781 | 874 | 1535 | 1184 | 1630 | 1514 | 1022 | 1045 | 1146 | 919 | 20879 | 1046 | 1066 |
| 4 | Irrigation and Flood Control | 55420 | 11220 | 9905 | 13444 | 10814 | 15672 | 14210 | 16234 | 13529 | 16528 | 14552 | 103315 | 16964 | 11965 |
| 5 | Energy | 222375 | 37111 | 31793 | 45134 | 35572 | 43273 | 35810 | 44856 | 36613 | 49103 | 37145 | 403927 | 53781 | 44710 |
| 6 | Industry and Minerals | 65148 | 13319 | 10306 | 13901 | 7979 | 10918 | 7248 | 11250 | 6866 | 10487 | 7942 | 58939 | 10608 | 8776 |
| 7 | Transport | 119373 | 20661 | 18101 | 25423 | 20347 | 27104 | 23463 | 31271 | 25734 | 35834 | 29918 | 225977 | 44258 | 35244 |
| 8 | Communications | 47280 | 13367 | 10131 | 14887 | 11376 | 16822 | 14039 | 19473 | 14183 | 20300 | 18083 | 98968 | 19551 | 13057 |
| 9 | Science, Technology and Environment | 18458 | 2431 | 2004 | 3289 | 2443 | 3471 | 2942 | 3927 | 3557 | 4042 | 3670 | 30424 | 4783 | 4160 |
| 10 | General Economic Services | 14580 | 2403 | 1811 | 3588 | 3071 | 4601 | 2452 | 4607 | 2538 | 8929 | 2948 | 38630 | 4557 | 4995 |
| 11 | Social Services | 183273 | 34260 | 26867 | 41359 | 38738 | 45154 | 38439 | 48368 | 41336 | 56373 | 46474 | 347391 | 63006 | 56954 |
| 12 | General Services | 12496 | 2221 | 1961 | 2458 | 1374 | 2569 | 1847 | 2721 | 1649 | 3608 | 2181 | 16328 | 3588 | 1868 |
| | Total | 859200 | 155905 | 129757 | 185907 | 151581 | 192263 | 160608 | 203359 | 164479 | 228893 | 186315 | 1525639 | 247897 | 210203 |

contd. ...

...contd. ...

| Sl. No. | Sector | 2003-04 | | 2004-05 | | 2005-06 | | 2006-07 | | 11th Plan | 2007-08(#) | | 2008-09(##) |
|---|---|---|---|---|---|---|---|---|---|---|---|---|---|
| | | BE | AE** | BE | AE** | BE | RE*** | BE | RE*** | Outlays | BE | RE*** | BE |
| 1 | 2 | 17 | 18 | 19 | 20 | 21 | 22 | 23 | 24 | 25 | 26 | 27 | 28 |
| 1 | Agriculture and Allied Activities | 9948 | 8818 | 11109 | 10778 | 13840 | 13440 | 16163 | 16987 | 136381 | 8558 | 8544 | 10075 |
| 2 | Rural Development | 16029 | 19858 | 17634 | 18579 | 23931 | 27214 | 30711 | 30458 | 301069 | 16706 | 17511 | 18972 |
| 3 | Special Area Programmes | 1848 | 1733 | 4939 | 2381 | 4547 | 4904 | 5522 | 5840 | 26329 | | | |
| 4 | Irrigation and Flood Control | 17751 | 14499 | 21111 | 19471 | 26957 | 25007 | 33189 | 32120 | 210326 | 507 | 411 | 411 |
| 5 | Energy | 59689 | 60174 | 65965 | 60375 | 77356 | 71306 | 90499 | 88296 | 854123 | 79158 | 93815 | 93815 |
| 6 | Industry and Minerals | 10101 | 7758 | 10756 | 10121 | 15303 | 13512 | 18213 | 17017 | 153600 | 20434 | 28836 | 28836 |
| 7 | Transport | 42309 | 36186 | 44481 | 38690 | 59992 | 58915 | 72055 | 74625 | 572443 | 71589 | 68930 | 84177 |
| 8 | Communications | 14914 | 12897 | 11764 | 9279 | 12226 | 17773 | 20366 | 18121 | 95380 | 25812 | 16599 | 21937 |
| 9 | Science, Technology and Environment | 5047 | 4417 | 6030 | 5513 | 7520 | 6411 | 8395 | 7258 | 87933 | 8816 | 7742 | 9283 |
| 10 | General Economic Services | 7123 | 4502 | 13462 | 5067 | 10080 | 7146 | 10026 | 8691 | 62523 | 3632 | 3043 | 6052 |
| 11 | Social Services | 68946 | 62621 | 77549 | 79211 | 106370 | 103308 | 131989 | 127738 | 1102327 | 83951 | 78798 | 100778 |
| 12 | General Services | 2179 | 3201 | 2270 | 3266 | 3119 | 2695 | 4159 | 4087 | 42283 | 829 | 533 | 1149 |
| | Total | 255883 | 236664 | 287070 | 263434 | 361239 | 351629 | 441285 | 431238 | 3644718 | 319992 | 324762 | 375485 |

*Note:* $: Indicates Ninth Plan realisation.

**: For the Centre some of the revised estimate figures have been taken in to account for Actual Expenditure figures as the figures for some of the sub-heads are not available.

***: Revised Expenditure has been taken into account for some Head of Development for Centre and State in place of actual Expenditure.

#: For Centre only, as figures for States & UTs are not available yet. Only Revised figures for centre has been given in place of Actual Expenditure.

##: For Centre only, as figures for States & UTs are not available yet.

*Source:* Planning Commission, New Delhi.

## Table 3.3

*Disaggregated Public Sector Outlays and Expenditure under Agriculture and Allied Programmes during Ninth Five Year during Ninth Five Year Plan (1997-98 to 2001-02), Tenth Plan (2002-03 to 2006-07) and Eleventh Plan (2007-12)*

(Rs. Crore)

| Sl. No. | Programme | 1997-98 | | 1998-99 | | 1999-2000 | | 2000-2001 | | 2001-02 | | 2002-03 | |
|---|---|---|---|---|---|---|---|---|---|---|---|---|---|
| | | BE | AE** | BE | AE** | BE | AE** | BE | RE** | BE | AE** | BE | AE** |
| 1 | 2 | 3 | 4 | 5 | 6 | 7 | 8 | 9 | 10 | 11 | 12 | 13 | 14 |
| 1 | Crop Husbandry | 2059 | 1694 | 2655 | 1845 | 2569 | 2175 | 2785 | 2503 | 3402 | 3146 | 3361 | 2597 |
| 2 | Horticulture | - | - | - | - | - | - | - | - | - | - | - | 59 |
| 3 | Soil & Water Conservation | 566 | 561 | 825 | 638 | 932 | 806 | 706 | 885 | 787 | 629 | 1315 | 544 |
| 4 | Animal Husbandry | 451 | 340 | 495 | 326 | 525 | 389 | 461 | 425 | 500 | 488 | 506 | 371 |
| 5 | Dairy Development | 141 | 70 | 133 | 65 | 145 | 61 | 97 | 85 | 101 | 89 | 87 | 105 |
| 6 | Fisheries | 341 | 284 | 397 | 291 | 410 | 323 | 331 | 319 | 346 | 268 | 314 | 269 |
| 7 | Forestry & Wildlife | 1334 | 1264 | 1832 | 2491 | 1966 | 1695 | 1741 | 1740 | 1920 | 1458 | 2032 | 1520 |
| 8 | Plantation | 206 | 136 | 168 | 189 | 156 | 118 | 182 | 175 | 147 | 193 | 177 | 211 |
| 9 | Food Storage & Warehousing | 122 | 117 | 108 | 29 | 136 | 107 | 170 | 26 | 145 | 191 | 186 | 218 |
| 10 | Agriculture, Research & Education | 575 | 541 | 819 | 675 | 888 | 810 | 922 | 851 | 1000 | 968 | 1026 | 939 |
| 11 | Agricultural Financial Institutions | 489 | 154 | 499 | 402 | 303 | 297 | 90 | 108 | 103 | 83 | 57 | 82 |
| 12 | Cooperation | 573 | 616 | 635 | 660 | 632 | 536 | 689 | 381 | 492 | 184 | 214 | 443 |
| 13 | Other Agricultural Programmes | 118 | 153 | 122 | 88 | 134 | 50 | 108 | 79 | 155 | 551 | 703 | 297 |
| | Total Agriculture and Allied Programmes | 6974 | 5929 | 8687 | 7698 | 8796 | 8248 | 8281 | 7577 | 9097 | 8248 | 9978 | 7655 |

contd. ...

*...contd. ...*

| | | 2003-04 | | 2004-05 | | 2005-06 | | 2006-07 | | 2007-08(#) | | 2008-09(##) |
|---|---|---|---|---|---|---|---|---|---|---|---|---|
| | | BE | AE** | BE | AE** | BE | AE** | BE | RE** | BE | RE** | BE |
| 1 | 2 | 15 | 16 | 17 | 18 | 19 | 20 | 21 | 22 | 23 | 24 | 25 |
| 1 | Crop Husbandry | 3726 | 3256 | 4240 | 3932 | 5624 | 4989 | 6196 | 6216 | 4657 | 4922 | 5902 |
| 2 | Horticulture | - | 105 | 118 | 166 | 189 | 177 | 335 | 314 | - | | |
| 3 | Soil & Water Conservation | 893 | 667 | 976 | 767 | 941 | 669 | 1072 | 979 | 21 | 21 | 11 |
| 4 | Animal Husbandry | 545 | 451 | 661 | 634 | 909 | 777 | 1071 | 1185 | 540 | 467 | 620 |
| 5 | Dairy Development | 93 | 69 | 116 | 110 | 154 | 181 | 187 | 216 | 80 | 101 | 89 |
| 6 | Fisheries | 297 | 254 | 375 | 350 | 473 | 446 | 660 | 576 | 284 | 171 | 291 |
| 7 | Forestry & Wildlife | 2070 | 1729 | 2048 | 1758 | 2296 | 2256 | 2769 | 2797 | 627 | 694 | 751 |
| 8 | Plantation | 245 | 177 | 301 | 276 | 345 | 293 | 378 | 383 | 274 | 261 | 313 |
| 9 | Food Storage & Warehousing | 186 | 186 | 126 | 138 | 233 | 231 | 332 | 272 | 333 | 269 | 225 |
| 10 | Agriculture, Research & Education | 1020 | 1029 | 1257 | 1166 | 1434 | 1379 | 1768 | 1911 | 1458 | 1291 | 1584 |
| 11 | Agricultural Financial Institutions | 50 | 60 | 82 | 33 | 66 | 50 | 124 | 109 | - | | |
| 12 | Cooperation | 583 | 389 | 494 | 454 | 618 | 512 | 829 | 530 | 124 | 142 | 137 |
| 13 | Other Agricultural Programmes | 232 | 446 | 315 | 995 | 557 | 800 | 442 | 1499 | 160 | 205 | 152 |
| | Total Agriculture and Allied Programmes | 9940 | 8372 | 11109 | 10777 | 13840 | 11960 | 15721 | 15488 | 8558 | 8544 | 10075 |

*Note*: (1) Totals may not tally due to rounding off of the figures.

$: Indicates Ninth Plan realisation.

**: For the Centre some of the revised estimate figures have been taken in to account for actual Expenditure figures as the figures for some of the sub-heads are not abailable.

***: Revised Expenditure has been taken into account for some Head of Development for Centre and State in place of actual Expenditure.

#: For Centre only, as figures for States & UTs are not available yet. Only Revised figures for centre has been given in place of Actual Expenditure.

##: For Centre only, as figures for States & UTs are not available yet.

*Source*: Planning Commission, New Delhi.

# Table 3.4

*Budget Estimate, Revised Estimate and Actual Expenditure of Department of Agriculture & Cooperation from 1991-92 to 2007-08*

(Rs.in Crore)

| Year | BE | RE | AE |
|---|---|---|---|
| 1 | 2 | 3 | 4 |
| 1991-92 | 1041.35 | 1016.93 | 957.86 |
| 1992-93 | 1050.00 | 1273.16 | 1214.88 |
| 1993-94 | 1330.00 | 1320.05 | 1183.50 |
| 1994-95 | 1405.00 | 1458.84 | 1279.43 |
| 1995-96 | 1490.00 | 1325.39 | 1246.42 |
| 1996-97 | 1471.25 | 1377.91 | 1378.61 |
| 1997-98 | 1431.00 | 1266.28 | 1222.80 |
| 1998-99 | 1956.00 | 1378.41 | 1358.89 |
| 1999-2000 | 1956.00 | 1492.00 | 1471.90 |
| 2000-01 | 1965.00 | 1692.00 | 1666.04 |
| 2001-02 | 1985.00 | 1985.00 | 1792.92 |
| 2002-03 | 2167.00 | 1667.00 | 1676.78 |
| 2003-04 | 2167.00 | 2120.00 | 2050.73 |
| 2004-05 | 2650.00 | 2945.00 | 2657.42 |
| 2005-06 | 4179.32 | 3890.00 | 3798.69 |
| 2006-07 | 4800.00 | 4860.00 | 4860.00 |
| 2007-08 | 5560.00* | 6927.94* | 7049.36* |

*Note*: BE: Budget Estimate, RE: Revised Estimate, AE: Actual Expenditure
* including State Plan Scheme

*Source*: Plan & Policy and Budget Divisions, Department of Agriculture & Cooperation.

## Table 3.5

*Budget Estimate and Expenditure of Department of Agriculture and Cooperation*

(Rs. Crore)

| Period | Centrally Sponsored Schemes | Central Sector Schemes | Externally Aided Projects | State Plan Schemes# | Total |
|---|---|---|---|---|---|
| 1 | 2 | 3 | 4 | 5 | 6 |
| IX Plan Outlays | 5533.65 | 3343.17 | 277.00 | | 9153.82 |
| 1997-98 | | | | | |
| BE | 913.65 | 505.35 | 12.00 | | 1431.00 |
| Expenditure | 804.53 | 399.93 | 18.34 | | 1222.80 |
| 1998-99 | | | | | |
| BE | 1348.08 | 564.19 | 43.73 | | 1956.00 |
| Expenditure | 957.67 | 372.85 | 28.37 | | 1358.89 |
| 1999-00 | | | | | |
| BE | 1166.35 | 735.64 | 54.01 | | 1956.00 |
| Expenditure | 876.91 | 559.05 | 35.94 | | 1471.90 |
| 2000-01 | | | | | |
| BE | 1063.76 | 847.24 | 54.00 | | 1965.00 |
| Expenditure | 583.59 | 1062.60 | 19.85 | | 1666.04 |
| 2001-02 | | | | | |
| BE | 1215.40 | 731.99 | 37.61 | | 1985.00 |
| Expenditure | 801.97 | 962.74 | 28.21 | | 1792.92 |
| **IX Plan Expenditure** | 4024.67 | 3357.17 | 130.71 | | 7512.55 |
| **X Plan Outlays Proposed** | 7236.73 | 5694.00 | 264.27 | | 13195.00 |
| 2002-03 | | | | | |
| BE | 1341.53 | 760.47 | 65.00 | | 2167.00 |
| Expenditure | 839.58 | 737.94 | 79.26 | | 1676.78 |
| 2003-04 | | | | | |
| BE | 1005.72 | 1149.25 | 12.03 | | 2167.00 |
| Expenditure | 961.41 | 1065.72 | 23.64 | | 2050.73 |
| 2004-05 | | | | | |
| BE | 1072.69 | 1543.81 | 33.5 | | 2650.00 |
| Expenditure | 1676.23 | 958.53 | 22.66 | | 2657.42 |
| 2005-06 | | | | | |
| BE | 2785.62 | 1390.61 | 3.09 | | 4179.32 |
| Expenditure | 2336.21 | 1460.87 | 1.61 | | 3798.69 |
| 2006-07 | | | | | |
| BE | 2556.90 | 2243.10 | 0.00 | | 4800.00 |
| Expenditure* | 3253.11 | 1606.89 | 0.00 | | 4860.00 |
| 2007-08 | | | | | |
| BE | 4093.40 | 1426.60 | 0.00 | 40.00 | 5560.00 |
| RE | 3444.49 | 2443.45 | 0.00 | 1040.00 | 6927.94 |
| Expenditure* | 3724.19 | 2038.28 | 0.00 | 1286.89 | 7049.36 |

* Provisional

# The State Plan Schemes introduced from 2007-08 are: (1) Rashtriya Krishi Vikas Yojana and (2) Watershed Development in shifting Cultivation area in North Eastern states

*Source*: Plan & Policy and Budget Divisions, Department of Agriculture & Cooperation

## Table 3.6(a)

*Gross Capital Formation in Agriculture & Allied Sector (At Current Prices)*

*(Rs. Crore)*

| *Year* | *GCF in Agriculture & Allied Sector* | | | *GCF of Economy (by Industry of Use)* | | | *Share of Agriculture & Allied Sector in Total GCF (%)* | | |
|---|---|---|---|---|---|---|---|---|---|
| | *Public Sector* | *Private Sector* | *Total* | *Public Sector* | *Private Sector* | *Total* | *Public Sector* | *Private Sector* | *Total* |
| *(1)* | (2) | (3) | (4) | (5) | (6) | (7) | (8) | (9) | *(10)* |
| 1999-2000 | 8670 | 41481 | 50151 | 144610 | 349389 | 493999 | 6.0 | 11.9 | 10.2 |
| 2000-01 | 8176 | 38558 | 46734 | 144637 | 348648 | 493285 | 5.7 | 11.1 | 9.5 |
| 2001-02 | 10353 | 51285 | 61638 | 156643 | 380311 | 536954 | 6.6 | 13.5 | 11.5 |
| 2002-03 | 9563 | 52319 | 61882 | 149390 | 456143 | 605533 | 6.4 | 11.5 | 10.2 |
| 2003-04 | 12218 | 49249 | 61467 | 174575 | 538325 | 712900 | 7.0 | 9.1 | 8.6 |
| 2004-05 | 16031 | 55662 | 71693 | 216963 | 737926 | 954889 | 7.4 | 7.5 | 7.5 |
| 2005-06 | 20632 | 63323 | 83955 | 272003 | 923405 | 1195408 | 7.6 | 6.9 | 7.0 |
| 2006-07 | 25439 | 71169 | 96608 | 321753 | 1120851 | 1442604 | 7.9 | 6.3 | 6.7 |

*Source*: Central Statisitcal Organisation, New Delhi.

## Table 3.6(b)

*Gross Capital Formation in Agriculture & Allied Sector (At 1999-2000 Prices)*

(Rs. Crore)

| Year | GCF in Agriculture & Allied Sector | | | GCF of Economy (by Industry of Use)** | | | Share of Agriculture & Allied Sector in Total GCF (%) | | |
|---|---|---|---|---|---|---|---|---|---|
| | Public Sector | Private Sector | Total | Public Sector | Private Sector | Total | Public Sector | Private Sector | Total |
| (1) | (2) | (3) | (4) | (5) | (6) | (7) | (8) | (9) | (10) |
| 1999-2000 | 8670 | 41481 | 501151 | 144610 | 349389 | 493999 | 6.0 | 11.9 | 10.2 |
| 2000-01 | 8084 | 37396 | 45480 | 140316 | 330477 | 470793 | 5.8 | 11.3 | 9.7 |
| 2001-02 | 9712 | 47267 | 56979 | 144540 | 344086 | 488626 | 6.7 | 13.7 | 11.7 |
| 2002-03 | 8734 | 46934 | 55668 | 134468 | 407893 | 542361 | 6.5 | 11.5 | 10.3 |
| 2003-04 | 10804 | 42737 | 53541 | 145577 | 465503 | 611080 | 7.4 | 9.2 | 8.8 |
| 2004-05 | 13021 | 44738 | 57759 | 166989 | 580721 | 747710 | 7.8 | 7.7 | 7.7 |
| 2005-06 | 15947 | 48564 | 64511 | 202884 | 687804 | 890688 | 7.9 | 7.1 | 7.2 |
| 2006-07 | 18712 | 52496 | 71208 | 227308 | 791133 | 1018441 | 8.2 | 6.6 | 7.0 |

*Source*: Central Statisitcal Organisation, New Delhi.

## Table 3.6 (c)

*Public and Private Investment in Agriculture and Allied Sectors in Total GDP at Market Prices (1999-2000 Prices)*

| *Year* | *Public Investment* | *Private Investment* | *Total* | *GDP at market price* | *Share (%) in Total* | | |
|---|---|---|---|---|---|---|---|
| | | | | | *Public* | *Private* | *Total* |
| *1* | 2 | 3 | 4 | 5 | 6 | 7 | 8 |
| 2002-03 | 8734 | 46934 | 55668 | 2271134 | 0.4 | 2.1 | 2.5 |
| 2003-04 | 10804 | 42737 | 53541 | 2402727 | 0.4 | 1.8 | 2.1 |
| 2004-05 | 13021 | 44738 | 57759 | 601630 | 0.6 | 7.4 | 9.6 |
| 2005-06 | 15947 | 48569 | 64511 | 2841967 | 0.6 | 1.7 | 2.3 |
| 2006-07 | 18712 | 52496 | 71208 | 3117371 | 0.6 | 1.7 | 2.3 |
| Tenth Plan (2002-03 to 2006-07) | 67218 | 235469 | 302687 | 13180829 | 0.5 | 1.8 | 2.3 |

*Source*: Central Statistical Organisation.

## Table 3.7

*Sectoral Growth Rates and ICOR under Different Sectors of the Economy*

| Sl. No. | Sector | Eighth plan | | Ninth Plan | | Tenth Plan | |
|---|---|---|---|---|---|---|---|
| | | Growth Rate (%) | ICOR | Growth Rate (%) | ICOR | Growth Rate (%)* | ICOR |
| 1 | 2 | 3 | 4 | 5 | 6 | 7 | 8 |
| 1 | Agriculture & Allied Sectors | 4.69 | | 2.06 | 4.05 | 3.97 | 1.99 |
| 2 | Mining & Quarrying | 3.59 | 10.74 | 3.81 | 5.44 | 4.3 | 7.99 |
| 3 | Manufacturing | 9.77 | 6.67 | 3.69 | 18.37 | 9.82 | 7.77 |
| 4 | Electricity, Gas & Water | 5.5 | 18 | 6.46 | 15.43 | 7.99 | 14.97 |
| 5 | Construction | 3.56 | 1.74 | 6.82 | 1 | 8.34 | 0.99 |
| 6 | Trade | 9.06 | 0.54 | 5.86 | 1.09 | 9.44 | 0.91 |
| 7 | Rail Transport | 1.95 | 27.94 | 4.7 | 9.87 | 5.4 | 14.66 |
| 8 | Other Transport | 8.42 | 4.41 | 5.63 | 6.09 | 7.54 | 5.37 |
| 9 | Communications | 14.31 | 7.25 | 17.14 | 5.28 | 15 | 8.33 |
| 10 | Financial Services | 10.21 | 2.23 | 8.93 | 1.35 | 11.69 | 1.56 |
| 11 | Public Administration | 3.91 | 7.82 | 9.21 | 4.09 | 6.43 | 5.45 |
| 12 | Other Services | 6.22 | 4.19 | 8.19 | 3.7 | 9.26 | 3.53 |
| | Total | 6.54 | 3.43 | 5.35 | 4.53 | 7.93 | 3.58 |

*Note*: These are implicit ICORs calculated over the Plan period. For the Tenth Plan, these are targets.
* Estimated

*Source*: Planning Commission, New Delhi.

## 4.1(a)

*Targets and Achievements of Production of Major Crops during Ninth Five Year Plan (1997-98 to 2001-02)*

(Million Tonnes)

| Crop | 1997-98 | | 1998-99 | | 1999-2000 | | 2000-2001 | | 2001-02 | | IXth Plan | |
|---|---|---|---|---|---|---|---|---|---|---|---|---|
| | Targets | Achievements | Targets | Achievements | Targets | Achievements | Targets | Achievements | Targets | Achievements | Targets | Achievements |
| 1 | 2 | 3 | 4 | 5 | 6 | 7 | 8 | 9 | 10 | 11 | 12 | 13 |
| Rice | 83.00 | 82.54 | 86.00 | 86.08 | 86.00 | 89.68 | 90.00 | 84.98 | 92.00 | 93.08 | 437.00 | 436.36 |
| Wheat | 70.00 | 66.35 | 74.00 | 71.29 | 74.00 | 76.37 | 74.00 | 69.68 | 78.00 | 71.81 | 370.00 | 355.50 |
| Coarse Cereals | 34.00 | 30.40 | 34.50 | 31.33 | 34.50 | 30.34 | 33.00 | 31.08 | 33.00 | 33.94 | 169.00 | 157.09 |
| Pulses | 15.00 | 12.97 | 15.50 | 14.91 | 15.50 | 13.41 | 15.00 | 11.07 | 15.00 | 13.19 | 76.00 | 65.55 |
| Foodgrains | 202.00 | 192.26 | 210.00 | 203.61 | 210.00 | 209.80 | 212.00 | 196.81 | 218.00 | 212.02 | 1052.00 | 1014.50 |
| Oilseeds | 25.50 | 21.32 | 27.00 | 24.75 | 27.00 | 20.71 | 28.00 | 18.44 | 28.00 | 20.80 | 135.50 | 106.02 |
| Sugarcane | 280.00 | 279.54 | 300.00 | 288.72 | 305.00 | 299.32 | 325.00 | 295.96 | 325.00 | 300.10 | 1535.00 | 1463.64 |
| Cotton # | 14.80 | 10.85 | 14.80 | 12.29 | 15.00 | 11.53 | 14.50 | 9.52 | 14.50 | 10.09 | 73.60 | 54.28 |
| Jute & Mesta@ | 9.75 | 11.02 | 10.25 | 9.81 | 11.00 | 10.55 | 10.00 | 10.56 | 11.00 | 11.64 | 52.00 | 53.58 |

\# Million Bales of 170 kg. each.

@ Million Bales of 180 kg. each.

# Table 4.1 (b)

*Targets and Achievements of Production of Major Crops during Tenth Five Year Plan (2002-03 to 2006-07) and 2007-08*

(Million Tonnes)

| Crop | 2002-03 | | 2003-04 | | 2004-05 | | 2005-06 | | 2006-07 | | Xth Plan | | 2007-08* | |
|---|---|---|---|---|---|---|---|---|---|---|---|---|---|---|
| | Targets | Achievements | Targets | Achievements | Targets | Achievements | Targets | Achievements | Targets | Achievements | Targets | Achievements | Targets | Achievements |
| 1 | 2 | 3 | 4 | 5 | 6 | 7 | 8 | 9 | 10 | 11 | 12 | 13 | 14 | 15 |
| Rice | 93.00 | 71.82 | 93.00 | 88.53 | 93.50 | 83.13 | 87.80 | 91.79 | 92.80 | 93.35 | 460.10 | 428.37 | 93.00 | 96.43 |
| Wheat | 78.00 | 65.76 | 78.00 | 72.15 | 79.50 | 68.64 | 75.53 | 69.35 | 75.53 | 75.81 | 386.56 | 351.67 | 75.50 | 78.40 |
| Coarse Cereals | 33.00 | 26.07 | 34.00 | 37.60 | 36.80 | 33.46 | 36.52 | 34.06 | 36.52 | 33.92 | 176.84 | 165.63 | 37.50 | 40.73 |
| Pulses | 16.00 | 11.13 | 15.00 | 14.91 | 15.30 | 13.13 | 15.15 | 13.39 | 15.15 | 14.20 | 76.60 | 66.79 | 15.50 | 15.11 |
| Foodgrains | 220.00 | 174.77 | 220.00 | 213.19 | 225.10 | 198.36 | 215.00 | 208.60 | 220.00 | 217.28 | 1100.10 | 1012.47 | 221.50 | 230.67 |
| Oilseeds | 27.00 | 14.84 | 24.70 | 25.29 | 26.20 | 24.35 | 26.58 | 27.98 | 29.40 | 24.29 | 133.88 | 116.75 | 30.00 | 28.83 |
| Sugarcane | 320.00 | 287.38 | 320.00 | 237.31 | 270.00 | 237.09 | 237.50 | 281.17 | 270.00 | 355.52 | 1417.50 | 1398.47 | 310.00 | 340.56 |
| Cotton # | 15.00 | 8.62 | 15.00 | 13.87 | 15.00 | 16.43 | 16.50 | 18.50 | 18.50 | 22.63 | 80.00 | 80.05 | 22.00 | 25.81 |
| Jute & Mesta@ | 12.00 | 11.28 | 12.00 | 11.23 | 11.80 | 10.27 | 11.28 | 10.84 | 11.28 | 11.27 | 58.36 | 54.89 | 11.00 | 11.18 |

* Advance Estimates as released on 09.07.2008

\# Million Bales of 170 kg. each.

@ Million Bales of 180 kg. each.

## Table 4.2

*Three Largest Producing States of Important Crops during 2006-07*

*Production : Million Tonnes*

| *Crop/Group of Crops* | *States* | *Production* |
|---|---|---|
| *1* | 2 | 3 |
| I. Foodgrains | | |
| Rice | West Bengal | 14.75 |
| | Andhra Pradesh | 11.87 |
| | Uttar Pradesh | 11.12 |
| | All-India | 93.35 |
| Wheat | Uttar Pradesh | 25.03 |
| | Punjab | 14.60 |
| | Haryana | 10.06 |
| | All-India | 75.81 |
| Maize | Karnataka | 2.72 |
| | Andhra Pradesh | 2.46 |
| | Bihar | 1.71 |
| | All-India | 15.10 |
| **Total Coarse Cereals** | Maharashtra | 6.14 |
| | Rajasthan | 5.50 |
| | Karnataka | 5.06 |
| | All-India | 33.92 |
| **Total Pulses** | Madhya Pradesh | 3.20 |
| | Maharashtra | 2.30 |
| | Uttar Pradesh | 1.98 |
| | All-India | 14.20 |
| **Total Foodgrains** | Uttar Pradesh | 41.21 |
| | Punjab | 25.31 |
| | Andhra Pradesh | 16.23 |
| | All-India | 217.28 |
| **II. Oilseeds** | | |
| Groundnut | Gujarat | 1.44 |
| | Tamil Nadu | 1.01 |
| | Andhra Pradesh | 0.74 |
| | All-India | 4.86 |
| Rapeseed & Mustard | Rajasthan | 3.81 |
| | Uttar Pradesh | 0.87 |
| | Haryana | 0.80 |
| | All-India | 7.44 |
| Soyabean | Madhya Pradesh | 4.78 |
| | Maharashtra | 2.89 |
| | Rajasthan | 0.77 |
| | All-India | 8.85 |
| Sunflower | Karnataka | 0.52 |
| | Andhra Pradesh | 0.33 |
| | Maharashtra | 0.20 |
| | All-India | 1.23 |

*contd. ...*

*...contd. ...*

| *1* | *2* | *3* |
|---|---|---|
| **Total Oilseeds** | Madhya Pradesh. | 5.81 |
| | Rajasthan | 5.17 |
| | Maharashtra | 3.72 |
| | All-India | 24.29 |
| III. Other Cash Crops | | |
| Sugarcane | Uttar Pradesh | 133.95 |
| | Maharashtra | 78.57 |
| | Tamil Nadu | 41.12 |
| | All-India | 355.52 |
| Cotton @ | Gujarat | 8.79 |
| | Maharashtra | 4.62 |
| | Punjab | 2.68 |
| | All-India | 22.63 |
| Jute & Mesta$ | West Bengal | 8.51 |
| | Bihar | 1.39 |
| | Assam | 0.58 |
| | All-India | 11.27 |
| Potato | Uttar Pradesh | 10.25 |
| | West Bengal | 5.05 |
| | Punjab | 1.22 |
| | All-India | 22.09 |
| Onion | Gujarat | 2.13 |
| | Maharashtra | 1.11 |
| | Karnataka | 0.85 |
| | All-India | 7.16 |

@: Production in million bales of 170 kgs. each.

$: Production in million bales of 180 kg. each.

# Table 4.3

*Normal (Average of 2002-03 to 2006-07) Area, Production and Yield of Major Crops in India*

*Area- Million Hectares*
*Production- Million Tonnes*
*Yield- Kg./Hectare*

| *Crop/Group of Crops* | | *Season* | *Area* | *Production* | *Yield* |
|---|---|---|---|---|---|
| 1 | | 2 | 3 | 4 | 5 |
| I. Foodgrains | | | | | |
| | Rice | *Kharif* | 38.91 | 74.47 | 1914 |
| | | *Rabi* | 3.72 | 11.25 | 3027 |
| | | Total | 42.63 | 85.72 | 2011 |
| | Wheat | *Rabi* | 26.53 | 70.34 | 2651 |
| | *Jowar* | *Kharif* | 4.06 | 4.18 | 1029 |
| | | *Rabi* | 4.91 | 2.96 | 604 |
| | | Total | 8.97 | 7.14 | 796 |
| | *Bajra* | *Kharif* | 9.33 | 8.17 | 876 |
| | Maize | *Kharif* | 6.58 | 11.44 | 1740 |
| | | *Rabi* | 0.80 | 2.58 | 3220 |
| | | Total | 7.38 | 14.02 | 1901 |
| **Total Coarse Cereals** | | *Kharif* | 22.55 | 26.18 | 1161 |
| | | *Rabi* | 6.36 | 6.84 | 1075 |
| | | Total | 28.91 | 33.02 | 1142 |
| | *Tur* | *Kharif* | 3.51 | 2.39 | 681 |
| | Gram | *Rabi* | 6.82 | 5.47 | 803 |
| **Total Pulses** | | *Kharif* | 10.86 | 4.94 | 455 |
| | | *Rabi* | 11.60 | 8.41 | 725 |
| | | Total | 22.46 | 13.35 | 594 |
| **Total Foodgrains** | | *Kharif* | 72.33 | 105.60 | 1460 |
| | | *Rabi* | 48.21 | 96.84 | 2009 |
| | | Total | 120.54 | 202.44 | 1679 |
| II. Oilseeds | | | | | |
| Groundnut | | Total | 6.18 | 6.38 | 1031 |
| Rapeseed & Mustard | | *Rabi* | 6.27 | 6.67 | 1063 |
| Soyabean | | *Kharif* | 7.25 | 7.29 | 1006 |
| Sunflower | | Total | 2.06 | 1.13 | 549 |
| **Nine Oilseeds** | | **Total** | **25.41** | **23.33** | **918** |
| III. Other Cash Crops | | | | | |
| Sugarcane | | Total | 4.29 | 279.01 | 64969 |
| Cotton@ | | Total | 8.38 | 15.98 | 324 |
| Jute & Mesta$ | | Total | 0.96 | 10.97 | 2063 |
| Potato | | Total | 1.37 | 23.19 | 16962 |
| Onion | | Total | 0.55 | 6.54 | 11987 |

*Note*: 'Normal' is worked out as simple average of estimates for 5 years, i.e. 2002-03 to 2006-07.
@: Production in million bales of 170 kg. each.
$: Production in million bales of 180 kg. each.

# Table 4.4

## *Season-wise Area, Production and Yield of Foodgrains from 1966-1967 to 2007-08*

*A- Area in Million Hectares*
*P- Production in Million Tonnes*
*Y- Yield in Kg./Hectare.*

| *Year* | *Kharif* | | | *Rabi* | | | *Total* | | |
|---|---|---|---|---|---|---|---|---|---|
| | A | P | Y | A | P | Y | A | P | Y |
| 1 | 2 | 3 | 4 | 5 | 6 | 7 | 8 | 9 | 10 |
| 1966-67 | 78.21 | 48.89 | 625 | 37.09 | 25.34 | 683 | 115.30 | 74.23 | 644 |
| 1967-68 | 81.49 | 60.76 | 746 | 39.93 | 34.29 | 859 | 121.42 | 95.05 | 783 |
| 1968-69 | 80.40 | 59.57 | 741 | 40.03 | 34.44 | 860 | 120.43 | 94.01 | 781 |
| 1969-70 | 82.30 | 62.35 | 758 | 41.27 | 37.15 | 900 | 123.57 | 99.50 | 805 |
| 1970-71 | 82.36 | 68.92 | 837 | 41.96 | 39.50 | 941 | 124.32 | 108.42 | 872 |
| 1971-72 | 79.22 | 62.99 | 795 | 43.40 | 42.18 | 972 | 122.62 | 105.17 | 858 |
| 1972-73 | 78.34 | 58.64 | 749 | 40.94 | 38.39 | 938 | 119.28 | 97.03 | 813 |
| 1973-74 | 84.12 | 67.84 | 806 | 42.42 | 36.83 | 868 | 126.54 | 104.67 | 827 |
| 1974-75 | 79.74 | 59.10 | 741 | 41.34 | 40.73 | 985 | 121.08 | 99.83 | 824 |
| 1975-76 | 83.15 | 73.89 | 889 | 45.03 | 47.14 | 1047 | 128.18 | 121.03 | 944 |
| 1976-77 | 81.18 | 66.53 | 820 | 43.18 | 44.64 | 1034 | 124.36 | 111.17 | 894 |
| 1977-78 | 82.88 | 77.72 | 938 | 44.64 | 48.69 | 1091 | 127.52 | 126.41 | 991 |
| 1978-79 | 82.85 | 78.08 | 942 | 46.16 | 53.82 | 1166 | 129.01 | 131.90 | 1022 |
| 1979-80 | 80.79 | 63.25 | 783 | 44.42 | 46.45 | 1046 | 125.21 | 109.70 | 876 |
| 1980-81 | 83.21 | 77.65 | 933 | 43.46 | 51.94 | 1195 | 126.67 | 129.59 | 1023 |
| 1981-82 | 83.93 | 79.38 | 946 | 45.21 | 53.92 | 1193 | 129.14 | 133.30 | 1032 |
| 1982-83 | 79.08 | 69.90 | 884 | 46.02 | 59.62 | 1296 | 125.10 | 129.52 | 1035 |
| 1983-84 | 84.14 | 89.23 | 1060 | 47.02 | 63.14 | 1343 | 131.16 | 152.37 | 1162 |
| 1984-85 | 81.18 | 84.52 | 1041 | 45.49 | 61.02 | 1341 | 126.67 | 145.54 | 1149 |
| 1985-86 | 81.80 | 85.25 | 1042 | 46.22 | 65.19 | 1410 | 128.02 | 150.44 | 1175 |
| 1986-87 | 81.46 | 80.20 | 985 | 45.74 | 63.22 | 1382 | 127.20 | 143.42 | 1128 |
| 1987-88 | 74.89 | 74.56 | 996 | 44.80 | 65.79 | 1469 | 119.69 | 140.35 | 1173 |
| 1988-89 | 82.03 | 95.64 | 1166 | 45.64 | 74.28 | 1628 | 127.67 | 169.92 | 1331 |
| 1989-90 | 81.40 | 100.99 | 1241 | 45.37 | 70.05 | 1544 | 126.77 | 171.04 | 1349 |
| 1990-91 | 80.78 | 99.44 | 1231 | 47.06 | 76.95 | 1635 | 127.84 | 176.39 | 1380 |
| 1991-92 | 78.02 | 91.59 | 1174 | 43.85 | 76.79 | 1751 | 121.87 | 168.38 | 1382 |
| 1992-93 | 77.92 | 101.47 | 1302 | 45.23 | 78.01 | 1725 | 123.15 | 179.48 | 1457 |
| 1993-94 | 75.81 | 100.40 | 1324 | 46.94 | 83.86 | 1787 | 122.75 | 184.26 | 1501 |
| 1994-95 | 75.19 | 101.09 | 1344 | 48.67 | 90.41 | 1858 | 123.86 | 191.50 | 1546 |
| 1995-96 | 73.60 | 95.12 | 1292 | 47.42 | 85.30 | 1799 | 121.02 | 180.42 | 1491 |
| 1996-97 | 75.34 | 103.92 | 1379 | 48.24 | 95.52 | 1980 | 123.58 | 199.44 | 1614 |
| 1997-98 | 74.15 | 101.58 | 1370 | 49.70 | 90.68 | 1825 | 123.85 | 192.26 | 1552 |
| 1998-99 | 73.99 | 102.91 | 1391 | 51.18 | 100.69 | 1967 | 125.17 | 203.60 | 1627 |
| 1999-00 | 73.24 | 105.51 | 1441 | 49.87 | 104.29 | 2091 | 123.11 | 209.80 | 1704 |
| 2000-01 | 75.22 | 102.09 | 1357 | 45.83 | 94.73 | 2067 | 121.05 | 196.81 | 1626 |
| 2001-02 | 74.23 | 112.07 | 1510 | 48.55 | 100.78 | 2076 | 122.78 | 212.85 | 1734 |
| 2002-03 | 68.56 | 87.22 | 1272 | 45.30 | 87.55 | 1933 | 113.86 | 174.77 | 1535 |
| 2003-04 | 75.44 | 117.00 | 1551 | 48.01 | 96.19 | 2004 | 123.45 | 213.19 | 1727 |
| 2004-05 | 72.26 | 103.31 | 1430 | 47.82 | 95.05 | 2004 | 120.08 | 198.36 | 1652 |
| 2005-06 | 72.72 | 109.87 | 1511 | 48.88 | 98.73 | 2020 | 121.60 | 208.60 | 1715 |
| 2006-07 | 72.67 | 110.58 | 1522 | 51.04 | 106.71 | 2091 | 123.71 | 217.28 | 1756 |
| 2007-08* | 73.58 | 120.97 | 1644 | 50.86 | 109.70 | 2157 | 124.44 | 230.67 | 1854 |

* Advance Estimates as released on 09.07.2008

# Table 4.5(a)

*All-India Area, Production and Yield of Foodgrains from 1950-51 to 2007-08 Alongwith Percentage Coverage under Irrigation*

*Area- Million Hectares*
*Production- Million Tonnes*
*Yield- Kg./Hectare*

| Year | Area | Production | Yield | % Coverage Under Irrigation |
|---|---|---|---|---|
| 1 | 2 | 3 | 4 | 5 |
| 1950-51 | 97.32 | 50.82 | 522 | 18.1 |
| 1951-52 | 96.96 | 51.99 | 536 | 18.4 |
| 1952-53 | 102.09 | 59.20 | 580 | 18.1 |
| 1953-54 | 109.07 | 69.82 | 640 | 18.1 |
| 1954-55 | 107.86 | 68.03 | 631 | 18.4 |
| 1955-56 | 110.56 | 66.85 | 605 | 18.5 |
| 1956-57 | 111.14 | 69.86 | 629 | 18.2 |
| 1957-58 | 109.48 | 64.31 | 587 | 19.3 |
| 1958-59 | 114.76 | 77.14 | 672 | 18.7 |
| 1959-60 | 115.82 | 76.67 | 662 | 18.8 |
| 1960-61 | 115.58 | 82.02 | 710 | 19.1 |
| 1961-62 | 117.23 | 82.71 | 706 | 19.1 |
| 1962-63 | 117.84 | 80.15 | 680 | 19.8 |
| 1963-64 | 117.42 | 80.64 | 687 | 19.8 |
| 1964-65 | 118.11 | 89.36 | 757 | 20.2 |
| 1965-66 | 115.10 | 72.35 | 629 | 20.9 |
| 1966-67 | 115.30 | 74.23 | 644 | 22.2 |
| 1967-68 | 121.42 | 95.05 | 783 | 21.6 |
| 1968-69 | 120.43 | 94.01 | 781 | 23.6 |
| 1969-70 | 123.57 | 99.50 | 805 | 23.7 |
| 1970-71 | 124.32 | 108.42 | 872 | 24.1 |
| 1971-72 | 122.62 | 105.17 | 858 | 24.5 |
| 1972-73 | 119.28 | 97.03 | 813 | 25.4 |
| 1973-74 | 126.54 | 104.67 | 827 | 24.5 |
| 1974-75 | 121.08 | 99.83 | 824 | 26.5 |
| 1975-76 | 128.18 | 121.03 | 944 | 26.5 |
| 1976-77 | 124.36 | 111.17 | 894 | 27.4 |
| 1977-78 | 127.52 | 126.41 | 991 | 27.7 |

contd. ...

*...contd. ...*

| 1 | 2 | 3 | 4 | 5 |
|---|---|---|---|---|
| 1978-79 | 129.01 | 131.90 | 1022 | 28.8 |
| 1979-80 | 125.21 | 109.70 | 876 | 30.3 |
| 1980-81 | 126.67 | 129.59 | 1023 | 29.7 |
| 1981-82 | 129.14 | 133.30 | 1032 | 29.6 |
| 1982-83 | 125.10 | 129.52 | 1035 | 30.8 |
| 1983-84 | 131.16 | 152.37 | 1162 | 30.9 |
| 1984-85 | 126.67 | 145.54 | 1149 | 31.9 |
| 1985-86 | 128.02 | 150.44 | 1175 | 31.4 |
| 1986-87 | 127.20 | 143.42 | 1128 | 32.6 |
| 1987-88 | 119.69 | 140.35 | 1173 | 33.5 |
| 1988-89 | 127.67 | 169.92 | 1331 | 34.4 |
| 1989-90 | 126.77 | 171.04 | 1349 | 35.0 |
| 1990-91 | 127.84 | 176.39 | 1380 | 35.1 |
| 1991-92 | 121.87 | 168.38 | 1382 | 37.4 |
| 1992-93 | 123.15 | 179.48 | 1457 | 37.4 |
| 1993-94 | 122.75 | 184.26 | 1501 | 38.7 |
| 1994-95 | 123.86 | 191.50 | 1546 | 39.6 |
| 1995-96 | 121.01 | 180.42 | 1491 | 40.1 |
| 1996-97 | 123.58 | 199.44 | 1614 | 40.0 |
| 1997-98 | 123.85 | 192.26 | 1552 | 40.8 |
| 1998-99 | 125.17 | 203.61 | 1627 | 42.4 |
| 1999-00 | 123.10 | 209.80 | 1704 | 43.9 |
| 2000-01 | 121.05 | 196.81 | 1626 | 43.4 |
| 2001-02 | 122.78 | 212.85 | 1734 | 43.0 |
| 2002-03 | 113.86 | 174.77 | 1535 | 42.8 |
| 2003-04 | 123.45 | 213.19 | 1727 | 42.2 |
| 2004-05 | 120.00 | 198.36 | 1652 | 44.2 |
| 2005-06 | 121.60 | 208.60 | 1715 | 45.5 |
| 2006-07 | 123.71 | 217.28 | 1756 | NA |
| 2007-08* | 124.44 | 230.67 | 1854 | NA |

* Advance Estimates as released on 09.07.2008.

*Note*: The yield rates given above have been worked out on the basis of production & area figures taken in '000 units.

*Total Foodgrains—All India Production*

*Million Tonnes*

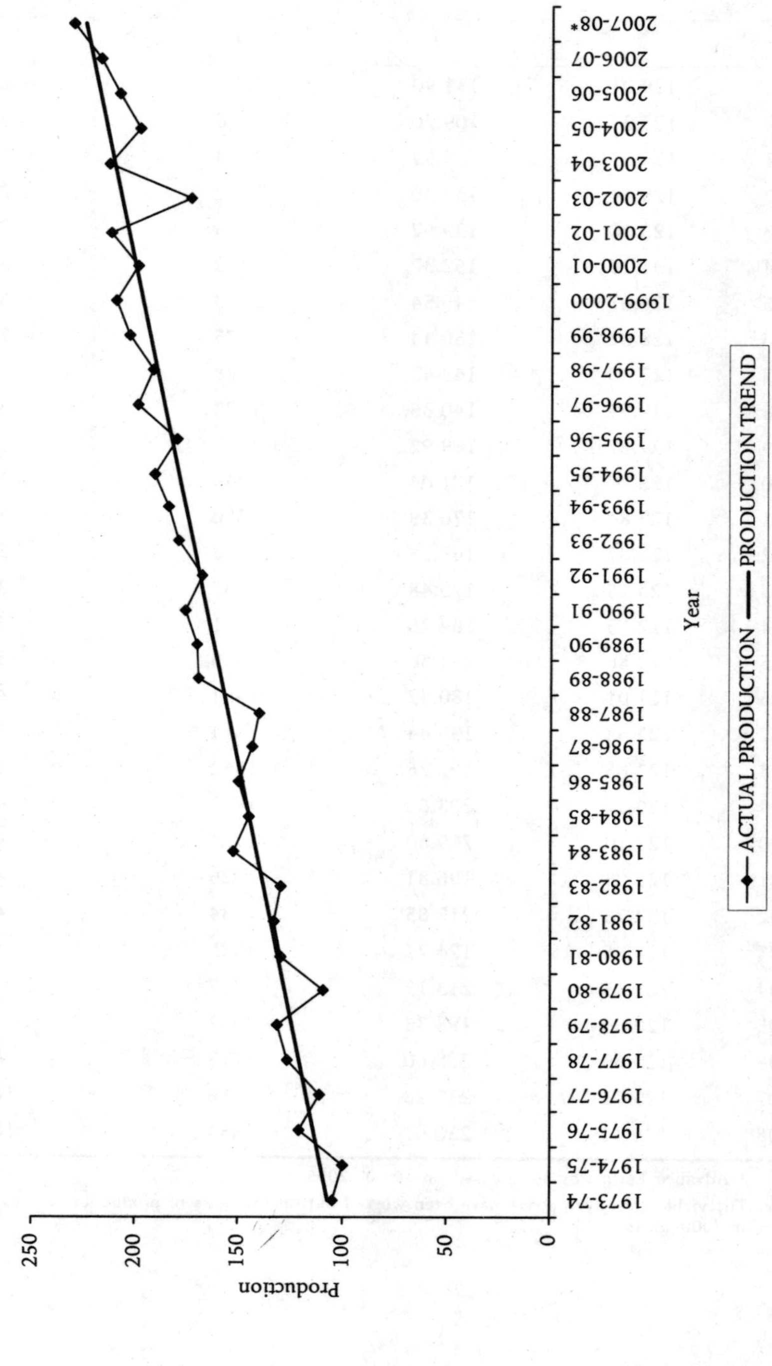


* Based on Advance estimate as on 09-07-08.

# Table 4.5 (b)

*Area, Production and Yield of Foodgrains during 2005-06 and 2006-07 in Respect of Major Foodgrains Producing States Alongwith Coverage under Irrigation*

*Area- Million Hectares*
*Production- Million Tonnes*
*Yield- Kg./Hectare*

| State | 2006-07 | | | | | | 2005-06 | | | | | | % Coverage Under Irrigation During 2005-06* |
|---|---|---|---|---|---|---|---|---|---|---|---|---|---|
| | Area | % to All-India | Production | % to All-India | Cumulative % | Yield | Area | % to All-India | Production | % to All-India | Cumulative % | Yield | |
| 1 | 2 | 3 | 4 | 5 | 6 | 7 | 8 | 9 | 10 | 11 | 12 | 13 | 14 |
| Uttar Pradesh | 20.04 | 16.20 | 41.21 | 18.97 | 18.97 | 2057 | 19.64 | 16.15 | 40.41 | 19.37 | 19.37 | 2057 | 71.6 |
| Punjab | 6.30 | 5.09 | 25.31 | 11.65 | 30.61 | 4017 | 6.32 | 5.20 | 25.18 | 12.07 | 31.44 | 3986 | 97.6 |
| Andhra Pradesh | 7.27 | 5.88 | 16.23 | 7.47 | 38.08 | 2231 | 7.17 | 5.90 | 16.95 | 8.13 | 39.57 | 2365 | 58.6 |
| West Bengal | 6.36 | 5.14 | 15.97 | 7.35 | 45.43 | 2511 | 6.44 | 5.30 | 15.61 | 7.48 | 47.05 | 2423 | 48.6 |
| Haryana | 4.35 | 3.52 | 14.76 | 6.79 | 52.23 | 3393 | 4.27 | 3.51 | 13.00 | 6.23 | 53.28 | 3045 | 85.2 |
| Rajasthan | 12.70 | 10.27 | 14.21 | 6.54 | 58.77 | 1119 | 12.45 | 10.24 | 11.45 | 5.49 | 58.77 | 919 | 25.0 |
| Madhya Pradesh | 11.78 | 9.52 | 13.75 | 6.33 | 65.10 | 1167 | 11.68 | 9.61 | 13.20 | 6.33 | 65.10 | 1130 | 39.7 |
| Maharashtra | 13.45 | 10.87 | 12.65 | 5.82 | 70.92 | 940 | 12.75 | 10.49 | 12.09 | 5.80 | 70.90 | 948 | 16.7 |
| Bihar | 6.70 | 5.42 | 11.10 | 5.11 | 76.03 | 1656 | 6.55 | 5.39 | 8.59 | 4.12 | 75.01 | 1311 | 59.3 |
| Karnataka | 7.45 | 6.02 | 9.60 | 4.42 | 80.44 | 1289 | 7.60 | 6.25 | 13.49 | 6.47 | 81.48 | 1776 | 26.0 |
| Tamil Nadu | 3.17 | 2.56 | 8.26 | 3.80 | 84.25 | 2610 | 3.32 | 2.73 | 6.13 | 2.94 | 84.42 | 1847 | 61.7 |
| Orissa | 5.40 | 4.37 | 7.34 | 3.38 | 87.62 | 1359 | 5.46 | 4.49 | 7.36 | 3.53 | 87.95 | 1349 | 37.0 |
| Gujarat | 4.57 | 3.69 | 6.50 | 2.99 | 90.62 | 1423 | 3.97 | 3.26 | 6.15 | 2.95 | 90.90 | 1551 | 37.1 |
| Chhattisgarh | 5.06 | 4.09 | 5.81 | 2.67 | 93.29 | 1148 | 5.15 | 4.24 | 5.71 | 2.74 | 93.63 | 1111 | 24.3 |
| Jharkhand | 2.38 | 1.92 | 3.69 | 1.70 | 94.99 | 1550 | 1.93 | 1.59 | 2.07 | 0.99 | 94.63 | 1073 | 4.3 |
| Assam | 2.38 | 1.92 | 3.06 | 1.41 | 96.40 | 1286 | 2.60 | 2.14 | 3.68 | 1.76 | 96.39 | 1416 | 7.0 |
| Uttarakhand | 0.99 | 0.80 | 1.74 | 0.80 | 97.20 | 1760 | 1.03 | 0.85 | 1.59 | 0.76 | 97.15 | 1548 | 39.6 |
| Others | 3.36 | 2.72 | 6.09 | 2.80 | 100.00 | @ | 3.27 | 2.69 | 5.94 | 2.85 | 100.00 | @ | 38.1 |
| All India | 123.71 | 100.00 | 217.28 | 100.00 | | 1756 | 121.60 | 100.01 | 208.60 | 100.00 | | 1715 | 45.5 |

@ Since area/production is low in individual states, yield rate is not worked out.

* Provisional

*Note:* States have been arranged in descending order of percentage share of production during 2006-07.

# Table 4.6 (a)

*All-India Area, Production and Yield of Rice from 1950-51 to 2007-08 Alongwith Percentage Coverage under Irrigation*

*Area- Million Hectares*
*Production- Million Tonnes*
*Yield- Kg./Hectare*

| Year | Area | Production | Yield | '% Coverage Under Irrigation |
|---|---|---|---|---|
| 1 | 2 | 3 | 4 | 5 |
| 1950-51 | 30.81 | 20.58 | 668 | 31.7 |
| 1951-52 | 29.83 | 21.30 | 714 | 31.7 |
| 1952-53 | 29.97 | 22.90 | 764 | 32.3 |
| 1953-54 | 31.29 | 28.21 | 902 | 33.6 |
| 1954-55 | 30.77 | 25.22 | 820 | 34.4 |
| 1955-56 | 31.52 | 27.56 | 874 | 34.9 |
| 1956-57 | 32.28 | 29.04 | 900 | 35.4 |
| 1957-58 | 32.30 | 25.53 | 790 | 36.4 |
| 1958-59 | 33.17 | 30.85 | 930 | 36.3 |
| 1959-60 | 33.82 | 31.68 | 937 | 35.8 |
| 1960-61 | 34.13 | 34.58 | 1013 | 36.8 |
| 1961-62 | 34.69 | 35.66 | 1028 | 37.5 |
| 1962-63 | 35.69 | 33.21 | 931 | 37.4 |
| 1963-64 | 35.81 | 37.00 | 1033 | 37.1 |
| 1964-65 | 36.46 | 39.31 | 1078 | 37.3 |
| 1965-66 | 35.47 | 30.59 | 862 | 36.5 |
| 1966-67 | 35.25 | 30.44 | 863 | 37.9 |
| 1967-68 | 36.44 | 37.61 | 1032 | 38.6 |
| 1968-69 | 36.97 | 39.76 | 1076 | 38.4 |
| 1969-70 | 37.68 | 40.43 | 1073 | 38.2 |
| 1970-71 | 37.59 | 42.22 | 1123 | 38.4 |
| 1971-72 | 37.76 | 43.07 | 1141 | 37.2 |
| 1972-73 | 36.69 | 39.24 | 1070 | 39.1 |
| 1973-74 | 38.29 | 44.05 | 1151 | 38.4 |
| 1974-75 | 37.89 | 39.58 | 1045 | 38.8 |
| 1975-76 | 39.48 | 48.74 | 1235 | 38.7 |
| 1976-77 | 38.51 | 41.92 | 1089 | 38.4 |
| 1977-78 | 40.28 | 52.67 | 1308 | 40.2 |

contd. ...

*...contd. ...*

| *1* | 2 | *3* | 4 | 5 |
|---|---|---|---|---|
| 1978-79 | 40.48 | 53.77 | 1328 | 41.6 |
| 1979-80 | 39.42 | 42.33 | 1074 | 42.8 |
| 1980-81 | 40.15 | 53.63 | 1336 | 40.7 |
| 1981-82 | 40.71 | 53.25 | 1308 | 41.5 |
| 1982-83 | 38.26 | 47.12 | 1231 | 42.0 |
| 1983-84 | 41.24 | 60.10 | 1457 | 42.7 |
| 1984-85 | 41.16 | 58.34 | 1417 | 43.7 |
| 1985-86 | 41.14 | 63.83 | 1552 | 42.9 |
| 1986-87 | 41.17 | 60.56 | 1471 | 44.1 |
| 1987-88 | 38.81 | 56.86 | 1465 | 43.6 |
| 1988-89 | 41.73 | 70.49 | 1689 | 45.8 |
| 1989-90 | 42.17 | 73.57 | 1745 | 46.1 |
| 1990-91 | 42.69 | 74.29 | 1740 | 45.5 |
| 1991-92 | 42.65 | 74.68 | 1751 | 47.3 |
| 1992-93 | 41.78 | 72.86 | 1744 | 48.0 |
| 1993-94 | 42.54 | 80.30 | 1888 | 48.6 |
| 1994-95 | 42.81 | 81.81 | 1911 | 49.8 |
| 1995-96 | 42.84 | 76.98 | 1797 | 49.9 |
| 1996-97 | 43.43 | 81.74 | 1882 | 51.0 |
| 1997-98 | 43.45 | 82.53 | 1900 | 50.8 |
| 1998-99 | 44.80 | 86.08 | 1921 | 52.3 |
| 1999-00 | 45.16 | 89.68 | 1936 | 53.9 |
| 2000-01 | 44.71 | 84.98 | 1901 | 53.6 |
| 2001-02 | 44.90 | 93.34 | 2079 | 53.2 |
| 2002-03 | 41.18 | 71.82 | 1744 | 50.2 |
| 2003-04 | 42.59 | 88.53 | 2077 | 52.6 |
| 2004-05 | 41.91 | 83.13 | 1984 | 54.7 |
| 2005-06 | 43.66 | 91.79 | 2102 | 56.0 |
| 2006-07 | 43.81 | 93.35 | 2131 | NA |
| 2007-08* | 43.77 | 96.43 | 2203 | NA |

* Advance Estimates as released on 09.07.2008.

*Note*: The yield rates given above have been worked out on the basis of production & area figures taken in '000 units.

*Rice—All India Production*

*Million Tonnes*

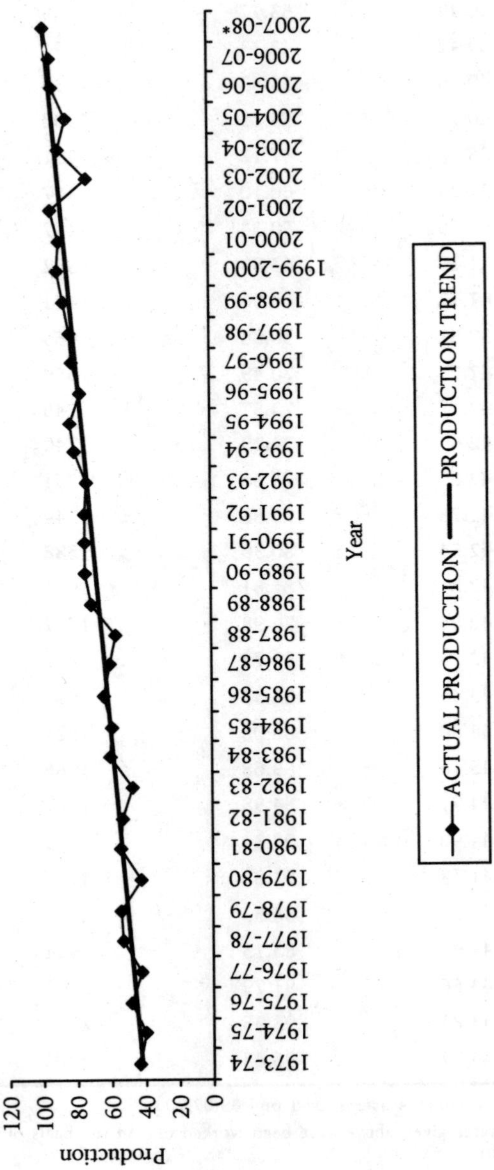


* Based on Advance estimate as on 09-07-08.

## Table 4.6 (b)

### *Area, Production and Yield of Rice during 2005-06 and 2006-07 in Respect of Major Rice Producing States Alongwith Coverage under Irrigation*

*Area- Million Hectares*
*Production- Million Tonnes*
*Yield- Kg./Hectare*

| State | 2006-07 | | | | | | 2005-06 | | | | | | % Coverage under Irrigation during 2005-06* |
|---|---|---|---|---|---|---|---|---|---|---|---|---|---|
| | Area | % to All-India | Production | % to All-India | Cumulative % | Yield | Area | % to All-India | Production | % to All-India | Cumulative % | Yield | |
| 1 | 2 | 3 | 4 | 5 | 6 | 7 | 8 | 9 | 10 | 11 | 12 | 13 | 14 |
| West Bengal | 5.69 | 12.99 | 14.75 | 15.80 | 15.80 | 2593 | 5.78 | 13.24 | 14.51 | 15.81 | 15.81 | 2509 | 49.8 |
| Andhra Pradesh | 3.98 | 9.08 | 11.87 | 12.71 | 28.51 | 2984 | 3.98 | 9.12 | 11.70 | 12.75 | 28.55 | 2939 | 96.6 |
| Uttar Pradesh | 5.92 | 13.51 | 11.12 | 11.91 | 40.42 | 1879 | 5.58 | 12.78 | 11.13 | 12.13 | 40.68 | 1996 | 73.0 |
| Punjab | 2.62 | 5.98 | 10.14 | 10.86 | 51.29 | 3868 | 2.64 | 6.05 | 10.19 | 11.10 | 51.78 | 3858 | 99.3 |
| Orissa | 4.45 | 10.16 | 6.82 | 7.31 | 58.59 | 1534 | 4.48 | 10.26 | 6.86 | 7.47 | 59.25 | 1531 | 42.6 |
| Tamil Nadu | 1.93 | 4.41 | 6.61 | 7.08 | 65.67 | 3423 | 2.05 | 4.70 | 5.22 | 5.69 | 64.94 | 2546 | 93.0 |
| Chattisgarh | 3.72 | 8.49 | 5.04 | 5.40 | 71.07 | 1354 | 3.75 | 8.59 | 5.01 | 5.46 | 70.40 | 1337 | 30.3 |
| Bihar | 3.36 | 7.67 | 4.99 | 5.34 | 76.41 | 1486 | 3.25 | 7.44 | 3.50 | 3.81 | 74.21 | 1075 | 51.1 |
| Karnataka | 1.40 | 3.20 | 3.45 | 3.70 | 80.11 | 2470 | 1.49 | 3.41 | 5.74 | 6.25 | 80.47 | 3868 | 75.8 |
| Haryana | 1.04 | 2.37 | 3.37 | 3.61 | 83.72 | 3238 | 1.05 | 2.40 | 3.21 | 3.50 | 83.96 | 3051 | 99.9 |
| Jharkhand | 1.62 | 3.70 | 2.97 | 3.18 | 86.90 | 1828 | 1.35 | 3.09 | 1.56 | 1.70 | 85.66 | 1150 | 5.6 |
| Assam | 2.19 | 5.00 | 2.92 | 3.13 | 90.03 | 1332 | 2.42 | 5.54 | 3.55 | 3.87 | 89.53 | 1468 | 4.6 |
| Maharashtra | 1.53 | 3.49 | 2.57 | 2.75 | 92.78 | 1680 | 1.52 | 3.48 | 2.70 | 2.94 | 92.47 | 1779 | 28.7 |
| Gujarat | 0.73 | 1.67 | 1.39 | 1.49 | 94.27 | 1894 | 0.67 | 1.53 | 1.30 | 1.42 | 93.89 | 1949 | 57.9 |
| Madhya Pradesh | 1.66 | 3.79 | 1.37 | 1.47 | 95.74 | 824 | 1.66 | 3.80 | 1.66 | 1.81 | 95.70 | 999 | 13.6 |
| Kerala | 0.26 | 0.59 | 0.63 | 0.67 | 96.41 | 2390 | 0.28 | 0.64 | 0.63 | 0.69 | 96.38 | 2284 | 66.3 |
| Others | 1.71 | 3.90 | 3.35 | 3.59 | 100.00 | @ | 1.71 | 3.92 | 3.32 | 3.62 | 100.00 | @ | - |
| All India | 43.81 | 100.00 | 93.36 | 100.00 | | 2131 | 43.66 | 100.00 | 91.79 | 100.00 | | 2102 | 56.0 |

@ Since area/production is low in individual states, yield rate is not worked out.

* Provisional

*Note*: States have been arranged in descending order of percentage share of production during 2006-07.

# Table 4.7(a)

*All-India Area, Production and Yield of Wheat from 1950-51 to 2007-08 Alongwith Percentage Coverage under Irrigation*

*Area- Million Hectares*
*Production- Million Tonnes*
*Yield- Kg./Hectare*

| Year | Area | Production | Yield | % Coverage under Irrigation |
|---|---|---|---|---|
| 1 | 2 | 3 | 4 | 5 |
| 1950-51 | 9.75 | 6.46 | 663 | 34.0 |
| 1951-52 | 9.47 | 6.18 | 653 | 35.8 |
| 1952-53 | 9.83 | 7.50 | 763 | 37.2 |
| 1953-54 | 10.68 | 8.02 | 750 | 36.2 |
| 1954-55 | 11.26 | 9.04 | 803 | 35.0 |
| 1955-56 | 12.37 | 8.76 | 708 | 32.7 |
| 1956-57 | 13.52 | 9.40 | 695 | 29.4 |
| 1957-58 | 11.73 | 7.99 | 682 | 33.3 |
| 1958-59 | 12.62 | 9.96 | 789 | 31.8 |
| 1959-60 | 13.38 | 10.32 | 772 | 31.8 |
| 1960-61 | 12.93 | 11.00 | 851 | 32.7 |
| 1961-62 | 13.57 | 12.07 | 890 | 31.9 |
| 1962-63 | 13.59 | 10.78 | 793 | 33.8 |
| 1963-64 | 13.50 | 9.85 | 730 | 34.9 |
| 1964-65 | 13.42 | 12.26 | 913 | 36.8 |
| 1965-66 | 12.57 | 10.40 | 827 | 43.1 |
| 1966-67 | 12.84 | 11.39 | 887 | 48.0 |
| 1967-68 | 14.99 | 16.54 | 1103 | 43.4 |
| 1968-69 | 15.96 | 18.65 | 1169 | 49.8 |
| 1969-70 | 16.63 | 20.09 | 1208 | 51.1 |
| 1970-71 | 18.24 | 23.83 | 1307 | 54.3 |
| 1971-72 | 19.14 | 26.41 | 1380 | 54.5 |
| 1972-73 | 19.46 | 24.74 | 1271 | 57.6 |
| 1973-74 | 18.58 | 21.78 | 1172 | 57.7 |
| 1974-75 | 18.01 | 24.10 | 1338 | 61.8 |
| 1975-76 | 20.45 | 28.84 | 1410 | 61.8 |
| 1976-77 | 20.92 | 29.01 | 1387 | 65.1 |
| 1977-78 | 21.46 | 31.75 | 1480 | 64.6 |

contd. ...

*...contd. ...*

| *1* | 2 | 3 | 4 | 5 |
|---|---|---|---|---|
| 1978-79 | 22.64 | 35.51 | 1568 | 66.0 |
| 1979-80 | 22.17 | 31.83 | 1436 | 68.3 |
| 1980-81 | 22.28 | 36.31 | 1630 | 76.5 |
| 1981-82 | 22.14 | 37.45 | 1691 | 70.7 |
| 1982-83 | 23.57 | 42.79 | 1816 | 72.5 |
| 1983-84 | 24.67 | 45.48 | 1843 | 73.0 |
| 1984-85 | 23.56 | 44.07 | 1870 | 74.5 |
| 1985-86 | 23.00 | 47.05 | 2046 | 74.6 |
| 1986-87 | 23.13 | 44.32 | 1916 | 76.3 |
| 1987-88 | 23.06 | 46.17 | 2002 | 76.8 |
| 1988-89 | 24.11 | 54.11 | 2244 | 79.2 |
| 1989-90 | 23.50 | 49.85 | 2121 | 80.3 |
| 1990-91 | 24.17 | 55.14 | 2281 | 81.1 |
| 1991-92 | 23.26 | 55.69 | 2394 | 83.7 |
| 1992-93 | 24.59 | 57.21 | 2327 | 84.2 |
| 1993-94 | 25.15 | 59.84 | 2380 | 84.8 |
| 1994-95 | 25.70 | 65.77 | 2559 | 85.2 |
| 1995-96 | 25.01 | 62.10 | 2483 | 85.8 |
| 1996-97 | 25.89 | 69.35 | 2679 | 86.2 |
| 1997-98 | 26.70 | 66.35 | 2485 | 85.8 |
| 1998-99 | 27.52 | 71.29 | 2590 | 85.8 |
| 1999-2000 | 27.49 | 76.37 | 2778 | 87.2 |
| 2000-01 | 25.73 | 69.68 | 2708 | 88.1 |
| 2001-02 | 26.34 | 72.77 | 2762 | 87.4 |
| 2002-03 | 25.20 | 65.76 | 2610 | 88.0 |
| 2003-04 | 26.60 | 72.16 | 2713 | 88.4 |
| 2004-05 | 26.38 | 68.64 | 2602 | 89.4 |
| 2005-06 | 26.48 | 69.35 | 2619 | 89.6 |
| 2006-07 | 27.99 | 75.80 | 2708 | NA |
| 2007-08* | 28.15 | 78.40 | 2785 | NA |

* Advance Estimates as released on 09.07.2008

*Note*: The yield rates given above have been worked out on the basis of production & area figures taken in '000 units.

*Wheat—All India Production*

*Million Tonnes*

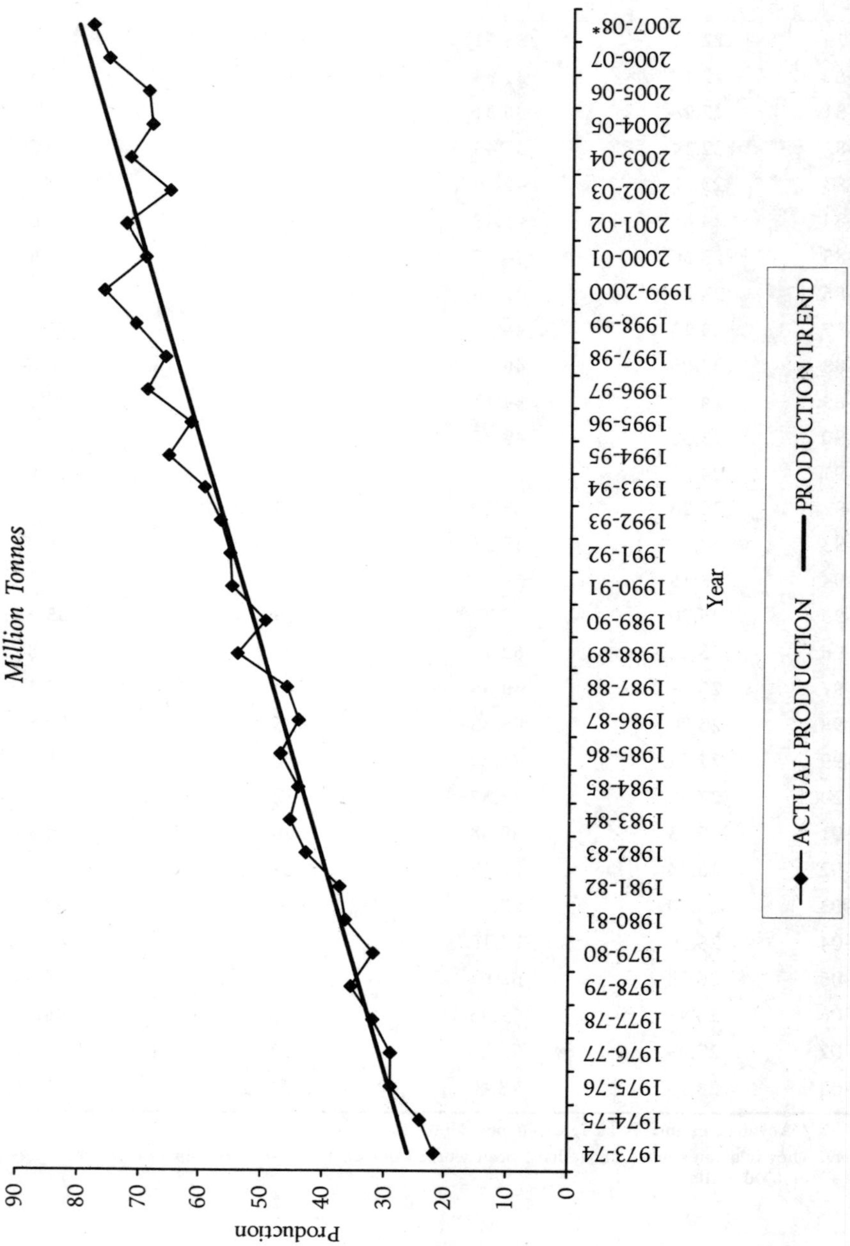


* Based on Advance estimate as on 09-07-08.

# Table 4.7 (b)

## *Area, Production and Yield of Wheat during 2005-06 and 2006-07 in Respect of Major Wheat Producing States Alongwith Coverage under Irrigation*

*Area- Million Hectares*
*Production- Million Tonnes*
*Yield- Kg./Hectare*

| State | 2006-07 | | | | | | 2005-06 | | | | | | % Coverage under Irrigation during 2005-06* |
|---|---|---|---|---|---|---|---|---|---|---|---|---|---|
| | Area | % to All-India | Production | % to All-India | Cumulative % | Yield | Area | % to All-India | Production | % to All-India | Cumulative % | Yield | |
| 1 | 2 | 3 | 4 | 5 | 6 | 7 | 8 | 9 | 10 | 11 | 12 | 13 | 14 |
| Uttar Pradesh | 9.20 | 32.87 | 25.03 | 33.02 | 33.02 | 2721 | 9.16 | 34.59 | 24.07 | 34.71 | 34.71 | 2627 | 97.5 |
| Punjab | 3.47 | 12.40 | 14.60 | 19.26 | 52.28 | 4210 | 3.47 | 13.10 | 14.49 | 20.89 | 55.60 | 4179 | 98.0 |
| Haryana | 2.38 | 8.50 | 10.06 | 13.27 | 65.55 | 4232 | 2.30 | 8.69 | 8.86 | 12.78 | 68.38 | 3844 | 99.0 |
| Madhya Pradesh | 3.99 | 14.26 | 7.33 | 9.67 | 75.21 | 1835 | 3.69 | 13.94 | 5.96 | 8.59 | 76.97 | 1613 | 78.3 |
| Rajasthan | 2.56 | 9.15 | 7.06 | 9.31 | 84.53 | 2751 | 2.12 | 8.01 | 5.87 | 8.46 | 85.44 | 2762 | 99.0 |
| Bihar | 2.05 | 7.32 | 3.91 | 5.16 | 89.68 | 1908 | 2.00 | 7.55 | 3.24 | 4.67 | 90.11 | 1617 | 90.7 |
| Gujarat | 1.20 | 4.29 | 3.00 | 3.96 | 93.64 | 2498 | 0.92 | 3.47 | 2.47 | 3.56 | 93.67 | 2700 | 87.5 |
| Maharashtra | 1.23 | 4.39 | 1.63 | 2.15 | 95.79 | 1325 | 0.93 | 3.51 | 1.30 | 1.87 | 95.54 | 1393 | 66.5 |
| Uttaranchal | 0.39 | 1.39 | 0.80 | 1.06 | 96.85 | 2049 | 0.40 | 1.51 | 0.65 | 0.94 | 96.48 | 1633 | 52.3 |
| West Bengal | 0.35 | 1.25 | 0.80 | 1.06 | 97.90 | 2282 | 0.37 | 1.40 | 0.77 | 1.11 | 97.59 | 2109 | 65.7 |
| Himachal Pradesh | 0.36 | 1.29 | 0.50 | 0.66 | 98.56 | 1385 | 0.36 | 1.36 | 0.68 | 0.98 | 98.57 | 1894 | 19.5 |
| Jammu & Kashmir | 0.26 | 0.93 | 0.49 | 0.65 | 99.21 | 1893 | 0.25 | 0.94 | 0.44 | 0.63 | 99.21 | 1790 | 28.2 |
| Karnataka | 0.27 | 0.96 | 0.21 | 0.28 | 99.49 | 762 | 0.25 | 0.94 | 0.22 | 0.32 | 99.52 | 858 | 51.6 |
| Jharkhand | 0.08 | 0.29 | 0.13 | 0.17 | 99.66 | 1529 | 0.06 | 0.23 | 0.08 | 0.12 | 99.64 | 1340 | 86.0 |
| Assam | 0.06 | 0.21 | 0.07 | 0.09 | 99.75 | 1117 | 0.05 | 0.19 | 0.05 | 0.07 | 99.71 | 1074 | 4.7 |
| Others | 0.14 | 0.50 | 0.19 | 0.25 | 100.00 | @ | 0.15 | 0.57 | 0.20 | 0.29 | 100.00 | @ | - |
| All India | 27.99 | 100.00 | 75.81 | 100.00 | | 2708 | 26.48 | 100.00 | 69.35 | 100.00 | | 2619 | 89.6 |

@ Since area/production is low in individual states, yield rate is not worked out.

* Provisional

*Note*: States have been arranged in descending order of percentage share of production during 2006-07.

# Table 4.8(a)

*All-India Area, Production and Yield of Coarse Cereals from 1950-51 to 2007-08 Alongwith Percentage Coverage under Irrigation*

*Area- Million Hectares*
*Production- Million Tonnes*
*Yield- Kg./Hectare*

| Year | Area | Production | Yield | % Coverage under Irrigation |
|---|---|---|---|---|
| 1 | 2 | 3 | 4 | 5 |
| 1950-51 | 37.67 | 15.38 | 408 | 7.9 |
| 1951-52 | 38.88 | 16.09 | 414 | 8.7 |
| 1952-53 | 42.45 | 19.61 | 462 | 8.1 |
| 1953-54 | 45.37 | 22.97 | 506 | 7.8 |
| 1954-55 | 43.92 | 22.82 | 520 | 8.1 |
| 1955-56 | 43.45 | 19.49 | 449 | 8.0 |
| 1956-57 | 42.02 | 19.87 | 473 | 7.6 |
| 1957-58 | 42.91 | 21.23 | 495 | 8.0 |
| 1958-59 | 44.66 | 23.18 | 519 | 7.5 |
| 1959-60 | 43.79 | 22.87 | 522 | 7.5 |
| 1960-61 | 44.96 | 23.74 | 528 | 7.7 |
| 1961-62 | 44.73 | 23.22 | 519 | 7.1 |
| 1962-63 | 44.29 | 24.63 | 556 | 7.3 |
| 1963-64 | 43.93 | 23.72 | 540 | 7.3 |
| 1964-65 | 44.35 | 25.37 | 514 | 7.3 |
| 1965-66 | 44.34 | 21.42 | 483 | 8.1 |
| 1966-67 | 45.09 | 24.05 | 533 | 8.5 |
| 1967-68 | 47.34 | 28.80 | 608 | 8.0 |
| 1968-69 | 46.24 | 25.18 | 545 | 9.6 |
| 1969-70 | 47.24 | 27.29 | 578 | 9.4 |
| 1970-71 | 45.95 | 30.55 | 665 | 8.3 |
| 1971-72 | 43.57 | 24.60 | 564 | 8.4 |
| 1972-73 | 42.21 | 23.14 | 548 | 8.6 |
| 1973-74 | 46.24 | 28.83 | 623 | 8.3 |
| 1974-75 | 43.15 | 26.13 | 606 | 10.9 |
| 1975-76 | 43.80 | 30.41 | 694 | 9.9 |
| 1976-77 | 41.94 | 28.88 | 689 | 9.7 |
| 1977-78 | 42.28 | 30.02 | 710 | 8.9 |

*contd. ...*

*...contd. ...*

| *1* | 2 | 3 | 4 | 5 |
|---|---|---|---|---|
| 1978-79 | 42.23 | 30.44 | 721 | 8.5 |
| 1979-80 | 41.36 | 26.97 | 652 | 9.4 |
| 1980-81 | 41.78 | 29.02 | 695 | 9.2 |
| 1981-82 | 42.45 | 31.09 | 733 | 9.0 |
| 1982-83 | 40.43 | 27.75 | 685 | 9.0 |
| 1983-84 | 41.71 | 33.90 | 813 | 7.8 |
| 1984-85 | 39.21 | 31.17 | 795 | 8.4 |
| 1985-86 | 39.47 | 26.20 | 664 | 8.4 |
| 1986-87 | 39.74 | 26.83 | 675 | 9.1 |
| 1987-88 | 36.55 | 26.36 | 721 | 9.8 |
| 1988-89 | 38.68 | 31.47 | 814 | 9.2 |
| 1989-90 | 37.69 | 34.76 | 922 | 9.8 |
| 1990-91 | 36.32 | 32.70 | 900 | 9.0 |
| 1991-92 | 33.42 | 25.99 | 778 | 10.7 |
| 1992-93 | 34.42 | 36.59 | 1063 | 10.1 |
| 1993-94 | 32.82 | 30.82 | 939 | 10.4 |
| 1994-95 | 32.17 | 29.88 | 929 | 10.4 |
| 1995-96 | 30.88 | 29.03 | 940 | 11.0 |
| 1996-97 | 31.81 | 34.10 | 1072 | 10.2 |
| 1997-98 | 30.83 | 30.40 | 986 | 11.0 |
| 1998-99 | 29.34 | 31.34 | 1068 | 12.0 |
| 1999-2000 | 29.34 | 30.33 | 1034 | 12.6 |
| 2000-01 | 30.26 | 31.08 | 1027 | 12.5 |
| 2001-02 | 29.52 | 33.38 | 1131 | 11.3 |
| 2002-03 | 26.99 | 26.07 | 966 | 11.0 |
| 2003-04 | 30.80 | 37.60 | 1221 | 6.6 |
| 2004-05 | 29.03 | 33.47 | 1153 | 6.6 |
| 2005-06 | 29.04 | 34.07 | 1172 | 13.0 |
| 2006-07 | 28.71 | 33.92 | 1182 | NA |
| 2007-08* | 28.72 | 40.73 | 1415 | NA |

* Advance Estimates as released on 09.07.2008

*Note*: The yield rates given above have been worked out on the basis of production & area figures taken in '000 units.

*Coarse Cereals—All India Production*

*Million Tonnes*

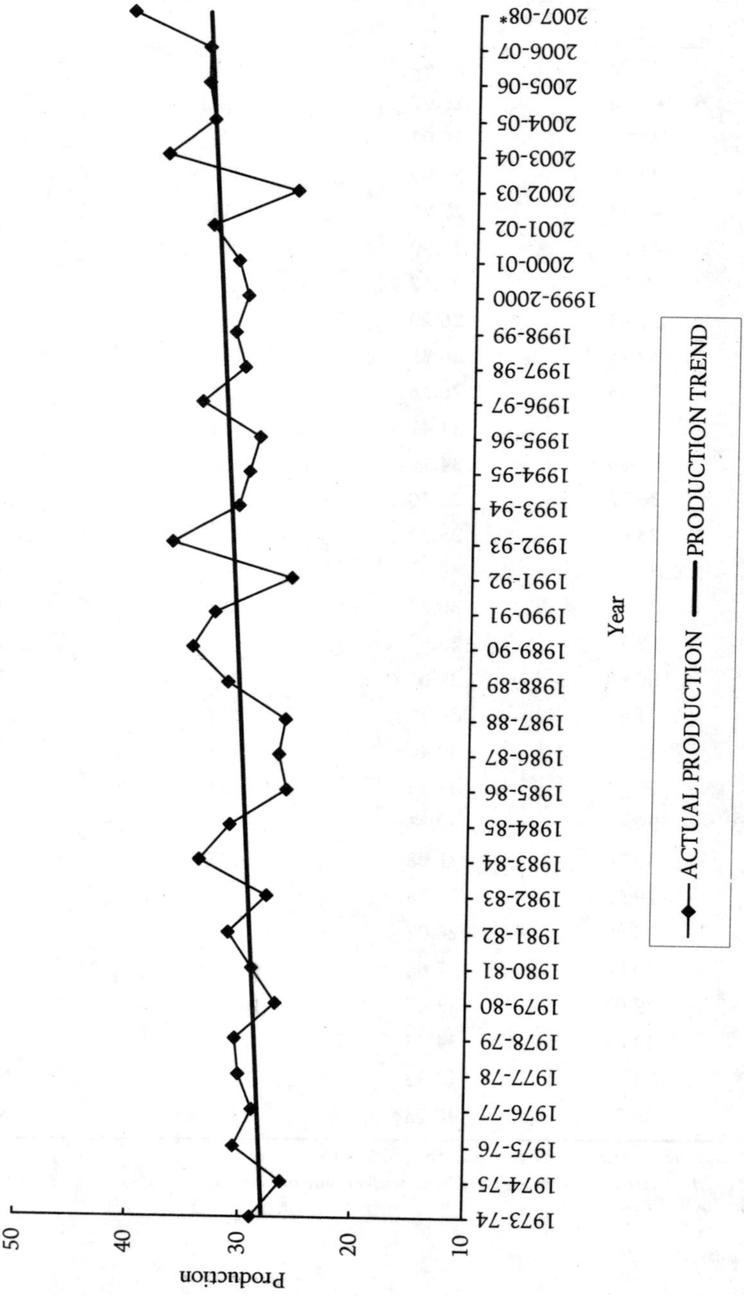


* Based on Advance estimate as on 09.07.08.

# Table 4.8 (b)

*Area, Production and Yield of Coarse Cereals during 2005-06 and 2006-07 in Respect of Major Coarse Cereals Producing States Alongwith Coverage under Irrigation*

*Area- Million Hectares*
*Production- Million Tonnes*
*Yield- Kg./Hectare*

| State | 2006-07 | | | | | | 2005-06 | | | | | | % Coverage under Irrigation during 2005-06* |
|---|---|---|---|---|---|---|---|---|---|---|---|---|---|
| | Area | % to All-India | Production | % to All-India | Cumulative % | Yield | Area | % to All-India | Production | % to All-India | Cumulative % | Yield | |
| 1 | 2 | 3 | 4 | 5 | 6 | 7 | 8 | 9 | 10 | 11 | 12 | 13 | 14 |
| Maharashtra | 6.86 | 23.89 | 6.14 | 18.10 | 18.10 | 895 | 6.87 | 23.64 | 6.09 | 17.87 | 17.87 | 887 | 9.2 |
| Rajasthan | 6.82 | 23.75 | 5.50 | 16.21 | 34.32 | 807 | 6.77 | 23.30 | 4.53 | 13.30 | 31.17 | 669 | 6.5 |
| Karnataka | 3.41 | 11.88 | 5.06 | 14.92 | 49.23 | 1481 | 3.88 | 13.35 | 6.56 | 19.25 | 50.43 | 1693 | 15.6 |
| Uttar Pradesh | 2.20 | 7.66 | 3.08 | 9.08 | 58.31 | 1404 | 2.15 | 7.40 | 2.97 | 8.72 | 59.14 | 1383 | 17.5 |
| Andhra pradesh | 1.30 | 4.53 | 3.00 | 8.84 | 67.16 | 2305 | 1.39 | 4.78 | 3.86 | 11.33 | 70.47 | 2772 | 25.4 |
| Madhya Pradesh | 2.02 | 7.04 | 1.85 | 5.45 | 72.61 | 918 | 2.04 | 7.02 | 2.35 | 6.90 | 77.37 | 1150 | 2.6 |
| Bihar | 0.69 | 2.40 | 1.76 | 5.19 | 77.80 | 2555 | 0.70 | 2.41 | 1.41 | 4.14 | 81.51 | 2021 | 57.1 |
| Gujarat | 1.63 | 5.68 | 1.52 | 4.48 | 82.28 | 928 | 1.61 | 5.54 | 1.84 | 5.40 | 86.91 | 1142 | 13.1 |
| Tamil Nadu | 0.70 | 2.44 | 1.36 | 4.01 | 86.29 | 1951 | 0.74 | 2.55 | 0.73 | 2.14 | 89.05 | 985 | 14.8 |
| Haryana | 0.76 | 2.65 | 1.20 | 3.54 | 89.83 | 1567 | 0.72 | 2.48 | 0.81 | 2.38 | 91.43 | 1124 | 37.2 |
| Himachal Pradesh | 0.33 | 1.15 | 0.73 | 2.15 | 91.98 | 2180 | 0.33 | 1.14 | 0.57 | 1.67 | 93.10 | 1715 | 8.4 |
| Punjab | 0.18 | 0.63 | 0.55 | 1.62 | 93.60 | 3033 | 0.18 | 0.62 | 0.47 | 1.38 | 94.48 | 2697 | 65.7 |
| Jammu & Kashmir | 0.37 | 1.29 | 0.51 | 1.50 | 95.11 | 1396 | 0.35 | 1.20 | 0.47 | 1.38 | 95.86 | 1344 | 10.4 |
| Uttarakhand | 0.26 | 0.91 | 0.34 | 1.00 | 96.11 | 1318 | 0.27 | 0.93 | 0.32 | 0.94 | 96.80 | 1188 | 0.7 |
| Jharkhand | 0.29 | 1.01 | 0.33 | 0.97 | 97.08 | 1126 | 0.22 | 0.76 | 0.26 | 0.76 | 97.56 | 1165 | 1.3 |
| West Bengal | 0.11 | 0.38 | 0.27 | 0.80 | 97.88 | 2589 | 0.07 | 0.24 | 0.15 | 0.44 | 98.00 | 2132 | 7.7 |
| Chattisgarh | 0.33 | 1.15 | 0.18 | 0.53 | 98.41 | 537 | 0.35 | 1.20 | 0.16 | 0.47 | 98.47 | 460 | 1.5 |
| Orissa | 0.16 | 0.56 | 0.16 | 0.47 | 98.88 | 1025 | 0.17 | 0.58 | 0.16 | 0.47 | 98.94 | 968 | 9.9 |
| Others | 0.29 | 1.01 | 0.38 | 1.12 | 100.00 | @ | 0.25 | 0.86 | 0.36 | 1.06 | 100.00 | @ | - |
| All India | 28.71 | 100.00 | 33.92 | 100.00 | | 1182 | 29.06 | 100.00 | 34.07 | 100.00 | | 1172 | 13.1 |

@ Since area/production is low in individual states, yield rate is not worked out.

* Provisional

*Note:* States have been arranged in descending order of percentage share of production during 2006-07.

# Table 4.9 (a)

*All-India Area, Production and Yield of Jowar from 1950-51 to 2007-08 Alongwith Percentage Coverage under Irrigation*

*Area- Million Hectares*
*Production- Million Tonnes*
*Yield- Kg./Hectare*

| Year | Area | Production | Yield | % Coverage under Irrigation |
|---|---|---|---|---|
| 1 | 2 | 3 | 4 | 5 |
| 1950-51 | 15.57 | 5.50 | 353 | 3.0 |
| 1951-52 | 15.94 | 6.08 | 381 | 3.5 |
| 1952-53 | 17.54 | 7.36 | 420 | 3.1 |
| 1953-54 | 17.76 | 8.08 | 455 | 3.1 |
| 1954-55 | 17.46 | 9.20 | 527 | 3.3 |
| 1955-56 | 17.36 | 6.73 | 387 | 3.6 |
| 1956-57 | 16.24 | 7.33 | 451 | 3.4 |
| 1957-58 | 17.31 | 8.64 | 499 | 3.5 |
| 1958-59 | 17.96 | 9.03 | 503 | 3.5 |
| 1959-60 | 17.71 | 8.58 | 484 | 3.6 |
| 1960-61 | 18.41 | 9.81 | 533 | 3.6 |
| 1961-62 | 18.25 | 8.03 | 440 | 3.7 |
| 1962-63 | 18.41 | 9.75 | 529 | 3.7 |
| 1963-64 | 18.38 | 9.20 | 501 | 3.9 |
| 1964-65 | 18.06 | 9.68 | 536 | 3.8 |
| 1965-66 | 17.68 | 7.58 | 429 | 4.1 |
| 1966-67 | 18.05 | 9.22 | 511 | 4.0 |
| 1967-68 | 18.42 | 10.05 | 545 | 3.9 |
| 1968-69 | 18.73 | 9.80 | 523 | 4.5 |
| 1969-70 | 18.61 | 9.72 | 522 | 4.1 |
| 1970-71 | 17.37 | 8.11 | 466 | 3.6 |
| 1971-72 | 16.78 | 7.72 | 460 | 4.4 |
| 1972-73 | 15.51 | 6.97 | 449 | 3.3 |
| 1973-74 | 16.72 | 9.10 | 544 | 4.0 |
| 1974-75 | 16.19 | 10.41 | 643 | 4.6 |
| 1975-76 | 16.09 | 9.50 | 591 | 4.9 |
| 1976-77 | 15.77 | 10.52 | 667 | 5.1 |
| 1977-78 | 16.32 | 12.06 | 739 | 4.9 |

*contd. ...*

*...contd. ...*

| *1* | 2 | 3 | 4 | 5 |
|---|---|---|---|---|
| 1978-79 | 16.15 | 11.44 | 708 | 4.8 |
| 1979-80 | 16.67 | 11.65 | 699 | 4.9 |
| 1980-81 | 15.81 | 10.43 | 660 | 4.7 |
| 1981-82 | 16.60 | 12.06 | 727 | 4.6 |
| 1982-83 | 16.38 | 10.75 | 657 | 4.3 |
| 1983-84 | 16.43 | 11.92 | 725 | 4.1 |
| 1984-85 | 15.94 | 11.40 | 715 | 4.4 |
| 1985-86 | 16.10 | 10.20 | 633 | 4.5 |
| 1986-87 | 15.95 | 9.19 | 576 | 4.9 |
| 1987-88 | 16.00 | 12.20 | 752 | 4.9 |
| 1988-89 | 14.60 | 10.17 | 697 | 5.8 |
| 1989-90 | 14.84 | 12.90 | 859 | 6.3 |
| 1990-91 | 14.36 | 11.68 | 814 | 5.6 |
| 1991-92 | 12.36 | 8.10 | 655 | 6.5 |
| 1992-93 | 13.04 | 12.81 | 982 | 6.1 |
| 1993-94 | 12.71 | 11.41 | 895 | 6.2 |
| 1994-95 | 11.51 | 8.97 | 779 | 6.7 |
| 1995-96 | 11.33 | 9.33 | 823 | 6.8 |
| 1996-97 | 11.43 | 10.93 | 956 | 6.7 |
| 1997-98 | 10.80 | 7.53 | 697 | 7.3 |
| 1998-99 | 9.79 | 8.42 | 859 | 8.1 |
| 1999-2000 | 10.25 | 8.68 | 847 | 7.7 |
| 2000-01 | 9.86 | 7.53 | 764 | 7.9 |
| 2001-02 | 9.80 | 7.56 | 771 | 8.3 |
| 2002-03 | 9.30 | 7.01 | 754 | 8.5 |
| 2003-04 | 9.33 | 6.68 | 716 | 7.5 |
| 2004-05 | 9.09 | 7.24 | 797 | 9.1 |
| 2005-06 | 8.67 | 7.24 | 880 | 9.0 |
| 2006-07 | 8.47 | 7.15 | 844 | NA |
| 2007-08* | 7.93 | 7.78 | 981 | NA |

* Advance Estimates as released on 09.07.2008

*Note*: The yield rates given above have been worked out on the basis of production & area figures taken in '000 units.

# Table 4.9(b)

*Area, Production and Yield of Jowar during 2005-06 and 2006-07 in Respect of Major Jowar Producing States Aalongwith Coverage under Irrigation*

*Area- Million Hectares*
*Production- Million Tonnes*
*Yield- Kg./Hectare*

| State | 2006-07 | | | | | | 2005-06 | | | | | | % Coverage under Irrigation during 2005-06* |
|---|---|---|---|---|---|---|---|---|---|---|---|---|---|
| | Area | % to All-India | Production | % to All-India | Cumulative % | Yield | Area | % to All-India | Production | % to All-India | Cumulative % | Yield | |
| 1 | 2 | 3 | 4 | 5 | 6 | 7 | 8 | 9 | 10 | 11 | 12 | 13 | 14 |
| Maharashtra | 4.62 | 54.55 | 3.77 | 52.73 | 52.73 | 817 | 4.74 | 54.67 | 3.90 | 51.11 | 51.11 | 824 | 10.5 |
| Karnataka | 1.42 | 16.77 | 1.31 | 18.32 | 71.05 | 925 | 1.52 | 17.53 | 1.67 | 21.89 | 73.00 | 1095 | 8.7 |
| Madhya Pradesh | 0.57 | 6.73 | 0.60 | 8.39 | 79.44 | 1039 | 0.58 | 6.69 | 0.63 | 8.26 | 81.26 | 1088 | 0.1 |
| Andhra Pradesh | 0.42 | 4.96 | 0.41 | 5.73 | 85.17 | 972 | 0.44 | 5.07 | 0.59 | 7.73 | 88.99 | 1324 | 7.9 |
| Rajasthan | 0.66 | 7.79 | 0.37 | 5.17 | 90.35 | 556 | 0.59 | 6.81 | 0.17 | 2.23 | 91.22 | 288 | 0.5 |
| Tamil Nadu | 0.29 | 3.42 | 0.29 | 4.06 | 94.41 | 999 | 0.32 | 3.69 | 0.23 | 3.01 | 94.23 | 732 | 9.4 |
| Uttar pradesh | 0.23 | 2.72 | 0.24 | 3.36 | 97.76 | 1030 | 0.23 | 2.65 | 0.24 | 3.15 | 97.38 | 1065 | 0.9 |
| Gujarat | 0.12 | 1.42 | 0.10 | 1.40 | 99.16 | 831 | 0.13 | 1.50 | 0.15 | 1.97 | 99.34 | 1138 | 7.6 |
| Haryana | 0.09 | 1.06 | 0.03 | 0.42 | 99.58 | 286 | 0.09 | 1.04 | 0.02 | 0.26 | 99.61 | 273 | 75.4 |
| Orissa | 0.01 | 0.12 | 0.01 | 0.14 | 99.72 | 611 | 0.01 | 0.12 | 0.01 | 0.13 | 99.74 | 600 | - |
| Others | 0.04 | 0.47 | 0.02 | 0.28 | 100.00 | @ | 0.02 | 0.23 | 0.02 | 0.26 | 100.00 | @ | - |
| All India | 8.47 | 100.00 | 7.15 | 100.00 | | 844 | 8.67 | 100.00 | 7.63 | 100.00 | | 880 | 9.0 |

@ Since area/production is low in individual states, yield rate is not worked out.

* Provisional

*Note*: States have been arranged in descending order of percentage share of production during 2006-07.

## Table 4.10(a)

*All-India Area, Production and Yield of Bajra from 1950-51 to 2007-08 Alongwith Percentage Coverage under Irrigation*

*Area- Million Hectares*
*Production- Million Tonnes*
*Yield- Kg./Hectare*

| Year | Area | Production | Yield | % Coverage under Irrigation |
|---|---|---|---|---|
| 1 | 2 | 3 | 4 | 5 |
| 1950-51 | 9.02 | 2.60 | 288 | 3.4 |
| 1951-52 | 9.52 | 2.35 | 246 | 3.8 |
| 1952-53 | 10.77 | 3.19 | 296 | 3.3 |
| 1953-54 | 12.20 | 4.55 | 373 | 3.1 |
| 1954-55 | 11.37 | 3.52 | 310 | 3.6 |
| 1955-56 | 11.34 | 3.43 | 302 | 3.6 |
| 1956-57 | 11.25 | 2.87 | 255 | 3.1 |
| 1957-58 | 11.17 | 3.62 | 324 | 2.9 |
| 1958-59 | 11.43 | 3.87 | 338 | 3.0 |
| 1959-60 | 10.70 | 3.49 | 327 | 2.5 |
| 1960-61 | 11.47 | 3.28 | 286 | 2.8 |
| 1961-62 | 11.28 | 3.65 | 323 | 2.6 |
| 1962-63 | 10.96 | 3.96 | 361 | 2.7 |
| 1963-64 | 11.10 | 3.88 | 349 | 2.3 |
| 1964-65 | 11.83 | 4.52 | 382 | 2.3 |
| 1965-66 | 11.97 | 3.75 | 314 | 2.8 |
| 1966-67 | 12.24 | 4.47 | 365 | 3.1 |
| 1967-68 | 12.81 | 5.19 | 405 | 3.0 |
| 1968-69 | 12.05 | 3.80 | 315 | 4.1 |
| 1969-70 | 12.49 | 5.33 | 426 | 4.2 |
| 1970-71 | 12.91 | 8.03 | 622 | 4.0 |
| 1971-72 | 11.77 | 5.32 | 452 | 3.7 |
| 1972-73 | 11.82 | 3.93 | 333 | 4.5 |
| 1973-74 | 13.93 | 7.52 | 540 | 4.2 |
| 1974-75 | 11.29 | 3.27 | 290 | 5.5 |
| 1975-76 | 11.57 | 5.74 | 496 | 5.1 |
| 1976-77 | 10.75 | 5.85 | 544 | 4.9 |
| 1977-78 | 11.10 | 4.73 | 425 | 4.4 |

*contd. ...*

...contd. ...

| 1 | 2 | 3 | 4 | 5 |
|---|---|---|---|---|
| 1978-79 | 11.39 | 5.57 | 489 | 4.4 |
| 1979-80 | 10.58 | 3.95 | 373 | 5.5 |
| 1980-81 | 11.66 | 5.34 | 458 | 5.5 |
| 1981-82 | 11.78 | 5.54 | 470 | 6.0 |
| 1982-83 | 10.94 | 5.13 | 469 | 6.0 |
| 1983-84 | 11.83 | 7.72 | 653 | 4.8 |
| 1984-85 | 10.62 | 6.05 | 569 | 5.2 |
| 1985-86 | 10.65 | 3.66 | 344 | 5.1 |
| 1986-87 | 11.27 | 4.51 | 401 | 6.0 |
| 1987-88 | 8.71 | 3.30 | 378 | 8.4 |
| 1988-89 | 12.04 | 7.78 | 646 | 5.3 |
| 1989-90 | 10.90 | 6.65 | 610 | 6.3 |
| 1990-91 | 10.48 | 6.89 | 658 | 5.1 |
| 1991-92 | 10.03 | 4.67 | 465 | 6.5 |
| 1992-93 | 10.62 | 8.88 | 836 | 5.8 |
| 1993-94 | 9.55 | 4.97 | 521 | 6.6 |
| 1994-95 | 10.22 | 7.16 | 700 | 5.5 |
| 1995-96 | 9.32 | 5.38 | 577 | 6.2 |
| 1996-97 | 9.98 | 7.87 | 788 | 5.3 |
| 1997-98 | 9.67 | 7.64 | 791 | 5.9 |
| 1998-99 | 9.30 | 6.96 | 748 | 7.0 |
| 1999-00 | 8.90 | 5.78 | 650 | 8.3 |
| 2000-01 | 9.83 | 6.76 | 688 | 8.0 |
| 2001-02 | 9.53 | 8.28 | 869 | 6.3 |
| 2002-03 | 7.74 | 4.72 | 610 | 9.0 |
| 2003-04 | 10.61 | 12.11 | 1141 | 6.3 |
| 2004-05 | 9.23 | 7.93 | 859 | 8.3 |
| 2005-06 | 9.58 | 7.68 | 802 | 8.9 |
| 2006-07 | 9.51 | 8.42 | 886 | NA |
| 2007-08* | 9.50 | 9.79 | 1030 | NA |

* Advance Estimates as released on 09.07.2008

*Note*: The yield rates given above have been worked out on the basis of production & area figures taken in '000 units.

## Table 4.10 (b)

*Area, Production and Yield of Bajra during 2005-06 and 2006-07 in Respect of Major Bajra Producing States Alongwith Coverage under Irrigation*

*Area- Million Hectares*
*Production- Million Tonnes*
*Yield- Kg./Hectare*

| State | 2006-07 | | | | | | 2005-06 | | | | | | % Coverage under Irrigation during 2005-06* |
|---|---|---|---|---|---|---|---|---|---|---|---|---|---|
| | Area | % to All-India | Production | % to All-India | Cumulative % | Yield | Area | % to All-India | Production | % to All-India | Cumulative % | Yield | |
| 1 | 2 | 3 | 4 | 5 | 6 | 7 | 8 | 9 | 10 | 11 | 12 | 13 | 14 |
| Rajasthan | 4.88 | 51.31 | 3.42 | 40.62 | 40.62 | 701 | 4.96 | 51.77 | 2.76 | 35.94 | 35.94 | 556 | 6.4 |
| Uttar Pradesh | 0.88 | 9.25 | 1.29 | 15.32 | 55.94 | 1455 | 0.87 | 9.08 | 1.25 | 16.28 | 52.21 | 1434 | 4.5 |
| Maharashtra | 1.45 | 15.25 | 1.06 | 12.59 | 68.53 | 729 | 1.43 | 14.93 | 1.03 | 13.41 | 65.63 | 720 | 6.3 |
| Gujarat | 0.94 | 9.88 | 1.02 | 12.11 | 80.64 | 1088 | 0.92 | 9.60 | 1.07 | 13.93 | 79.56 | 1169 | 16.9 |
| Haryana | 0.62 | 6.52 | 1.02 | 12.11 | 92.76 | 1649 | 0.59 | 6.16 | 0.68 | 8.85 | 88.41 | 1147 | 29.8 |
| Madhya Pradesh | 0.19 | 2.00 | 0.25 | 2.97 | 95.72 | 1365 | 0.18 | 1.88 | 0.27 | 3.52 | 91.93 | 1491 | 0.1 |
| Karnataka | 0.39 | 4.10 | 0.19 | 2.26 | 97.98 | 483 | 0.43 | 4.49 | 0.42 | 5.47 | 97.40 | 977 | 6.3 |
| Tamil Nadu | 0.07 | 0.74 | 0.10 | 1.19 | 99.17 | 1511 | 0.08 | 0.84 | 0.09 | 1.17 | 98.57 | 1158 | 8.4 |
| Andhra Pradesh | 0.06 | 0.63 | 0.05 | 0.59 | 99.76 | 770 | 0.08 | 0.84 | 0.08 | 1.04 | 99.61 | 1012 | 22.1 |
| Jammu & Kashmir | 0.02 | 0.21 | 0.01 | 0.12 | 99.88 | 591 | 0.01 | 0.10 | 0.01 | 0.13 | 99.74 | 600 | 0.5 |
| Others | 0.01 | 0.11 | 0.01 | 0.12 | 100.00 | @ | 0.03 | 0.31 | 0.02 | 0.26 | 100.00 | @ | - |
| All India | 9.51 | 100.00 | 8.42 | 100.00 | | 886 | 9.58 | 100.00 | 7.68 | 100.00 | | 802 | 8.9 |

@ Since area/production is low in individual states, yield rate is not worked out.

* Provisional

*Note:* States have been arranged in descending order of percentage share of production during 2006-07.

# Table 4.11 (a)

*All-India Area, Production and Yield of Maize from 1950-51 to 2007-08 Alongwith Percentage Coverage under Irrigation*

*Area- Million Hectares*
*Production- Million Tonnes*
*Yield- Kg./Hectare*

| Year | Area | Production | Yield | % Coverage under Irrigation |
|---|---|---|---|---|
| 1 | 2 | 3 | 4 | 5 |
| 1950-51 | 3.16 | 1.73 | 547 | 11.4 |
| 1951-52 | 3.31 | 2.08 | 627 | 16.4 |
| 1952-53 | 3.61 | 2.87 | 796 | 14.0 |
| 1953-54 | 3.87 | 3.04 | 785 | 11.5 |
| 1954-55 | 3.75 | 2.98 | 794 | 14.6 |
| 1955-56 | 3.70 | 2.60 | 704 | 11.8 |
| 1956-57 | 3.76 | 3.08 | 819 | 13.1 |
| 1957-58 | 4.08 | 3.15 | 772 | 13.4 |
| 1958-59 | 4.27 | 3.46 | 812 | 10.5 |
| 1959-60 | 4.34 | 4.07 | 938 | 9.9 |
| 1960-61 | 4.41 | 4.08 | 926 | 12.6 |
| 1961-62 | 4.51 | 4.31 | 957 | 9.5 |
| 1962-63 | 4.64 | 4.61 | 992 | 11.5 |
| 1963-64 | 4.58 | 4.56 | 995 | 11.4 |
| 1964-65 | 4.62 | 4.66 | 1010 | 12.1 |
| 1965-66 | 4.80 | 4.82 | 1005 | 16.1 |
| 1966-67 | 5.07 | 4.89 | 964 | 15.6 |
| 1967-68 | 5.58 | 6.27 | 1123 | 11.9 |
| 1968-69 | 5.72 | 5.70 | 997 | 19.5 |
| 1969-70 | 5.86 | 5.67 | 968 | 18.2 |
| 1970-71 | 5.85 | 7.49 | 1279 | 15.9 |
| 1971-72 | 5.67 | 5.10 | 900 | 14.3 |
| 1972-73 | 5.84 | 6.39 | 1094 | 18.8 |
| 1973-74 | 6.02 | 5.80 | 965 | 14.7 |
| 1974-75 | 5.86 | 5.56 | 948 | 21.0 |
| 1975-76 | 6.03 | 7.26 | 1203 | 16.2 |
| 1976-77 | 6.00 | 6.36 | 1060 | 17.7 |
| 1977-78 | 5.68 | 5.97 | 1051 | 16.3 |

*contd. ...*

*...contd. ...*

| *1* | 2 | *3* | *4* | *5* |
|---|---|---|---|---|
| 1978-79 | 5.76 | 6.20 | 1076 | 16.3 |
| 1979-80 | 5.72 | 5.60 | 979 | 24.0 |
| 1980-81 | 6.01 | 6.96 | 1159 | 20.1 |
| 1981-82 | 5.94 | 6.90 | 1162 | 19.8 |
| 1982-83 | 5.72 | 6.55 | 1145 | 21.7 |
| 1983-84 | 5.86 | 7.92 | 1352 | 16.9 |
| 1984-85 | 5.80 | 8.44 | 1456 | 17.5 |
| 1985-86 | 5.80 | 6.64 | 1146 | 18.7 |
| 1986-87 | 5.92 | 7.59 | 1282 | 21.2 |
| 1987-88 | 5.56 | 5.72 | 1029 | 21.2 |
| 1988-89 | 5.90 | 8.23 | 1395 | 21.0 |
| 1989-90 | 5.92 | 9.65 | 1632 | 20.8 |
| 1990-91 | 5.90 | 8.96 | 1518 | 19.7 |
| 1991-92 | 5.86 | 8.06 | 1376 | 22.5 |
| 1992-93 | 5.96 | 9.99 | 1676 | 21.5 |
| 1993-94 | 6.00 | 9.60 | 1602 | 22.4 |
| 1994-95 | 6.14 | 8.88 | 1570 | 20.5 |
| 1995-96 | 5.98 | 9.53 | 1595 | 22.6 |
| 1996-97 | 6.26 | 10.77 | 1720 | 20.6 |
| 1997-98 | 6.32 | 10.82 | 1711 | 20.6 |
| 1998-99 | 6.20 | 11.15 | 1797 | 21.7 |
| 1999-00 | 6.42 | 11.51 | 1792 | 22.9 |
| 2000-01 | 6.61 | 12.04 | 1822 | 22.4 |
| 2001-02 | 6.58 | 13.16 | 2000 | 20.5 |
| 2002-03 | 6.64 | 11.15 | 1681 | 19.5 |
| 2003-04 | 7.34 | 14.98 | 2041 | 19.1 |
| 2004-05 | 7.43 | 14.17 | 1907 | 20.5 |
| 2005-06 | 7.59 | 14.71 | 1938 | 21.1 |
| 2006-07 | 7.89 | 15.09 | 1912 | NA |
| 2007-08* | 8.26 | 19.30 | 2337 | NA |

* Advance Estimates as released on 09.07.2008

*Note*: The yield rates given above have been worked out on the basis of production & area figures taken in '000 units.

## Table 4.11 (b)

*Area, Production and Yield of Maize during 2005-06 and 2006-07 in Respect of Major Maize Producing States Alongwith Coverage under Irrigation*

*Area- Million Hectares*
*Production- Million Tonnes*
*Yield- Kg./Hectare*

| State | 2006-07 | | | | | | 2005-06 | | | | | | % Coverage under Irrigation during 2005-06* |
|---|---|---|---|---|---|---|---|---|---|---|---|---|---|
| | Area | % to All-India | Production | % to All-India | Cumulative % | Yield | Area | % to All-India | Production | % to All-India | Cumulative % | Yield | |
| 1 | 2 | 3 | 4 | 5 | 6 | 7 | 8 | 9 | 10 | 11 | 12 | 13 | 14 |
| Karnataka | 0.96 | 12.17 | 2.72 | 18.01 | 18.01 | 2829 | 0.94 | 12.38 | 2.73 | 18.56 | 18.56 | 2915 | 41.7 |
| Andhra Pradesh | 0.73 | 9.25 | 2.46 | 16.29 | 34.30 | 3396 | 0.76 | 10.01 | 3.09 | 21.01 | 39.56 | 4073 | 34.3 |
| Bihar | 0.64 | 8.11 | 1.72 | 11.39 | 45.70 | 2671 | 0.65 | 8.56 | 1.36 | 9.25 | 48.81 | 2098 | 57.3 |
| Uttar Pradesh | 0.87 | 11.03 | 1.16 | 7.68 | 53.38 | 1335 | 0.81 | 10.67 | 1.05 | 7.14 | 55.95 | 1295 | 28.4 |
| Maharashtra | 0.58 | 7.35 | 1.15 | 7.62 | 60.99 | 1983 | 0.47 | 6.19 | 1.00 | 6.80 | 62.75 | 2106 | 9.7 |
| Rajasthan | 1.03 | 13.05 | 1.12 | 7.42 | 68.41 | 1086 | 1.00 | 13.18 | 1.10 | 7.48 | 70.22 | 1098 | 4.7 |
| Madhya Pradesh | 0.86 | 10.90 | 0.84 | 5.56 | 73.97 | 976 | 0.86 | 11.33 | 1.25 | 8.50 | 78.72 | 1450 | 1.9 |
| Tamil Nadu | 0.20 | 2.53 | 0.76 | 5.03 | 79.01 | 3838 | 0.20 | 2.64 | 0.24 | 1.63 | 80.35 | 1189 | 29.5 |
| Himachal Pradesh | 0.30 | 3.80 | 0.70 | 4.64 | 83.64 | 2326 | 0.30 | 3.95 | 0.54 | 3.67 | 84.02 | 1839 | 8.0 |
| Jammu & Kashmir | 0.32 | 4.06 | 0.49 | 3.25 | 86.89 | 1505 | 0.32 | 4.22 | 0.45 | 3.06 | 87.08 | 1413 | 8.0 |
| Punjab | 0.15 | 1.90 | 0.48 | 3.18 | 90.07 | 3123 | 0.15 | 1.98 | 0.40 | 2.72 | 89.80 | 2723 | 60.4 |
| Gujarat | 0.52 | 6.59 | 0.36 | 2.38 | 92.45 | 698 | 0.50 | 6.59 | 0.56 | 3.81 | 93.61 | 1124 | 6.7 |
| Jharkhand | 0.24 | 3.04 | 0.30 | 1.99 | 94.44 | 1230 | 0.18 | 2.37 | 0.24 | 1.63 | 95.24 | 1315 | 1.8 |
| West Bengal | 0.09 | 1.14 | 0.25 | 1.66 | 96.09 | 2968 | 0.05 | 0.66 | 0.13 | 0.88 | 96.13 | 2533 | 7.0 |
| Others | 0.40 | 5.07 | 0.59 | 3.91 | 100.00 | @ | 0.40 | 5.27 | 0.57 | 3.87 | 100.00 | @ | - |
| All India | 7.89 | 100.00 | 15.10 | 100.00 | | 1912 | 7.59 | 100.00 | 14.71 | 100.00 | | 1938 | 21.1 |

@ Since area/production is low in individual states, yield rate is not worked out.

* Provisional

*Note*: States have been arranged in descending order of percentage share of production during 2006-07.

# Table 4.12 (a)

*All-India Area, Production and Yield of Total Pulses from 1950-51 to 2007-08 Alongwith Percentage Coverage under Irrigation*

*Area- Million Hectares*
*Production- Million Tonnes*
*Yield- Kg./Hectare*

| Year | Area | Production | Yield | % Coverage under Irrigation |
|---|---|---|---|---|
| 1 | 2 | 3 | 4 | 5 |
| 1950-51 | 19.09 | 8.41 | 441 | 9.4 |
| 1951-52 | 18.78 | 8.42 | 448 | 9.7 |
| 1952-53 | 19.84 | 9.19 | 463 | 9.8 |
| 1953-54 | 21.73 | 10.62 | 489 | 9.2 |
| 1954-55 | 21.91 | 10.95 | 500 | 8.8 |
| 1955-56 | 23.22 | 11.04 | 476 | 8.4 |
| 1956-57 | 23.32 | 11.55 | 495 | 7.3 |
| 1957-58 | 22.54 | 9.56 | 424 | 9.1 |
| 1958-59 | 24.31 | 13.15 | 541 | 8.4 |
| 1959-60 | 24.83 | 11.80 | 475 | 8.5 |
| 1960-61 | 23.56 | 12.70 | 539 | 8.0 |
| 1961-62 | 24.24 | 11.76 | 485 | 8.1 |
| 1962-63 | 24.27 | 11.53 | 475 | 8.9 |
| 1963-64 | 24.18 | 10.07 | 416 | 8.9 |
| 1964-65 | 23.88 | 12.42 | 520 | 9.2 |
| 1965-66 | 22.72 | 9.94 | 438 | 9.4 |
| 1966-67 | 22.12 | 8.35 | 377 | 10.9 |
| 1967-68 | 22.65 | 12.10 | 534 | 8.7 |
| 1968-69 | 21.26 | 10.42 | 490 | 9.8 |
| 1969-70 | 22.02 | 11.69 | 531 | 9.4 |
| 1970-71 | 22.54 | 11.82 | 524 | 8.8 |
| 1971-72 | 22.15 | 11.09 | 501 | 8.8 |
| 1972-73 | 20.92 | 9.91 | 474 | 8.1 |
| 1973-74 | 23.43 | 10.01 | 427 | 7.9 |
| 1974-75 | 22.03 | 10.02 | 455 | 8.1 |
| 1975-76 | 24.45 | 13.04 | 533 | 7.9 |
| 1976-77 | 22.98 | 11.36 | 494 | 7.5 |
| 1977-78 | 23.50 | 11.97 | 510 | 7.1 |

contd. ...

*...contd. ...*

| 1 | 2 | 3 | 4 | 5 |
|---|---|---|---|---|
| 1978-79 | 23.66 | 12.18 | 515 | 7.9 |
| 1979-80 | 22.26 | 8.57 | 385 | 8.8 |
| 1980-81 | 22.46 | 10.63 | 473 | 9.0 |
| 1981-82 | 23.84 | 11.51 | 483 | 8.5 |
| 1982-83 | 22.83 | 11.86 | 519 | 8.2 |
| 1983-84 | 23.54 | 12.89 | 548 | 7.5 |
| 1984-85 | 22.74 | 11.96 | 526 | 7.9 |
| 1985-86 | 24.42 | 13.36 | 547 | 8.5 |
| 1986-87 | 23.16 | 11.71 | 506 | 9.6 |
| 1987-88 | 21.27 | 10.96 | 515 | 9.4 |
| 1988-89 | 23.15 | 13.85 | 598 | 9.3 |
| 1989-90 | 23.41 | 12.86 | 549 | 10.0 |
| 1990-91 | 24.66 | 14.26 | 578 | 10.5 |
| 1991-92 | 22.54 | 12.02 | 533 | 10.7 |
| 1992-93 | 22.36 | 12.82 | 573 | 10.4 |
| 1993-94 | 22.25 | 13.30 | 598 | 11.3 |
| 1994-95 | 23.03 | 14.04 | 610 | 12.7 |
| 1995-96 | 22.28 | 12.31 | 552 | 12.9 |
| 1996-97 | 22.45 | 14.24 | 635 | 12.7 |
| 1997-98 | 22.87 | 12.98 | 567 | 11.3 |
| 1998-99 | 23.50 | 14.91 | 634 | 12.1 |
| 1999-00 | 21.12 | 13.42 | 635 | 16.1 |
| 2000-01 | 20.35 | 11.08 | 544 | 12.5 |
| 2001-02 | 22.01 | 13.37 | 607 | 13.3 |
| 2002-03 | 20.50 | 11.13 | 543 | 14.4 |
| 2003-04 | 23.46 | 14.91 | 635 | 13.6 |
| 2004-05 | 22.76 | 13.13 | 577 | 13.9 |
| 2005-06 | 22.39 | 13.39 | 598 | 15.0 |
| 2006-07 | 23.19 | 14.20 | 612 | NA |
| 2007-08* | 23.86 | 15.12 | 638 | NA |

* Advance Estimates as released on 09.07.2008

*Note*: The yield rates given above have been worked out on the basis of production and area figures taken in '000 units.

*Total Pulses—All India Production*

*Million Tonnes*

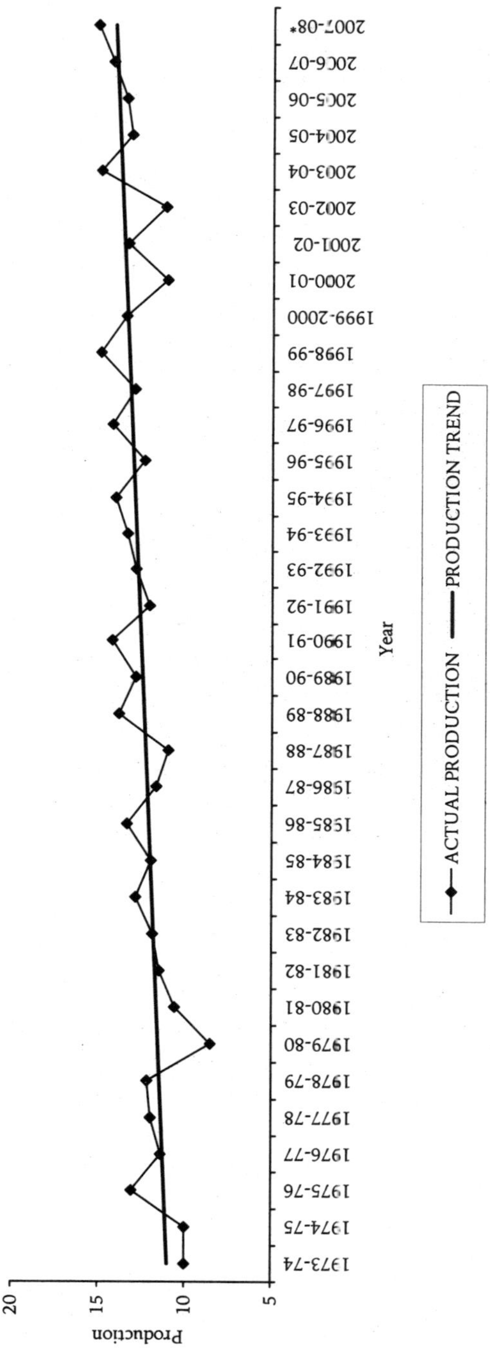


* Based on Advance estimate as on 09-07-08.

# Table 4.12 (b)

*Area, Production and Yield of Total Pulses during 2005-06 and 2006-07 in Respect of Major Pulses Producing States Alongwith Coverage under Irrigation*

*Area- Million Hectares*
*Production- Million Tonnes*
*Yield- Kg./Hectare*

| State | 2006-07 | | | | | | 2005-06 | | | | | | % Coverage under Irrigation during 2005-06* |
|---|---|---|---|---|---|---|---|---|---|---|---|---|---|
| | Area | % to All-India | Production | % to All-India | Cumulative % | Yield | Area | % to All-India | Production | % to All-India | Cumulative % | Yield | |
| 1 | 2 | 3 | 4 | 5 | 6 | 7 | 8 | 9 | 10 | 11 | 12 | 13 | 14 |
| Madhya Pradesh | 4.11 | 17.72 | 3.20 | 22.54 | 22.54 | 780 | 4.28 | 19.12 | 3.23 | 24.14 | 24.14 | 754 | 33.5 |
| Maharashtra | 3.83 | 16.52 | 2.30 | 16.20 | 38.73 | 602 | 3.43 | 15.32 | 2.01 | 15.02 | 39.16 | 584 | 23.4 |
| Uttar Pradesh | 2.72 | 11.73 | 1.98 | 13.94 | 52.68 | 725 | 2.75 | 12.28 | 2.23 | 16.67 | 55.83 | 811 | 11.3 |
| Rajasthan | 3.21 | 13.84 | 1.48 | 10.42 | 63.10 | 462 | 3.44 | 15.36 | 0.90 | 6.73 | 62.56 | 261 | 1.1 |
| Andhra Pradesh | 1.98 | 8.54 | 1.35 | 9.51 | 72.61 | 679 | 1.78 | 7.95 | 1.38 | 10.31 | 72.87 | 772 | 5.4 |
| Karnataka | 2.37 | 10.22 | 0.89 | 6.27 | 78.87 | 377 | 1.98 | 8.84 | 0.96 | 7.17 | 80.04 | 487 | 12.0 |
| Gujarat | 1.00 | 4.31 | 0.59 | 4.15 | 83.03 | 593 | 0.78 | 3.48 | 0.55 | 4.11 | 84.16 | 704 | 10.2 |
| Chhattisgarh | 0.91 | 3.92 | 0.49 | 3.45 | 86.48 | 543 | 0.95 | 4.24 | 0.45 | 3.36 | 87.52 | 477 | 3.6 |
| Bihar | 0.61 | 2.63 | 0.44 | 3.10 | 89.58 | 722 | 0.60 | 2.68 | 0.45 | 3.36 | 90.88 | 449 | 4.7 |
| Orissa | 0.79 | 3.41 | 0.35 | 2.46 | 92.04 | 445 | 0.81 | 3.62 | 0.34 | 2.54 | 93.42 | 416 | 8.3 |
| Tamil Nadu | 0.54 | 2.33 | 0.29 | 2.04 | 94.08 | 541 | 0.53 | 2.37 | 0.18 | 1.35 | 94.77 | 337 | 5.5 |
| Jharkhand | 0.38 | 1.64 | 0.26 | 1.83 | 95.92 | 686 | 0.29 | 1.30 | 0.17 | 1.27 | 96.04 | 592 | 7.6 |
| West Bengal | 0.22 | 0.95 | 0.15 | 1.06 | 96.97 | 703 | 0.22 | 0.98 | 0.17 | 1.27 | 97.31 | 785 | 2.1 |
| Haryana | 0.17 | 0.73 | 0.14 | 0.99 | 97.96 | 824 | 0.19 | 0.85 | 0.12 | 0.90 | 98.21 | 622 | 31.9 |
| Others | 0.35 | 1.51 | 0.29 | 2.04 | 100.00 | @ | 0.36 | 1.61 | 0.24 | 1.79 | 100.00 | @ | - |
| All India | 23.19 | 100.00 | 14.20 | 100.00 | | 612 | 22.39 | 100.00 | 13.38 | 100.00 | | 598 | 15.1 |

@ Since area/production is low in individual states, yield rate is not worked out.

* Provisional

*Note*: States have been arranged in descending order of percentage share of production during 2006-07.

# Table 4.13(a)

*All-India Area, Production and Yield of Gram from 1950-51 to 2007-08 Alongwith Percentage Coverage under Irrigation*

*Area- Million Hectares*
*Production- Million Tonnes*
*Yield- Kg./Hectare*

| *Year* | *Area* | *Production* | *Yield* | *% Coverage under Irrigation* |
|---|---|---|---|---|
| *1* | 2 | 3 | 4 | 5 |
| 1950-51 | 7.57 | 3.65 | 482 | 12.5 |
| 1951-52 | 6.83 | 3.39 | 496 | 15.1 |
| 1952-53 | 7.26 | 4.21 | 580 | 14.6 |
| 1953-54 | 7.97 | 4.83 | 606 | 14.0 |
| 1954-55 | 9.25 | 5.62 | 608 | 12.8 |
| 1955-56 | 9.78 | 5.42 | 554 | 12.1 |
| 1956-57 | 9.67 | 6.23 | 644 | 10.8 |
| 1957-58 | 9.09 | 4.89 | 538 | 12.5 |
| 1958-59 | 10.08 | 7.02 | 697 | 12.0 |
| 1959-60 | 10.33 | 5.62 | 544 | 12.4 |
| 1960-61 | 9.28 | 6.25 | 674 | 11.9 |
| 1961-62 | 9.57 | 5.79 | 605 | 12.2 |
| 1962-63 | 9.19 | 5.36 | 583 | 13.7 |
| 1963-64 | 9.35 | 4.50 | 481 | 13.7 |
| 1964-65 | 8.87 | 5.78 | 651 | 15.8 |
| 1965-66 | 8.02 | 4.22 | 527 | 16.4 |
| 1966-67 | 8.00 | 3.62 | 453 | 18.8 |
| 1967-68 | 8.26 | 5.97 | 723 | 15.6 |
| 1968-69 | 7.11 | 4.31 | 607 | 18.8 |
| 1969-70 | 7.75 | 5.55 | 715 | 17.3 |
| 1970-71 | 7.84 | 5.20 | 663 | 15.6 |
| 1971-72 | 7.91 | 5.08 | 642 | 15.0 |
| 1972-73 | 6.97 | 4.54 | 651 | 15.6 |
| 1973-74 | 7.76 | 4.10 | 528 | 15.8 |
| 1974-75 | 7.04 | 4.02 | 570 | 17.8 |
| 1975-76 | 8.32 | 5.88 | 707 | 16.5 |
| 1976-77 | 7.97 | 5.42 | 680 | 15.2 |
| 1977-78 | 7.97 | 5.41 | 678 | 14.7 |

*contd. ...*

...contd. ...

| 1 | 2 | 3 | 4 | 5 |
|---|---|---|---|---|
| 1978-79 | 7.71 | 5.74 | 745 | 15.6 |
| 1979-80 | 6.99 | 3.36 | 481 | 18.6 |
| 1980-81 | 6.58 | 4.33 | 657 | 20.6 |
| 1981-82 | 7.87 | 4.64 | 590 | 17.9 |
| 1982-83 | 7.40 | 5.29 | 715 | 15.6 |
| 1983-84 | 7.16 | 4.75 | 663 | 14.5 |
| 1984-85 | 6.91 | 4.56 | 661 | 14.8 |
| 1985-86 | 7.80 | 5.79 | 742 | 15.6 |
| 1986-87 | 6.98 | 4.53 | 649 | 19.4 |
| 1987-88 | 5.77 | 3.63 | 629 | 19.2 |
| 1988-89 | 6.81 | 5.13 | 753 | 18.3 |
| 1989-90 | 6.47 | 4.22 | 652 | 21.0 |
| 1990-91 | 7.52 | 5.36 | 712 | 20.5 |
| 1991-92 | 5.58 | 4.12 | 739 | 24.2 |
| 1992-93 | 6.45 | 4.42 | 684 | 22.0 |
| 1993-94 | 6.36 | 4.98 | 783 | 24.0 |
| 1994-95 | 7.54 | 6.44 | 853 | 25.3 |
| 1995-96 | 7.12 | 4.98 | 700 | 26.0 |
| 1996-97 | 6.85 | 5.57 | 813 | 25.1 |
| 1997-98 | 7.56 | 6.13 | 811 | 21.8 |
| 1998-99 | 8.47 | 6.80 | 803 | 21.0 |
| 1999-2000 | 6.15 | 5.12 | 833 | 29.1 |
| 2000-01 | 5.19 | 3.86 | 744 | 30.9 |
| 2001-02 | 6.42 | 5.47 | 853 | 30.4 |
| 2002-03 | 5.91 | 4.24 | 717 | 32.1 |
| 2003-04 | 7.05 | 5.72 | 811 | 31.0 |
| 2004-05 | 6.71 | 5.47 | 815 | 31.4 |
| 2005-06 | 6.93 | 5.60 | 808 | 31.1 |
| 2006-07 | 7.49 | 6.33 | 845 | NA |
| 2007-08* | 7.58 | 6.91 | 780 | NA |

* Advance Estimates as released on 09.07.2008

*Note*: The yield rates given above have been worked out on the basis of production & area figures taken in '000 units.

# Table 4.13 (b)

*Area, Production and Yield of Gram during 2005-06 and 2006-07 in Respect of Major Gram Producing States Alongwith Coverage under Irrigation*

*Area- Million Hectares*
*Production- Million Tonnes*
*Yield- Kg./Hectare*

| State | 2006-07 | | | | | | 2005-06 | | | | | | % Coverage under Irrigation during 2005-06* |
|---|---|---|---|---|---|---|---|---|---|---|---|---|---|
| | Area | % to All-India | Production | % to All-India | Cumulative % | Yield | Area | % to All-India | Production | % to All-India | Cumulative % | Yield | |
| 1 | 2 | 3 | 4 | 5 | 6 | 7 | 8 | 9 | 10 | 11 | 12 | 13 | 14 |
| Madhya Pradesh | 2.46 | 32.84 | 2.41 | 38.07 | 38.07 | 980 | 2.56 | 36.94 | 2.37 | 42.32 | 42.32 | 926 | 47.5 |
| Maharashtra | 1.31 | 17.49 | 0.92 | 14.53 | 52.61 | 706 | 1.02 | 14.72 | 0.71 | 12.68 | 55.00 | 691 | 26.8 |
| Rajasthan | 1.01 | 13.48 | 0.87 | 13.74 | 66.35 | 863 | 1.08 | 15.58 | 0.48 | 8.57 | 63.57 | 443 | 31.8 |
| Andhra Pradesh | 0.60 | 8.01 | 0.65 | 10.27 | 76.62 | 1085 | 0.39 | 5.63 | 0.63 | 11.25 | 74.82 | 1591 | 1.2 |
| Uttar Pradesh | 0.68 | 9.08 | 0.50 | 7.90 | 84.52 | 742 | 0.74 | 10.68 | 0.66 | 11.79 | 86.61 | 893 | 15.1 |
| Karnataka | 0.65 | 8.68 | 0.31 | 4.90 | 89.42 | 473 | 0.42 | 6.06 | 0.23 | 4.11 | 90.71 | 548 | 16.9 |
| Gujarat | 0.25 | 3.34 | 0.21 | 3.32 | 92.73 | 870 | 0.17 | 2.45 | 0.14 | 2.50 | 93.21 | 850 | 28.7 |
| Chattisgarh | 0.21 | 2.80 | 0.18 | 2.84 | 95.58 | 843 | 0.23 | 3.32 | 0.16 | 2.86 | 96.07 | 706 | 13.6 |
| Haryana | 0.11 | 1.47 | 0.09 | 1.42 | 97.00 | 843 | 0.13 | 1.88 | 0.07 | 1.25 | 97.32 | 554 | 17.2 |
| Bihar | 0.06 | 0.80 | 0.05 | 0.79 | 97.79 | 818 | 0.06 | 0.87 | 0.06 | 1.07 | 98.39 | 902 | 15.5 |
| West Bengal | 0.03 | 0.40 | 0.02 | 0.32 | 98.10 | 769 | 0.04 | 0.58 | 0.04 | 0.71 | 99.11 | 913 | 31.6 |
| Orissa | 0.04 | 0.53 | 0.02 | 0.32 | 98.42 | 652 | 0.04 | 0.58 | 0.02 | 0.36 | 99.46 | 646 | - |
| Others | 0.08 | 1.07 | 0.10 | 1.58 | 100.00 | @ | 0.05 | 0.72 | 0.03 | 0.54 | 100.00 | @ | - |
| All India | 7.49 | 100.00 | 6.33 | 100.00 | | 845 | 6.93 | 100.00 | 5.60 | 100.00 | | 808 | 31.1 |

@ Since area/production is low in individual states, yield rate is not worked out.

* Provisional

*Note:* States have been arranged in descending order of percentage share of production during 2006-07.

# Table 4.14 (a)

*All-India Area, Production and Yield of Tur (Arhar) from 1950-51 to 2007-08 Alongwith Percentage Coverage under Irrigation*

*Area- Million Hectares*
*Production- Million Tonnes*
*Yield- Kg./Hectare*

| Year | Area | Production | Yield | % Coverage under Irrigation |
|---|---|---|---|---|
| 1 | 2 | 3 | 4 | 5 |
| 1950-51 | 2.18 | 1.72 | 788 | 0.5 |
| 1951-52 | 2.45 | 1.83 | 748 | 0.5 |
| 1952-53 | 2.40 | 1.70 | 709 | 0.6 |
| 1953-54 | 2.40 | 1.86 | 775 | 0.6 |
| 1954-55 | 2.40 | 1.72 | 715 | 0.5 |
| 1955-56 | 2.29 | 1.86 | 814 | 0.4 |
| 1956-57 | 2.29 | 1.99 | 867 | 0.3 |
| 1957-58 | 2.36 | 1.48 | 625 | 0.5 |
| 1958-59 | 2.47 | 1.70 | 688 | 0.5 |
| 1959-60 | 2.43 | 1.70 | 700 | 0.5 |
| 1960-61 | 2.43 | 2.07 | 849 | 0.5 |
| 1961-62 | 2.45 | 1.37 | 559 | 0.5 |
| 1962-63 | 2.45 | 1.58 | 645 | 0.4 |
| 1963-64 | 2.52 | 1.38 | 549 | 0.4 |
| 1964-65 | 2.58 | 1.89 | 733 | 0.5 |
| 1965-66 | 2.56 | 1.73 | 678 | 0.4 |
| 1966-67 | 2.52 | 1.13 | 448 | 0.7 |
| 1967-68 | 2.67 | 1.74 | 653 | 0.5 |
| 1968-69 | 2.53 | 1.82 | 718 | 0.5 |
| 1969-70 | 2.67 | 1.84 | 690 | 0.3 |
| 1970-71 | 2.66 | 1.88 | 709 | 0.3 |
| 1971-72 | 2.35 | 1.68 | 718 | 0.3 |
| 1972-73 | 2.42 | 1.93 | 795 | 0.4 |
| 1973-74 | 2.65 | 1.41 | 532 | 0.4 |
| 1974-75 | 2.53 | 1.83 | 725 | 0.5 |
| 1975-76 | 2.67 | 2.10 | 786 | 0.4 |
| 1976-77 | 2.57 | 1.73 | 672 | 0.5 |
| 1977-78 | 2.63 | 1.93 | 735 | 0.5 |

*contd. ...*

*...contd. ...*

| 1 | 2 | 3 | 4 | 5 |
|---|---|---|---|---|
| 1978-79 | 2.64 | 1.89 | 716 | 0.8 |
| 1979-80 | 2.73 | 1.76 | 643 | 1.8 |
| 1980-81 | 2.84 | 1.96 | 689 | 2.6 |
| 1981-82 | 3.00 | 2.24 | 745 | 2.1 |
| 1982-83 | 2.93 | 1.99 | 680 | 2.4 |
| 1983-84 | 3.22 | 2.58 | 801 | 3.0 |
| 1984-85 | 3.16 | 2.58 | 819 | 3.9 |
| 1985-86 | 3.18 | 2.44 | 767 | 4.3 |
| 1986-87 | 3.15 | 2.27 | 722 | 4.2 |
| 1987-88 | 3.33 | 2.28 | 685 | 4.9 |
| 1988-89 | 3.49 | 2.72 | 779 | 4.7 |
| 1989-90 | 3.60 | 2.75 | 763 | 5.1 |
| 1990-91 | 3.59 | 2.41 | 673 | 5.5 |
| 1991-92 | 3.63 | 2.13 | 588 | 5.2 |
| 1992-93 | 3.58 | 2.33 | 652 | 4.9 |
| 1993-94 | 3.53 | 2.69 | 762 | 5.0 |
| 1994-95 | 3.31 | 2.14 | 644 | 5.5 |
| 1995-96 | 3.45 | 2.31 | 670 | 5.2 |
| 1996-97 | 3.51 | 2.66 | 756 | 5.4 |
| 1997-98 | 3.36 | 1.85 | 551 | 5.5 |
| 1998-99 | 3.44 | 2.71 | 787 | 4.2 |
| 1999-00 | 3.43 | 2.69 | 786 | 4.4 |
| 2000-01 | 3.63 | 2.25 | 618 | 4.2 |
| 2001-02 | 3.33 | 2.26 | 679 | 4.7 |
| 2002-03 | 3.36 | 2.19 | 651 | 4.6 |
| 2003-04 | 3.52 | 2.35 | 670 | 4.7 |
| 2004-05 | 3.52 | 2.35 | 667 | 4.4 |
| 2005-06 | 3.58 | 2.74 | 765 | 4.4 |
| 2006-07 | 3.56 | 2.31 | 650 | NA |
| 2007-08* | 3.79 | 3.09 | 824 | NA |

* Advance Estimates as released on 09.07.2008

*Note*: The yield rates given above have been worked out on the basis of production & area figures taken in '000 units.

# Table 4.14 (b)

*Area, Production and Yield of Tur (Arhar) during 2005-06 and 2006-07 in Respect of Major Tur Producing States Alongwith Coverage under Irrigation*

*Area- Million Hectares*
*Production- Million Tonnes*
*Yield- Kg./Hectare*

| State | 2006-07 | | | | | | 2005-06 | | | | | | % Coverage under Irrigation during 2005-06* |
|---|---|---|---|---|---|---|---|---|---|---|---|---|---|
| | Area | % to All-India | Production | % to All-India | Cumulative % | Yield | Area | % to All-India | Production | % to All-India | Cumulative % | Yield | |
| 1 | 2 | 3 | 4 | 5 | 6 | 7 | 8 | 9 | 10 | 11 | 12 | 13 | 14 |
| Maharashtra | 1.12 | 31.46 | 0.82 | 35.50 | 35.50 | 726 | 1.10 | 30.73 | 0.79 | 28.83 | 28.83 | 720 | 1.7 |
| Uttar Pradesh | 0.41 | 11.52 | 0.30 | 12.99 | 48.48 | 749 | 0.38 | 10.61 | 0.38 | 13.87 | 42.70 | 987 | 14.3 |
| Karnataka | 0.60 | 16.85 | 0.28 | 12.12 | 60.61 | 470 | 0.60 | 16.76 | 0.44 | 16.06 | 58.76 | 727 | 3.3 |
| Gujarat | 0.29 | 8.15 | 0.22 | 9.52 | 70.13 | 751 | 0.25 | 6.98 | 0.28 | 10.22 | 68.98 | 1102 | 9.2 |
| Madhya Pradesh | 0.32 | 8.99 | 0.22 | 9.52 | 79.65 | 683 | 0.32 | 8.94 | 0.24 | 8.76 | 77.74 | 739 | 1.2 |
| Andhra Pradesh | 0.40 | 11.24 | 0.16 | 6.93 | 86.58 | 401 | 0.49 | 13.69 | 0.30 | 10.95 | 88.69 | 609 | 0.4 |
| Orissa | 0.13 | 3.65 | 0.11 | 4.76 | 91.34 | 803 | 0.13 | 3.63 | 0.10 | 3.65 | 92.34 | 742 | - |
| Jharkhand | 0.07 | 1.97 | 0.06 | 2.60 | 93.94 | 645 | 0.09 | 2.51 | 0.05 | 1.82 | 94.16 | 633 | - |
| Bihar | 0.04 | 1.12 | 0.04 | 1.73 | 95.67 | 989 | 0.03 | 0.84 | 0.04 | 1.46 | 95.62 | 1291 | 1.2 |
| Tamil Nadu | 0.03 | 0.84 | 0.02 | 0.87 | 96.54 | 732 | 0.04 | 1.12 | 0.02 | 0.73 | 96.35 | 540 | 3.2 |
| Others | 0.15 | 4.21 | 0.08 | 3.46 | 100.00 | @ | 0.15 | 4.19 | 0.10 | 3.65 | 100.00 | @ | - |
| All-India | 3.56 | 100.00 | 2.31 | 100.00 | | 650 | 3.58 | 100.00 | 2.74 | 100.00 | | 765 | 4.4 |

@ Since area/production is low in individual states, yield rate is not worked out.

* Provisional

*Note*: States have been arranged in descending order of percentage share of production during 2006-07.

# Table 4.15(a)

*All-India Area, Production and Yield of Lentil (Masur) from 1970-71 to 2006-07*

*Area- Million Hectares*
*Production- Million Tonnes*
*Yield- Kg./Hectare*

| Year | Area | Production | Yield |
|---|---|---|---|
| 1 | 2 | 3 | 4 |
| 1970-71 | 0.75 | 0.37 | 497 |
| 1971-72 | 0.63 | 0.30 | 481 |
| 1972-73 | 0.83 | 0.37 | 447 |
| 1973-74 | 0.93 | 0.41 | 440 |
| 1974-75 | 0.95 | 0.46 | 480 |
| 1975-76 | 0.94 | 0.46 | 492 |
| 1976-77 | 0.85 | 0.39 | 461 |
| 1977-78 | 0.88 | 0.38 | 436 |
| 1978-79 | 1.01 | 0.45 | 443 |
| 1979-80 | 0.85 | 0.32 | 377 |
| 1980-81 | 0.93 | 0.47 | 498 |
| 1981-82 | 0.95 | 0.50 | 525 |
| 1982-83 | 1.00 | 0.49 | 491 |
| 1983-84 | 0.94 | 0.53 | 566 |
| 1984-85 | 0.98 | 0.55 | 557 |
| 1985-86 | 1.09 | 0.66 | 607 |
| 1986-87 | 1.09 | 0.66 | 607 |
| 1987-88 | 1.04 | 0.66 | 635 |
| 1988-89 | 1.08 | 0.73 | 679 |
| 1989-90 | 1.11 | 0.71 | 635 |
| 1990-91 | 1.19 | 0.85 | 717 |
| 1991-92 | 1.19 | 0.80 | 674 |
| 1992-93 | 1.20 | 0.79 | 657 |
| 1993-94 | 1.18 | 0.75 | 632 |
| 1994-95 | 1.16 | 0.78 | 678 |
| 1995-96 | 1.26 | 0.71 | 569 |
| 1996-97 | 1.37 | 0.96 | 702 |
| 1997-98 | 1.29 | 0.80 | 623 |
| 1998-99 | 1.39 | 0.94 | 675 |
| 1999-00 | 1.46 | 1.08 | 738 |
| 2000-01 | 1.48 | 0.92 | 619 |
| 2001-02 | 1.47 | 0.97 | 664 |
| 2002-03 | 1.38 | 0.87 | 634 |
| 2003-04 | 1.39 | 1.04 | 743 |
| 2004-05 | 1.47 | 0.99 | 675 |
| 2005-06 | 1.51 | 0.95 | 629 |
| 2006-07 | 1.47 | 0.91 | 621 |

*Note*: The yield rates given above have been worked out on the basis of production & area figures taken in '000 units.

## Table 4.15 (b)

*Area, Production and Yield of Lentil (Masur) during 2005-06 and 2006-07 in Respect of Major Lentil Producing States*

*Area- Million Hectares*
*Production- Million Tonnes*
*Yield- Kg./Hectare*

| State | 2006-07 | | | | | | 2005-06 | | | | | |
|---|---|---|---|---|---|---|---|---|---|---|---|---|
| | Area | % to All-India | Production | % to All-India | Cumulative % | Yield | Area | % to All-India | Production | % to All-India | Cumulative % | Yield |
| 1 | 2 | 3 | 4 | 5 | 6 | 7 | 8 | 9 | 10 | 11 | 12 | 13 |
| Uttar Pradesh | 0.62 | 31.00 | 0.43 | 47.25 | 47.25 | 705 | 0.61 | 40.40 | 0.43 | 45.26 | 45.26 | 715 |
| Madhya Pradesh | 0.52 | 26.00 | 0.26 | 28.57 | 75.82 | 497 | 0.56 | 37.09 | 0.29 | 30.53 | 75.79 | 510 |
| Bihar | 0.17 | 8.50 | 0.12 | 13.19 | 89.01 | 705 | 0.16 | 10.60 | 0.11 | 11.58 | 87.37 | 705 |
| West Bengal | 0.06 | 3.00 | 0.04 | 4.40 | 93.41 | 665 | 0.06 | 3.97 | 0.05 | 5.26 | 92.63 | 764 |
| Rajasthan | 0.01 | 0.50 | 0.01 | 1.10 | 94.51 | 921 | 0.02 | 1.32 | 0.02 | 2.11 | 94.74 | 908 |
| Assam | 0.02 | 1.00 | 0.01 | 1.10 | 95.60 | 550 | 0.02 | 1.32 | 0.01 | 1.05 | 95.79 | 532 |
| Others | 0.60 | 30.00 | 0.04 | 4.40 | 100.00 | @ | 0.08 | 5.30 | 0.04 | 4.21 | 100.00 | @ |
| All-India | 2.00 | 100.00 | 0.91 | 100.00 | | 621 | 1.51 | 100.00 | 0.95 | 100.00 | | 629 |

@ - Since area/production is low in individual states, yield rate is not worked out.

*Note*: States have been arranged in descending order of percentage share of production during 2006-07.

# Table 4.16 (a)

*All-India Area, Production and Yield of Nine Oilseeds from 1950-51 to 2007-08 Alongwith Percentage Coverage under Irrigation*

*Area- Million Hectares*
*Production- Million Tonnes*
*Yield- Kg./Hectare*

| *Year* | *Area* | *Production* | *Yield* | *% Coverage under Irrigation* |
|---|---|---|---|---|
| *1* | *2* | *3* | *4* | *5* |
| 1950-51 | 10.73 | 5.16 | 481 | NA |
| 1951-52 | 11.69 | 5.03 | 430 | NA |
| 1952-53 | 11.18 | 4.73 | 424 | 0.8 |
| 1953-54 | 10.99 | 5.37 | 488 | 1.0 |
| 1954-55 | 12.52 | 6.40 | 511 | 1.1 |
| 1955-56 | 12.09 | 5.73 | 474 | 2.4 |
| 1956-57 | 12.49 | 6.36 | 509 | 2.7 |
| 1957-58 | 12.66 | 6.35 | 502 | 3.5 |
| 1958-59 | 13.00 | 7.30 | 561 | 3.2 |
| 1959-60 | 13.95 | 6.56 | 470 | 3.1 |
| 1960-61 | 13.77 | 6.98 | 507 | 3.3 |
| 1961-62 | 14.77 | 7.28 | 493 | 3.6 |
| 1962-63 | 15.34 | 7.39 | 482 | 3.3 |
| 1963-64 | 14.82 | 7.13 | 481 | 3.7 |
| 1964-65 | 15.26 | 8.56 | 561 | 3.4 |
| 1965-66 | 15.25 | 6.40 | 419 | 3.7 |
| 1966-67 | 15.00 | 6.43 | 428 | 5.1 |
| 1967-68 | 15.67 | 8.30 | 530 | 5.0 |
| 1968-69 | 14.47 | 6.85 | 473 | 4.8 |
| 1969-70 | 14.81 | 7.73 | 522 | 6.1 |
| 1970-71 | 16.64 | 9.63 | 579 | 7.4 |
| 1971-72 | 17.27 | 9.08 | 526 | 7.7 |
| 1972-73 | 15.79 | 7.14 | 452 | 7.2 |
| 1973-74 | 16.90 | 9.39 | 555 | 8.8 |
| 1974-75 | 17.31 | 9.15 | 529 | 9.2 |
| 1975-76 | 16.92 | 10.61 | 627 | 7.9 |
| 1976-77 | 16.47 | 8.43 | 512 | 7.6 |

*contd. ...*

*...contd. ...*

| 1 | 2 | 3 | 4 | 5 |
|---|---|---|---|---|
| 1977-78 | 17.17 | 9.66 | 563 | 10.4 |
| 1978-79 | 17.71 | 10.10 | 570 | 11.0 |
| 1979-80 | 16.94 | 8.74 | 516 | 12.6 |
| 1980-81 | 17.60 | 9.37 | 532 | 14.5 |
| 1981-82 | 18.91 | 12.08 | 639 | 15.4 |
| 1982-83 | 17.76 | 10.00 | 563 | 15.6 |
| 1983-84 | 18.69 | 12.69 | 679 | 17.0 |
| 1984-85 | 18.92 | 12.95 | 684 | 19.6 |
| 1985-86 | 19.02 | 10.83 | 570 | 17.3 |
| 1986-87 | 18.63 | 11.27 | 605 | 17.9 |
| 1987-88 | 20.13 | 12.65 | 629 | 20.6 |
| 1988-89 | 21.90 | 18.03 | 824 | 22.3 |
| 1989-90 | 22.80 | 16.92 | 742 | 22.1 |
| 1990-91 | 24.15 | 18.61 | 771 | 22.9 |
| 1991-92 | 25.89 | 18.60 | 719 | 25.5 |
| 1992-93 | 25.24 | 20.11 | 797 | 24.1 |
| 1993-94 | 26.90 | 21.50 | 799 | 22.8 |
| 1994-95 | 25.30 | 21.34 | 843 | 25.0 |
| 1995-96 | 25.96 | 22.11 | 851 | 26.0 |
| 1996-97 | 26.34 | 24.38 | 926 | 26.3 |
| 1997-98 | 26.12 | 21.32 | 816 | 24.3 |
| 1998-99 | 26.23 | 24.75 | 944 | 23.2 |
| 1999-00 | 24.28 | 20.72 | 853 | 25.2 |
| 2000-01 | 22.77 | 18.44 | 810 | 23.0 |
| 2001-02 | 22.64 | 20.66 | 913 | 24.3 |
| 2002-03 | 21.49 | 14.84 | 691 | 22.7 |
| 2003-04 | 23.66 | 25.19 | 1064 | 24.5 |
| 2004-05 | 27.52 | 24.35 | 885 | 26.6 |
| 2005-06 | 27.86 | 27.98 | 1004 | 28.0 |
| 2006-07 | 26.51 | 24.29 | 916 | NA |
| 2007-08* | 26.54 | 28.83 | 1086 | NA |

* Advance Estimates as released on 09.07.2008

*Note*: 1. The yield rates given above have been worked out on the basis of production & area figures taken in '000 units.

2. Data for 1950-51 to 1969-70 relate to total of five major oilseeds *viz.* groundnut, castorseed, sesamum, rapeseed & mustard and linseed.

*Nine Oilseeds - All India Production*

*Million Tonnes*

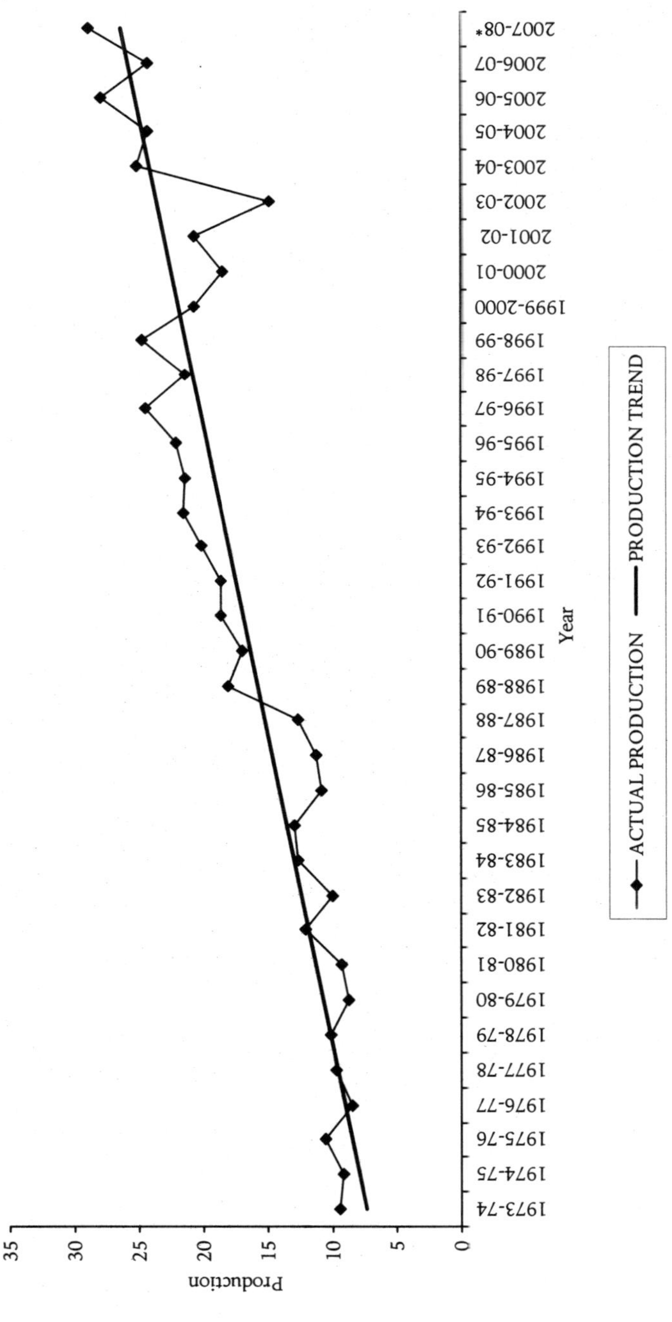


* Based on Advance estimate as on 09-07-08.

# Table 4.16(b)

## *Area, Production and Yield of Nine Oilseeds during 2005-06 and 2006-07 in Respect of Major Oilseeds Producing States Alongwith Coverage under Irrigation*

*Area- Million Hectares*
*Production- Million Tonnes*
*Yield- Kg./Hectare*

| State | 2006-07 | | | | | | 2005-06 | | | | | | |
|---|---|---|---|---|---|---|---|---|---|---|---|---|---|
| | Area | % to All-India | 2006-07 | % to All-India | Cumulative % | Yield | Area | % to All-India | Production | % to All-India | Cumulative % | Yield | % Coverage under Irrigation during 2005-06* |
| 1 | 2 | 3 | 4 | 5 | 6 | 7 | 8 | 9 | 10 | 11 | 12 | 13 | 14 |
| Madhya Pradesh | 6.09 | 22.97 | 5.81 | 5.81 | 11.81 | 955 | 5.67 | 20.35 | 5.72 | 20.44 | 20.44 | 1009 | 7.7 |
| Rajasthan | 4.51 | 17.01 | 5.17 | 5.17 | 16.98 | 1146 | 5.26 | 18.88 | 5.96 | 21.30 | 41.74 | 1134 | 63.6 |
| Maharashtra | 3.86 | 14.56 | 3.72 | 3.72 | 20.70 | 963 | 3.65 | 13.10 | 3.37 | 12.04 | 53.79 | 925 | 7.7 |
| Gujarat | 2.83 | 10.68 | 2.57 | 2.57 | 2.57 | 908 | 3.03 | 10.88 | 4.68 | 16.73 | 70.51 | 1544 | 12.8 |
| Andhra Pradesh | 2.24 | 8.45 | 1.36 | 1.36 | 3.93 | 609 | 2.92 | 10.48 | 2.04 | 7.29 | 77.81 | 698 | 16.5 |
| Karnataka | 2.35 | 8.86 | 1.13 | 1.13 | 5.06 | 478 | 2.86 | 10.27 | 1.72 | 6.15 | 83.95 | 600 | 21.1 |
| Tamil Nadu | 0.59 | 2.23 | 1.08 | 1.08 | 6.14 | 1829 | 0.71 | 2.55 | 1.15 | 4.11 | 88.06 | 1624 | 49.4 |
| Uttar Pradesh | 1.23 | 4.64 | 1.03 | 1.03 | 7.17 | 837 | 1.07 | 3.84 | 1.07 | 3.82 | 91.89 | 993 | 57.0 |
| Haryana | 0.62 | 2.34 | 0.83 | 0.83 | 8.00 | 1344 | 0.73 | 2.62 | 0.83 | 2.97 | 94.85 | 1124 | 74.3 |
| West Bengal | 0.70 | 2.64 | 0.65 | 0.65 | 8.65 | 918 | 0.64 | 2.30 | 0.61 | 2.18 | 97.03 | 952 | 58.8 |
| Orissa | 0.32 | 1.21 | 0.18 | 0.18 | 8.83 | 550 | 0.33 | 1.18 | 0.19 | 0.68 | 97.71 | 565 | 14.3 |
| Bihar | 0.14 | 0.53 | 0.15 | 0.15 | 8.98 | 1031 | 0.14 | 0.50 | 0.14 | 0.50 | 98.21 | 982 | 37.1 |
| Assam | 0.27 | 1.02 | 0.13 | 0.13 | 9.11 | 495 | 0.24 | 0.86 | 0.11 | 0.39 | 98.61 | 465 | 2.5 |
| Punjab | 0.07 | 0.26 | 0.08 | 0.08 | 9.19 | 1111 | 0.08 | 0.29 | 0.09 | 0.32 | 98.93 | 1097 | 87.4 |
| Others | 0.69 | 2.60 | 0.40 | 0.40 | 9.59 | @ | 0.53 | 1.90 | 0.30 | 1.07 | 100.00 | @ | - |
| All India | 26.51 | 100.00 | 24.29 | 24.29 | | 916 | 27.86 | 100.00 | 27.98 | 100.00 | | 1004 | 28.2 |

@ Since area/production is low in individual states, yield rate is not worked out.

* Provisional

*Note*: States have been arranged in descending order of percentage share of production during 2006-07.

# Table 4.17 (a)

## *All-India Area, Production and Yield of Groundnut from 1950-51 to 2007-08 Alongwith Percentage Coverage under Irrigation*

*Area- Million Hectares*
*Production- Million Tonnes*
*Yield- Kg./Hectare*

| *Year* | *Area* | *Production* | *Yield* | *% Coverage under Irrigation* |
|---|---|---|---|---|
| *1* | *2* | *3* | *4* | *5* |
| 1950-51 | 4.49 | 3.48 | 775 | NA |
| 1951-52 | 4.92 | 3.19 | 649 | NA |
| 1952-53 | 4.80 | 2.93 | 611 | 1.2 |
| 1953-54 | 4.25 | 3.45 | 811 | 1.5 |
| 1954-55 | 5.54 | 4.25 | 766 | 1.7 |
| 1955-56 | 5.13 | 3.86 | 752 | 1.7 |
| 1956-57 | 5.53 | 4.37 | 783 | 1.8 |
| 1957-58 | 6.42 | 4.71 | 734 | 2.9 |
| 1958-59 | 6.25 | 5.18 | 828 | 2.5 |
| 1959-60 | 6.44 | 4.56 | 708 | 2.5 |
| 1960-61 | 6.46 | 4.81 | 745 | 3.0 |
| 1961-62 | 6.89 | 4.99 | 725 | 3.4 |
| 1962-63 | 7.28 | 5.06 | 695 | 2.6 |
| 1963-64 | 6.89 | 5.30 | 769 | 3.0 |
| 1964-65 | 7.38 | 6.00 | 814 | 2.9 |
| 1965-66 | 7.70 | 4.26 | 554 | 3.4 |
| 1966-67 | 7.30 | 4.41 | 604 | 4.8 |
| 1967-68 | 7.55 | 5.73 | 759 | 5.4 |
| 1968-69 | 7.09 | 4.63 | 653 | 5.1 |
| 1969-70 | 7.13 | 5.13 | 720 | 5.8 |
| 1970-71 | 7.33 | 6.11 | 834 | 7.5 |
| 1971-72 | 7.51 | 6.18 | 823 | 7.3 |
| 1972-73 | 6.99 | 4.09 | 585 | 6.6 |
| 1973-74 | 7.02 | 5.93 | 845 | 9.1 |
| 1974-75 | 7.06 | 5.11 | 724 | 8.2 |
| 1975-76 | 7.22 | 6.76 | 935 | 6.9 |
| 1976-77 | 7.04 | 5.26 | 747 | 5.9 |
| 1977-78 | 7.03 | 6.09 | 866 | 8.1 |
| 1978-79 | 7.43 | 6.21 | 835 | 9.6 |
| 1979-80 | 7.17 | 5.77 | 805 | 12.1 |

*contd. ...*

*...contd. ...*

| *1* | 2 | 3 | 4 | 5 |
|---|---|---|---|---|
| 1980-81 | 6.80 | 5.01 | 736 | 13.3 |
| 1981-82 | 7.43 | 7.22 | 972 | 14.2 |
| 1982-83 | 7.22 | 5.28 | 732 | 14.8 |
| 1983-84 | 7.54 | 7.09 | 940 | 16.0 |
| 1984-85 | 7.17 | 6.44 | 898 | 16.1 |
| 1985-86 | 7.12 | 5.12 | 719 | 14.8 |
| 1986-87 | 6.98 | 5.88 | 841 | 15.1 |
| 1987-88 | 6.84 | 5.85 | 855 | 19.0 |
| 1988-89 | 8.53 | 9.66 | 1132 | 18.6 |
| 1989-90 | 8.71 | 8.10 | 930 | 17.0 |
| 1990-91 | 8.31 | 7.51 | 904 | 18.6 |
| 1991-92 | 8.67 | 7.09 | 818 | 19.1 |
| 1992-93 | 8.17 | 8.56 | 1049 | 19.7 |
| 1993-94 | 8.32 | 7.83 | 941 | 19.4 |
| 1994-95 | 7.85 | 8.06 | 1027 | 19.9 |
| 1995-96 | 7.52 | 7.58 | 1007 | 18.1 |
| 1996-97 | 7.60 | 8.64 | 1138 | 17.9 |
| 1997-98 | 7.09 | 7.37 | 1040 | 19.5 |
| 1998-99 | 7.40 | 8.98 | 1214 | 19.4 |
| 1999-00 | 6.87 | 5.26 | 766 | 19.0 |
| 2000-01 | 6.56 | 6.41 | 977 | 17.6 |
| 2001-02 | 6.24 | 7.03 | 1127 | 17.3 |
| 2002-03 | 5.94 | 4.12 | 694 | 16.5 |
| 2003-04 | 5.99 | 8.13 | 1357 | 16.6 |
| 2004-05 | 6.64 | 6.77 | 1020 | 17.8 |
| 2005-06 | 6.74 | 7.99 | 1187 | 19.6 |
| 2006-07 | 5.61 | 4.86 | 866 | NA |
| 2007-08* | 6.41 | 9.36 | 1460 | NA |

* Advance Estimates as released on 09.07.2008

*Note*: The yield rates given above have been worked out on the basis of production & area figures taken in '000 units.

## Table 4.17 (b)

*Area, Production and Yield of Groundnut during 2005-06 and 2006-07 in Respect of Major Groundnut Producing States Alongwith Coverage under Irrigation*

*Area- Million Hectares*
*Production- Million Tonnes*
*Yield- Kg./Hectare*

| State | 2006-07 | | | | | | 2005-06 | | | | | | % Coverage under Irrigation during 2005-06* |
|---|---|---|---|---|---|---|---|---|---|---|---|---|---|
| | Area | % to All-India | Production | % to All-India | Cumulative % | Yield | Area | % to All-India | Production | % to All-India | Cumulative % | Yield | |
| 1 | 2 | 3 | 4 | 5 | 6 | 7 | 8 | 9 | 10 | 11 | 12 | 13 | 14 |
| Gujarat | 1.77 | 31.49 | 1.44 | 29.63 | 29.63 | 809 | 1.95 | 28.93 | 3.39 | 42.43 | 42.43 | 1734 | 7.5 |
| Tamil Nadu | 0.51 | 9.07 | 1.01 | 20.78 | 50.41 | 1981 | 0.62 | 9.20 | 1.10 | 13.77 | 56.20 | 1775 | 15.0 |
| Andhra Pradesh | 1.33 | 23.67 | 0.74 | 15.23 | 65.64 | 557 | 1.88 | 27.89 | 1.37 | 17.15 | 73.34 | 728 | 30.7 |
| Rajasthan | 0.30 | 5.34 | 0.40 | 8.23 | 73.87 | 1310 | 0.32 | 4.75 | 0.49 | 6.13 | 79.47 | 645 | 20.2 |
| Maharashtra | 0.45 | 8.01 | 0.40 | 8.23 | 82.10 | 889 | 0.43 | 6.38 | 0.41 | 5.13 | 84.61 | 958 | 70.6 |
| Karnataka | 0.76 | 13.52 | 0.38 | 7.82 | 89.92 | 497 | 1.04 | 15.43 | 0.67 | 8.39 | 92.99 | 645 | 36.2 |
| Madhya Pradesh | 0.20 | 3.56 | 0.19 | 3.91 | 93.83 | 948 | 0.21 | 3.12 | 0.23 | 2.88 | 95.87 | 1126 | 9.0 |
| Orissa | 0.08 | 1.42 | 0.09 | 1.85 | 95.68 | 1113 | 0.09 | 1.34 | 0.11 | 1.38 | 97.25 | 1171 | 37.5 |
| Uttar Pradesh | 0.10 | 1.78 | 0.07 | 1.44 | 97.12 | 730 | 0.11 | 1.63 | 0.09 | 1.13 | 98.37 | 851 | 1.0 |
| Others | 0.12 | 2.14 | 0.14 | 2.88 | 100.00 | @ | 0.09 | 1.34 | 0.13 | 1.63 | 100.00 | @ | - |
| All India | 5.62 | 100.00 | 4.86 | 100.00 | | 866 | 6.74 | 100.00 | 7.99 | 100.00 | | 1187 | 19.6 |

@ Since area/production is low in individual states, yield rate is not worked out.

* Provisional

*Note*: States have been arranged in descending order of percentage share of production during 2006-07.

# Table 4.18 (a)

*All-India Area, Production and Yield of Rapeseed & Mustard from 1950-51 to 2007-08 Alongwith Percentage Coverage under Irrigation*

*Area- Million Hectares*
*Production- Million Tonnes*
*Yield- Kg./Hectare*

| Year | Area | Production | Yield | % Coverage under Irrigation |
|---|---|---|---|---|
| 1 | 2 | 3 | 4 | 5 |
| 1950-51 | 2.07 | 0.76 | 368 | NA |
| 1951-52 | 2.40 | 0.94 | 393 | NA |
| 1952-53 | 2.11 | 0.86 | 408 | NA |
| 1953-54 | 2.24 | 0.87 | 389 | NA |
| 1954-55 | 2.44 | 1.04 | 425 | NA |
| 1955-56 | 2.56 | 0.86 | 336 | 10.4 |
| 1956-57 | 2.54 | 1.04 | 411 | 13.5 |
| 1957-58 | 2.41 | 0.93 | 387 | 13.0 |
| 1958-59 | 2.45 | 1.04 | 426 | 13.0 |
| 1959-60 | 2.91 | 1.06 | 365 | 12.3 |
| 1960-61 | 2.88 | 1.35 | 467 | 12.1 |
| 1961-62 | 3.17 | 1.35 | 425 | 13.2 |
| 1962-63 | 3.13 | 1.30 | 417 | 13.3 |
| 1963-64 | 3.05 | 0.92 | 300 | 16.1 |
| 1964-65 | 2.91 | 1.47 | 507 | 15.2 |
| 1965-66 | 2.91 | 1.30 | 446 | 15.8 |
| 1966-67 | 3.01 | 1.23 | 408 | 20.3 |
| 1967-68 | 3.24 | 1.57 | 483 | 14.8 |
| 1968-69 | 2.87 | 1.35 | 469 | 18.4 |
| 1969-70 | 3.17 | 1.56 | 493 | 23.5 |
| 1970-71 | 3.32 | 1.98 | 594 | 25.2 |
| 1971-72 | 3.61 | 1.43 | 396 | 28.6 |
| 1972-73 | 3.32 | 1.81 | 545 | 26.7 |
| 1973-74 | 3.46 | 1.70 | 493 | 30.4 |
| 1974-75 | 3.68 | 2.25 | 612 | 35.4 |
| 1975-76 | 3.34 | 1.94 | 580 | 31.2 |
| 1976-77 | 3.13 | 1.55 | 496 | 34.4 |
| 1977-78 | 3.58 | 1.65 | 460 | 39.7 |

contd. ...

...*contd.* ...

| *1* | 2 | 3 | 4 | 5 |
|---|---|---|---|---|
| 1978-79 | 3.54 | 1.86 | 525 | 39.7 |
| 1979-80 | 3.47 | 1.43 | 411 | 41.9 |
| 1980-81 | 4.11 | 2.30 | 560 | 43.7 |
| 1981-82 | 4.40 | 2.38 | 541 | 44.9 |
| 1982-83 | 3.83 | 2.21 | 577 | 44.0 |
| 1983-84 | 3.87 | 2.61 | 673 | 46.6 |
| 1984-85 | 3.99 | 3.07 | 771 | 53.3 |
| 1985-86 | 3.98 | 2.68 | 674 | 51.3 |
| 1986-87 | 3.72 | 2.60 | 700 | 51.8 |
| 1987-88 | 4.62 | 3.45 | 748 | 54.7 |
| 1988-89 | 4.83 | 4.38 | 906 | 60.0 |
| 1989-90 | 4.97 | 4.13 | 831 | 61.6 |
| 1990-91 | 5.78 | 5.23 | 904 | 59.8 |
| 1991-92 | 6.55 | 5.86 | 895 | 63.9 |
| 1992-93 | 6.19 | 4.80 | 776 | 60.0 |
| 1993-94 | 6.29 | 5.33 | 847 | 59.4 |
| 1994-95 | 6.01 | 5.76 | 958 | 62.4 |
| 1995-96 | 6.55 | 6.00 | 916 | 65.8 |
| 1996-97 | 6.55 | 6.66 | 1017 | 69.1 |
| 1997-98 | 7.04 | 4.70 | 668 | 60.0 |
| 1998-99 | 6.51 | 5.66 | 869 | 58.3 |
| 1999-00 | 6.03 | 5.79 | 960 | 63.2 |
| 2000-01 | 4.48 | 4.19 | 935 | 66.1 |
| 2001-02 | 5.07 | 5.08 | 1002 | 68.3 |
| 2002-03 | 4.54 | 3.88 | 854 | 69.2 |
| 2003-04 | 5.43 | 6.29 | 1159 | 67.0 |
| 2004-05 | 7.32 | 7.59 | 1038 | 69.3 |
| 2005-06 | 7.28 | 8.13 | 1117 | 72.1 |
| 2006-07 | 6.79 | 7.44 | 1095 | NA |
| 2007-08* | 5.75 | 5.80 | 1009 | NA |

* Advance Estimates as released on 09.07.2008

*Note*: The yield rates given above have been worked out on the basis of production & area figures taken in '000 units.

# Table 4.18 (b)

*Area, Production and Yield of Rapeseed & Mustard during 2005-06 and 2006-07 in Respect of Major Rapeseed & Mustard Producing States Alongwith Coverage under Irrigation*

*Area- Million Hectares*
*Production- Million Tonnes*
*Yield- Kg./Hectare*

| State | 2006-07 | | | | | | 2005-06 | | | | | | |
|---|---|---|---|---|---|---|---|---|---|---|---|---|---|
| | Area | % to All-India | Production | % to All-India | Cumulative % | Yield | Area | % to All-India | Production | % to All-India | Cumulative % | Yield | % Coverage under Irrigation during 2005-06* |
| 1 | 2 | 3 | 4 | 5 | 6 | 7 | 8 | 9 | 10 | 11 | 12 | 13 | 14 |
| Rajasthan | 3.21 | 47.28 | 3.81 | 51.21 | 51.21 | 1185 | 3.67 | 50.41 | 4.42 | 54.37 | 54.37 | 1205 | 82.5 |
| Uttar Pradesh | 0.83 | 12.22 | 0.87 | 11.69 | 62.90 | 1057 | 0.79 | 10.85 | 0.91 | 11.19 | 65.56 | 1149 | 75.6 |
| Haryana | 0.60 | 8.84 | 0.80 | 10.75 | 73.66 | 1343 | 0.71 | 9.75 | 0.79 | 9.72 | 75.28 | 1117 | 74.2 |
| Madhya Pradesh | 0.69 | 10.16 | 0.69 | 9.27 | 82.93 | 999 | 0.81 | 11.13 | 0.85 | 10.46 | 85.73 | 1047 | 50.2 |
| Gujarat | 0.36 | 5.30 | 0.50 | 6.72 | 89.65 | 1396 | 0.34 | 4.67 | 0.46 | 5.66 | 91.39 | 1349 | 98.0 |
| West Bengal | 0.42 | 6.19 | 0.34 | 4.57 | 94.22 | 803 | 0.42 | 5.77 | 0.38 | 4.67 | 96.06 | 909 | 72.6 |
| Assam | 0.24 | 3.53 | 0.12 | 1.61 | 95.83 | 487 | 0.21 | 2.88 | 0.10 | 1.23 | 97.29 | 456 | 1.9 |
| Bihar | 0.09 | 1.33 | 0.09 | 1.21 | 97.04 | 1029 | 0.08 | 1.10 | 0.08 | 0.98 | 98.28 | 926 | 38.3 |
| Punjab | 0.04 | 0.59 | 0.05 | 0.67 | 97.72 | 1122 | 0.05 | 0.69 | 0.05 | 0.62 | 98.89 | 1102 | 92.4 |
| Others | 0.31 | 4.57 | 0.17 | 2.28 | 100.00 | @ | 0.20 | 2.75 | 0.09 | 1.11 | 100.00 | @ | - |
| All India | 6.79 | 100.00 | 7.44 | 100.00 | | 1095 | 7.28 | 100.00 | 8.13 | 100.00 | | 1117 | 72.1 |

@ Since area/production is low in individual states, yield rate is not worked out.

* Provisional

*Note:* States have been arranged in descending order of percentage share of production during 2006-07.

# Table 4.19 (a)

## *All-India Area, Production and Yield of Soyabean from 1970-71 to 2007-08 Alongwith Percentage Coverage under Irrigation*

*Area- Million Hectares*
*Production- Million Tonnes*
*Yield- Kg./Hectare*

| *Year* | *Area* | *Production* | *Yield* | *% Coverage under Irrigation* |
|---|---|---|---|---|
| 1 | 2 | 3 | 4 | 5 |
| 1970-71 | 0.03 | 0.01 | 426 | NA |
| 1971-72 | 0.03 | 0.01 | 426 | NA |
| 1972-73 | 0.03 | 0.03 | 319 | NA |
| 1973-74 | 0.05 | 0.04 | 329 | NA |
| 1974-75 | 0.07 | 0.05 | 768 | NA |
| 1975-76 | 0.09 | 0.09 | 975 | NA |
| 1976-77 | 0.13 | 0.12 | 988 | NA |
| 1977-78 | 0.20 | 0.18 | 940 | NA |
| 1978-79 | 0.31 | 0.30 | 975 | NA |
| 1979-80 | 0.50 | 0.28 | 568 | NA |
| 1980-81 | 0.61 | 0.44 | 728 | NA |
| 1981-82 | 0.48 | 0.35 | 741 | NA |
| 1982-83 | 0.77 | 0.49 | 637 | NA |
| 1983-84 | 0.84 | 0.61 | 735 | NA |
| 1984-85 | 1.24 | 0.95 | 768 | NA |
| 1985-86 | 1.34 | 1.02 | 764 | NA |
| 1986-87 | 1.53 | 0.89 | 584 | NA |
| 1987-88 | 1.54 | 0.90 | 582 | NA |
| 1988-89 | 1.73 | 1.55 | 892 | NA |
| 1989-90 | 2.25 | 1.81 | 801 | NA |
| 1990-91 | 2.56 | 2.60 | 1015 | NA |
| 1991-92 | 3.18 | 2.49 | 782 | NA |
| 1992-93 | 3.79 | 3.39 | 894 | 2.6 |
| 1993-94 | 4.37 | 4.75 | 1086 | 2.5 |
| 1994-95 | 4.32 | 3.93 | 911 | 2.9 |
| 1995-96 | 5.04 | 5.10 | 1012 | 3.7 |
| 1996-97 | 5.45 | 5.38 | 987 | 2.7 |
| 1997-98 | 5.99 | 6.46 | 1079 | 2.6 |
| 1998-99 | 6.49 | 7.14 | 1100 | 2.5 |
| 1999-2000 | 6.22 | 7.08 | 1138 | 1.6 |
| 2000-01 | 6.42 | 5.28 | 822 | 1.4 |
| 2001-02 | 6.34 | 5.96 | 940 | 1.7 |
| 2002-03 | 6.11 | 4.65 | 762 | 0.8 |
| 2003-04 | 6.56 | 7.82 | 1193 | 1.4 |
| 2004-05 | 7.57 | 6.87 | 908 | 1.8 |
| 2005-06 | 7.71 | 8.27 | 1073 | 1.7 |
| 2006-07 | 8.32 | 8.85 | 1063 | NA |
| 2007-08* | 8.88 | 9.99 | 1124 | NA |

* Advance Estimates as released on 09.07.2008

*Note*: The yield rates given above have been worked out on the basis of production & area figures taken in '000 units.

# Table 4.19 (b)

*Area, Production and Yield of Soyabean during 2005-06 and 2006-07 in Respect of Major Soyabean Producing States Alongwith Coverage under Irrigation*

*Area- Million Hectares*
*Production- Million Tonnes*
*Yield- Kg./Hectare*

| State | 2006-07 | | | | | | 2005-06 | | | | | | % Coverage under Irrigation during 2005-06* |
|---|---|---|---|---|---|---|---|---|---|---|---|---|---|
| | Area | % to All-India | Production | % to All-India | Cumulative % | Yield | Area | % to All-India | Production | % to All-India | Cumulative % | Yield | |
| 1 | 2 | 3 | 4 | 5 | 6 | 7 | 8 | 9 | 10 | 11 | 12 | 13 | 14 |
| Madhya Pradesh | 4.76 | 57.14 | 4.78 | 54.01 | 54.01 | 1006 | 4.26 | 55.25 | 4.50 | 54.41 | 54.41 | 1058 | 0.5 |
| Maharashtra | 2.52 | 30.25 | 2.89 | 32.66 | 86.67 | 1147 | 2.35 | 30.48 | 2.53 | 30.59 | 85.01 | 1077 | 0.5 |
| Rajasthan | 0.64 | 7.68 | 0.77 | 8.70 | 95.37 | 1203 | 0.74 | 9.60 | 0.86 | 10.40 | 95.41 | 1150 | 11.4 |
| Andhra Pradesh | 0.10 | 1.20 | 0.16 | 1.81 | 97.18 | 1515 | 0.10 | 1.30 | 0.19 | 2.30 | 97.70 | 1949 | 2.6 |
| Karnataka | 0.13 | 1.56 | 0.09 | 1.02 | 98.19 | 718 | 0.13 | 1.69 | 0.07 | 0.85 | 98.55 | 534 | 10.9 |
| Others | 0.18 | 2.16 | 0.16 | 1.81 | 100.00 | @ | 0.13 | 1.69 | 0.12 | 1.45 | 100.00 | @ | - |
| All India | 8.33 | 100.00 | 8.85 | 100.00 | | 1063 | 7.71 | 100.00 | 8.27 | 100.00 | | 1073 | 1.7 |

@ Since area/production is low in individual states, yield rate is not worked out.

* Provisional

*Note*: States have been arranged in descending order of percentage share of production during 2006-07.

# Table 4.20 (a)

## All-India area, Production and Yield of Sunflower from 1970-71 to 2007-08 Alongwith Percentage Coverage under Irrigation

*Area- Million Hectares*
*Production- Million Tonnes*
*Yield- Kg./Hectare*

| Year | Area | Production | Yield | % Coverage under Irrigation |
|---|---|---|---|---|
| 1 | 2 | 3 | 4 | 5 |
| 1970-71 | 0.12 | 0.08 | 653 | NA |
| 1971-72 | 0.12 | 0.08 | 653 | NA |
| 1972-73 | 0.12 | 0.08 | 653 | NA |
| 1973-74 | 0.24 | 0.17 | 712 | NA |
| 1974-75 | 0.34 | 0.23 | 671 | NA |
| 1975-76 | 0.32 | 0.22 | 686 | NA |
| 1976-77 | 0.26 | 0.14 | 541 | NA |
| 1977-78 | 0.27 | 0.14 | 523 | NA |
| 1978-79 | 0.18 | 0.10 | 543 | NA |
| 1979-80 | 0.06 | 0.03 | 519 | NA |
| 1980-81 | 0.12 | 0.07 | 555 | NA |
| 1981-82 | 0.28 | 0.16 | 564 | NA |
| 1982-83 | 0.46 | 0.23 | 497 | NA |
| 1983-84 | 0.70 | 0.30 | 431 | NA |
| 1984-85 | 0.84 | 0.44 | 527 | NA |
| 1985-86 | 0.75 | 0.28 | 374 | NA |
| 1986-87 | 1.02 | 0.42 | 411 | NA |
| 1987-88 | 1.65 | 0.64 | 385 | NA |
| 1988-89 | 1.10 | 0.37 | 335 | NA |
| 1989-90 | 1.19 | 0.63 | 529 | NA |
| 1990-91 | 1.63 | 0.87 | 535 | NA |
| 1991-92 | 2.11 | 1.19 | 565 | NA |
| 1992-93 | 2.09 | 1.18 | 567 | 19.6 |
| 1993-94 | 2.67 | 1.35 | 505 | 17.7 |
| 1994-95 | 2.00 | 1.22 | 610 | 18.2 |
| 1995-96 | 2.12 | 1.26 | 593 | 21.0 |
| 1996-97 | 1.93 | 1.25 | 646 | 21.1 |
| 1997-98 | 1.74 | 0.89 | 548 | 21.6 |
| 1998-99 | 1.82 | 0.94 | 517 | 20.1 |
| 1999-00 | 1.29 | 0.69 | 538 | 23.3 |
| 2000-01 | 1.07 | 0.65 | 602 | 27.6 |
| 2001-02 | 1.18 | 0.68 | 577 | 24.1 |
| 2002-03 | 1.64 | 0.87 | 531 | 22.9 |
| 2003-04 | 2.01 | 0.93 | 464 | 15.4 |
| 2004-05 | 2.17 | 1.19 | 549 | 26.0 |
| 2005-06 | 2.34 | 1.44 | 615 | 24.9 |
| 2006-07 | 2.16 | 1.23 | 567 | NA |
| 2007-08* | 1.88 | 1.44 | 694 | NA |

* Advance Estimates as released on 09.07.2008

*Note*: The yield rates given above have been worked out on the basis of production & area figures taken in '000 units.

# Table 4.20 (b)

*Area, Production and Yield of Sunflower during 2005-06 and 2006-07 in Respect of Major Sunflower Producing States Alongwith Coverage under Irrigation*

*Area- Million Hectares*
*Production- Million Tonnes*
*Yield- Kg./Hectare*

| State | 2006-07 | | | | | | 2005-06 | | | | | | % Coverage under Irrigation during 2005-06* |
|---|---|---|---|---|---|---|---|---|---|---|---|---|---|
| | Area | % to All-India | Production | % to All-India | Cumulative % | Yield | Area | % to All-India | Production | % to All-India | Cumulative % | Yield | |
| 1 | 2 | 3 | 4 | 5 | 6 | 7 | 8 | 9 | 10 | 11 | 12 | 13 | 14 |
| Karnataka | 1.23 | 56.94 | 0.52 | 42.28 | 42.28 | 420 | 1.43 | 61.11 | 0.79 | 54.86 | 54.86 | 552 | 17.1 |
| Andhra Pradesh | 0.45 | 20.83 | 0.33 | 26.83 | 69.11 | 738 | 0.44 | 18.80 | 0.30 | 20.83 | 75.69 | 671 | 31.7 |
| Maharashtra | 0.36 | 16.67 | 0.20 | 16.26 | 85.37 | 548 | 0.36 | 15.38 | 0.21 | 14.58 | 90.28 | 580 | 28.8 |
| Tamil Nadu | 0.03 | 1.39 | 0.05 | 4.07 | 89.43 | 1828 | 0.02 | 0.85 | 0.02 | 1.39 | 91.67 | 1240 | 86.5 |
| Bihar | 0.02 | 0.93 | 0.03 | 2.44 | 91.87 | 1332 | 0.02 | 0.85 | 0.03 | 2.08 | 93.75 | 1345 | 90.9 |
| Haryana | 0.02 | 0.93 | 0.03 | 2.44 | 94.31 | 1667 | 0.02 | 0.85 | 0.03 | 2.08 | 95.83 | 1667 | 68.0 |
| Uttar Pradesh | 0.01 | 0.51 | 0.02 | 1.63 | 95.93 | 1500 | 0.01 | 0.47 | 0.02 | 1.39 | 97.22 | 1278 | 98.0 |
| Others | 0.04 | 1.81 | 0.05 | 4.07 | 100.00 | @ | 0.04 | 1.67 | 0.04 | 2.78 | 100.00 | @ | - |
| All India | 2.16 | 100.00 | 1.23 | 100.00 | | 567 | 2.34 | 100.00 | 1.44 | 100.00 | | 615 | 24.9 |

@ Since area/production is low in individual states, yield rate is not worked out.

* Provisional

*Note*: States have been arranged in descending order of percentage share of production during 2006-07.

## Table 4.21 (a)

*All-India Area, Production and Yield of Cotton from 1950-51 to 2007-08 Alongwith Percentage Coverage under Irrigation*

*Area- Million Hectares*
*Production- Million Tonnes*
*Yield- Kg./Hectare*

| *Year* | *Area* | *Production* | *Yield* | *% Coverage under Irrigation* |
|---|---|---|---|---|
| *1* | 2 | 3 | 4 | 5 |
| 1950-51 | 5.88 | 3.04 | 83 | 8.2 |
| 1951-52 | 6.56 | 3.28 | 85 | 9.1 |
| 1952-53 | 6.36 | 3.34 | 89 | 8.5 |
| 1953-54 | 6.99 | 4.13 | 100 | 8.4 |
| 1954-55 | 7.55 | 4.45 | 100 | 9.8 |
| 1955-56 | 8.09 | 4.18 | 88 | 10.0 |
| 1956-57 | 8.02 | 4.92 | 104 | 11.0 |
| 1957-58 | 8.01 | 4.96 | 105 | 12.7 |
| 1958-59 | 7.96 | 4.88 | 104 | 12.5 |
| 1959-60 | 7.30 | 3.68 | 86 | 12.9 |
| 1960-61 | 7.61 | 5.60 | 125 | 12.7 |
| 1961-62 | 7.98 | 4.85 | 103 | 13.0 |
| 1962-63 | 7.73 | 5.54 | 122 | 14.1 |
| 1963-64 | 8.22 | 5.75 | 119 | 15.3 |
| 1964-65 | 8.37 | 6.01 | 122 | 15.5 |
| 1965-66 | 7.96 | 4.85 | 104 | 15.9 |
| 1966-67 | 7.84 | 5.27 | 114 | 16.1 |
| 1967-68 | 8.00 | 5.78 | 123 | 16.7 |
| 1968-69 | 7.60 | 5.45 | 122 | 16.5 |
| 1969-70 | 7.73 | 5.56 | 122 | 16.4 |
| 1970-71 | 7.61 | 4.76 | 106 | 17.3 |
| 1971-72 | 7.80 | 6.95 | 151 | 20.3 |
| 1972-73 | 7.68 | 5.74 | 127 | 21.0 |
| 1973-74 | 7.57 | 6.31 | 142 | 22.1 |
| 1974-75 | 7.56 | 7.16 | 161 | 22.9 |
| 1975-76 | 7.35 | 5.95 | 138 | 23.5 |
| 1976-77 | 6.89 | 5.84 | 144 | 24.6 |
| 1977-78 | 7.87 | 7.24 | 157 | 26.2 |
| 1978-79 | 8.12 | 7.96 | 167 | 27.2 |
| 1979-80 | 8.13 | 7.65 | 160 | 27.5 |

*contd. ...*

...contd. ...

| 1 | 2 | 3 | 4 | 5 |
|---|---|---|---|---|
| 1980-81 | 7.82 | 7.01 | 152 | 27.3 |
| 1981-82 | 8.06 | 7.88 | 166 | 27.7 |
| 1982-83 | 7.87 | 7.53 | 163 | 29.0 |
| 1983-84 | 7.72 | 6.39 | 141 | 29.9 |
| 1984-85 | 7.38 | 8.51 | 196 | 28.5 |
| 1985-86 | 7.53 | 8.73 | 197 | 30.2 |
| 1986-87 | 6.95 | 6.91 | 169 | 31.1 |
| 1987-88 | 6.46 | 6.38 | 168 | 32.0 |
| 1988-89 | 7.34 | 8.74 | 202 | 33.0 |
| 1989-90 | 7.69 | 11.42 | 252 | 34.2 |
| 1990-91 | 7.44 | 9.84 | 225 | 32.9 |
| 1991-92 | 7.66 | 9.71 | 216 | 33.3 |
| 1992-93 | 7.54 | 11.40 | 257 | 34.6 |
| 1993-94 | 7.32 | 10.74 | 249 | 34.7 |
| 1994-95 | 7.87 | 11.89 | 257 | 34.2 |
| 1995-96 | 9.04 | 12.86 | 242 | 35.0 |
| 1996-97 | 9.12 | 14.23 | 265 | 35.8 |
| 1997-98 | 8.87 | 10.85 | 208 | 36.8 |
| 1998-99 | 9.34 | 12.29 | 224 | 34.9 |
| 1999-00 | 8.71 | 11.53 | 225 | 35.2 |
| 2000-01 | 8.53 | 9.52 | 190 | 34.3 |
| 2001-02 | 9.13 | 10.00 | 186 | 34.0 |
| 2002-03 | 7.67 | 8.62 | 191 | 33.1 |
| 2003-04 | 7.60 | 13.73 | 307 | 27.1 |
| 2004-05 | 8.79 | 16.43 | 318 | 36.9 |
| 2005-06 | 8.68 | 18.50 | 362 | 36.1 |
| 2006-07 | 9.14 | 22.63 | 421 | NA |
| 2007-08* | 9.43 | 25.81 | 466 | NA |

* Advance Estimates as released on 09.07.2008

*Note*: The yield rates given above have been worked out on the basis of production & area figures taken in '000 units.

*Cotton - All India Production*

*Million Tonnes*

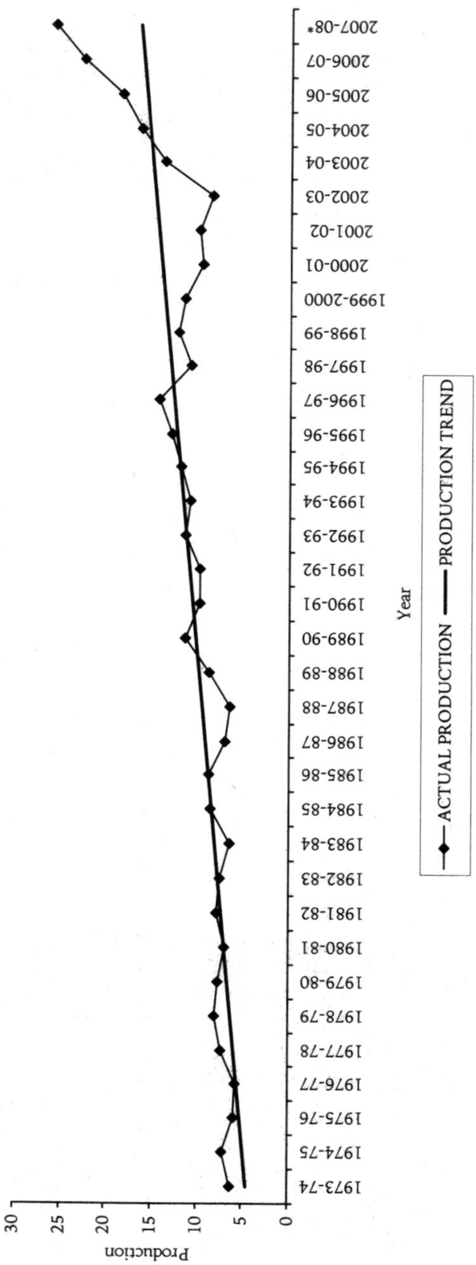


* Based on Advance estimate as on 09-07-08.

# Table 4.21 (b)

*Area, Production and Yield of Cotton during 2005-06 and 2006-07 in Respect of Major Cotton Producing States Alongwith Coverage under Irrigation*

*Area- Million Hectares*
*Production- Million Tonnes*
*Yield- Kg./Hectare*

| State | 2006-07 | | | | | | 2005-06 | | | | | | % Coverage under Irrigation during 2005-06* |
|---|---|---|---|---|---|---|---|---|---|---|---|---|---|
| | Area | % to All-India | Production | % to All-India | Cumulative % | Yield | Area | % to All-India | Production | % to All-India | Cumulative % | Yield | |
| 1 | 2 | 3 | 4 | 5 | 6 | 7 | 8 | 9 | 10 | 11 | 12 | 13 | 14 |
| Gujarat | 2.39 | 26.15 | 8.79 | 38.84 | 38.84 | 625 | 1.91 | 22.00 | 6.77 | 36.59 | 36.59 | 604 | 44.7 |
| Maharashtra | 3.11 | 34.03 | 4.62 | 20.42 | 59.26 | 253 | 2.88 | 33.18 | 3.16 | 17.08 | 53.68 | 187 | 4.8 |
| Punjab | 0.61 | 6.67 | 2.68 | 11.84 | 71.10 | 750 | 0.56 | 6.45 | 2.40 | 12.97 | 66.65 | 731 | 99.8 |
| Andhra Pradesh | 0.97 | 10.61 | 2.18 | 9.63 | 80.73 | 381 | 1.03 | 11.87 | 2.11 | 11.41 | 78.05 | 347 | 20.1 |
| Haryana | 0.53 | 5.80 | 1.81 | 8.00 | 88.73 | 582 | 0.58 | 6.68 | 1.50 | 8.11 | 86.16 | 437 | 94.8 |
| Madhya Pradesh | 0.64 | 7.00 | 0.83 | 3.67 | 92.40 | 220 | 0.62 | 7.14 | 0.75 | 4.05 | 90.22 | 204 | 94.4 |
| Rajasthan | 0.35 | 3.83 | 0.75 | 3.31 | 95.71 | 363 | 0.47 | 5.41 | 0.88 | 4.76 | 94.97 | 317 | 36.4 |
| Karnataka | 0.38 | 4.16 | 0.61 | 2.70 | 98.41 | 276 | 0.41 | 4.72 | 0.55 | 2.97 | 97.95 | 228 | 14.7 |
| Tamil Nadu | 0.10 | 1.09 | 0.22 | 0.97 | 99.38 | 374 | 0.14 | 1.61 | 0.21 | 1.14 | 99.08 | 258 | 40.1 |
| Others | 0.06 | 0.66 | 0.14 | 0.62 | 100.00 | @ | 0.08 | 0.92 | 0.17 | 0.92 | 100.00 | @ | - |
| All India | 9.14 | 100.00 | 22.63 | 100.00 | | 421 | 8.68 | 100.00 | 18.50 | 100.00 | | 362 | 36.1 |

@ Since area/production is low in individual states, yield rate is not worked out.

* Provisional

*Note:* States have been arranged in descending order of percentage share of production during 2006-07.

# Table 4.21 (c)

*Staple-wise Estimates of Area and Production of Cotton from 1997-98 to 1999-2000*

*Area: Lakh Hectares*
*Production: Lakh Bales of 170 kgs. each*

| *Staple Length* | *Area* | | | *Production* | | |
|---|---|---|---|---|---|---|
| | *1997-98* | *1998-99* | *1999-2000* | *1997-98* | *1998-99* | *1999-2000* |
| *1* | *2* | *3* | *4* | *5* | *6* | *7* |
| I. Long | | | | | | |
| i. Superior Long (27 mm & above) | 22.0 | 25.8 | 24.0 | 37.5 | 47.9 | 35.8 |
| ii. Long (24.5 mm to 26 mm) | 14.2 | 15.7 | 14.9 | 10.9 | 14.5 | 15.3 |
| Sub-Total (Long) | 36.2 | 41.5 | 38.9 | 48.4 | 62.4 | 51.1 |
| II. Medium | | | | | | |
| i. Superior Medium (22 mm to 24 mm) | 36.7 | 35.5 | 32.7 | 43.9 | 43.2 | 44.6 |
| ii. Medium (20 mm to 21.5 mm) | 2.6 | 2.0 | 2.0 | 3.2 | 2.7 | 2.5 |
| Sub-Total (Medium) | 39.3 | 37.5 | 34.7 | 47.1 | 45.9 | 47.1 |
| III. Short | | | | | | |
| i. Short (19 mm & below) | 13.2 | 14.5 | 13.5 | 13.1 | 14.5 | 17.1 |
| Total | 88.7 | 93.5 | 87.1 | 108.6 | 122.8 | 115.3 |

# Table 4.22 (a)

*All-India Area, Production and Yield of Jute and Mesta from 1950-51 to 2007-08*

*Area- Million Hectares*
*Production- Million Bales of 180 kgs. each*
*Yield- Kg./Hectare*

| Year | Area | Production | Yield |
|---|---|---|---|
| 1 | 2 | 3 | 4 |
| 1950-51 | 0.57 | 3.31 | 1043 |
| 1951-52 | 0.79 | 4.72 | 1074 |
| 1952-53 | 0.93 | 5.32 | 1028 |
| 1953-54 | 0.68 | 3.77 | 992 |
| 1954-55 | 0.68 | 3.86 | 1021 |
| 1955-56 | 0.94 | 5.39 | 1038 |
| 1956-57 | 1.07 | 5.81 | 977 |
| 1957-58 | 1.02 | 5.33 | 944 |
| 1958-59 | 1.10 | 6.91 | 1130 |
| 1959-60 | 0.98 | 5.69 | 1049 |
| 1960-61 | 0.90 | 5.26 | 1049 |
| 1961-62 | 1.34 | 8.24 | 1104 |
| 1962-63 | 1.24 | 7.19 | 1041 |
| 1963-64 | 1.27 | 7.98 | 1130 |
| 1964-65 | 1.21 | 7.66 | 1136 |
| 1965-66 | 1.11 | 5.78 | 936 |
| 1966-67 | 1.12 | 6.58 | 1058 |
| 1967-68 | 1.20 | 7.59 | 1137 |
| 1968-69 | 0.81 | 3.84 | 855 |
| 1969-70 | 1.09 | 6.79 | 1120 |
| 1970-71 | 1.08 | 6.19 | 1032 |
| 1971-72 | 1.11 | 6.84 | 1107 |
| 1972-73 | 0.99 | 6.09 | 1104 |
| 1973-74 | 1.16 | 7.68 | 1188 |
| 1974-75 | 0.98 | 5.83 | 1068 |
| 1975-76 | 0.91 | 5.91 | 1164 |
| 1976-77 | 1.09 | 7.10 | 1173 |
| 1977-78 | 1.16 | 7.15 | 1108 |
| 1978-79 | 1.27 | 8.33 | 1186 |

*contd. ...*

*...contd. ...*

| *1* | *2* | *3* | *4* |
|---|---|---|---|
| 1979-80 | 1.22 | 7.96 | 1177 |
| 1980-81 | 1.30 | 8.16 | 1130 |
| 1981-82 | 1.15 | 8.37 | 1311 |
| 1982-83 | 1.02 | 7.17 | 1265 |
| 1983-84 | 1.05 | 7.72 | 1320 |
| 1984-85 | 1.13 | 7.79 | 1242 |
| 1985-86 | 1.50 | 12.65 | 1524 |
| 1986-87 | 1.07 | 8.62 | 1454 |
| 1987-88 | 0.96 | 6.78 | 1274 |
| 1988-89 | 0.92 | 7.86 | 1540 |
| 1989-90 | 0.91 | 8.29 | 1646 |
| 1990-91 | 1.02 | 9.23 | 1634 |
| 1991-92 | 1.11 | 10.29 | 1662 |
| 1992-93 | 0.93 | 8.59 | 1658 |
| 1993-94 | 0.89 | 8.43 | 1713 |
| 1994-95 | 0.93 | 9.08 | 1760 |
| 1995-96 | 0.93 | 8.81 | 1712 |
| 1996-97 | 1.10 | 11.13 | 1818 |
| 1997-98 | 1.11 | 11.02 | 1792 |
| 1998-99 | 1.03 | 9.81 | 1722 |
| 1999-00 | 1.04 | 10.56 | 1836 |
| 2000-01 | 1.02 | 10.56 | 1867 |
| 2001-02 | 1.05 | 11.68 | 2007 |
| 2002-03 | 1.04 | 11.28 | 1960 |
| 2003-04 | 1.00 | 11.17 | 2008 |
| 2004-05 | 0.92 | 10.27 | 2019 |
| 2005-06 | 0.90 | 10.84 | 2173 |
| 2006-07 | 0.94 | 11.27 | 2170 |
| 2007-08* | 0.96 | 11.18 | 2093 |

*Note*: Figures for 1950-51 & 1951-52 relate to Jute crop only.

* Advance Estimates as released on 09.07.2008

*Note*: The yield rates given above have been worked out on the basis of production & area figures taken in '000 units.

*Jute & Mesta - All India Production*

*Million Tonnes*

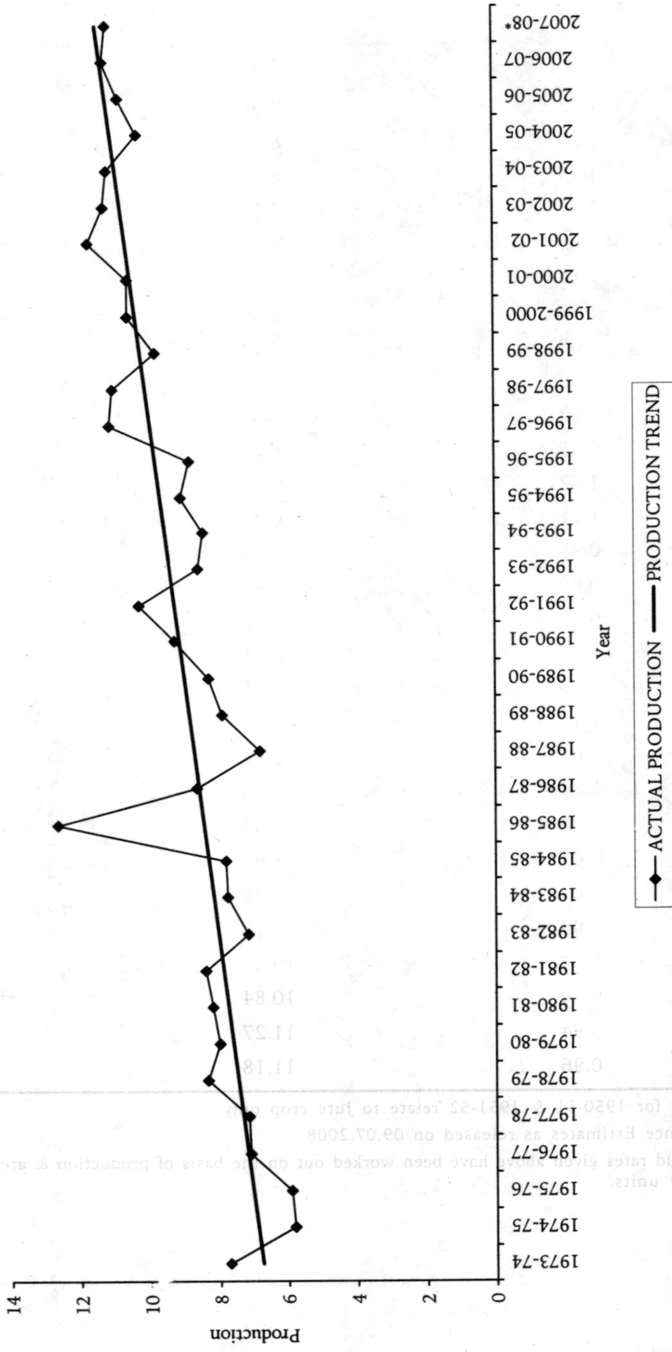


* Based on Advance estimate as on 09-07-08.

# Table 4.22 (b)

*Area, Production and Yield of Jute & Mesta during 2005-06 and 2006-07 in Respect of Major Jute & Mesta Producing States*

*Area- Million Hectares*
*Production- Million Bales of 180 kgs. Each*
*Yield- Kg./Hectare*

| State | 2006-07 | | | | | | 2005-06 | | | | | |
|---|---|---|---|---|---|---|---|---|---|---|---|---|
| | Area | % to All-India | Production | % to All-India | Cumulative % | Yield | Area | % to All-India | Production | % to All-India | Cumulative % | Yield |
| 1 | 2 | 3 | 4 | 5 | 6 | 7 | 8 | 9 | 10 | 11 | 12 | 13 |
| West Bengal | 0.61 | 64.89 | 8.51 | 75.51 | 75.51 | 2528 | 0.57 | 63.33 | 8.11 | 74.82 | 74.82 | 2566 |
| Bihar | 0.14 | 14.89 | 1.39 | 12.33 | 87.84 | 1772 | 0.15 | 16.67 | 1.39 | 12.82 | 87.64 | 1692 |
| Assam | 0.06 | 6.38 | 0.58 | 5.15 | 92.99 | 1667 | 0.06 | 6.67 | 0.60 | 5.54 | 93.17 | 1733 |
| Andhra Pradesh | 0.06 | 6.38 | 0.54 | 4.79 | 97.78 | 1579 | 0.05 | 5.56 | 0.46 | 4.24 | 97.42 | 1638 |
| Orissa | 0.03 | 3.45 | 0.13 | 1.15 | 98.94 | 893 | 0.03 | 3.60 | 0.14 | 1.29 | 98.71 | 991 |
| Meghalaya | 0.01 | 1.06 | 0.06 | 0.53 | 99.47 | 1181 | 0.01 | 1.11 | 0.06 | 0.55 | 99.26 | 1194 |
| Maharashtra | 0.02 | 2.13 | 0.03 | 0.27 | 99.73 | 297 | 0.02 | 2.22 | 0.04 | 0.37 | 99.63 | 270 |
| Others | 0.01 | 0.81 | 0.03 | 0.27 | 100.00 | @ | 0.01 | 0.84 | 0.04 | 0.37 | 100.00 | @ |
| All India | 0.94 | 100.00 | 11.27 | 100.00 | | 2170 | 0.90 | 100.00 | 10.84 | 100.00 | | 2173 |

@ - Since area/production is low in individual states, yield rate is not worked out.

*Note:* States have been arranged in descending order of percentage share of production during 2006-07.

# Table 4.23 (a)

*All-India Area, Production and Yield of Sugarcane (Cane) from 1950-51 to 2007-08 Alongwith Percentage Coverage under Irrigation*

*Area- Million Hectares*
*Production- Million Tonnes*
*Yield- Kg./Hectare*

| Year | Area | Production | Yield | % Coverage under Irrigation |
|---|---|---|---|---|
| 1 | 2 | 3 | 4 | 5 |
| 1950-51 | 1.71 | 57.05 | 33422 | 67.3 |
| 1951-52 | 1.94 | 61.63 | 31786 | 68.8 |
| 1952-53 | 1.73 | 51.00 | 29495 | 66.3 |
| 1953-54 | 1.41 | 44.41 | 31497 | 67.7 |
| 1954-55 | 1.62 | 58.74 | 36303 | 68.8 |
| 1955-56 | 1.85 | 60.54 | 32779 | 67.2 |
| 1956-57 | 2.05 | 69.05 | 33683 | 64.9 |
| 1957-58 | 2.07 | 71.16 | 34325 | 65.2 |
| 1958-59 | 1.95 | 73.36 | 37658 | 67.4 |
| 1959-60 | 2.14 | 77.82 | 36414 | 67.9 |
| 1960-61 | 2.42 | 110.00 | 45549 | 69.3 |
| 1961-62 | 2.46 | 103.97 | 42349 | 68.0 |
| 1962-63 | 2.24 | 91.91 | 40996 | 67.8 |
| 1963-64 | 2.25 | 104.23 | 46353 | 69.6 |
| 1964-65 | 2.60 | 121.91 | 46838 | 71.5 |
| 1965-66 | 2.84 | 123.99 | 43717 | 71.1 |
| 1966-67 | 2.30 | 92.83 | 40336 | 71.0 |
| 1967-68 | 2.05 | 95.50 | 40665 | 74.1 |
| 1968-69 | 2.53 | 124.68 | 49236 | 76.9 |
| 1969-70 | 2.75 | 135.02 | 49121 | 75.5 |
| 1970-71 | 2.62 | 126.37 | 48322 | 72.4 |
| 1971-72 | 2.39 | 113.57 | 47511 | 71.8 |
| 1972-73 | 2.45 | 124.87 | 50933 | 75.0 |
| 1973-74 | 2.75 | 140.81 | 51163 | 76.5 |
| 1974-75 | 2.89 | 144.29 | 49855 | 77.9 |
| 1975-76 | 2.76 | 140.60 | 50903 | 78.0 |
| 1976-77 | 2.87 | 153.01 | 53383 | 77.2 |
| 1977-78 | 3.15 | 176.97 | 56160 | 78.1 |
| 1978-79 | 3.09 | 151.66 | 49114 | 77.8 |
| 1979-80 | 2.61 | 128.83 | 49358 | 77.2 |

contd. ...

*...contd. ...*

| *1* | *2* | *3* | *4* | *5* |
|---|---|---|---|---|
| 1980-81 | 2.67 | 154.25 | 57844 | 81.2 |
| 1981-82 | 3.19 | 186.36 | 58359 | 82.3 |
| 1982-83 | 3.36 | 189.51 | 56441 | 80.5 |
| 1983-84 | 3.11 | 174.08 | 55978 | 80.3 |
| 1984-85 | 2.95 | 170.32 | 57673 | 83.6 |
| 1985-86 | 2.85 | 170.65 | 59889 | 84.5 |
| 1986-87 | 3.08 | 186.09 | 60444 | 85.4 |
| 1987-88 | 3.28 | 196.74 | 60006 | 85.6 |
| 1988-89 | 3.33 | 203.04 | 60992 | 86.2 |
| 1989-90 | 3.44 | 225.57 | 65612 | 86.9 |
| 1990-91 | 3.69 | 241.05 | 65395 | 86.9 |
| 1991-92 | 3.84 | 254.00 | 66069 | 88.0 |
| 1992-93 | 3.57 | 228.03 | 63843 | 88.3 |
| 1993-94 | 3.42 | 229.66 | 67120 | 88.8 |
| 1994-95 | 3.87 | 275.54 | 71254 | 87.9 |
| 1995-96 | 4.15 | 281.10 | 67787 | 87.4 |
| 1996-97 | 4.17 | 277.56 | 66496 | 88.1 |
| 1997-98 | 3.93 | 279.54 | 71134 | 91.3 |
| 1998-99 | 4.05 | 288.72 | 71203 | 91.7 |
| 1999-2000 | 4.22 | 299.32 | 70935 | 92.0 |
| 2000-01 | 4.32 | 295.96 | 68577 | 92.1 |
| 2001-02 | 4.41 | 297.21 | 67370 | 91.6 |
| 2002-03 | 4.52 | 287.38 | 63576 | 91.3 |
| 2003-04 | 3.93 | 233.86 | 59380 | 90.7 |
| 2004-05 | 3.66 | 237.08 | 64752 | 91.2 |
| 2005-06 | 4.20 | 281.17 | 66928 | 92.5 |
| 2006-07 | 5.15 | 355.52 | 69022 | NA |
| 2007-08* | 5.04 | 340.56 | 67531 | NA |

* Advance Estimates as released on 09.07.2008.

*Note*: The yield rates given above have been worked out on the basis of production & area figures taken in '000 units.

*Sugarcane - All India Production*

*Million Tonnes*

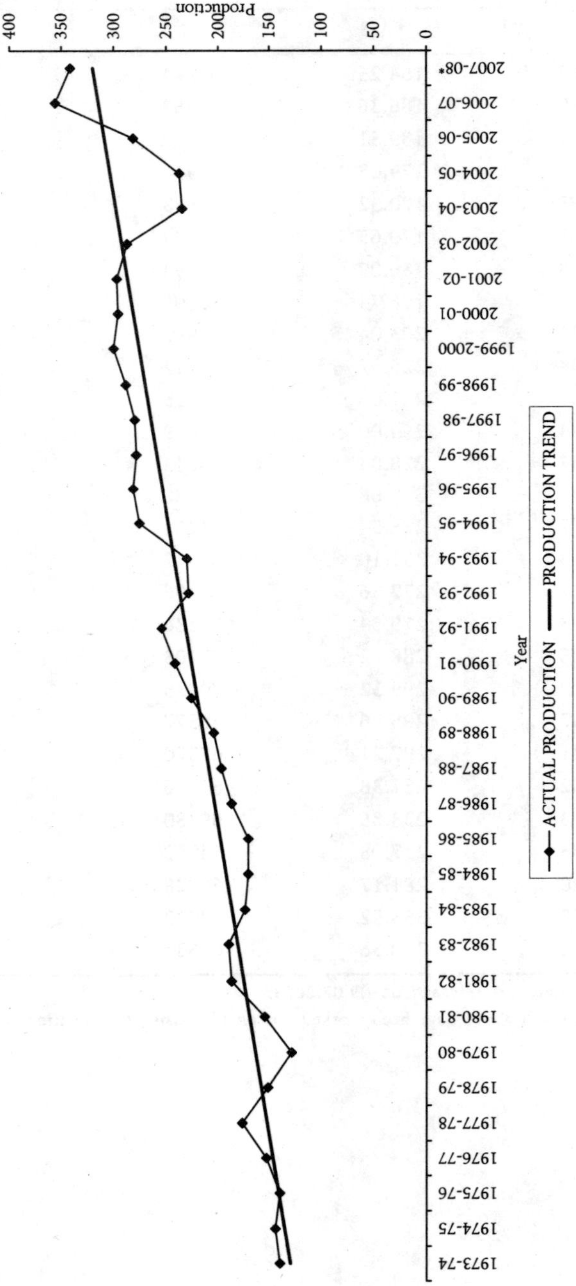


* Based on Advance Estimates As on 09.07.08.

# Table 4.23 (b)

*Area, Production and Yield of Sugarcane during 2005-06 and 2006-07 in Respect of Major Sugarcane Producing States Alongwith Coverage under Irrigation*

*Area- Million Hectares*
*Production- Million Tonnes*
*Yield- Kg./Hectare*

| State | 2006-07 | | | | | | 2005-06 | | | | | | % Coverage under Irrigation during 2005-06* |
|---|---|---|---|---|---|---|---|---|---|---|---|---|---|
| | Area | % to All-India | Production | % to All-India | Cumulative % | Yield | Area | % to All-India | Production | % to All-India | Cumulative % | Yield | |
| 1 | 2 | 3 | 4 | 5 | 6 | 7 | 8 | 9 | 10 | 11 | 12 | 13 | 14 |
| Uttar Pradesh | 2.25 | 43.69 | 133.95 | 37.68 | 37.68 | 59626 | 2.16 | 51.43 | 125.47 | 44.62 | 44.62 | 58201 | 91.0 |
| Maharashtra | 1.05 | 20.39 | 78.57 | 22.10 | 59.78 | 74898 | 0.50 | 11.90 | 38.85 | 13.82 | 58.44 | 77551 | 100.0 |
| Tamil Nadu | 0.39 | 7.57 | 41.12 | 11.57 | 71.34 | 105123 | 0.34 | 8.10 | 35.11 | 12.49 | 70.93 | 104671 | 100.0 |
| Karnataka | 0.33 | 6.41 | 28.67 | 8.06 | 79.41 | 87944 | 0.22 | 5.24 | 18.27 | 6.50 | 77.43 | 83411 | 100.0 |
| Andhra Pradesh | 0.26 | 5.05 | 21.69 | 6.10 | 85.51 | 82167 | 0.23 | 5.48 | 17.66 | 6.28 | 83.71 | 76765 | 93.5 |
| Gujarat | 0.21 | 4.08 | 15.63 | 4.40 | 89.90 | 73037 | 0.20 | 4.76 | 14.58 | 5.19 | 88.89 | 74010 | 100.0 |
| Haryana | 0.14 | 2.72 | 9.58 | 2.69 | 92.60 | 68429 | 0.13 | 3.10 | 8.18 | 2.91 | 91.80 | 64409 | 99.2 |
| Uttarakhand | 0.12 | 2.33 | 6.10 | 1.72 | 94.32 | 50413 | 0.10 | 2.38 | 6.13 | 2.18 | 93.98 | 60733 | 97.2 |
| Punjab | 0.10 | 1.94 | 6.02 | 1.69 | 96.01 | 60808 | 0.08 | 1.90 | 4.86 | 1.73 | 95.71 | 57857 | 96.5 |
| Bihar | 0.13 | 2.52 | 5.96 | 1.68 | 97.69 | 45953 | 0.10 | 2.38 | 4.34 | 1.54 | 97.25 | 42822 | 25.7 |
| Madhya Pradesh | 0.06 | 1.17 | 2.81 | 0.79 | 98.48 | 43639 | 0.06 | 1.43 | 2.43 | 0.86 | 98.12 | 43694 | 99.8 |
| West Bengal | 0.02 | 0.39 | 1.27 | 0.36 | 98.83 | 76307 | 0.02 | 0.48 | 1.25 | 0.44 | 98.56 | 83180 | 20.0 |
| Orissa | 0.02 | 0.39 | 1.27 | 0.36 | 99.19 | 63403 | 0.02 | 0.48 | 1.07 | 0.38 | 98.94 | 65828 | 100.0 |
| Assam | 0.03 | 0.58 | 1.06 | 0.30 | 99.49 | 39074 | 0.02 | 0.48 | 0.87 | 0.31 | 99.25 | 37231 | 16.7 |
| Others | 0.04 | 0.78 | 1.82 | 0.51 | 100.00 | @ | 0.02 | 0.48 | 2.10 | 0.75 | 100.00 | @ | - |
| All India | 5.15 | 100.00 | 355.52 | 100.00 | | 69022 | 4.20 | 100.00 | 281.17 | 100.00 | | 66928 | 91.2 |

@ Since area/production is low in individual states, yield rate is not worked out.

* Provisional

*Note:* States have been arranged in descending order of percentage share of production during 2006-07.

# Table 4.24

*Production of Oilseeds/Oils and Net Domestic Availability of Edible Oils during years 1995-96 to 2006-2007*

*(In Lakh MT)*

| Name of the Oil | 1995-96 | | 1996-97 | | 1997-98 | | 1998-99 | | 1999-2000 | | 2000-01 | | 2001-02 | |
|---|---|---|---|---|---|---|---|---|---|---|---|---|---|---|
| | Oilseeds | Oils | Oilseeds | Oils | Oilseeds | Oils | Oilseeds | Oils | Oilseeds | Oils | Oilseeds | Oils | Oilseeds | Oils |
| 1 | 2 | 3 | 4 | 5 | 6 | 7 | 8 | 9 | 10 | 11 | 12 | 13 | 14 | 15 |
| A. Primary Source | | | | | | | | | | | | | | |
| Groundnut | 75.80 | 17.44 | 86.40 | 19.87 | 73.70 | 16.95 | 89.82 | 20.66 | 52.58 | 12.09 | 64.10 | 14.74 | 70.27 | 16.16 |
| Rapeseed & Mustard | 60.00 | 18.60 | 66.60 | 20.65 | 47.00 | 14.57 | 56.64 | 17.56 | 57.88 | 17.94 | 41.90 | 12.99 | 50.83 | 15.76 |
| Soyabean | 50.90 | 8.14 | 53.80 | 8.61 | 64.60 | 10.34 | 71.43 | 11.43 | 70.81 | 11.33 | 52.80 | 8.45 | 59.63 | 9.54 |
| Sunflower | 12.60 | 4.16 | 12.50 | 4.13 | 8.90 | 2.94 | 9.44 | 3.12 | 6.94 | 2.29 | 6.50 | 2.15 | 6.79 | 2.24 |
| Sesamum | 5.30 | 1.64 | 6.40 | 1.98 | 5.70 | 1.77 | 5.27 | 1.63 | 4.80 | 1.49 | 5.20 | 1.61 | 6.98 | 2.16 |
| Nigerseed | 1.90 | 0.57 | 1.50 | 0.45 | 1.40 | 0.42 | 1.41 | 0.42 | 1.48 | 0.44 | 1.10 | 0.33 | 1.30 | 0.39 |
| Safflower | 3.80 | 1.14 | 4.50 | 1.35 | 1.20 | 0.36 | 2.42 | 0.73 | 2.60 | 0.78 | 2.00 | 0.60 | 2.21 | 0.66 |
| Castor | 7.80 | 3.12 | 9.00 | 3.60 | 8.30 | 3.32 | 8.40 | 3.36 | 7.65 | 3.06 | 8.80 | 3.52 | 6.53 | 2.61 |
| Linseed | 2.90 | 0.87 | 3.10 | 0.93 | 2.40 | 0.72 | 2.65 | 0.80 | 2.41 | 0.72 | 2.00 | 0.60 | 2.09 | 0.63 |
| Sub Total | 221.00 | 55.68 | 243.80 | 61.57 | 213.20 | 51.38 | 247.48 | 59.70 | 207.15 | 50.15 | 184.40 | 44.99 | 206.63 | 50.16 |
| B. Secondary Source | | | | | | | | | | | | | | |
| Coconut | | 4.20 | | 4.50 | | 4.50 | | 4.90 | | 4.50 | | 5.60 | | 5.50 |
| Cottonseed | | 4.00 | | 4.80 | | 4.20 | | 4.80 | | 5.00 | | 4.60 | | 4.70 |
| Ricebran | | 4.50 | | 4.60 | | 4.80 | | 5.00 | | 5.00 | | 4.80 | | 5.50 |
| Solvent Extracted Oils | | 4.90 | | 4.30 | | 4.20 | | 3.70 | | 2.50 | | 2.00 | | 2.80 |
| Tree & Forest Origin | | 1.40 | | 1.50 | | 1.50 | | 1.50 | | 1.00 | | 1.00 | | 0.80 |
| Sub Total | | 19.00 | | 19.70 | | 19.20 | | 19.90 | | 18.00 | | 18.00 | | 19.30 |
| Total (A+B) | | 74.68 | | 81.27 | | 70.58 | | 79.60 | | 68.15 | | 62.99 | | 69.46 |
| C. Less: Export & Industrial Use | | 10.00 | | 10.00 | | 10.00 | | 10.00 | | 8.00 | | 8.00 | | 8.00 |
| D. Net Domestic Availability of Edible Oils | | 64.68 | | 71.27 | | 60.58 | | 69.60 | | 60.15 | | 54.99 | | 61.46 |
| E. Shortfall (Import) | | 11.61 | | 14.06 | | 12.66 | | 26.22 | | 41.96 | | 41.77 | | 43.22 |
| F. Actual Consumption | | 76.29 | | 85.33 | | 73.24 | | 95.82 | | 102.11 | | 96.76 | | 104.68 |

*contd. ...*

...contd. ...

| Name of the Oil | 2002-03 | | 2003-04 | | 2004-05 | | 2005-06 | | 2006-07* | |
|---|---|---|---|---|---|---|---|---|---|---|
| | Oilseeds | Oils | Oilseeds | Oils | Oilseeds | Oils | Oilseeds | Oils | Oilseeds | Oils |
| 1 | 16 | 17 | 18 | 19 | 20 | 21 | 22 | 23 | 24 | 25 |
| A. Primary Source | | | | | | | | | | |
| Groundnut | 41.21 | 9.48 | 81.27 | 18.69 | 67.74 | 15.58 | 79.93 | 18.38 | 49.09 | 11.29 |
| Rapeseed & Mustard | 38.80 | 12.03 | 62.91 | 19.50 | 75.93 | 23.54 | 81.31 | 25.21 | 70.97 | 22.00 |
| Soyabean | 46.55 | 7.45 | 78.18 | 12.51 | 68.77 | 11.00 | 82.74 | 13.24 | 88.57 | 14.17 |
| Sunflower | 8.73 | 2.88 | 9.30 | 3.07 | 11.87 | 3.92 | 14.39 | 4.75 | 11.78 | 3.89 |
| Sesamum | 4.41 | 1.37 | 7.82 | 2.42 | 6.74 | 2.09 | 6.41 | 1.99 | 5.86 | 1.82 |
| Nigerseed | 0.85 | 0.26 | 1.09 | 0.33 | 1.12 | 0.34 | 1.08 | 0.32 | 0.71 | 0.21 |
| Safflower | 1.79 | 0.54 | 1.35 | 0.40 | 1.74 | 0.52 | 2.29 | 0.69 | 2.24 | 0.67 |
| Castor | 4.28 | 1.71 | 7.97 | 3.19 | 7.93 | 3.17 | 9.91 | 3.96 | 7.95 | 3.18 |
| Linseed | 1.77 | 0.53 | 1.97 | 0.59 | 1.70 | 0.51 | 1.73 | 0.52 | 1.67 | 0.50 |
| Sub Total | 148.39 | 36.24 | 251.86 | 60.70 | 243.54 | 60.67 | 279.79 | 69.06 | 238.84 | 57.73 |
| B. Secondary Source | | | | | | | | | | |
| Coconut | | 5.50 | | 5.50 | | 5.50 | | 4.20 | | 4.50 |
| Cottonseed | | 4.30 | | 4.30 | | 4.30 | | 5.70 | | 6.30 |
| Ricebran | | 6.00 | | 6.00 | | 6.20 | | 6.80 | | 7.00 |
| Solvent Extracted Oils | | 2.00 | | 3.30 | | 3.50 | | 4.30 | | 3.50 |
| Tree & Forest Origin | | 0.80 | | 0.80 | | 0.80 | | 1.30 | | 1.20 |
| Sub Total | | 18.60 | | 19.90 | | 20.30 | | 22.30 | | 22.50 |
| Total (A+B) | | 54.84 | | 80.60 | | 80.97 | | 91.36 | | 80.23 |
| C. Less: Export & Industrial Use | | 8.20 | | 9.20 | | 8.50 | | 8.20 | | 7.80 |
| D. Net Domestic Availability of Edible Oils | | 46.64 | | 71.40 | | 72.47 | | 83.16 | | 72.43 |
| E. Shortfall (Import) | | 43.65 | | 52.90 | | 45.42 | | 42.88 | | 42.17 |
| F. Actual Consumption | | 90.29 | | 124.30 | | 117.89 | | 126.04 | | 114.60 |

* Fourth advance Estimates (declared by Ministry of Agriculture on 19.7.2007).

*Source*: Directorate of Vanaspati, Vegetable Oils and Fats; Department of Food and Public Distribution for oils.

# Table 4.25

## *State-wise, Year-wise Quantity of Crude Palm Oil (CPO) Obtained under Oil Palm Development Programme (OPDP)*

*(Tonnes)*

| Sl. No. | Name of the State | 1993-94 | 1994-95 | 1995-96 | 1996-97 | 1997-98 | 1998-99 | 1999-2000 | 2000-01 | 2001-02 | 2002-03 | 2003-04 | 2004-05 | 2005-06 | 2006-07 |
|---|---|---|---|---|---|---|---|---|---|---|---|---|---|---|---|
| | 1 | 2 | 3 | 4 | 5 | 6 | 7 | 8 | 9 | 10 | 11 | 12 | 13 | 14 | 15 |
| 1 | Andhra Pradesh | 150.90 | 525.90 | 862.21 | 1795.71 | 3710.90 | 5298.40 | 9684.00 | 15729.00 | 18974.00 | 18960.00 | 21457.00 | 23905.00 | 43500.00 | 35509.00 |
| 2 | Karnataka | 28.92 | 163.85 | 347.91 | 432.35 | 567.21 | 535.51 | 738.76 | 731.97 | 573.58 | 606.64 | 646.00 | 681.01 | 793.00 | 974.00 |
| 3 | Tamil Nadu | 0.00 | 0.00 | 0.00 | 0.00 | 2.21 | 30.10 | 82.09 | 86.06 | 0.00 | 0.00 | NR | 110.49 | 178.26 | 249.00 |
| 4 | Gujarat | 0.00 | 0.00 | 0.00 | 0.00 | 0.00 | 0.00 | 0.06 | 3.56 | 2.94 | 2.94 | 3.00 | NR | NR | Nil |
| 5 | Orissa | 0.00 | 0.00 | 0.00 | 0.00 | 11.19 | 19.18 | 21.71 | 42.35 | 0.00 | 0.00 | 0.00 | 0.00 | 0.00 | 0.00 |
| 6 | Goa | 0.00 | 0.00 | 0.00 | 0.00 | 54.83 | 102.95 | 150.00 | 207.75 | 243.00 | 330.85 | 324.00 | 348.93 | 379.00 | 345.00 |
| 7 | Tripura | 0.00 | 0.00 | 0.00 | 0.00 | 0.00 | 0.00 | 8.00 | 20.64 | 0.00 | 0.00 | NR | NR | NR | NR |
| 8 | Kerala | 954.00 | 0.00 | 3955.00 | 4261.00 | 4428.00 | 3812.00 | 5159.15 | 6667.26 | 980.00 | 6572.00 | 6387.00 | 5792.94 | 6478.00 | 6773.00 |
| 9 | Andaman & Nicobar | 0.00 | 2314.00 | 0.00 | 1426.00 | 1284.00 | 1142.00 | 2314.00 | 1800.76 | 1840.32 | 1696.00 | 0.00 | 0.00 | 0.00 | 0.00 |
| | Total | 1133.82 | 3003.75 | 5165.12 | 7915.06 | 10058.34 | 10940.14 | 18157.77 | 25289.35 | 22613.84 | 28168.43 | 28817.00 | 30838.37 | 51328.26 | 43850.00 |

*Note*: Information Based on inputs provided by State Governments and updated up to July 2007. NR= Not Reported

*Source*: Technology Mission on Oilseeds, Pulses and Maize, Ministry of Agriculture, New Delhi.

## Table 4.26(a)

*State-wise Production of Oil Palm Fresh Fruit Bunches under Oil Palm Development Programme (OPDP)*

*(Tonnes)*

| Sl.No. | Name of the State | 1992-93 | 1993-94 | 1994-95 | 1995-96 | 1996-97 | 1997-98 | 1998-99 | 1999-2000 |
|---|---|---|---|---|---|---|---|---|---|
| 1 | 2 | 3 | 4 | 5 | 6 | 7 | 8 | 9 | 10 |
| 1 | Andhra Pradesh | 58.00 | 983.00 | 3855.00 | 6966.00 | 11374.00 | 33763.00 | 56152.00 | 125000.00 |
| 2 | Karnataka | 0.00 | 237.00 | 1197.00 | 2143.68 | 2665.00 | 3439.00 | 3346.95 | 4588.54 |
| 3 | Tamil Nadu | 0.00 | 0.00 | 0.00 | 0.00 | 0.00 | 12.80 | 188.18 | 850.00 |
| 4 | Gujarat | 0.00 | 0.00 | 0.00 | 0.00 | 0.00 | 0.00 | 0.00 | 0.00 |
| 5 | Orissa | 0.00 | 0.00 | 0.00 | 0.00 | 0.00 | 70.00 | 120.00 | 200.00 |
| 6 | Goa | 0.00 | 0.00 | 0.00 | 28.00 | 193.00 | 357.00 | 700.00 | 819.00 |
| 7 | Tripura | 0.00 | 0.00 | 0.00 | 0.00 | 0.00 | 0.00 | 0.00 | 0.00 |
| 8 | Kerala | 14051.00 | 14051.00 | 0.00 | 0.00 | 0.00 | 0.00 | 0.00 | 0.00 |
| 0 | Andaman & Nicobar | 7124.00 | 9347.00 | 0.00 | 0.00 | 0.00 | 9771.10 | 6864.18 | 12970.88 |
|  | Total | 21233.00 | 24618.00 | 5052.00 | 9137.68 | 14232.00 | 47412.90 | 67371.31 | 144428.42 |

*Contd. ...*

...*Contd.* ...

*(Tonnes)*

| *Sl.No.* | *Name of the State* | *2000-01* | *2001-02* | *2002-03* | *2003-04* | *2004-05* | *2005-06* | *2006-07* | *2007-08** |
|---|---|---|---|---|---|---|---|---|---|
| *1* | *2* | *11* | *12* | *13* | *14* | *15* | *16* | *17* | *18* |
| 1 | Andhra Pradesh | 125000.00 | 108757.00 | 109688.61 | 129841.00 | 138929.00 | 203000.00 | 208262.00 | 194000.00 |
| 2 | Karnataka | 3998.13 | 3562.50 | 3676.65 | 3483.00 | 4127.34 | 4528.00 | 5415.00 | 530400.00 |
| 3 | Tamil Nadu | 2250.00 | 229.58 | 434.00 | 415.00 | 680.72 | 1802.00 | 1584.02 | 1692.69 |
| 4 | Gujarat | 0.00 | 24.90 | 29.42 | 28.00 | 6.10 | 20.00 | 10.25 | 10.72 |
| 5 | Orissa | 265.00 | 0.00 | 0.00 | 0.00 | 0.00 | 0.00 | 0.00 | NR |
| 6 | Goa | 1300.00 | 1603.00 | 1879.05 | 1851.00 | 1993.93 | 2143.00 | 1951.00 | NR |
| 7 | Tripura | 0.00 | 75.00 | 80.00 | 503.00 | 465.00 | NR | 447.12 | NR |
| 8 | Kerala | 0.00 | 4820.43 | 32991.64 | 32295.00 | 29939.63 | 33795.00 | 33899.00 | 29329.81 |
| 9 | Andaman & Nicobar | 10010.82 | 9800.92 | 8957.00 | 0.00 | 0.00 | 0.00 | 0.00 | NR |
|  | Total | 142823.95 | 128873.33 | 157736.37 | 168416.00 | 176141.72 | 245288.00 | 251568.39 | 230337.22 |

*Note*: Information Based on inputs provided by State Governments. NR= Not Reported

* Data/information updated on July 2007.

*Source*: Technology Mission on Oilseeds, Pulses and Maize, Ministry of Agriculture, New Delhi.

## Table 4.26(b)

*State-wise & Year-wise Area Target and Achievement under Oil Palm Development Programme OPDP/ISOPOM during 8th Plan, 9th Plan & Years 2002-03, 2003-04, 2004-05, 2005-06, 2006-07 and 10th Plan*

(Area in Hectare)

| Sl.No. | Name of the State | Prior to OPDP | Eighth Plan | | Ninth Plan | | 2002-03 | | 2003-04 | | 2004-05 | |
|---|---|---|---|---|---|---|---|---|---|---|---|---|
| | | | Targets | Achievements | Targets | Achievements | Targets | Achievements | Targets | Achievements | Targets | Achievements |
| 1 | 2 | 3 | 4 | 5 | 6 | 7 | 8 | 9 | 10 | 11 | 12 | 13 |
| 1 | Andhra Pradesh | 1232 | 47000 | 18888 | 48000 | 12851 | 3200 | 1948 | 3800 | 2780 | 4200 | 5998 |
| 2 | Karnataka | 1114 | 18000 | 6431 | 10198 | 1421 | 1200 | 335 | 1400 | 452 | 1500 | 1046 |
| 3 | Tamil Nadu | 0 | 8000 | 6104 | 12258 | 5193 | 1800 | 570 | 1800 | 722 | 1900 | 1091 |
| 4 | Gujarat | 0 | 3100 | 300 | 2550 | 194 | 300 | 193 | 400 | 8 | 450 | 3 |
| 5 | Orissa | 0 | 1700 | 893 | 3444 | 591 | 400 | 0 | 400 | 0 | 400 | 0 |
| 6 | Goa | 0 | 1000 | 655 | 800 | 140 | 200 | 48 | 300 | 28 | 300 | 14 |
| 7 | Tripura | 0 | 500 | 85 | 200 | 229 | 200 | 0 | 200 | 0 | 300 | 25 |
| 8 | Assam | 0 | 400 | 10 | 650 | 0 | 200 | 0 | 200 | 0 | 250 | 0 |
| 9 | West Bengal@ | 0 | 300 | 0 | 450 | 0 | 0 | 0 | 0 | 0 | 0 | 0 |
| 10 | Kerala | 3646 | 0 | 0 | 1000 | 161 | 300 | 277 | 300 | 438 | 400 | 132 |
| 11 | Maharashtra@ | 1000 | 0 | 0 | 450 | 0 | 0 | 0 | 0 | 0 | 0 | 0 |
| 12 | Andaman & Nicobar Islands@ | 1593 | 0 | 0 | 0 | 0 | 0 | 0 | 0 | 0 | 0 | 0 |
| 13 | Mizoram | 0 | 0 | 0 | 0 | 0 | 200 | 0 | 200 | 0 | 300 | 12 |
| | Total | 8585 | 80000 | 33366 | 80000 | 20780 | 8000 | 3371 | 9000 | 4428 | 10000 | 8321 |

contd. ...

...contd. ...

| Sl.No. | Name of the State | 2005-06 | | 2006-07 | | 10th Plan | | 2007-08 | | Area | Net |
|---|---|---|---|---|---|---|---|---|---|---|---|
| | | Target | Achievement | Target | Achievement | Target | Achievement | Target | Achievement# | Uprooted | Area |
| 1 | 2 | 14 | 15 | 16 | 17 | 18 | 19 | 20 | 21 | 22 | 23 |
| 1 | Andhra Pradesh | 4800 | 9563 | 5200 | 11882 | 21200 | 32171 | 15000 | 11049 | 5457 | 59685 |
| 2 | Karnataka | 1500 | 1591 | 1800 | 3714 | 7400 | 7138 | 5000 | 3429 | 6842 | 9262 |
| 3 | Tamil Nadu | 2000 | 1210 | 2200 | 1746 | 9700 | 5339 | 3500 | 890 | 5384 | 10551 |
| 4 | Gujarat | 500 | 24 | 550 | 57 | 2200 | 285 | 1000 | 328 | 219 | 561 |
| 5 | Orissa | 500 | 0 | 500 | 300 | 2200 | 300 | 1000 | 0 | 0 | 1734 |
| 6 | Goa | 350 | 7 | 400 | 3 | 1550 | 100 | 100 | 0 | 42 | 853 |
| 7 | Tripura | 300 | 55 | 300 | 19 | 1300 | 216 | 0 | 0 | 0 | 530 |
| 8 | Assam | 300 | 0 | 300 | 0 | 1250 | 0 | 0 | 0 | 0 | 10 |
| 9 | West Bengal @ | 0 | 0 | 0 | 0 | 0 | 0 | 0 | 0 | 0 | 0 |
| 10 | Kerala | 400 | 188 | 400 | 270 | 1800 | 1305 | 480 | 182 | 0 | 5112 |
| 11 | Maharashtra @ | 0 | 0 | 0 | 0 | 0 | 0 | 0 | 0 | 0 | 0 |
| 12 | Andaman & Nicobar Islands@ | | 0 | 0 | 0 | 0 | 0 | 0 | 0 | 0 | 0 |
| 0 | | | | | | | | | | | |
| 13 | Mizoram | 350 | 24 | 350 | 185 | 1400 | 221 | 35 | 1614 | 0 | 221 |
| | Total | 11000 | 12662 | 12000 | 18176 | 50000 | 47075 | 29580 | 17492 | 17944 | 88519 |

@ OPDP scheme not implemented so far. # Up to February 2008.

ISOPOM: Integrated scheme Oil seeds, Pulses,Oil Palm and Maize.

*Note*: Net Area = Gross Area - Area Uprooted

*Source*: Technology Mission on Oilseeds, Pulses and Maize, Ministry of Agriculture, New Delhi.

# Table 4.1.1

## *Area under High Yielding Varieties of Rice in Major States*

*('000 hectare)*

| *State* | *Season* | *1998-1999* | *1999-2000* | *2000-2001* | *2001-2002* | *2002-2003* | *2003-2004* | *2004-2005* |
|---|---|---|---|---|---|---|---|---|
| *1* | *2* | *3* | *4* | *5* | *6* | *7* | *8* | *9* |
| Andhra Pradesh | *Kharif* | - | - | - | - | - | - | - |
| | *Rabi* | - | - | - | - | - | - | - |
| Assam | Autumn | - | - | - | - | - | - | - |
| | Winter | - | - | - | - | - | | |
| | Summer | - | - | - | - | - | - | - |
| Bihar | *Bhadai* | - | - | - | - | 144 | - | - |
| | *Aghani* | 1851 | 1881 | 1649 | 1608 | 1583 | 1556 | 1411 |
| | *Garma* | 124 | 103 | 88 | 94 | 119 | 88 | 80 |
| Chhattisgarh | | - | - | 581 | 637 | 981 | 736 | 732 |
| Goa | *Kharif* | 33 | 33 | 33 | 30 | 31 | 35 | 35 |
| | *Rabi* | 15 | 16 | 16 | 16 | 16 | 17 | 17 |
| Gujarat | *Kharif* | 439 | 498 | 424 | 576 | 374 | 458 | 518 |
| | Summer | - | - | - | 3 | 9 | 22 | 10 |
| Haryana | | - | - | - | - | - | 690 | 741 |
| Himachal Pradesh | | - | - | - | - | - | - | - |
| Jammu & Kashmir | | 103 | - | 183 | 140 | 121 | 141 | 162 |
| Jharkhand | *Aghani* | - | - | - | - | - | - | - |
| Karnataka | *Kharif* | 818 | 849 | 875 | 867 | 805 | 743 | 877 |
| | *Rabi* | 44 | 45 | 45 | 47 | 44 | 50 | 41 |
| | Summer | 294 | 299 | 310 | 274 | 72 | 74 | 209 |
| Kerala | Autumn | 62 | 71 | 84 | 79 | 98 | 93 | 98 |
| | Winter | 67 | 86 | 89 | 103 | 112 | 110 | 116 |
| | Summer | 49 | 55 | 53 | 40 | 40 | 36 | 35 |
| Madhya Pradesh | | 740 | 792 | 154 | 182 | 124 | 123 | 136 |
| Maharashtra | *Kharif* | 1379 | 1407 | 1412 | 1325 | 1282 | 1312 | 1342 |
| | Summer | - | - | - | - | - | - | - |
| Meghalaya | Autumn | - | - | - | - | - | - | - |
| | Winter | - | - | - | - | - | - | - |
| Orissa | Autumn | 306 | 336 | 400 | 432 | 404 | 450 | 434 |
| | Winter | 1613 | 1868 | 2022 | 2153 | 2086 | 2185 | 2276 |
| | Summer | 268 | 382 | 207 | 272 | 178 | 253 | 293 |
| Punjab | | - | - | - | - | - | - | 2549 |
| Rajasthan | | 94 | 119 | 121 | 114 | 59 | 90 | 91 |
| Tamil Nadu# | K/K/S | - | - | - | - | - | - | - |
| | S/T/P | - | - | - | - | - | - | - |
| | N/K | - | - | - | - | - | - | - |
| Uttar Pradesh | | - | - | - | - | - | - | - |
| West Bengal | Aus. | - | - | - | - | - | - | - |
| | Aman | - | - | - | - | - | - | - |
| | Summer | - | - | - | - | - | - | - |
| Dadra & Nagar Haveli | | 4 | 7 | 10 | 7 | 6 | 9 | 9 |
| Delhi | | 6 | 8 | 6 | 6 | 6 | 7 | 9 |
| Daman & Diu | | 1 | 1 | 1 | 1 | 1 | 1 | 1 |
| Pondicherry* | I | 5 | 5 | 6 | 6 | 5 | 6 | 6 |
| | II | 14 | 14 | 14 | 14 | 13 | 12 | 13 |
| | III | 6 | 6 | 5 | 6 | 6 | 6 | 6 |

# K/K/S- Kar/Kuruvai/Sornaveri (Early *Kharif*), S/T/P- Samba/Thalady/Pishanam (Late *Kharif*), N/K- Navarai/Kodai (*Rabi*)* I- *Kharif*, II- *Rabi*-I, III- *Rabi*-II (Summer). Blank (-) Not available with State Agricultural Statistics Authority (SASA)

*Source*: Consolidated results of Crop Estimation Survey on Principal Crops (from 1998-99 to 2004-05 individual years) Published by NSSO, Ministry of Statistics and Programme Implementation.

## Table 4.1.2

*Area under High Yielding Varieties of Jowar in Major States*

('000 hectare)

| State | Season | 1998-1999 | 1999-2000 | 2000-2001 | 2001-2002 | 2002-2003 | 2003-2004 | 2004-2005 |
|---|---|---|---|---|---|---|---|---|
| 1 | 2 | 3 | 4 | 5 | 6 | 7 | 8 | 9 |
| Andhra Pradesh | *Kharif* | - | - | - | - | - | - | - |
| | *Rabi* | - | - | - | - | - | - | - |
| Chhattisgarh | | - | - | - | - | - | - | - |
| Gujarat | *Kharif* | 38 | 32 | 40 | 53 | 41 | 55 | 31 |
| | Summer | 8 | 13 | 8 | 7 | 4 | 38 | 18 |
| Karnataka | *Kharif* | 312 | 309 | 306 | 234 | 245 | 189 | 304 |
| | *Rabi* | 492 | 624 | 400 | 677 | 856 | 848 | 830 |
| | Summer | - | 8 | 6 | 6 | - | - | 16 |
| Madhya Pradesh | | - | - | - | - | - | - | - |
| Maharashtra | *Kharif* | 1904 | - | - | 1717 | 1677 | 1570 | 1518 |
| | *Rabi* | - | - | - | - | - | - | - |
| Rajasthan | | 102 | 125 | 147 | 131 | 117 | 292 | 174 |
| Tamil Nadu | | - | - | - | - | - | - | - |
| Uttar Pradesh | | - | - | - | - | - | - | - |

Blank (-) *Not available with State Agricultural Statistics Authority (SASA)*

*Source*: Consolidated results of Crop Estimation Survey on Principal Crops (from 1998-99 to 2004-05 individual years) - Published by NSSO, Ministry of Statistics and Programme Implementation.

# Table 4.1.3

*Area under High Yielding Varieties of Bajra Major States*

('000 hectare)

| State | Season | 1998-1999 | 1999-2000 | 2000-2001 | 2001-2002 | 2002-2003 | 2003-2004 | 2004-2005 |
|---|---|---|---|---|---|---|---|---|
| 1 | 2 | 3 | 4 | 5 | 6 | 7 | 8 | 9 |
| Andhra Pradesh | | - | - | - | - | - | - | - |
| Gujarat | *Kharif* | 226 | 716 | 832 | 779 | 784 | - | 750 |
| | Summer | 177 | 174 | 140 | 134 | 137 | 163 | 135 |
| Haryana | | - | - | - | - | - | 560 | 480 |
| Karnataka | *Kharif* | 333 | 377 | 419 | 192 | 286 | 279 | 426 |
| Madhya Pradesh | | - | - | - | - | - | - | - |
| Maharashtra | | 1716 | 1684 | 1651 | 1291 | 1377 | 1214 | 1423 |
| Punjab | | - | - | - | - | - | - | - |
| Rajasthan | | 1376 | 1589 | 1770 | 1498 | 2461 | 2935 | 2194 |
| Tamil Nadu | | - | - | - | - | - | - | - |
| Uttar Pradesh | | - | - | - | - | - | - | - |
| Delhi | | 3 | - | 3 | 3 | 2 | 2 | 2 |
| Daman & Diu | | - | - | - | - | - | - | - |

Blank (-) Not available with State Agricultural Statistics Authority (SASA)

*Source:* Consolidated results of Crop Estimation Survey on Principal Crops (from 1998-99 to 2004-05 individual years) - Published by NSSO, Ministry of Statistics and Programme Implementation.

# Table 4.1.4

*Area under High Yielding Varieties of Maize in Major States*

*('000 hectare)*

| *State* | *Season* | *1998-1999* | *1999-2000* | *2000-2001* | *2001-2002* | *2002-2003* | *2003-2004* | *2004-2005* |
|---|---|---|---|---|---|---|---|---|
| *1* | *2* | *3* | *4* | *5* | *6* | *7* | *8* | *9* |
| Andhra Pradesh | *Kharif* | - | - | - | - | - | - | - |
| | *Rabi* | - | - | - | - | - | - | - |
| Bihar | *Bhadai* | 124 | 124 | 106 | 92 | 104 | 105 | 105 |
| | *Rabi* | - | - | - | - | 188 | - | - |
| | *Garma* | 108 | 118 | 105 | 113 | - | 103 | 111 |
| Chhattisgarh | | - | - | - | - | - | - | - |
| Gujarat | | 197 | 208 | 227 | 268 | 222 | 249 | 320 |
| Haryana | | - | - | - | - | - | 8 | 10 |
| Himachal Pradesh | | - | - | - | - | - | - | - |
| Jammu & Kashmir | | 40 | - | 82 | 46 | 90 | 71 | 47 |
| Karnataka | *Kharif* | 446 | 518 | 595 | 500 | 574 | 544 | 779 |
| | *Rabi* | 48 | 61 | 51 | 61 | 50 | 47 | 55 |
| | Summer | - | 13 | 11 | 9 | - | - | 9 |
| Madhya Pradesh | | - | - | - | - | - | - | - |
| Maharashtra | *Kharif* | - | - | - | - | - | - | - |
| | *Rabi* | - | - | - | - | - | - | - |
| Orissa | Autumn | 23 | 27 | 22 | 26 | 14 | 26 | 31 |
| | Winter | 0.22 | 0.03 | 0.13 | 0.04 | 0.07 | 0.62 | 2 |
| | Summer | 0.23 | 1.00 | 1.00 | 0.40 | 1.00 | 2.00 | 2 |
| Punjab | | - | - | - | - | - | - | 100 |
| Rajasthan | *Kharif* | 33 | 205 | 313 | 386 | 428 | 498 | 444 |
| | *Rabi* | - | - | - | - | - | - | - |
| Uttar Pradesh | | - | - | - | - | - | - | - |
| West Bengal | | - | - | - | - | - | - | - |

**Blank** (-) Not available with State Agricultural Statistics Authority (SASA)

*Source*: Consolidated results of Crop Estimation Survey on Principal Crops (from 1998-99 to 2004-05 individual years) - Published by NSSO, Ministry of Statistics and Programme Implementation.

## Table 4.1.5

*Area under High Yielding Varieties of Wheat in Major States*

*('000 hectare)*

| State | Season | 1998-1999 | 1999-2000 | 2000-2001 | 2001-2002 | 2002-2003 | 2003-2004 | 2004-05 |
|---|---|---|---|---|---|---|---|---|
| 1 | 2 | 3 | 4 | 5 | 6 | 7 | 8 | 9 |
| Assam | | - | - | - | - | - | - | - |
| Bihar | | 1602 | 1636 | 1516 | 1637 | 1582 | 1470 | 1473 |
| Chhattisgarh | | - | - | 21 | 57 | 62 | 67 | 41 |
| Gujarat | | 641 | 442 | 260 | 409 | 390 | 731 | 641 |
| Haryana | | - | - | - | - | - | 2290 | 2288 |
| Himachal Pradesh | | - | - | - | - | - | - | - |
| Jammu & Kashmir | | 173 | | 170 | 149 | 110 | 143 | 140 |
| Jharkhand | | - | 173 | 107 | - | - | - | - |
| Karnataka | | 102 | 105 | - | 134 | 123 | 101 | 120 |
| Madhya Pradesh | | 2364 | 2286 | 1220 | 1751 | 1603 | 2269 | 2405 |
| Maharashtra | | 904 | 885 | 689 | 680 | 710 | 622 | 724 |
| Orissa | Summer | 3 | 5 | 8 | 7 | 4 | 5 | 4 |
| Punjab | | - | - | - | - | - | - | 3482 |
| Rajasthan | | 2059 | 1989 | 1821 | 1859 | 1450 | 1528 | 1350 |
| Uttar Pradesh | | - | - | - | - | - | - | - |
| West Bengal | | - | - | - | - | - | - | - |
| Delhi | | - | 34 | 28 | 26 | 20 | 19 | 13 |

Blank (-) Not available with State Agricultural Statistics Authority (SASA)

*Source*: Consolidated results of Crop Estimation Survey on Principal Crops (from 1998-99 to 2004-05 individual years) - Published by NSSO, Ministry of Statistics and Programme Implementation.

# Table 4.1.6

## *Yield of Local Variety of Rice in Major States*

*(Kgs. Per hectare)*

| *State* | *Season* | *1998-1999* | *1999-2000* | *2000-2001* | *2001-2002* | *2002-2003* | *2003-2004* | *2004-2005* |
|---|---|---|---|---|---|---|---|---|
| *1* | *2* | *3* | *4* | *5* | *6* | *7* | *8* | *9* |
| Andhra Pradesh | *Kharif* | - | - | - | - | - | - | - |
| | *Rabi* | - | - | - | - | - | - | - |
| Assam | Autumn | 938 | 1138 | 1045 | 960 | 963 | 1012 | 665 |
| | Winter | 1350 | 1470 | 1488 | 1516 | 1459 | 1554 | 1421 |
| | Summer | 1497 | 1952 | 1783 | 1636 | 1677 | 1399 | 1704 |
| Bihar | *Bhadai* | 1082 | 1157 | 1339 | 1240 | 1095 | 1234 | 912 |
| | *Aghani* | 1043 | 1261 | 1202 | 1177 | 1150 | 1217 | 551 |
| | *Garma* | 1080 | 1251 | 1385 | 1659 | 2307 | 1331 | 1174 |
| Chhattisgarh | | - | - | 602 | 1354 | 582 | 958 | 1084 |
| Goa | *Kharif* | 2376 | 2004 | 1823 | 2063 | 1823 | - | - |
| | *Rabi* | - | - | - | - | - | - | - |
| Gujarat | *Kharif* | 1073 | 683 | 590 | 782 | 677 | 1438 | 966 |
| | Summer | - | - | - | 3374 | 2489 | - | 2765 |
| Haryana | | - | 1690 | 1670 | - | - | 2126 | 2208 |
| Himachal Pradesh | | 1455 | 1386 | 1554 | 1738 | 1036 | 1478 | 1306 |
| Jammu & Kashmir | | 2044 | 2044 | 1343 | 1575 | 1699 | 1780 | 1596 |
| Jharkhand | *Aghani* | - | - | - | - | - | - | - |
| Karnataka | *Kharif* | 1489 | 1375 | 1605 | 1025 | 880 | 1059 | 1463 |
| | *Rabi* | 1386 | 1398 | 1507 | 1573 | 1676 | 2300 | 2509 |
| | Summer | - | - | - | - | - | - | - |
| Kerala | Autumn | 1925 | 1922 | 1912 | 1988 | 1441 | 1549 | 1426 |
| | Winter | 1885 | 1965 | 1818 | 1944 | 1784 | 1729 | 1873 |
| | Summer | 2344 | 2211 | 1842 | 2044 | 1628 | 1391 | 1417 |
| Madhya Pradesh | | 982 | 1161 | 521 | 883 | 585 | 987 | 724 |
| Maharashtra | *Kharif* | 1235 | 1406 | 1078 | 1325 | 369 | 1069 | 1191 |
| | Summer | - | - | - | - | - | - | - |
| Meghalaya | Autumn | 881 | 1191 | 1191 | 1422 | - | 1285 | 1277 |
| | Winter | 1582 | 1409 | 1409 | 1377 | - | 1487 | 1378 |
| Orissa | Autumn | 520 | 757 | 488 | 783 | 326 | 765 | 677 |
| | Winter | 994 | 821 | 905 | 1303 | 608 | 1276 | 1249 |
| | Summer | - | - | - | - | - | - | - |
| Punjab | | 2761 | - | - | 3369 | 3121 | 3396 | 3537 |
| Rajasthan | | 828 | 595 | 232 | 845 | 324 | 1140 | 1130 |
| Tamil Nadu# | K/K/S | 3429 | - | - | 3124 | - | 3333 | - |
| | S/T/P | 3042 | 1247 | 1790 | 1319 | 2626 | 2552 | 1738 |
| | N/K | 4651 | 2501 | 2830 | 2571 | 2557 | 3082 | 1613 |
| Uttar Pradesh | | - | - | - | - | - | - | - |
| West Bengal | Aus. | 1171 | 1451 | 2181 | 1529 | 1401 | 1525 | 1395 |
| | *Aman* | 1226 | 1493 | 2568 | 1738 | 1621 | 1744 | 1746 |
| | Summer | 2308 | 2672 | 4197 | 1872 | 2338 | 2526 | 2733 |
| Dadra & Nagar Haveli | | 1468 | 1856 | 1895 | 2002 | 2627 | 1572 | 1576 |
| Delhi | | - | - | - | - | - | - | - |
| Daman & Diu | | 1110 | 1243 | 1772 | 990 | 1491 | 1843 | 1715 |
| Pondicherry* | I | - | - | - | - | - | - | - |
| | II | - | - | - | - | - | - | - |
| | III | - | - | - | - | - | - | - |

# K/K/S- Kar/Kuruvai/Sornaveri (Early *Kharif*), S/T/P- Samba/Thalady/Pishanam (Late *Kharif*), N/K- Navarai/Kodai (*Rabi*)* I- *Kharif*, II- *Rabi*-I, III- *Rabi*-II (Summer). Blank (-) Not available with State Agricultural Statistics Authority (SASA)

*Source*: Consolidated results of Crop Estimation Survey on Principal Crops (from 1998-99 to 2004-05 individual years) Published by NSSO, Ministry of Statistics and Programme Implementation.

## Table 4.1.7

*Yield of Local Variety of Jowar in Major States*

*(Kgs. Per hectare)*

| *State* | *Season* | *1998-1999* | *1999-2000* | *2000-2001* | *2001-2002* | *2002-2003* | *2003-2004* | *2004-2005* |
|---|---|---|---|---|---|---|---|---|
| 1 | 2 | 3 | 4 | 5 | 6 | 7 | 8 | 9 |
| Andhra Pradesh | *Kharif* | - | - | - | - | - | - | - |
| | *Rabi* | - | - | - | - | - | - | - |
| Chhattisgarh | | - | - | 665 | 973 | 740 | 1001 | 667 |
| Gujarat | *Kharif* | 660 | 1249 | 425 | 832 | 585 | 828 | 1242 |
| | Summer | 835 | 716 | 506 | 752 | 657 | 835 | 932 |
| Karnataka | *Kharif* | 655 | 582 | 700 | 612 | 453 | 332 | 537 |
| | *Rabi* | 740 | 734 | 652 | 584 | 442 | 136 | 610 |
| | Summer | - | - | - | - | - | - | - |
| Madhya Pradesh | | - | - | - | - | 760 | 1197 | 953 |
| Maharashtra | *Kharif* | 738 | - | - | 819 | 536 | 892 | 740 |
| | *Rabi* | - | - | - | - | - | - | - |
| Rajasthan | | 281 | 244 | 109 | 350 | 86 | 671 | 412 |
| Tamil Nadu | | 912 | 613 | 484 | 902 | 704 | 587 | 715 |
| Uttar Pradesh | | - | - | - | - | - | - | - |

Blank (-) Not available with State Agricultural Statistics Authority (SASA)

*Source*: Consolidated results of Crop Estimation Survey on Principal Crops (from 1998-99 to 2004-05 individual years) - Published by NSSO, Ministry of Statistics and Programme Implementation.

# Table 4.1.8

## *Yield of Local Variety of Bajra in Major States*

*(Kgs. Per hectare)*

| *State* | *Season* | *1998-1999* | *1999-2000* | *2000-2001* | *2001-2002* | *2002-2003* | *2003-2004* | *2004-2005* |
|---|---|---|---|---|---|---|---|---|
| *1* | 2 | 3 | 4 | 5 | 6 | 7 | 8 | 9 |
| Andhra Pradesh | | - | - | - | - | - | - | - |
| Gujarat | *Kharif* | 860 | 461 | 695 | 829 | 961 | 1320 | 1001 |
| | Summer | 2451 | 2537 | 458 | 1713 | - | - | 3198 |
| Haryana | | - | - | - | - | - | 1071 | 756 |
| Karnataka | *Kharif* | 493 | 364 | 460 | 424 | 385 | 300 | 391 |
| Madhya Pradesh | | - | - | - | - | 486 | 822 | 795 |
| Maharashtra | | 588 | 537 | 566 | 341 | 438 | 466 | 512 |
| Punjab | | 718 | 632 | 808 | - | - | - | - |
| Rajasthan | | 283 | 132 | 243 | 610 | 116 | 942 | 399 |
| Tamil Nadu | | 794 | 702 | 746 | 871 | 636 | 826 | 1050 |
| Uttar Pradesh | | - | - | - | - | - | - | - |
| Delhi | | - | - | - | - | - | - | - |
| Daman & Diu | | 1420 | 1218 | 1992 | 780 | 780 | 780 | 1454 |

Blank (-) Not available with State Agricultural Statistics Authority (SASA)

*Source*: Consolidated results of Crop Estimation Survey on Principal Crops (from 1998-99 to 2004-05 individual years) - Published by NSSO, Ministry of Statistics and Programme Implementation.

# Table 4.1.9

*Yield of Local Variety of Maize in Major States*

(Kgs. Per hectare)

| State | Season | 1998-1999 | 1999-2000 | 2000-2001 | 2001-2002 | 2002-2003 | 2003-2004 | 2004-2005 |
|---|---|---|---|---|---|---|---|---|
| 1 | 2 | 3 | 4 | 5 | 6 | 7 | 8 | 9 |
| Andhra Pradesh | *Kharif* | - | - | - | - | - | - | - |
| | *Rabi* | - | - | - | - | - | - | - |
| Bihar | *Bhadai* | 1080 | 1453 | 1704 | 1920 | 1537 | 1650 | 1496 |
| | *Rabi* | 3005 | 3003 | 2919 | 3186 | - | 3006 | 3041 |
| | *Garma* | 2009 | 2252 | 2541 | 2309 | - | 2953 | 2600 |
| Chhattisgarh | | - | - | 1353 | 743 | 1305 | 1370 | 1430 |
| Gujarat | | 1758 | 1309 | 734 | 1907 | 1702 | 1661 | 899 |
| Haryana | | - | - | - | - | - | 2449 | 2302 |
| Himachal Pradesh | | 2158 | 2332 | 2323 | 2559 | 1670 | 2551 | 2227 |
| Jammu & Kashmir | | 1648 | 1648 | 1560 | 1632 | 1199 | 1555 | 1473 |
| Karnataka | *Kharif* | 1516 | 1824 | 2283 | 1931 | 1430 | 1500 | 1999 |
| | *Rabi* | - | 1622 | - | 1672 | 1374 | 670 | 1691 |
| | Summer | - | - | - | - | - | - | - |
| Madhya Pradesh | | - | - | - | - | - | 1557 | 1185 |
| Maharashtra | *Kharif* | - | - | - | - | - | - | - |
| | *Rabi* | - | - | - | - | - | - | - |
| Orissa | Autumn | 957 | 982 | 873 | 740 | 950 | 891 | 1006 |
| | Winter | 771 | 673 | 650 | 683 | 667 | 1291 | 899 |
| | Summer | 771 | 660 | 637 | 875 | 769 | 740 | 837 |
| Punjab | | 2069 | 2213 | 2212 | 2128 | 1638 | 2287 | 2395 |
| Rajasthan | *Kharif* | 1026 | 890 | 979 | 1399 | 696 | 1796 | 1225 |
| | *Rabi* | - | - | - | - | - | - | - |
| Uttar Pradesh | | - | - | - | - | - | - | - |
| West Bengal | | - | - | - | - | - | - | 2029 |

Blank (-) Not available with State Agricultural Statistics Authority (SASA)

*Source*: Consolidated results of Crop Estimation Survey on Principal Crops (from 1998-99 to 2004-05 individual years) - Published by NSSO, Ministry of Statistics and Programme Implementation.

# Table 4.1.10

## *Yield of Local Variety of Wheat in Major States*

*(Kgs. Per hectare)*

| State | Season | 1998-1999 | 1999-2000 | 2000-2001 | 2001-2002 | 2002-2003 | 2003-2004 | 2004-05 |
|---|---|---|---|---|---|---|---|---|
| 1 | 2 | 3 | 4 | 5 | 6 | 7 | 8 | 9 |
| Assam | | 1001 | 1218 | 1207 | 1162 | 1133 | 1040 | 1084 |
| Bihar | | 1621 | 1791 | 1812 | 1732 | 1648 | 1552 | 1393 |
| Chhattisgarh | | - | - | 461 | 667 | 651 | 784 | 652 |
| Gujarat | | 1267 | 1067 | 1799 | 2078 | 1367 | 697 | 1910 |
| Haryana | | - | - | - | - | - | 2240 | 2419 |
| Himachal Pradesh | | 1225 | 1506 | 568 | 1622 | 1408 | 1360 | 1795 |
| Jammu & Kashmir | | 1292 | 1292 | 429 | 1401 | 1435 | 1437 | 1567 |
| Jharkhand | | - | - | - | - | - | - | - |
| Karnataka | | 624 | 553 | 751 | 564 | 339 | 216 | 596 |
| Madhya Pradesh | | 1384 | 1342 | 1077 | 1241 | 1109 | 1346 | 1298 |
| Maharashtra | | 898 | 957 | 1008 | 911 | 695 | 708 | 864 |
| Orissa | Summer | 1120 | 1145 | 1165 | - | 970 | 1359 | - |
| Punjab | | 2149 | - | - | 1417 | 2048 | 3582 | - |
| Rajasthan | | 1794 | 1884 | 2029 | 2596 | 2469 | 2331 | 2780 |
| Uttar Pradesh | | - | - | - | - | - | - | - |
| West Bengal | | 1338 | 1797 | 2588 | 1728 | 1808 | 2052 | 1873 |
| Delhi | | - | - | - | - | - | - | - |

Blank (-) Not available with State Agricultural Statistics Authority (SASA)

*Source*: Consolidated results of Crop Estimation Survey on Principal Crops (from 1998-99 to 2004-05 individual years) - Published by NSSO, Ministry of Statistics and Programme Implementation.

# Table 4.1.11

## *Yield of High Yielding Variety of Rice in Major States*

*(Kgs. Per hectare)*

| *State* | *Season* | *1998-1999* | *1999-2000* | *2000-2001* | *2001-2002* | *2002-2003* | *2003-2004* | *2004-2005* |
|---|---|---|---|---|---|---|---|---|
| *1* | *2* | *3* | *4* | *5* | *6* | *7* | *8* | *9* |
| Andhra Pradesh | *Kharif* | - | - | - | - | - | - | - |
| | *Rabi* | - | - | - | - | - | - | - |
| Assam | Autumn | 1157 | 1752 | 1804 | 1720 | 1839 | 1778 | 1551 |
| | Winter | 1752 | 2068 | 2154 | 2308 | 2262 | 2263 | 2013 |
| | Summer | 2580 | 1693 | 2459 | 2445 | 2518 | 2220 | 2309 |
| Bihar | *Bhadai* | - | - | - | - | 1401 | - | - |
| | *Aghani* | 1737 | 1778 | 1745 | 1734 | 1703 | 1810 | 942 |
| | *Garma* | 2204 | 2157 | 1833 | 2017 | 1641 | 1502 | 1521 |
| Chhattisgarh | | - | - | 903 | 2035 | 1163 | 1267 | 1840 |
| Goa | *Kharif* | 2540 | 2355 | 2369 | 2312 | 2369 | 3182 | 2852 |
| | *Rabi* | 2877 | 2610 | 2726 | 2724 | 2584 | 3038 | 2597 |
| Gujarat | *Kharif* | 1810 | 1717 | 1442 | 1672 | 1330 | 2054 | 1893 |
| | Summer | - | - | - | 2965 | 3104 | 3106 | 2993 |
| Haryana | | - | 2869 | 3037 | - | - | 3042 | 3218 |
| Himachal Pradesh | | 2116 | 1741 | 1837 | 1981 | 1283 | 1825 | 1643 |
| Jammu & Kashmir | | 2401 | - | 1796 | 1787 | 1862 | 2078 | 2172 |
| Jharkhand | *Aghani* | - | - | - | - | - | - | - |
| Karnataka | *Kharif* | 2869 | 2878 | 2910 | 2647 | 2503 | 2688 | 2856 |
| | *Rabi* | 2298 | 1905 | 2255 | 2017 | 2198 | 3120 | 3550 |
| | Summer | 3398 | 3089 | 3216 | 2832 | 2642 | 4107 | 4709 |
| Kerala | Autumn | 2052 | 2242 | 2058 | 2189 | 2165 | 2244 | 2365 |
| | Winter | 2214 | 2479 | 2275 | 2408 | 2351 | 1875 | 2365 |
| | Summer | 2588 | 2551 | 2840 | 2424 | 2748 | 2287 | 2525 |
| Madhya Pradesh | | 1818 | 1855 | 1304 | 1908 | 1221 | 1872 | 1087 |
| Maharashtra | *Kharif* | 1677 | 1688 | 1274 | 1792 | 869 | 1957 | 1434 |
| | Summer | - | - | - | - | - | - | - |
| | Autumn | - | - | - | - | - | - | - |
| | Winter | - | - | - | - | - | - | - |
| Orissa | Autumn | 786 | 1189 | 624 | 1212 | 423 | 1202 | 949 |
| | Winter | 1634 | 1243 | 1208 | 1924 | 844 | 1795 | 1668 |
| | Summer | 1908 | 2413 | 2157 | 2148 | 2125 | 2133 | 2230 |
| Punjab | | 3157 | 3185 | 3246 | 3530 | 3501 | 3681 | 3942 |
| Rajasthan | | 1524 | 1742 | 1244 | 1347 | 947 | 1373 | 1551 |
| Tamil Nadu# | K/K/S | 4088 | 4158 | 4104 | 3999 | 3495 | 3347 | 3542 |
| | S/T/P | 3539 | 3487 | 3534 | 3148 | 2845 | 2749 | 3033 |
| | N/K | 1724 | 3704 | 3544 | 3556 | 2988 | 2969 | 3100 |
| Uttar Pradesh | | - | - | - | - | - | - | - |
| West Bengal | Aus. | 2074 | 2075 | 3218 | 2233 | 2184 | 2194 | 2242 |
| | *Aman* | 2167 | 2155 | 3193 | 2500 | 2431 | 2422 | 2464 |
| | Summer | 2716 | 2923 | 5155 | 2860 | 2940 | 3076 | 3089 |
| Dadra & Nagar Haveli | | 2378 | 2125 | 1837 | 2419 | 3017 | 2058 | 2275 |
| Delhi | | 2112 | 2828 | 2830 | 2825 | 2117 | 2635 | 2956 |
| Daman & Diu | | 1025 | 947 | 1876 | 1477 | 1910 | 1957 | 1561 |
| Pondicherry* | I | 2304 | 2477 | 2677 | 2555 | 2757 | 2759 | 2720 |
| | II | 2123 | 2325 | 2275 | 2517 | 1979 | 2609 | 2147 |
| | III | 2603 | 2400 | 2525 | 2573 | 2789 | 2714 | 2731 |

# K/K/S- Kar/Kuruvai/Sornaveri (Early *Kharif*), S/T/P- Samba/Thalady/Pishanam (Late *Kharif*), N/K- Navarai/Kodai (*Rabi*)* I- *Kharif*, II- *Rabi*-I, III- *Rabi*-II (Summer). Blank (-) Not available with State Agricultural Statistics Authority (SASA)

*Source*: Consolidated results of Crop Estimation Survey on Principal Crops (from 1998-99 to 2004-05 individual years) - Published by NSSO, Ministry of Statistics and Programme Implementation.

## Table 4.1.12

*Yield of High Yielding Variety of Jowar in Major States*

(*Kgs. Per hectare*)

| *State* | *Season* | *1998-1999* | *1999-2000* | *2000-2001* | *2001-2002* | *2002-2003* | *2003-2004* | *2004-2005* |
|---|---|---|---|---|---|---|---|---|
| *1* | 2 | 3 | 4 | 5 | 6 | 7 | 8 | 9 |
| Andhra Pradesh | *Kharif* | - | - | - | - | - | - | - |
| | *Rabi* | - | - | - | - | - | - | - |
| Chhattisgarh | | - | - | - | - | - | - | - |
| Gujarat | *Kharif* | 1379 | 1500 | 1048 | 1347 | 1125 | 1407 | 1379 |
| | Summer | 815 | 966 | 462 | 708 | 821 | 798 | 976 |
| Karnataka | *Kharif* | 1849 | 1653 | 1805 | 1521 | 1417 | 1234 | 1703 |
| | *Rabi* | 753 | 847 | 901 | 831 | 730 | 552 | 711 |
| | Summer | 2792 | 1198 | 2137 | 2065 | - | - | - |
| Madhya Pradesh | | - | - | - | - | 1387 | 1161 | 1066 |
| Maharashtra | *Kharif* | 1443 | - | - | 1177 | 1434 | 1335 | 1173 |
| | *Rabi* | - | - | - | - | - | - | - |
| Rajasthan | | 338 | 431 | 876 | 759 | 250 | 745 | 523 |
| Tamil Nadu | | 1401 | 1246 | 1248 | 1250 | 1075 | 897 | 935 |
| Uttar Pradesh | | - | - | - | - | - | - | - |

Blank (-) Not available with State Agricultural Statistics Authority (SASA)

*Source*: Consolidated results of Crop Estimation Survey on Principal Crops (from 1998-99 to 2004-05 individual years) - Published by NSSO, Ministry of Statistics and Programme Implementation.

# Table 4.1.13

*Yield of High Yielding Variety of Bajra in Major States*

(Kgs. Per hectare)

| State | Season | 1998-1999 | 1999-2000 | 2000-2001 | 2001-2002 | 2002-2003 | 2003-2004 | 2004-2005 |
|---|---|---|---|---|---|---|---|---|
| 1 | 2 | 3 | 4 | 5 | 6 | 7 | 8 | 9 |
| Andhra Pradesh | | - | - | - | - | - | - | - |
| Gujarat | *Kharif* | 1192 | 634 | 666 | 1297 | 874 | - | 1043 |
| | Summer | 2308 | 2130 | 1936 | 1798 | 1648 | 2084 | 1998 |
| Haryana | | - | - | - | - | - | 1701 | 1417 |
| Karnataka | *Kharif* | 769 | 639 | 809 | 574 | 416 | 678 | 637 |
| Madhya Pradesh | | - | - | - | - | 964 | 1638 | 1523 |
| Maharashtra | | 854 | 657 | 607 | 614 | 778 | 695 | 750 |
| Punjab | | 1537 | 843 | 1034 | - | - | - | - |
| Rajasthan | | 711 | 570 | 768 | 1060 | 288 | 1416 | 858 |
| Tamil Nadu | | 2102 | 2170 | 1921 | 1900 | 1406 | 1338 | 1620 |
| Uttar Pradesh | | - | - | - | - | - | - | - |
| Delhi | | - | 1379 | 1486 | 1953 | 1807 | 2473 | 1628 |
| Daman & Diu | | - | - | - | - | - | - | - |

Blank (-) Not available with State Agricultural Statistics Authority (SASA)

*Source:* Consolidated results of Crop Estimation Survey on Principal Crops (from 1998-99 to 2004-05 individual years) - Published by NSSO, Ministry of Statistics and Programme Implementation.

# Table 4.1.14

## *Yield of High Yielding Variety of Maize in Major States*

(Kgs. Per hectare)

| State | Season | 1998-1999 | 1999-2000 | 2000-2001 | 2001-2002 | 2002-2003 | 2003-2004 | 2004-2005 |
|---|---|---|---|---|---|---|---|---|
| 1 | 2 | 3 | 4 | 5 | 6 | 7 | 8 | 9 |
| Andhra Pradesh | *Kharif* | - | - | - | - | - | - | - |
| | *Rabi* | - | - | - | - | - | - | - |
| Bihar | *Bhadai* | 1156 | 1822 | 1836 | 1820 | 1907 | 1695 | 1909 |
| | *Rabi* | - | - | - | - | 2610 | - | - |
| | *Garma* | 2521 | 2644 | 3151 | 2804 | - | 2709 | 2928 |
| Chhattisgarh | | - | - | - | - | - | - | - |
| Gujarat | | 1602 | 1276 | 774 | 1978 | 1663 | 1852 | 913 |
| Haryana | | - | - | - | - | - | 2409 | 2695 |
| Himachal Pradesh | | 3238 | 3028 | 2818 | 3154 | 1946 | 3292 | 2763 |
| Jammu & Kashmir | | 2136 | - | 1692 | 1745 | 1982 | 2020 | 1837 |
| Karnataka | *Kharif* | 3628 | 2782 | 3453 | 2665 | 2125 | 2059 | 3109 |
| | *Rabi* | 2159 | 2205 | 2652 | 2530 | 2694 | 2425 | 3201 |
| | Summer | - | 543 | 2960 | 2769 | - | - | 3220 |
| Madhya Pradesh | | - | - | - | - | - | 2330 | 1703 |
| Maharashtra | *Kharif* | - | - | - | - | - | - | - |
| | *Rabi* | - | - | - | - | - | - | - |
| Orissa | Autumn | 1746 | 2025 | 1257 | 1047 | 1005 | 1892 | 2178 |
| | Winter | 689 | 661 | 967 | 1753 | 1179 | 1986 | 1979 |
| | Summer | 894 | 1342 | 1358 | 1688 | 1635 | 1999 | 2066 |
| Punjab | | 2446 | 3008 | 3277 | 3204 | 2467 | 3385 | 3291 |
| Rajasthan | *Kharif* | 1091 | 1177 | 1079 | 1395 | 1174 | 2143 | 1099 |
| | *Rabi* | - | - | - | - | - | - | - |
| Uttar Pradesh | | - | - | - | - | - | - | - |
| West Bengal | | - | - | - | - | - | - | 3039 |

Blank (-) Not available with State Agricultural Statistics Authority (SASA)

*Source*: Consolidated results of Crop Estimation Survey on Principal Crops (from 1998-99 to 2004-05 individual years) Published by NSSO, Ministry of Statistics and Programme Implementation.

## Table 4.1.15

*Yield of High Yielding Variety of Wheat in Major States*

*(Kgs. Per hectare)*

| State | Season | 1998-1999 | 1999-2000 | 2000-2001 | 2001-2002 | 2002-2003 | 2003-2004 | 2004-2005 |
|---|---|---|---|---|---|---|---|---|
| 1 | 2 | 3 | 4 | 5 | 6 | 7 | 8 | 9 |
| Assam | | 910 | 1444 | 1229 | 1219 | 1199 | 1049 | 965 |
| Bihar | | 2208 | 2316 | 2265 | 2169 | 1979 | 1860 | 1706 |
| Chhattisgarh | | - | - | 1500 | 1373 | 1266 | 1226 | 1092 |
| Gujarat | | 2571 | 2219 | 2258 | 2449 | 1974 | 2712 | 2560 |
| Haryana | | - | - | - | - | - | 3975 | 3921 |
| Himachal Pradesh | | 1525 | 1855 | 882 | 2049 | 1556 | 1563 | 2098 |
| Jammu & Kashmir | | 1631 | | 584 | 1276 | 1904 | 2107 | 2175 |
| Jharkhand | | - | 1621 | 1341 | - | - | - | - |
| Karnataka | | 1249 | 1185 | - | 1030 | 918 | 723 | 1575 |
| Madhya Pradesh | | 2286 | 2376 | 1973 | 2096 | 1827 | 2165 | 2146 |
| Maharashtra | | 1337 | 1445 | 1287 | 1457 | 1335 | 1201 | 1366 |
| Orissa | Summer | 1159 | 1313 | 1490 | 1515 | 1263 | 1471 | 1311 |
| Punjab | | 4348 | 4711 | 4548 | 4523 | 4200 | 4203 | 4222 |
| Rajasthan | | 2775 | 2855 | 2596 | 2894 | 2681 | 2848 | 2846 |
| Uttar Pradesh | | - | - | - | - | - | - | - |
| West Bengal | | 1994 | 2274 | 3868 | 2186 | 2011 | 2261 | 2067 |
| Delhi | | - | 3643 | 3518 | 3510 | 4014 | 4052 | 3745 |

Blank (-) Not available with State Agricultural Statistics Authority (SASA)

*Source:* Consolidated results of Crop Estimation Survey on Principal Crops (from 1998-99 to 2004-05 individual years) Published by NSSO, Ministry of Statistics and Programme Implementation.

# Table 5.1(a)

*All-India Area, Production and Yield of Potato from 1950-51 to 2006-07*

*Area- Million Hectares*
*Production- Million Tonnes*
*Yield- Kg./Hectare*

| *Year* | *Area* | *Production* | *Yield* |
|---|---|---|---|
| *1* | 2 | 3 | 4 |
| 1950-51 | 0.24 | 1.66 | 6917 |
| 1951-52 | 0.25 | 1.71 | 6848 |
| 1952-53 | 0.26 | 1.99 | 7812 |
| 1953-54 | 0.26 | 1.96 | 7611 |
| 1954-55 | 0.27 | 1.76 | 6632 |
| 1955-56 | 0.28 | 1.86 | 6639 |
| 1956-57 | 0.29 | 1.72 | 6028 |
| 1957-58 | 0.32 | 2.00 | 6243 |
| 1958-59 | 0.34 | 2.35 | 6947 |
| 1959-60 | 0.36 | 2.73 | 7550 |
| 1960-61 | 0.38 | 2.72 | 7251 |
| 1961-62 | 0.37 | 2.45 | 6704 |
| 1962-63 | 0.41 | 3.37 | 8152 |
| 1963-64 | 0.42 | 2.59 | 6249 |
| 1964-65 | 0.43 | 3.61 | 8395 |
| 1965-66 | 0.48 | 4.08 | 8513 |
| 1966-67 | 0.47 | 3.52 | 7440 |
| 1967-68 | 0.50 | 4.23 | 8441 |
| 1968-69 | 0.52 | 4.73 | 9011 |
| 1969-70 | 0.50 | 3.91 | 7888 |
| 1970-71 | 0.48 | 4.81 | 9976 |
| 1971-72 | 0.49 | 4.83 | 9810 |
| 1972-73 | 0.51 | 4.45 | 8821 |
| 1973-74 | 0.54 | 4.86 | 8946 |
| 1974-75 | 0.59 | 6.23 | 10598 |
| 1975-76 | 0.62 | 7.31 | 11738 |
| 1976-77 | 0.62 | 7.17 | 11573 |
| 1977-78 | 0.67 | 8.14 | 12228 |
| 1978-79 | 0.81 | 10.13 | 12555 |
| 1979-80 | 0.69 | 8.33 | 12152 |

*contd. ...*

*...contd. ...*

| *1* | *2* | *3* | *4* |
|---|---|---|---|
| 1980-81 | 0.73 | 9.67 | 13258 |
| 1981-82 | 0.76 | 9.91 | 12996 |
| 1982-83 | 0.74 | 9.96 | 13549 |
| 1983-84 | 0.79 | 12.15 | 15299 |
| 1984-85 | 0.85 | 12.57 | 14806 |
| 1985-86 | 0.84 | 10.42 | 12364 |
| 1986-87 | 0.83 | 12.74 | 15322 |
| 1987-88 | 0.89 | 14.05 | 15869 |
| 1988-89 | 0.93 | 14.86 | 15929 |
| 1989-90 | 0.94 | 14.77 | 15714 |
| 1990-91 | 0.94 | 15.21 | 16254 |
| 1991-92 | 1.03 | 16.39 | 15904 |
| 1992-93 | 1.05 | 15.23 | 14458 |
| 1993-94 | 1.05 | 17.39 | 16610 |
| 1994-95 | 1.07 | 17.40 | 16262 |
| 1995-96 | 1.11 | 18.84 | 16973 |
| 1996-97 | 1.25 | 24.22 | 19376 |
| 1997-98 | 1.21 | 17.65 | 14587 |
| 1998-99 | 1.32 | 23.61 | 17886 |
| 1999-00 | 1.34 | 24.71 | 18443 |
| 2000-01 | 1.22 | 22.49 | 18404 |
| 2001-02 | 1.21 | 23.92 | 19806 |
| 2002-03 | 1.35 | 23.27 | 17300 |
| 2003-04 | 1.29 | 23.06 | 17887 |
| 2004-05 | 1.32 | 23.63 | 17923 |
| 2005-06 | 1.40 | 23.91 | 17058 |
| 2006-07 | 1.48 | 22.09 | 14903 |

# Table 5.1 (b)

## *Area, Production and Yield of Potato during 2005-06 and 2006-07 in Respect of Major Potato Producing States*

*Area- '000 Hectares*
*Production- '000 Tonnes*
*Yield- Kg./ Hectare*

| State | 2006-07 | | | | | | 2005-06 | | | | | |
|---|---|---|---|---|---|---|---|---|---|---|---|---|
| | Area | % to All-India | Production | % to All-India | Cumulative % | Yield | Area | % to All-India | Production | % to All-India | Cumulative % | Yield |
| 1 | 2 | 3 | 4 | 5 | 6 | 7 | 8 | 9 | 10 | 11 | 12 | 13 |
| Uttar Pradesh | 469.0 | 31.64 | 10248.0 | 46.39 | 46.39 | 21851 | 445.2 | 31.77 | 9986.8 | 41.78 | 41.78 | 22432 |
| West Bengal | 407.9 | 27.52 | 5052.0 | 22.87 | 69.26 | 12385 | 354.5 | 25.30 | 7462.5 | 31.22 | 72.99 | 21051 |
| Punjab | 75.0 | 5.06 | 1223.0 | 5.54 | 74.80 | 16307 | 75.0 | 5.35 | 1223.0 | 5.12 | 78.11 | 16307 |
| Gujarat | 44.0 | 2.97 | 1210.0 | 5.48 | 80.27 | 27500 | 44.0 | 3.14 | 1210.0 | 5.06 | 83.17 | 27500 |
| Bihar | 151.4 | 10.21 | 1178.1 | 5.33 | 85.61 | 7781 | 142.3 | 10.15 | 1232.7 | 5.16 | 88.33 | 8663 |
| Madhya Pradesh | 46.6 | 3.14 | 683.3 | 3.09 | 88.70 | 14663 | 42.8 | 3.05 | 562.4 | 2.35 | 90.68 | 13140 |
| Karnataka | 63.0 | 4.25 | 546.0 | 2.47 | 91.17 | 8667 | 73.0 | 5.21 | 325.0 | 1.36 | 92.04 | 4452 |
| Assam | 78.0 | 5.26 | 505.0 | 2.29 | 93.46 | 6474 | 69.6 | 4.97 | 353.7 | 1.48 | 93.52 | 5082 |
| Haryana | 21.5 | 1.45 | 341.6 | 1.55 | 95.00 | 15888 | 19.0 | 1.36 | 374.4 | 1.57 | 95.09 | 19705 |
| Meghalaya | 18.2 | 1.23 | 181.0 | 0.82 | 95.82 | 9945 | 18.2 | 1.30 | 148.7 | 0.62 | 95.71 | 8170 |
| Himachal Pradesh | 12.7 | 0.86 | 118.3 | 0.54 | 96.36 | 9315 | 14.0 | 1.00 | 95.9 | 0.40 | 96.11 | 6850 |
| Tamil Nadu | 5.2 | 0.35 | 81.2 | 0.37 | 96.73 | 15615 | 5.0 | 0.36 | 74.6 | 0.31 | 96.42 | 14920 |
| Tripura | 5.4 | 0.36 | 80.9 | 0.37 | 97.09 | 14981 | 5.3 | 0.38 | 93.5 | 0.39 | 96.81 | 17642 |
| Orissa | 8.2 | 0.55 | 79.3 | 0.36 | 97.45 | 9671 | 8.3 | 0.59 | 75.1 | 0.31 | 97.13 | 9048 |
| Others | 76.2 | 5.14 | 563.0 | 2.55 | 100.00 | @ | 85.2 | 6.08 | 687.0 | 2.87 | 100.00 | @ |
| All India | 1482.3 | 100.00 | 22090.7 | 100.00 | | 14903 | 1401.4 | 100.00 | 23905.3 | 100.00 | | 17058 |

@ - Since Area/Production is low in individual states, yield rate is not worked out.

*Note*: States have been arranged in descending order of percentage share of production during 2006-07.

## Table 5.2(a)

*All-India Area, Production and Yield of Onion from 1978-79 to 2006-07*

*Area- Million Hectares*
*Production- Million Tonnes*
*Yield- Kg./Hectare*

| Year | Area | Production | Yield |
|---|---|---|---|
| 1 | 2 | 3 | 4 |
| 1978-79 | 0.21 | 2.20 | 10403 |
| 1979-80 | 0.24 | 2.50 | 10232 |
| 1980-81 | 0.25 | 2.50 | 9961 |
| 1981-82 | 0.25 | 2.65 | 10562 |
| 1982-83 | 0.24 | 2.43 | 10330 |
| 1983-84 | 0.27 | 2.70 | 9982 |
| 1984-85 | 0.28 | 3.10 | 11139 |
| 1985-86 | 0.28 | 2.86 | 10202 |
| 1986-87 | 0.26 | 2.53 | 9659 |
| 1987-88 | 0.27 | 2.70 | 9857 |
| 1988-89 | 0.32 | 3.35 | 10620 |
| 1989-90 | 0.30 | 3.07 | 10176 |
| 1990-91 | 0.30 | 3.23 | 10686 |
| 1991-92 | 0.32 | 3.58 | 11088 |
| 1992-93 | 0.32 | 3.49 | 10791 |
| 1993-94 | 0.37 | 4.01 | 10902 |
| 1994-95 | 0.38 | 4.04 | 10661 |
| 1995-96 | 0.40 | 4.08 | 10316 |
| 1996-97 | 0.40 | 4.18 | 10348 |
| 1997-98 | 0.40 | 3.62 | 9091 |
| 1998-99 | 0.47 | 5.33 | 11391 |
| 1999-00 | 0.49 | 4.90 | 9932 |
| 2000-01 | 0.42 | 4.55 | 10786 |
| 2001-02 | 0.45 | 4.83 | 10686 |
| 2002-03 | 0.42 | 4.21 | 9912 |
| 2003-04 | 0.50 | 5.92 | 11784 |
| 2004-05 | 0.55 | 6.43 | 11718 |
| 2005-06 | 0.66 | 8.68 | 13118 |
| 2006-07 | 0.58 | 7.16 | 12357 |

# Table 5.2(b)

## *Area, Production and Yield of Onion during 2005-06 and 2006-07 in Respect of Major Onion Producing States*

*Area- '000 Hectares*
*Production- '000 Tonnes*
*Yield- Kg./Hectare*

| State | 2006-07 | | | | | | 2005-06 | | | | | |
|---|---|---|---|---|---|---|---|---|---|---|---|---|
| | Area | % to All-India | Production | % to All-India | Cumulative % | Yield | Area | % to All-India | Production | % to All-India | Cumulative % | Yield |
| 1 | 2 | 3 | 4 | 5 | 6 | 7 | 8 | 9 | 10 | 11 | 12 | 13 |
| Gujarat | 76.0 | 13.1 | 2128.0 | 29.73 | 45.26 | 28000 | 76.0 | 11.48 | 2128.0 | 24.51 | 24.51 | 13949 |
| Maharashtra | 109.0 | 18.8 | 1112.0 | 15.53 | 15.53 | 10202 | 177.0 | 26.74 | 2469.0 | 28.44 | 52.94 | 28000 |
| Karnataka | 151.0 | 26.1 | 845.0 | 11.80 | 57.07 | 5596 | 151.0 | 22.81 | 870.0 | 10.02 | 62.97 | 5762 |
| Andhra Pradesh | 31.0 | 5.4 | 579.0 | 8.09 | 65.15 | 18677 | 36.0 | 5.44 | 696.0 | 8.02 | 70.98 | 19333 |
| Madhya Pradesh | 36.2 | 6.2 | 617.4 | 8.62 | 73.78 | 17055 | 37.8 | 5.71 | 572.0 | 6.59 | 77.57 | 15132 |
| Rajasthan | 38.6 | 6.7 | 370.3 | 5.17 | 78.95 | 9593 | 44.5 | 6.72 | 427.2 | 4.92 | 82.49 | 9600 |
| Haryana | 16.4 | 2.8 | 314.9 | 4.40 | 87.50 | 19201 | 15.5 | 2.34 | 317.7 | 3.66 | 86.15 | 12871 |
| Uttar Pradesh | 21.0 | 3.6 | 297.0 | 4.15 | 83.10 | 14143 | 25.5 | 3.85 | 328.2 | 3.78 | 89.93 | 20497 |
| Orissa | 28.5 | 4.9 | 260.0 | 3.63 | 91.13 | 9123 | 26.8 | 4.05 | 243.9 | 2.81 | 92.74 | 9101 |
| Tamil Nadu | 29.6 | 5.1 | 258.3 | 3.61 | 94.74 | 8726 | 29.2 | 4.41 | 233.8 | 2.69 | 95.43 | 8007 |
| Others | 42.0 | 7.3 | 376.5 | 5.26 | 100.00 | @ | 42.6 | 6.44 | 396.8 | 4.57 | 100.00 | @ |
| All India | 579.3 | 100.0 | 7158.4 | 100.00 | | 12357 | 661.9 | 100.00 | 8682.6 | 100.00 | | 13118 |

@ - Since Area/Production is low in individual states, yield rate is not worked out.

*Note*: States have been arranged in descending order of percentage share of production during 2006-07.

## Table 5.3(a)

*All-India Area, Production and Yield of Tobacco from 1950-51 to 2006-07 Alongwith Percentage Coverage under Irrigation*

*Area- Million Hectares*
*Production- Million Tonnes*
*Yield- Kg./Hectare*

| Year | Area | Production | Yield | % Coverage under Irrigation |
|---|---|---|---|---|
| 1 | 2 | 3 | 4 | 5 |
| 1950-51 | 0.36 | 0.26 | 731 | NA |
| 1951-52 | 0.29 | 0.21 | 723 | NA |
| 1952-53 | 0.36 | 0.25 | 675 | 9.7 |
| 1953-54 | 0.37 | 0.27 | 737 | 9.7 |
| 1954-55 | 0.35 | 0.26 | 737 | 10.8 |
| 1955-56 | 0.41 | 0.30 | 739 | 16.9 |
| 1956-57 | 0.42 | 0.31 | 728 | 20.4 |
| 1957-58 | 0.36 | 0.24 | 669 | 22.5 |
| 1958-59 | 0.38 | 0.32 | 836 | 21.0 |
| 1959-60 | 0.41 | 0.29 | 716 | 22.4 |
| 1960-61 | 0.40 | 0.31 | 766 | 21.9 |
| 1961-62 | 0.42 | 0.34 | 811 | 17.9 |
| 1962-63 | 0.41 | 0.34 | 842 | 18.2 |
| 1963-64 | 0.44 | 0.36 | 817 | 15.4 |
| 1964-65 | 0.41 | 0.36 | 876 | 17.0 |
| 1965-66 | 0.38 | 0.29 | 778 | 16.9 |
| 1966-67 | 0.42 | 0.35 | 834 | 17.1 |
| 1967-68 | 0.42 | 0.37 | 871 | 17.4 |
| 1968-69 | 0.44 | 0.36 | 821 | 20.4 |
| 1969-70 | 0.44 | 0.34 | 770 | 19.5 |
| 1970-71 | 0.45 | 0.36 | 810 | 23.7 |
| 1971-72 | 0.46 | 0.42 | 914 | 24.5 |
| 1972-73 | 0.45 | 0.37 | 837 | 26.3 |
| 1973-74 | 0.46 | 0.46 | 1001 | 24.2 |
| 1974-75 | 0.38 | 0.36 | 954 | 21.0 |
| 1975-76 | 0.37 | 0.35 | 950 | 29.8 |
| 1976-77 | 0.43 | 0.42 | 969 | 30.2 |
| 1977-78 | 0.50 | 0.49 | 979 | 23.3 |

*contd. ...*

*...contd. ...*

| *1* | 2 | 3 | 4 | 5 |
|---|---|---|---|---|
| 1978-79 | 0.41 | 0.45 | 1109 | 31.5 |
| 1979-80 | 0.43 | 0.44 | 1031 | 32.6 |
| 1980-81 | 0.45 | 0.48 | 1065 | 33.7 |
| 1981-82 | 0.44 | 0.52 | 1172 | 33.9 |
| 1982-83 | 0.50 | 0.58 | 1157 | 29.6 |
| 1983-84 | 0.44 | 0.49 | 1120 | 34.0 |
| 1984-85 | 0.44 | 0.49 | 1113 | 37.7 |
| 1985-86 | 0.40 | 0.44 | 1111 | 39.9 |
| 1986-87 | 0.39 | 0.46 | 1187 | 40.7 |
| 1987-88 | 0.32 | 0.37 | 1155 | 45.9 |
| 1988-89 | 0.38 | 0.49 | 1307 | 42.4 |
| 1989-90 | 0.41 | 0.55 | 1335 | 43.2 |
| 1990-91 | 0.41 | 0.56 | 1353 | 43.2 |
| 1991-92 | 0.43 | 0.58 | 1369 | 39.0 |
| 1992-93 | 0.42 | 0.60 | 1425 | 40.4 |
| 1993-94 | 0.38 | 0.56 | 1463 | 43.7 |
| 1994-95 | 0.38 | 0.57 | 1486 | 45.3 |
| 1995-96 | 0.39 | 0.54 | 1356 | 46.5 |
| 1996-97 | 0.43 | 0.62 | 1444 | 43.8 |
| 1997-98 | 0.46 | 0.64 | 1394 | 39.8 |
| 1998-99 | 0.51 | 0.74 | 1451 | 43.1 |
| 1999-00 | 0.43 | 0.52 | 1211 | 46.0 |
| 2000-01 | 0.26 | 0.34 | 1318 | 52.8 |
| 2001-02 | 0.35 | 0.55 | 1565 | 43.6 |
| 2002-03 | 0.33 | 0.49 | 1506 | 46.2 |
| 2003-04 | 0.37 | 0.55 | 1486 | 49.9 |
| 2004-05 | 0.37 | 0.55 | 1498 | 42.3 |
| 2005-06 | 0.37 | 0.55 | 1481 | NA |
| 2006-07 | 0.37 | 0.52 | 1409 | NA |

# Table 5.3(b)

*Area, Production and Yield of Tobacco during 2005-06 and 2006-07 in Respect of Major Tobacco Producing States Alongwith Coverage under Irrigation*

*Area- '000 Hectares*
*Production- '000 Tonnes*
*Yield- Kg./Hectare*

| State | 2006-07 | | | | | | 2005-06 | | | | | | % Coverage under Irrigation during 2004-05* |
|---|---|---|---|---|---|---|---|---|---|---|---|---|---|
| | Area | % to All-India | Production | % to All-India | Cumulative % | Yield | Area | % to All-India | Production | % to All-India | Cumulative % | Yield | |
| 1 | 2 | 3 | 4 | 5 | 6 | 7 | 8 | 9 | 10 | 11 | 12 | 13 | 14 |
| Andhra Pradesh | 127.0 | 34.46 | 182.0 | 35.04 | 35.04 | 1433 | 134.0 | 35.94 | 198.0 | 35.86 | 35.86 | 1478 | 24.5 |
| Uttar Pradesh | 23.0 | 6.24 | 125.0 | 24.07 | 59.11 | 5435 | 20.2 | 5.42 | 121.5 | 22.00 | 57.86 | 6015 | 100.0 |
| Gujarat | 71.3 | 19.35 | 113.9 | 21.93 | 81.04 | 1597 | 71.3 | 19.13 | 113.9 | 20.63 | 78.49 | 1597 | 89.0 |
| Karnataka | 103.0 | 27.95 | 46.0 | 8.86 | 89.89 | 947 | 101.0 | 27.09 | 62.0 | 11.23 | 89.71 | 614 | 3.3 |
| Bihar | 14.1 | 3.83 | 16.1 | 3.10 | 92.99 | 1142 | 14.9 | 4.00 | 17.3 | 3.13 | 92.85 | 1161 | 70.2 |
| Tamil Nadu | 5.0 | 1.36 | 7.7 | 1.48 | 94.47 | 1540 | 5.1 | 1.37 | 7.8 | 1.41 | 94.26 | 1529 | 98.8 |
| Maharashtra | 6.0 | 1.63 | 7.0 | 1.35 | 95.82 | 1167 | 6.0 | 1.61 | 7.0 | 1.27 | 95.53 | 1167 | 13.0 |
| Others | 19.1 | 5.18 | 5.7 | 4.18 | 100.00 | @ | 20.3 | 5.45 | 24.7 | 4.47 | 100.00 | @ | - |
| All India | 368.5 | 100.00 | 519.4 | 100.00 | | 1409 | 372.8 | 100.00 | 552.2 | 100.00 | | 1481 | 42.3 |

@ - Since Area/Production is low, yield rate is not worked out.

* - Provisional

*Note:* States have been arranged in descending order of percentage share of production during 2006-07.

## Table 5.4(a)

*All-India Area, Production and Yield of Coconut from 1950-51 to 2006-07*

*Area- Million Hectares*
*Production- '00 Million Nuts*
*Yield- Nuts/Hectare*

| Year | Area | Production | Yield |
|---|---|---|---|
| 1 | 2 | 3 | 4 |
| 1950-51 | 0.62 | 35.82 | 5759 |
| 1951-52 | 0.63 | 36.06 | 5724 |
| 1952-53 | 0.65 | 44.98 | 6909 |
| 1953-54 | 0.66 | 46.49 | 7012 |
| 1954-55 | 0.64 | 44.09 | 6878 |
| 1955-56 | 0.65 | 42.26 | 6532 |
| 1956-57 | 0.66 | 43.83 | 6671 |
| 1957-58 | 0.67 | 44.55 | 6689 |
| 1958-59 | 0.69 | 45.89 | 6651 |
| 1959-60 | 0.72 | 47.34 | 6621 |
| 1960-61 | 0.72 | 46.39 | 6470 |
| 1961-62 | 0.72 | 44.78 | 6194 |
| 1962-63 | 0.80 | 50.17 | 6288 |
| 1963-64 | 0.80 | 47.25 | 5920 |
| 1964-65 | 0.85 | 50.43 | 5950 |
| 1965-66 | 0.88 | 50.35 | 5698 |
| 1966-67 | 0.89 | 51.92 | 5814 |
| 1967-68 | 0.92 | 53.21 | 5760 |
| 1968-69 | 0.99 | 55.46 | 5613 |
| 1969-70 | 1.03 | 58.59 | 5670 |
| 1970-71 | 1.05 | 60.75 | 5811 |
| 1971-72 | 1.09 | 61.24 | 5626 |
| 1972-73 | 1.10 | 59.97 | 5456 |
| 1973-74 | 1.10 | 58.51 | 5309 |
| 1974-75 | 1.12 | 60.30 | 5401 |
| 1975-76 | 1.07 | 58.29 | 5449 |
| 1976-77 | 1.08 | 57.65 | 5366 |
| 1977-78 | 1.06 | 54.13 | 5123 |
| 1978-79 | 1.06 | 57.30 | 5431 |
| 1979-80 | 1.08 | 56.36 | 5239 |

*contd. ...*

*...contd. ...*

| 1 | 2 | 3 | 4 |
|---|---|---|---|
| 1980-81 | 1.08 | 57.20 | 5280 |
| 1981-82 | 1.09 | 55.73 | 5109 |
| 1982-83 | 1.15 | 63.56 | 5531 |
| 1983-84 | 1.17 | 58.08 | 4983 |
| 1984-85 | 1.18 | 69.13 | 5842 |
| 1985-86 | 1.23 | 67.70 | 5524 |
| 1986-87 | 1.23 | 63.77 | 5179 |
| 1987-88 | 1.35 | 72.70 | 5401 |
| 1988-89 | 1.43 | 85.41 | 5992 |
| 1989-90 | 1.47 | 93.59 | 6357 |
| 1990-91 | 1.48 | 97.30 | 6595 |
| 1991-92 | 1.53 | 100.50 | 6593 |
| 1992-93 | 1.54 | 112.41 | 7310 |
| 1993-94 | 1.64 | 119.75 | 7324 |
| 1994-95 | 1.71 | 133.00 | 7760 |
| 1995-96 | 1.83 | 129.52 | 7066 |
| 1996-97 | 1.89 | 130.61 | 6908 |
| 1997-98 | 1.86 | 127.17 | 6834 |
| 1998-99 | 1.75 | 125.36 | 7145 |
| 1999-00 | 1.77 | 121.29 | 6860 |
| 2000-01 | 1.84 | 125.97 | 6847 |
| 2001-02 | 1.93 | 129.63 | 6709 |
| 2002-03 | 1.92 | 125.35 | 6337 |
| 2003-04 | 1.93 | 121.78 | 6310 |
| 2004-05 | 1.94 | 128.33 | 6615 |
| 2005-06 | 1.95 | 148.11 | 7608 |
| 2006-07 | 1.94 | 158.40 | 8165 |

# Table 5.4(b)

*Area, Production and Yield of Coconut during 2005-06 and 2006-07 in Respect of Major Coconut Producing States*

*Area- 'Million Hectares*
*Production- '00 Million Nuts*
*Yield- Nuts/Hectare*

| State | 2006-07 | | | | | | 2005-06 | | | | | |
|---|---|---|---|---|---|---|---|---|---|---|---|---|
| | Area | % to All-India | Production | % to All-India | Cumulative % | Yield | Area | % to All-India | Production | % to All-India | Cumulative % | Yield |
| 1 | 2 | 3 | 4 | 5 | 6 | 7 | 8 | 9 | 10 | 11 | 12 | 13 |
| Kerala | 0.87 | 44.85 | 6.05 | 38.19 | 38.19 | 6951 | 0.90 | 46.15 | 6.33 | 42.74 | 42.74 | 7046 |
| Tamil Nadu | 0.37 | 19.07 | 5.43 | 34.28 | 72.47 | 14495 | 0.37 | 18.97 | 4.87 | 32.88 | 75.62 | 13137 |
| Karnataka | 0.40 | 20.62 | 1.63 | 10.29 | 82.77 | 4052 | 0.39 | 20.00 | 1.21 | 8.17 | 83.79 | 3139 |
| Andhra Pradesh | 0.10 | 5.15 | 1.33 | 8.40 | 91.16 | 12629 | 0.10 | 5.13 | 0.89 | 6.01 | 89.80 | 8577 |
| West Bengal | 0.02 | 1.03 | 0.36 | 2.27 | 93.43 | 14307 | 0.02 | 1.03 | 0.32 | 2.16 | 91.96 | 12992 |
| Orissa | 0.05 | 2.58 | 0.28 | 1.77 | 95.20 | 5408 | 0.05 | 2.56 | 0.27 | 1.82 | 93.79 | 5406 |
| Maharashtra | 0.02 | 1.03 | 0.18 | 1.14 | 96.34 | 8338 | 0.02 | 1.03 | 0.27 | 1.82 | 95.61 | 15189 |
| Assam | 0.02 | 1.03 | 0.15 | 0.95 | 97.29 | 8053 | 0.02 | 1.03 | 0.20 | 1.35 | 96.96 | 10728 |
| Goa | 0.03 | 1.55 | 0.13 | 0.82 | 98.11 | 4969 | 0.03 | 1.54 | 0.13 | 0.88 | 97.84 | 4953 |
| Others | 0.06 | 3.09 | 0.30 | 1.89 | 100.00 | @ | 0.05 | 2.56 | 0.32 | 2.16 | 100.00 | @ |
| All India | 1.94 | 100.00 | 15.84 | 100.00 | | 8165 | 1.95 | 100.00 | 14.81 | 100.00 | | 7608 |

@ - Since Area/ Production is low, yield rate is not worked out.

*Note:* States have been arranged in descending order of percentage share of production during 2006-07.

## Table 5.5(a)

*All-India Area, Production and Yield of Cashewnut from 1990-91 to 2006-07*

*Area- Thousand Hectares*
*Production- Thousand Tonnes*
*Yield- Kg./Hectare*

| *Year* | *Area* | *Productive Area* | *Production* | *Yield* |
|---|---|---|---|---|
| *1* | 2 | 3 | 4 | 5 |
| 1990-91 | 532 | 464 | 295 | 636 |
| 1991-92 | 534 | 482 | 305 | 633 |
| 1992-93 | 560 | 508 | 349 | 687 |
| 1993-94 | 565 | 517 | 348 | 673 |
| 1994-95 | 577 | 510 | 322 | 631 |
| 1995-96 | 635 | 485 | 418 | 862 |
| 1996-97 | 659 | 515 | 430 | 835 |
| 1997-98 | 675 | 512 | 360 | 703 |
| 1998-99 | 706 | 573 | 460 | 803 |
| 1999-00 | 686 | 601 | 520 | 865 |
| 2000-01 | 720 | 627 | 450 | 710 |
| 2001-02 | 750 | 666 | 470 | 710 |
| 2002-03 | 770 | 675 | 500 | 760 |
| 2003-04 | 780 | 684 | 535 | 800 |
| 2004-05 | 820 | 700 | 544 | 810 |
| 2005-06 | 837 | 760 | 573 | 815 |
| 2006-07 | 854 | 765 | 620 | 820 |

*Source*: Directorate of Cashewnut & Cocoa Development, Cochin, Kerala.

## Table 5.5(b)

*Area, Production and Yield of Cashewnut in India for the year 2005-06 and 2006-07*

*Area- Thousand Hectares*
*Production- Thousand Tonnes*
*Yield- Kg./Hectare*

| State | 2006-07 | | | | | | 2005-06 | | | | | |
|---|---|---|---|---|---|---|---|---|---|---|---|---|
| | Area | % to All-India | Production | % to All-India | Cumulative % | Yield | Area | % to All-India | Production | % to All-India | Cumulative % | Yield |
| 1 | 2 | 3 | 4 | 5 | 6 | 7 | 8 | 9 | 10 | 11 | 12 | 13 |
| Maharashtra | 164.00 | 19.20 | 197.00 | 31.77 | 31.77 | 1500 | 160.00 | 19.10 | 183.00 | 31.90 | 31.90 | 1300 |
| Andhra Pradesh | 171.00 | 20.02 | 99.00 | 15.97 | 47.74 | 890 | 170.00 | 20.30 | 92.00 | 16.10 | 48.00 | 880 |
| Orissa | 125.00 | 14.64 | 84.00 | 13.54 | 61.28 | 860 | 120.00 | 14.30 | 78.00 | 13.60 | 61.60 | 860 |
| Kerala | 80.00 | 9.37 | 72.00 | 11.61 | 72.89 | 900 | 80.00 | 9.60 | 67.00 | 11.70 | 73.30 | 900 |
| Tamil Nadu | 123.00 | 14.41 | 60.00 | 9.68 | 82.57 | 670 | 121.00 | 14.50 | 56.00 | 9.80 | 83.10 | 640 |
| Karnataka | 102.00 | 11.94 | 52.00 | 8.40 | 90.97 | 700 | 100.00 | 11.90 | 45.00 | 7.90 | 91.00 | 700 |
| Goa | 55.00 | 6.44 | 29.00 | 4.68 | 95.65 | 690 | 55.00 | 6.60 | 27.00 | 4.70 | 95.70 | 690 |
| West Bengal | 10.00 | 1.17 | 10.00 | 1.61 | 97.26 | 1000 | 10.00 | 1.20 | 10.00 | 1.70 | 97.40 | 950 |
| Others | 24.00 | 2.81 | 17.00 | 2.74 | 100.00 | 700 | 21.00 | 2.50 | 15.00 | 2.60 | 100.00 | 650 |
| All India | 854.00 | 100.00 | 620.00 | 100.00 | | 820 | 837.00 | 100.00 | 573.00 | 100.00 | | 815 |

*Note*: States have been arranged in descending order of percentage share of production during 2006-07.

*Source*: The Directorate of Cashewnut and Cocoa Development, Cochin, Kerala.

# Table 5.6

## *Area and Producation of Horticulture Crops during 2005-06 to 2007-08*

*Area (In '000 Hectares)*
*Production (In '000 Tonnes)*

| *Crops* | *2005-06* | | *2006-07* | | *2007-08* | |
|---|---|---|---|---|---|---|
| | *Area* | *Production* | *Area* | *Production* | *Area* | *Production* |
| *1* | *2* | *3* | *4* | *5* | *6* | *7* |
| Fruits | | | | | | |
| Mango | 2081 | 12663 | 2143 | 13501 | | |
| Apple | 227 | 1814 | 252 | 1622 | | |
| Banana | 570 | 18888 | 600 | 20858 | | |
| Citrus | 742 | 6139 | 796 | 7097 | | |
| Guava | 167 | 1737 | 179 | 1856 | | |
| Grapes | 66 | 1650 | 64 | 1668 | | |
| *Litchi* | 63 | 392 | 65 | 404 | | |
| Papaya | 68 | 2139 | 72 | 2405 | | |
| Pineapple | 82 | 1263 | 86 | 1308 | | |
| Pomegranate | 113 | 820 | 118 | 853 | | |
| Sapota | 140 | 1114 | 146 | 1191 | | |
| Others | 1006 | 6737 | 986 | 6159 | | |
| Total | 5324 | 55356 | 5507 | 58922 | 5807 | 62858 |
| Vegetables | | | | | | |
| Potato | 1569 | 29175 | 1640 | 30195 | | |
| Onion | 704 | 9432 | 757 | 10655 | | |
| Tomato | 546 | 9820 | 594 | 9878 | | |
| Brinjal | 560 | 9365 | 564 | 9301 | | |
| Cabbage | 253 | 5637 | 254 | 5724 | | |
| Cauliflower | 289 | 5323 | 310 | 5524 | | |
| Okra | 392 | 3975 | 398 | 4099 | | |
| Peas | 286 | 2270 | 287 | 2370 | | |
| Tapioca | 245 | 7855 | 255 | 8226 | | |
| Sweet Potato | 123 | 1067 | 97 | 880 | | |
| Others | 2247 | 27481 | 2336 | 29179 | | |
| Total | 7213 | 111399 | 7492 | 116033 | 7728 | 122256 |
| Aromatic | 262 | 202 | 280 | 174 | 362 | 270 |
| Almond/Walnut | 130 | 149 | 132 | 150 | 132 | 177 |
| Flowers Loose | 130 | 656 | 146 | 778 | 154 | 805 |
| Flowers Cut* | | 29206 | | 36497 | | 37721 |
| Plantation Crops | | | | | | |
| Coconut | 2029 | 10190 | 1940 | 10894 | 1940 | 10669 |
| Cashewnut | 843 | 579 | 854 | 620 | 868 | 665 |
| Other | 411 | 493 | 427 | 569 | 431 | 570 |
| Total | 3283 | 11263 | 3221 | 12083 | 3239 | 11904 |
| Mushroom | | 35 | | 37 | | 37 |
| Honey | | 52 | | 51 | | 65 |
| Spices | 2366 | 3705 | 2462 | 3982 | 2663 | 4307 |
| Grand Total | 18707 | 182817 | 19241 | 192210 | 20085 | 202678 |

* Lakh Numbers

*Note*: Totals may not exactly match due to rounding of decimals.

# Table 5.7

## *State-wise Area under Fruits & Vegetables during 1990-91, 2004-05 and 2005-06 as Per Land Use Statistics*

*(Thousand Hectares)*

| State/ Uts. | 1990-91 | 2004-05(P) | 2005-06(P) |
|---|---|---|---|
| Andhra Pradesh | 527 | 896 | 921 |
| Arunachal Pradesh | 33 | 28 | 28 |
| Assam | 209 | 282 | 279 |
| Bihar | 442 | 414** | 413** |
| Chhatisgarh | * | 114 | 117 |
| Goa | 61 | 73 | 74 |
| Gujarat | 200 | 384 | 421 |
| Haryana | 62 | 60 | 71 |
| Himachal Pradesh | 73 | 102 | 102 |
| Jammu & Kashmir | 60 | 77 | 76 |
| Jharkhand | * | 101 | 101 |
| Karnataka | 272 | 549 | 587 |
| Kerala | 574 | 599 | 585 |
| Madhya Pradesh | 217 | 233** | 242** |
| Maharashtra | 508 | 1121 | 769 |
| Manipur | 24 | 31 | 34 |
| Meghalaya | 57 | 73 | 73 |
| Mizoram | 9 | 4 | 4 |
| Nagaland | 9 | 29 | 29 |
| Orissa | 996 | 975 | 975 |
| Punjab | 83 | 151 | 167 |
| Rajasthan | 75 | 122 | 126 |
| Sikkim | 7 | 8 | 8 |
| Tamil Nadu | 358 | 556 | 598 |
| Tripura | 84 | 11 | 11 |
| Uttaranchal | * | 49 | 51 |
| Uttar Pradesh | 867 | 973** | 986** |
| West Bengal | 871 | 1373 | 1455 |
| A.&N.Islands | 2 | 3 | 3 |
| D.& N.Haveli | 2 | 2 | 2 |
| Delhi | 6 | 2 | 2 |
| Lakshyadweep | 1 | - | - |
| Pondicherry | 1 | 2 | 2 |
| ALL INDIA | 6690 | 9396 | 9313 |

P: Provisional

* The figures for Chhattisgarh, Jharkhand and Uttaranchal for 1990-91 are not shown as they did not exist in 1990-91.

** The figures for Madhya Pradesh, Bihar and Uttar Pradesh for 2004-05 & 2005-06 exclude Chhattisgarh, Jharkhand and Uttaranchal respectively.

## Table 5.8(a)

### *All-India Area, Production and Yield of Arecanut, Banana and Black Pepper from 1997-98 to 2006-07*

*Area- '000 Hectares*
*Production- '000 Tonnes*
*Yield- Kg/Hectare*

| *Year* | *Area* | *Production* | *Yield* |
|---|---|---|---|
| Arecanut | | | |
| 1997-98 | 273.7 | 333.5 | 1218 |
| 1998-99 | 277.5 | 309.8 | 1116 |
| 1999-00 | 289.0 | 334.4 | 1157 |
| 2000-01 | 315.2 | 373.1 | 1184 |
| 2001-02 | 340.9 | 403.1 | 1182 |
| 2002-03 | 349.4 | 400.8 | 1147 |
| 2003-04 | 365.0 | 439.2 | 1203 |
| 2004-05 | 364.3 | 452.7 | 1243 |
| 2005-06 | 381.1 | 483.1 | 1268 |
| 2006-07 | 396.8 | 559.2 | 1409 |
| Banana | | | |
| 1997-98 | 446.7 | 12642.5 | 28302 |
| 1998-99 | 445.7 | 12425.4 | 27878 |
| 1999-00 | 464.6 | 11123.2 | 23941 |
| 2000-01 | 458.5 | 12898.1 | 28131 |
| 2001-02 | 432.4 | 12519.2 | 28953 |
| 2003-04 | 390.5 | 11388.0 | 29163 |
| 2004-05 | 404.2 | 11710.3 | 28972 |
| 2005-06 | 423.6 | 12104.5 | 28575 |
| 2006-07 | 462.5 | 15922.9 | 34428 |
| Black Pepper | | | |
| 1997-98 | 181.5 | 57.3 | 316 |
| 1998-99 | 239.8 | 75.7 | 316 |
| 1999-00 | 195.6 | 59.0 | 302 |
| 2000-01 | 213.9 | 63.7 | 298 |
| 2001-02 | 219.4 | 62.4 | 285 |
| 2002-03 | 216.0 | 64.0 | 296 |
| 2003-04 | 233.4 | 73.2 | 314 |
| 2004-05 | 228.3 | 73.0 | 320 |
| 2005-06 | 260.2 | 92.9 | 357 |
| 2006-07 | 246.0 | 69.0 | 280 |

## Table 5.8(b)

*All-India Area, Production and Yield of Cardamom, Chillies and Coriander from 1997-98 to 2006-07*

*Area- '000 Hectares*
*Production- '000 Tonnes*
*Yield- Kg/Hectare*

| *Year* | *Area* | *Production* | *Yield* |
|---|---|---|---|
| Cardamom | | | |
| 1997-98 | 94.3 | 11.7 | 124 |
| 1998-99 | 85.0 | 9.0 | 106 |
| 1999-00 | 84.8 | 10.7 | 126 |
| 2000-01 | 92.4 | 14.5 | 157 |
| 2001-02 | 91.6 | 15.2 | 166 |
| 2002-03 | 88.5 | 15.4 | 174 |
| 2003-04 | 95.6 | 16.7 | 175 |
| 2004-05 | 95.2 | 16.9 | 178 |
| 2005-06 | 95.8 | 17.8 | 186 |
| 2006-07 | 98.2 | 15.7 | 160 |
| Chillies | | | |
| 1997-98 | 840.6 | 870.1 | 1035 |
| 1998-99 | 891.2 | 1043.2 | 1171 |
| 1999-00 | 959.2 | 1052.8 | 1098 |
| 2000-01 | 836.5 | 983.7 | 1176 |
| 2001-02 | 880.0 | 1069.0 | 1215 |
| 2002-03 | 828.6 | 896.9 | 1082 |
| 2003-04 | 774.3 | 1235.7 | 1596 |
| 2004-05 | 737.5 | 1185.5 | 1607 |
| 2005-06 | 654.0 | 1014.6 | 1551 |
| 2006-07 | 758.0 | 1234.1 | 1628 |
| Coriander | | | |
| 1997-98 | 656.5 | 337.7 | 514 |
| 1998-99 | 525.8 | 292.4 | 556 |
| 1999-00 | 415.1 | 239.3 | 576 |
| 2000-01 | 353.6 | 231.3 | 654 |
| 2001-02 | 429.2 | 319.4 | 744 |
| 2002-03 | 243.3 | 164.9 | 678 |
| 2003-04 | 483.5 | 389.8 | 806 |
| 2004-05 | 348.9 | 248.0 | 711 |
| 2005-06 | 340.4 | 223.4 | 656 |
| 2006-07 | 320.8 | 233.2 | 727 |

## Table 5.8(c)

### *All-India Area, Production and Yield of Garlic, Ginger and Guarseed from 1997-98 to 2006-07*

*Area- '000 Hectares*
*Production- '000 Tonnes*
*Yield- Kg/Hectare*

| *Year* | *Area* | *Production* | *Yield* |
|---|---|---|---|
| Garlic | | | |
| 1997-98 | 108.8 | 484.4 | 4452 |
| 1998-99 | 123.2 | 570.7 | 4632 |
| 1999-00 | 124.6 | 524.6 | 4210 |
| 2000-01 | 69.2 | 247.1 | 3571 |
| 2001-02 | 92.2 | 385.3 | 4190 |
| 2002-03 | 98.2 | 411.4 | 4189 |
| 2003-04 | 138.9 | 691.1 | 4976 |
| 2004-05 | 144.1 | 646.6 | 4487 |
| 2005-06 | 134.9 | 598.2 | 4434 |
| 2006-07 | 149.7 | 707.4 | 4725 |
| Ginger | | | |
| 1997-98 | 75.6 | 252.1 | 3335 |
| 1998-99 | 77.6 | 263.2 | 3392 |
| 1999-00 | 80.8 | 282.6 | 3498 |
| 2000-01 | 86.2 | 288.0 | 3341 |
| 2001-02 | 90.8 | 318.0 | 3502 |
| 2003-04 | 85.1 | 301.9 | 3548 |
| 2004-05 | 95.3 | 359.0 | 3767 |
| 2005-06 | 110.6 | 391.2 | 3537 |
| 2006-07 | 105.9 | 370.3 | 3497 |
| Guarseed | | | |
| 1997-98 | 2301.2 | 962.7 | 418 |
| 1998-99 | 1922.1 | 488.5 | 254 |
| 1999-00 | 2933.9 | 375.1 | 128 |
| 2000-01 | 3497.4 | 658.8 | 188 |
| 2001-02 | 2903.1 | 1089.9 | 375 |
| 2002-03 | 974.1 | 202.6 | 208 |
| 2003-04 | 2854.0 | 1513.4 | 530 |
| 2004-05 | 2867.4 | 903.3 | 315 |
| 2005-06 | 2955.5 | 1059.0 | 358 |
| 2006-07 | 3343.7 | 1169.3 | 350 |

## Table 5.8(d)

*All-India Area, Production and Yield of Sunhemp, Sweet Potato and Tapioca from 1997-98 to 2006-07*

*Area- '000 Hectares*
*Production- '000 Tonnes*
*Yield- Kg/Hectare*

| *Year* | *Area* | *Production* | *Yield* |
|---|---|---|---|
| Sunhemp | | | |
| 1997-98 | 54.8 | 28.2 | 515 |
| 1998-99 | 48.4 | 22.4 | 463 |
| 1999-00 | 49.7 | 25.2 | 453 |
| 2000-01 | 55.8 | 24.5 | 439 |
| 2001-02 | 51.4 | 23.3 | 434 |
| 2002-03 | 45.9 | 18.2 | 397 |
| 2003-04 | 37.2 | 24.3 | 653 |
| 2004-05 | 35.6 | 25.1 | 705 |
| 2005-06 | 34.3 | 19.2 | 560 |
| 2006-07 | 34.4 | 19.6 | 570 |
| Sweet Potato | | | |
| 1997-98 | 128.8 | 1171.0 | 9092 |
| 1998-99 | 134.4 | 1152.4 | 8574 |
| 1999-2000 | 122.9 | 1110.6 | 9037 |
| 2000-01 | 101.3 | 838.3 | 8275 |
| 2001-02 | 96.7 | 834.6 | 8631 |
| 2002-03 | 104.6 | 954.7 | 9127 |
| 2003-04 | 106.4 | 937.8 | 8814 |
| 2004-05 | 107.2 | 932.5 | 8699 |
| 2005-06 | 106.8 | 955.1 | 8943 |
| 2006-07 | 106.6 | 963.7 | 9040 |
| Tapioca | | | |
| 1997-98 | 264.6 | 7007.8 | 26485 |
| 1998-99 | 243.4 | 5830.4 | 23954 |
| 1999-2000 | 223.5 | 6014.1 | 26909 |
| 2000-01 | 251.8 | 7123.8 | 28292 |
| 2001-02 | 247.6 | 6834.0 | 27601 |
| 2002-03 | 207.0 | 5426.2 | 26214 |
| 2003-04 | 219.0 | 5945.3 | 27147 |
| 2004-05 | 229.4 | 5854.8 | 25522 |
| 2005-06 | 242.4 | 7620.2 | 31436 |
| 2006-07 | 256.4 | 8429.0 | 32874 |

## Table 5.8(e)

*All-India Area, Production and Yield of Turmeric from 1997-98 to 2006-07*

*Area- '000 Hectares*
*Production- '000 Tonnes*
*Yield- Kg/Hectare*

| *Year* | *Area* | *Production* | *Yield* |
|---|---|---|---|
| *Turmeric* | | | |
| 1997-98 | 139.7 | 549.2 | 3931 |
| 1998-99 | 160.7 | 597.9 | 3721 |
| 1999-00 | 141.2 | 668.5 | 4734 |
| 2000-01 | 191.7 | 714.3 | 3726 |
| 2001-02 | 167.1 | 562.3 | 3368 |
| 2002-03 | 154.2 | 573.9 | 3722 |
| 2003-04 | 150.1 | 564.9 | 3763 |
| 2004-05 | 158.7 | 718.1 | 4525 |
| 2005-06 | 172.0 | 851.7 | 4952 |
| 2006-07 | 186.0 | 837.0 | 4500 |

# Table 6.1

*Index Numbers of Area, Production and Yield of Foodgrain and Non-Foodgrain Crops in India from 1949-50 to 2007-08 (Base: Triennium Ending 1981-82 = 100)*

*A- Index Numbers of Area*
*P- Index Numbers of Production*
*Y- Index Numbers of Yield*

| Year | Foodgrains (Weight 62.92) | | | Non-Foodgrains (Weight 37.08) | | | All Principal Crops | | |
|---|---|---|---|---|---|---|---|---|---|
| | A | P | Y | A | P | Y | A | P | Y |
| 1 | 2 | 3 | 4 | 5 | 6 | 7 | 8 | 9 | 10 |
| 1949-50 | 78.0 | 51.5 | 70.3 | 67.8 | 44.6 | 80.9 | 74.0 | 48.9 | 73.6 |
| 1950-51 | 76.4 | 46.5 | 64.2 | 66.6 | 45.8 | 75.1 | 74.1 | 46.2 | 67.8 |
| 1951-52 | 76.5 | 47.0 | 65.8 | 72.5 | 46.9 | 71.2 | 75.6 | 46.9 | 67.7 |
| 1952-53 | 80.5 | 51.7 | 70.1 | 69.4 | 44.1 | 70.1 | 77.9 | 48.9 | 70.1 |
| 1953-54 | 86.0 | 60.7 | 77.7 | 68.6 | 46.6 | 77.6 | 81.9 | 55.5 | 77.7 |
| 1954-55 | 85.0 | 59.0 | 75.7 | 76.5 | 53.6 | 79.4 | 83.0 | 57.0 | 77.0 |
| 1955-56 | 87.0 | 59.0 | 73.3 | 77.6 | 51.6 | 73.7 | 84.8 | 56.2 | 73.4 |
| 1956-57 | 87.5 | 61.7 | 75.2 | 80.1 | 56.5 | 77.5 | 85.8 | 59.8 | 76.0 |
| 1957-58 | 86.2 | 55.7 | 69.4 | 78.4 | 56.1 | 76.5 | 84.4 | 55.8 | 71.9 |
| 1958-59 | 90.3 | 66.7 | 79.6 | 80.8 | 60.8 | 82.0 | 88.1 | 64.6 | 80.4 |
| 1959-60 | 91.5 | 64.9 | 75.6 | 82.3 | 59.6 | 77.5 | 89.3 | 63.0 | 76.4 |
| 1960-61 | 90.9 | 69.6 | 81.9 | 83.8 | 67.4 | 84.0 | 89.2 | 68.8 | 82.7 |
| 1961-62 | 92.0 | 69.4 | 80.2 | 88.4 | 65.9 | 79.2 | 91.2 | 68.1 | 79.9 |
| 1962-63 | 92.9 | 67.3 | 76.6 | 88.9 | 67.5 | 81.5 | 92.0 | 67.4 | 78.3 |
| 1963-64 | 92.5 | 67.9 | 77.5 | 89.4 | 69.1 | 82.6 | 91.8 | 68.3 | 79.3 |
| 1964-65 | 91.6 | 75.4 | 87.8 | 92.2 | 78.2 | 89.2 | 91.7 | 76.4 | 88.3 |
| 1965-66 | 90.6 | 60.6 | 71.0 | 91.0 | 68.9 | 76.8 | 90.7 | 63.7 | 73.2 |
| 1966-67 | 90.8 | 60.8 | 71.2 | 90.0 | 66.6 | 77.5 | 90.6 | 63.0 | 73.6 |
| 1967-68 | 95.7 | 78.3 | 86.6 | 91.1 | 75.2 | 87.2 | 94.6 | 77.1 | 86.8 |
| 1968-69 | 94.8 | 76.3 | 84.2 | 87.1 | 73.7 | 84.3 | 93.0 | 75.3 | 84.2 |
| 1969-70 | 97.3 | 81.6 | 87.5 | 90.3 | 78.3 | 86.3 | 95.7 | 80.4 | 87.1 |
| 1970-71 | 97.9 | 87.9 | 93.2 | 91.1 | 82.6 | 91.4 | 96.3 | 85.9 | 92.6 |
| 1971-72 | 96.6 | 86.1 | 91.3 | 93.3 | 83.7 | 92.2 | 95.8 | 85.2 | 91.6 |
| 1972-73 | 93.9 | 79.1 | 85.8 | 87.9 | 76.6 | 86.3 | 92.5 | 78.2 | 86.0 |
| 1973-74 | 99.6 | 85.3 | 89.0 | 93.7 | 88.6 | 94.7 | 98.2 | 86.5 | 91.1 |
| 1974-75 | 95.3 | 81.0 | 87.0 | 93.2 | 89.2 | 94.0 | 94.8 | 84.0 | 89.6 |
| 1975-76 | 100.8 | 98.8 | 99.7 | 95.4 | 91.1 | 97.7 | 99.5 | 96.2 | 99.0 |
| 1976-77 | 97.8 | 89.6 | 92.4 | 94.1 | 88.8 | 94.3 | 96.9 | 89.3 | 93.1 |
| 1977-78 | 100.3 | 103.0 | 103.2 | 99.3 | 99.4 | 99.0 | 100.1 | 101.6 | 101.6 |
| 1978-79 | 101.5 | 107.0 | 105.7 | 102.0 | 101.6 | 101.8 | 101.6 | 105.0 | 104.2 |

contd. ...

*...contd. ...*

| *1* | *2* | *3* | *4* | *5* | *6* | *7* | *8* | *9* | *10* |
|---|---|---|---|---|---|---|---|---|---|
| 1979-80 | 98.5 | 87.5 | 88.7 | 96.6 | 90.8 | 93.9 | 98.1 | 88.7 | 90.6 |
| 1980-81 | 99.8 | 104.9 | 105.1 | 99.4 | 97.4 | 99.2 | 99.7 | 102.1 | 102.9 |
| 1981-82 | 101.7 | 107.6 | 105.9 | 104.1 | 111.8 | 106.4 | 102.2 | 109.2 | 106.1 |
| 1982-83 | 98.6 | 103.7 | 104.9 | 101.8 | 106.6 | 101.3 | 99.3 | 104.8 | 103.6 |
| 1983-84 | 103.4 | 122.8 | 117.8 | 104.4 | 111.5 | 105.7 | 103.6 | 118.6 | 113.3 |
| 1984-85 | 99.8 | 117.5 | 115.5 | 103.4 | 118.6 | 113.6 | 100.6 | 117.9 | 114.8 |
| 1985-86 | 100.9 | 123.4 | 120.6 | 104.2 | 113.0 | 108.3 | 101.6 | 119.5 | 116.0 |
| 1986-87 | 100.2 | 116.9 | 114.9 | 100.7 | 112.5 | 108.3 | 100.3 | 115.2 | 112.4 |
| 1987-88 | 94.3 | 113.5 | 117.2 | 101.7 | 118.3 | 110.2 | 96.0 | 115.3 | 114.4 |
| 1988-89 | 100.6 | 138.1 | 134.2 | 112.8 | 143.2 | 124.8 | 103.4 | 140.0 | 130.5 |
| 1989-90 | 99.9 | 139.1 | 135.5 | 115.8 | 149.7 | 126.7 | 103.5 | 143.0 | 131.9 |
| 1990-91 | 100.7 | 143.7 | 137.8 | 120.0 | 156.3 | 128.0 | 105.2 | 148.4 | 133.8 |
| 1991-92 | 96.0 | 137.6 | 136.5 | 124.8 | 158.8 | 123.7 | 102.7 | 145.5 | 131.0 |
| 1992-93 | 97.0 | 144.3 | 142.0 | 123.2 | 164.0 | 130.2 | 103.1 | 151.6 | 137.2 |
| 1993-94 | 96.7 | 150.2 | 146.5 | 127.3 | 169.5 | 132.7 | 103.8 | 157.3 | 140.7 |
| 1994-95 | 97.6 | 155.9 | 150.4 | 126.2 | 180.9 | 138.9 | 104.2 | 165.2 | 145.5 |
| 1995-96 | 95.3 | 146.1 | 143.1 | 131.8 | 185.5 | 135.7 | 103.8 | 160.7 | 139.8 |
| 1996-97 | 97.4 | 160.9 | 154.5 | 134.6 | 200.9 | 143.8 | 106.0 | 175.7 | 149.8 |
| 1997-98 | 97.6 | 155.7 | 148.4 | 133.6 | 181.6 | 132.3 | 105.9 | 165.3 | 141.4 |
| 1998-99 | 98.6 | 165.2 | 154.0 | 134.8 | 200.2 | 141.3 | 107.0 | 178.2 | 148.4 |
| 1999-00 | 97.0 | 169.7 | 159.8 | 130.7 | 189.0 | 136.4 | 104.8 | 176.9 | 149.6 |
| 2000-01 | 95.4 | 158.4 | 152.8 | 126.9 | 178.2 | 133.2 | 102.7 | 165.7 | 144.3 |
| 2001-02 | 96.7 | 172.5 | 164.1 | 128.1 | 190.3 | 138.6 | 104.0 | 179.1 | 153.0 |
| 2002-03 | 89.7 | 140.4 | 143.2 | 116.1 | 166.4 | 124.7 | 95.8 | 150.1 | 135.0 |
| 2003-04 | 97.3 | 172.0 | 165.3 | 125.7 | 201.5 | 151.2 | 103.8 | 182.9 | 159.2 |
| 2004-05 | 94.6 | 159.9 | 156.5 | 137.9 | 206.7 | 147.9 | 104.6 | 177.3 | 152.6 |
| 2005-06 | 95.8 | 169.2 | 162.3 | 140.8 | 230.1 | 157.0 | 106.2 | 191.8 | 159.9 |
| 2006-07 | 97.5 | 175.9 | 180.4 | 142.5 | 241.2 | 169.3 | 107.9 | 200.1 | 185.5 |
| 2007-08* | 98.0 | 186.7 | 190.4 | 143.1 | 237.6 | 166.0 | 108.5 | 201.8 | 186.0 |

* Based on Fourth Advance Estimates, 2007-08 as released on 09.07.2008.

# Table 6.2

## *Compound Growth-Rates of Area, Production and Yield of Principal Crops since 1949-50 (Base: T.E.1981-82=100)*

*A- Growth rates of Area*
*P- Growth rates of Production*
*Y- Growth rates of Yield*

*(% per annum)*

| Crop | 1949-50 to 2007-08** | | | 1949-50 to 1964-65 | | | 1967-68 to 2007-08* | | |
|---|---|---|---|---|---|---|---|---|---|
| | A | P | Y | A | P | Y | A | P | Y |
| 1 | 2 | 3 | 4 | 5 | 6 | 7 | 8 | 9 | 10 |
| Rice | 0.67 | 2.53 | 1.85 | 1.21 | 3.50 | 2.25 | 0.50 | 2.48 | 2.01 |
| Wheat | 1.88 | 4.83 | 2.89 | 2.69 | 3.98 | 1.27 | 1.22 | 3.75 | 2.50 |
| *Jowar* | -1.20 | 0.14 | 1.36 | 0.99 | 2.51 | 1.49 | -1.99 | -0.67 | 1.35 |
| *Bajra* | -0.33 | 1.68 | 2.01 | 1.08 | 2.34 | 1.24 | -0.82 | 1.36 | 2.20 |
| Maize | 1.12 | 2.61 | 1.47 | 2.67 | 3.88 | 1.18 | 0.56 | 2.55 | 1.97 |
| Ragi | -0.84 | 0.65 | 1.50 | 0.84 | 3.08 | 2.22 | -1.56 | -0.11 | 1.47 |
| Small Millets | -2.89 | -2.66 | 0.23 | -0.30 | -0.20 | 0.09 | -4.38 | -3.88 | 0.52 |
| Barley | -3.37 | -1.52 | 1.92 | -0.64 | -0.28 | 0.36 | -4.31 | -2.15 | 2.26 |
| Coarse Cereals | -0.80 | 0.96 | 1.64 | 0.90 | 2.25 | 1.23 | -1.41 | 0.62 | 1.91 |
| Total Cereals | 0.32 | 2.69 | 2.02 | 1.25 | 3.21 | 1.77 | -0.07 | 2.48 | 2.18 |
| Gram | -0.56 | 0.17 | 0.74 | 1.64 | 2.66 | 1.00 | -0.39 | 0.35 | 0.75 |
| *Tur* | 0.89 | 0.80 | -0.09 | 0.57 | -1.34 | -1.90 | 1.03 | 1.04 | 0.01 |
| Other Pulses | 0.30 | 0.84 | 0.54 | 2.06 | 1.28 | -0.77 | 0.00 | 1.02 | 1.01 |
| Total Pulses | 0.08 | 0.56 | 0.50 | 1.72 | 1.41 | -0.18 | 0.02 | 0.75 | 0.70 |
| Total Foodgrains | 0.27 | 2.35 | 1.77 | 1.35 | 2.82 | 1.36 | -0.05 | 2.25 | 1.99 |
| Sugarcane | 1.78 | 2.87 | 1.07 | 3.28 | 4.26 | 0.95 | 1.71 | 2.72 | 0.99 |
| Groundnut | 0.47 | 1.35 | 0.87 | 4.01 | 4.34 | 0.31 | -0.25 | 0.85 | 1.10 |
| Sesamum | -0.58 | 0.86 | 1.45 | 0.14 | -0.32 | -0.46 | -1.08 | 1.10 | 2.20 |
| R & M | 1.96 | 4.10 | 2.10 | 2.97 | 3.35 | 0.37 | 2.01 | 4.47 | 2.41 |
| Sunflower @ | | | | | | | 0.00 | 0.00 | 0.00 |
| Soyabean @ | | | | | | | 0.00 | 0.00 | 0.00 |
| Nine Oilseeds | 1.33 | 2.74 | 1.29 | 2.53 | 3.12 | 0.00 | 1.45 | 3.16 | 1.67 |
| Total Oilseeds | 1.15 | 2.72 | 1.25 | 2.67 | 3.20 | 0.30 | 1.19 | 3.11 | 1.63 |
| Cotton | 0.35 | 2.66 | 2.30 | 2.47 | 4.55 | 2.04 | 0.40 | 2.95 | 2.54 |
| Jute | 0.41 | 1.78 | 1.36 | 3.00 | 3.50 | 0.49 | 0.28 | 2.16 | 1.87 |
| Mesta | -0.91 | -0.11 | 0.80 | 6.21 | 7.97 | 1.66 | -2.35 | -1.07 | 1.33 |
| Jute & Mesta | 0.11 | 1.49 | 1.30 | 3.88 | 4.31 | 0.70 | -0.35 | 1.69 | 1.88 |
| Total Fibres | 0.30 | 2.49 | 2.16 | 2.71 | 4.56 | 1.88 | 0:27 | 2.77 | 2.46 |
| Potato | 3.42 | 5.35 | 1.87 | 4.38 | 4.28 | -0.11 | 3.04 | 4.89 | 1.80 |
| Tobacco | 0.00 | 1.53 | 1.53 | 1.66 | 2.79 | 1.10 | -0.52 | 1.07 | 1.60 |
| Non Foodgrains | 1.22 | 2.92 | 1.40 | 2.44 | 3.74 | 0.89 | 1.26 | 3.06 | 1.61 |
| All Crops | 0.51 | 2.57 | 1.64 | 1.58 | 3.15 | 1.21 | 0.29 | 2.56 | 1.86 |

*Notes:* * Growth rates are based on Forth advance estimates for 2007-08.

$ Growth rates for the period from 1970-71

Nine oilseeds include groundnut, castorseed, sesamum, rapeseed & mustard, Linseed, nigerseed, safflower, sunflower and soyabean.

Total oilseeds include nine oilseeds, cottonseed and coconut.

## Table 6.3

### *Compound Growth Rates of Area, Production and Yield of Principal Crops during 1980-1990, 1990-2000 and 2000-2007 (Base: T.E.1981-82=100)*

*A- Growth rates of Area*
*P- Growth rates of Production*
*Y- Growth rates of Yield*

*(% per annum)*

| *Crop* | *1980-81 to 1989-90* | | | *1990-91 to 1999-2000* | | | *2000-01 to 2007-08** | | |
|---|---|---|---|---|---|---|---|---|---|
| | *A* | *P* | *Y* | *A* | *P* | *Y* | *A* | *P* | *Y* |
| *1* | *2* | *3* | *4* | *5* | *6* | *7* | *8* | *9* | *10* |
| Rice | 0.41 | 3.62 | 3.19 | 0.68 | 2.02 | 1.34 | -0.14 | 1.87 | 2.00 |
| Wheat | 0.46 | 3.57 | 3.10 | 1.72 | 3.57 | 1.83 | 1.29 | 1.36 | 0.08 |
| *Jowar* | -0.99 | 0.28 | 1.29 | -3.53 | -3.07 | 0.48 | -2.90 | 0.34 | 3.37 |
| *Bajra* | -1.05 | 0.03 | 1.09 | -1.46 | 0.95 | 2.44 | 0.31 | 4.53 | 4.21 |
| Maize | -0.20 | 1.89 | 2.09 | 0.94 | 3.28 | 2.32 | 3.50 | 5.84 | 2.26 |
| Ragi | -1.23 | -0.10 | 1.14 | -2.85 | -0.80 | 2.10 | -3.28 | -2.68 | 0.63 |
| Small Millets | -4.32 | -3.23 | 1.14 | -5.40 | -5.88 | -0.51 | -5.40 | -2.63 | 2.90 |
| Barley | -6.03 | -3.48 | 2.72 | -2.62 | -0.64 | 2.03 | -2.52 | -2.27 | 0.25 |
| Coarse Cereals | -1.34 | 0.40 | 1.62 | -2.12 | -0.02 | 1.82 | -0.41 | 3.25 | 4.25 |
| Total Cereals | -0.26 | 3.03 | 2.90 | 0.04 | -0.02 | 1.59 | 0.16 | 1.95 | 3.10 |
| Gram | -1.41 | -0.81 | 0.61 | 1.26 | 2.96 | 1.68 | 4.70 | 5.51 | 0.77 |
| *Tur* | 2.30 | 2.87 | 0.56 | -0.66 | 0.89 | 1.55 | 0.91 | 3.66 | 2.74 |
| Other Pulses | 0.02 | 3.05 | 3.03 | -1.61 | -1.58 | 0.04 | 0.76 | 1.59 | 0.82 |
| Total Pulses | -0.09 | 1.52 | 1.61 | -0.60 | 0.59 | 0.93 | 1.91 | 3.42 | 1.65 |
| Total Food grains | -0.23 | 2.85 | 2.74 | -0.07 | 2.02 | 1.52 | 0.48 | 2.09 | 2.82 |
| Sugarcane | 1.44 | 2.70 | 1.24 | -0.07 | 2.73 | 1.05 | 1.89 | 2.20 | 0.30 |
| Groundnut | 1.67 | 3.76 | 2.06 | -2.31 | -1.25 | 1.08 | -0.23 | 3.17 | 3.41 |
| Sesamum | -0.52 | 3.20 | 3.74 | -5.52 | -4.84 | 0.72 | 1.03 | 3.98 | 2.91 |
| R & M | 1.95 | 7.28 | 5.22 | 0.71 | 0.78 | 0.07 | 6.03 | 8.17 | 2.02 |
| Sunflower | 25.69 | 21.32 | -3.47 | -2.97 | -3.20 | -0.24 | 10.14 | 12.76 | 2.37 |
| Soyabean | 17.10 | 17.96 | 0.73 | 10.23 | 13.06 | 2.56 | 5.48 | 10.04 | 4.32 |
| Nine Oilseeds | 2.47 | 5.36 | 2.49 | 0.17 | 1.42 | 1.42 | 3.38 | 7.23 | 3.70 |
| Total Oilseeds | 1.51 | 5.20 | 2.43 | -0.86 | 1.63 | 1.15 | 2.71 | 6.69 | 4.91 |
| Cotton | -1.25 | 2.80 | 4.10 | 2.71 | 2.29 | -0.41 | 1.45 | 17.48 | 15.79 |
| Jute | -2.38 | 0.91 | 3.37 | 1.48 | 2.32 | 0.83 | -1.25 | 0.38 | 1.65 |
| Mesta | -4.24 | -3.59 | 0.67 | -2.47 | -2.08 | 0.40 | -4.19 | -3.15 | 1.09 |
| Jute & Mesta | -2.90 | 0.16 | 2.96 | 1.81 | 1.81 | 0.87 | -1.75 | 0.04 | 2.51 |
| Total Fibres | -1.50 | 2.46 | 3.98 | 2.45 | 2.21 | -0.27 | 1.09 | 15.23 | 14.14 |
| Potato | 2.90 | 5.17 | 2.20 | 3.84 | 5.44 | 1.54 | 3.05 | -0.50 | -3.44 |
| Tobacco | -2.79 | -1.05 | 1.79 | 1.56 | 1.00 | -0.55 | 3.71 | 3.54 | -0.17 |
| Non Foodgrains | 1.12 | 3.77 | 2.31 | 1.18 | 2.69 | 1.09 | 2.46 | 5.09 | 3.87 |
| All Principal Crops | 0.10 | 3.19 | 2.56 | 0.27 | 2.29 | 1.33 | 1.06 | 3.17 | 3.8 |

*Notes:* * Growth rates are based on fourth advance estimates for 2007-08 except potato and tobacco for which final estimates for 2006-07 have been used.

Nine oilseeds include groundnut, castorseed, sesamum, rapeseed and mustard, linseed, nigerseed, safflower, sunflower and soysabean.

Total oilseeds include nine oilseeds, cottonseed and coconut.

# Table 6.4

*Index Numbers of Area, Production and Yield of Foodgrain and Non-Foodgrain Crops in India from 1993-94 to 2007-08 (Base: Triennium Ending 1993-94 = 100)*

*A- Index Numbers of Area*
*P- Index Numbers of Production*
*Y- Index Numbers of Yield*

| Crops | Weight | 1993-94 | | | 1994-95 | | | 1995-96 | | | 1996-97 | | | 1997-98 | | |
|---|---|---|---|---|---|---|---|---|---|---|---|---|---|---|---|---|
| | | A | P | Y | A | P | Y | A | P | Y | A | P | Y | A | P | Y |
| 1 | 2 | 3 | 4 | 5 | 6 | 7 | 8 | 9 | 10 | 11 | 12 | 13 | 14 | 15 | 16 | 17 |
| Rice | 19.427 | 163.6 | 172.7 | 105.5 | 164.7 | 175.9 | 106.8 | 164.8 | 165.5 | 100.5 | 167.0 | 175.7 | 105.2 | 167.0 | 177.5 | 106.2 |
| Wheat | 17.100 | 103.3 | 103.9 | 100.6 | 105.6 | 114.2 | 108.1 | 102.8 | 107.8 | 104.9 | 106.4 | 120.4 | 113.2 | 109.7 | 115.2 | 105.0 |
| *Jowar* | 1.406 | 184.9 | 201.8 | 109.2 | 164.1 | 155.9 | 95.0 | 161.4 | 162.2 | 100.5 | 162.9 | 190.2 | 116.8 | 153.9 | 130.9 | 85.1 |
| *Bajra* | 1.607 | 94.9 | 80.6 | 84.9 | 101.6 | 116.0 | 114.2 | 92.6 | 87.2 | 94.1 | 99.2 | 127.4 | 128.5 | 96.1 | 123.8 | 128.8 |
| Maize | 1.839 | 116.9 | 126.4 | 108.1 | 121.4 | 117.0 | 96.4 | 118.3 | 125.5 | 106.1 | 124.0 | 141.8 | 114.4 | 125.0 | 142.4 | 113.9 |
| Ragi | 0.568 | 96.3 | 100.9 | 104.7 | 90.2 | 91.0 | 100.9 | 90.7 | 97.2 | 107.2 | 91.2 | 91.2 | 100.0 | 84.7 | 81.1 | 95.7 |
| Small Millets | 0.212 | 95.0 | 103.1 | 108.5 | 90.2 | 89.7 | 99.5 | 83.7 | 87.6 | 104.6 | 80.6 | 81.9 | 101.6 | 77.0 | 72.5 | 94.2 |
| Barlely | 0.455 | 89.4 | 87.1 | 97.3 | 100.3 | 114.5 | 114.2 | 92.8 | 100.2 | 108.0 | 85.3 | 97.0 | 113.7 | 96.6 | 111.4 | 115.3 |
| Coarse Cereals | 6.086 | 122.4 | 125.6 | 102.6 | 119.8 | 122.2 | 101.9 | 114.5 | 118.0 | 103.0 | 118.0 | 139.0 | 117.8 | 114.3 | 124.4 | 108.8 |
| Cereals | 42.613 | 130.3 | 138.4 | 106.2 | 130.4 | 143.5 | 110.0 | 127.7 | 135.6 | 106.2 | 130.8 | 148.3 | 113.3 | 130.6 | 144.9 | 110.9 |
| Gram | 3.727 | 103.7 | 110.5 | 106.6 | 123.0 | 142.8 | 116.1 | 116.1 | 110.5 | 95.2 | 111.7 | 123.5 | 110.6 | 123.4 | 136.1 | 110.3 |
| *Arhar* | 1.814 | 98.8 | 112.8 | 114.2 | 92.7 | 89.8 | 96.9 | 96.4 | 96.8 | 100.4 | 98.3 | 111.3 | 113.2 | 94.0 | 77.5 | 82.4 |
| Other Pulses | 2.479 | 131.1 | 131.7 | 100.5 | 133.5 | 133.9 | 100.3 | 123.7 | 116.9 | 94.5 | 131.4 | 146.8 | 111.7 | 130.1 | 125.3 | 96.3 |
| Pulses | 8.020 | 115.6 | 117.6 | 101.7 | 122.1 | 128.1 | 104.9 | 115.8 | 109.4 | 94.4 | 118.3 | 127.9 | 108.1 | 120.8 | 119.5 | 98.9 |
| Foodgrains | 50.633 | 127.4 | 135.1 | 106.0 | 128.8 | 141.0 | 109.5 | 125.4 | 131.4 | 104.8 | 128.4 | 145.1 | 113.0 | 128.7 | 140.9 | 109.5 |
| Groundnut | 3.201 | 184.0 | 185.5 | 100.8 | 173.5 | 191.0 | 110.1 | 166.4 | 179.6 | 107.9 | 167.9 | 204.8 | 121.9 | 156.7 | 174.6 | 111.4 |
| Sesamum | 0.641 | 96.5 | 87.0 | 90.1 | 85.8 | 90.5 | 105.6 | 79.5 | 82.0 | 103.2 | 86.7 | 98.9 | 114.0 | 72.3 | 87.7 | 121.3 |
| Rapeseed and Mustard | 4.312 | 99.1 | 101.2 | 102.1 | 95.5 | 107.3 | 112.4 | 103.2 | 111.8 | 108.4 | 103.2 | 124.1 | 120.3 | 111.0 | 87.7 | 79.0 |
| Sunflower | 0.592 | 215.6 | 177.1 | 82.1 | 161.5 | 160.0 | 99.1 | 171.5 | 165.2 | 96.4 | 156.1 | 164.2 | 105.2 | 140.9 | 116.3 | 82.5 |
| Soyabean | 2.131 | 116.9 | 135.8 | 116.1 | 115.5 | 112.5 | 97.4 | 134.7 | 145.8 | 108.2 | 145.7 | 153.8 | 105.6 | 160.2 | 184.9 | 115.4 |
| Nine Oilseeds | 11.971 | 127.9 | 134.1 | 104.8 | 120.4 | 133.8 | 111.1 | 123.6 | 137.3 | 111.1 | 125.2 | 151.7 | 121.1 | 124.2 | 130.3 | 104.9 |
| Total Oilseeds | 15.479 | 119.2 | 127.8 | 107.2 | 116.0 | 130.2 | 112.3 | 122.5 | 132.8 | 108.4 | 124.2 | 144.7 | 116.5 | 122.5 | 125.9 | 102.8 |
| Cotton (Lint) | 6.399 | 97.4 | 101.1 | 103.8 | 104.8 | 111.9 | 106.9 | 120.2 | 121.1 | 100.7 | 121.4 | 134.0 | 110.4 | 118.0 | 102.2 | 86.6 |
| Jute | 0.368 | 90.8 | 157.2 | 173.2 | 96.5 | 170.8 | 177.0 | 96.3 | 163.9 | 170.3 | 117.2 | 212.8 | 181.5 | 118.4 | 212.7 | 179.6 |
| Mesta | 0.059 | 90.1 | 91.3 | 101.3 | 89.7 | 92.2 | 102.8 | 89.3 | 96.7 | 108.3 | 96.5 | 99.9 | 103.5 | 94.7 | 90.3 | 95.4 |
| Jute and Mesta | 0.428 | 90.6 | 148.0 | 163.4 | 95.0 | 159.9 | 168.3 | 94.8 | 154.6 | 163.1 | 112.8 | 197.1 | 174.8 | 113.3 | 195.7 | 172.7 |
| Fibres | 6.846 | 96.7 | 104.1 | 107.6 | 103.6 | 114.9 | 110.9 | 117.2 | 123.1 | 105.1 | 120.2 | 137.8 | 114.6 | 117.3 | 108.0 | 92.1 |
| Sugarcane | 11.980 | 94.7 | 96.8 | 102.2 | 107.0 | 116.1 | 108.5 | 114.8 | 118.5 | 103.2 | 115.5 | 117.0 | 101.3 | 108.8 | 122.6 | 108.3 |
| Potato | 2.478 | 100.3 | 106.5 | 106.1 | 102.4 | 106.5 | 104.0 | 106.2 | 115.3 | 108.6 | 119.3 | 148.2 | 124.2 | 115.5 | 124.5 | 93.5 |
| Tobacco | 0.857 | 90.5 | 93.4 | 103.2 | 94.5 | 100.0 | 105.9 | 95.1 | 91.0 | 95.7 | 103.2 | 105.1 | 101.8 | 112.1 | 117.8 | 96.8 |
| Non-Foodgrains | 49.367 | 111.2 | 110.7 | 99.5 | 114.3 | 118.8 | 103.9 | 118.3 | 122.0 | 103.2 | 120.2 | 130.3 | 108.4 | 118.8 | 176.8 | 101.5 |
| All Crops | 100.000 | 121.9 | 123.0 | 101.0 | 123.8 | 130.1 | 105.0 | 123.0 | 126.8 | 103.1 | 125.6 | 137.8 | 109.7 | 125.3 | 120.6 | 104.4 |

*contd. ...*

*...contd. ...*

| *Crops* | *1998-99* | | | *1999-2000* | | | *2000-01* | | | *2001-02* | | | *2002-03* | | |
|---|---|---|---|---|---|---|---|---|---|---|---|---|---|---|---|
| | *A* | *P* | *Y* | *A* | *P* | *Y* | *A* | *P* | *Y* | *A* | *P* | *Y* | *A* | *P* | *Y* |
| *1* | *18* | *19* | *20* | *21* | *22* | *23* | *24* | *25* | *26* | *27* | *28* | *29* | *30* | *31* | *32* |
| Rice | 172.3 | 185.1 | 107.4 | 173.7 | 192.9 | 111.1 | 171.9 | 182.7 | 106.3 | 172.7 | 200.7 | 116.2 | 158.3 | 154.5 | 97.5 |
| Wheat | 113.1 | 123.8 | 109.4 | 112.9 | 132.6 | 117.4 | 105.7 | 121.0 | 114.4 | 108.3 | 126.3 | 116.7 | 103.5 | 114.2 | 110.3 |
| *Jowar* | 139.6 | 146.3 | 104.9 | 146.1 | 151.0 | 103.4 | 140.4 | 130.9 | 93.2 | 139.6 | 131.4 | 94.1 | 132.5 | 121.9 | 92.0 |
| *Bajra* | 92.4 | 112.6 | 121.9 | 88.4 | 93.6 | 105.9 | 97.7 | 109.5 | 112.0 | 94.7 | 134.2 | 141.6 | 76.9 | 76.4 | 99.3 |
| Maize | 122.7 | 146.8 | 119.6 | 127.0 | 151.5 | 119.3 | 130.7 | 158.6 | 121.3 | 130.1 | 173.3 | 133.1 | 131.2 | 146.8 | 111.9 |
| Ragi | 89.9 | 101.4 | 112.8 | 83.6 | 89.0 | 106.5 | 89.9 | 106.2 | 118.0 | 84.2 | 92.3 | 109.6 | 72.3 | 51.1 | 70.7 |
| Small Millets | 75.2 | 75.4 | 100.2 | 71.0 | 69.5 | 97.8 | 71.7 | 66.0 | 92.0 | 66.0 | 64.8 | 98.3 | 60.4 | 51.6 | 85.4 |
| Barley | 89.3 | 102.0 | 114.3 | 81.6 | 96.0 | 117.6 | 87.6 | 94.9 | 108.4 | 74.3 | 94.5 | 127.2 | 79.0 | 93.4 | 118.2 |
| Coarse Cereals | 108.8 | 127.6 | 117.3 | 108.8 | 123.3 | 113.4 | 112.2 | 126.4 | 112.7 | 109.4 | 136.1 | 124.3 | 100.1 | 106.3 | 106.2 |
| Cereals | 131.5 | 152.3 | 115.8 | 131.9 | 158.8 | 120.4 | 130.2 | 149.9 | 115.1 | 130.3 | 161.7 | 124.0 | 120.7 | 131.4 | 108.8 |
| Gram | 138.2 | 150.9 | 109.2 | 100.3 | 113.6 | 113.3 | 84.6 | 85.6 | 101.1 | 104.7 | 121.5 | 116.0 | 96.4 | 94.0 | 97.6 |
| *Arhar* | 96.3 | 113.5 | 117.9 | 95.9 | 112.9 | 117.7 | 101.7 | 94.1 | 92.6 | 93.1 | 94.7 | 101.7 | 94.0 | 91.6 | 97.4 |
| Other Pulses | 125.5 | 135.4 | 107.9 | 124.9 | 140.6 | 112.5 | 124.8 | 124.7 | 99.9 | 132.7 | 141.3 | 106.5 | 121.5 | 233.3 | 192.0 |
| Pulses | 123.7 | 137.6 | 111.2 | 111.2 | 121.8 | 109.5 | 107.1 | 99.6 | 93.0 | 115.9 | 121.5 | 104.9 | 107.9 | 136.5 | 126.5 |
| Food Grains | 130.0 | 150.0 | 115.4 | 127.8 | 152.9 | 119.6 | 125.7 | 141.9 | 112.9 | 127.5 | 155.3 | 121.8 | 118.2 | 132.2 | 111.8 |
| Groundnut | 163.5 | 212.7 | 130.1 | 151.8 | 124.5 | 82.0 | 145.0 | 151.8 | 104.7 | 137.9 | 166.4 | 120.7 | 131.2 | 97.6 | 74.4 |
| Sesamum | 70.1 | 81.4 | 116.2 | 68.0 | 74.1 | 109.0 | 74.9 | 79.9 | 106.7 | 72.8 | 107.7 | 148.0 | 62.9 | 68.1 | 108.3 |
| Rapeseed & Mustard | 102.7 | 105.6 | 102.9 | 95.0 | 107.9 | 113.6 | 70.6 | 78.1 | 110.6 | 80.0 | 94.8 | 118.5 | 71.6 | 72.4 | 101.0 |
| Sunflower | 147.5 | 124.0 | 84.1 | 104.1 | 91.1 | 87.6 | 86.8 | 84.9 | 97.9 | 95.1 | 89.3 | 93.9 | 132.7 | 114.7 | 86.4 |
| Soyabean | 173.0 | 204.4 | 117.7 | 166.4 | 202.6 | 121.7 | 171.6 | 150.9 | 88.0 | 169.6 | 170.6 | 100.6 | 163.3 | 133.2 | 81.6 |
| Nine Oilseeds | 124.7 | 151.3 | 121.3 | 115.4 | 125.7 | 108.9 | 108.3 | 113.0 | 104.4 | 107.6 | 126.9 | 117.9 | 102.2 | 90.7 | 88.8 |
| Cotton (Lint) | 124.3 | 115.7 | 93.1 | 115.9 | 108.6 | 93.7 | 113.6 | 89.7 | 79.0 | 121.5 | 94.2 | 77.5 | 102.1 | 82.2 | 80.5 |
| Jute | 110.8 | 188.7 | 170.3 | 110.6 | 201.3 | 182.1 | 108.2 | 199.0 | 184.0 | 114.1 | 226.0 | 198.2 | 112.9 | 219.4 | 194.3 |
| Mesta | 83.6 | 83.2 | 99.5 | 89.1 | 96.5 | 108.3 | 89.6 | 105.9 | 118.2 | 82.2 | 93.5 | 113.7 | 80.7 | 85.6 | 106.1 |
| Jute & Mesta | 105.0 | 173.7 | 165.5 | 106.0 | 186.4 | 175.9 | 104.2 | 185.7 | 178.2 | 107.2 | 207.2 | 193.3 | 106.0 | 200.4 | 189.1 |
| Fibres | 121.8 | 119.2 | 97.9 | 114.5 | 113.3 | 99.0 | 112.4 | 95.6 | 85.1 | 119.6 | 101.1 | 84.5 | 102.3 | 89.5 | 87.5 |
| Sugarcane | 112.3 | 121.7 | 108.4 | 116.8 | 126.1 | 108.0 | 119.5 | 124.7 | 104.4 | 122.1 | 125.2 | 102.5 | 125.1 | 121.1 | 96.8 |
| Potato | 126.5 | 144.5 | 114.2 | 128.3 | 151.2 | 117.8 | 117.0 | 137.6 | 117.6 | 115.7 | 146.4 | 126.5 | 128.1 | 141.7 | 110.7 |
| Tobacco | 122.9 | 125.5 | 102.0 | 104.3 | 88.8 | 85.2 | 63.0 | 58.4 | 92.7 | 84.0 | 92.5 | 110.1 | 78.7 | 83.4 | 105.9 |
| Non-Food Grains | 120.1 | 131.9 | 109.9 | 117.5 | 123.8 | 105.4 | 114.3 | 118.2 | 103.4 | 114.9 | 124.6 | 108.5 | 102.4 | 110.8 | 108.2 |
| All Crops | 126.6 | 141.1 | 111.4 | 124.3 | 138.5 | 111.4 | 121.8 | 130.2 | 106.9 | 122.7 | 140.1 | 114.3 | 113.1 | 121.6 | 107.6 |

*contd. ...*

*...contd. ...*

| *Crops* | *2003-04* | | | *2004-05* | | | *2005-06* | | | *2006-07* | | | *2007-08** | | |
|---|---|---|---|---|---|---|---|---|---|---|---|---|---|---|---|
| | *A* | *P* | *Y* | *A* | *P* | *Y* | *A* | *P* | *Y* | *A* | *P* | *Y* | *A* | *P* | *Y* |
| *1* | *33* | *34* | *35* | *36* | *37* | *38* | *39* | *40* | *41* | *42* | *43* | *44* | *45* | *46* | *47* |
| Rice | 163.8 | 190.4 | 116.2 | 161.2 | 178.8 | 110.9 | 167.9 | 197.4 | 117.6 | 168.5 | 200.1 | 119.2 | 168.3 | 207.4 | 123.2 |
| Wheat | 109.3 | 125.3 | 114.6 | 108.4 | 119.2 | 109.9 | 108.8 | 120.4 | 110.7 | 115.0 | 131.6 | 114.4 | 115.7 | 136.1 | 117.7 |
| *Jowar* | 132.9 | 116.2 | 87.4 | 129.6 | 126.0 | 97.2 | 123.5 | 132.7 | 107.4 | 120.7 | 124.3 | 103.0 | 112.9 | 135.2 | 119.7 |
| *Bajra* | 105.5 | 196.1 | 185.9 | 91.8 | 128.4 | 140.0 | 95.2 | 124.4 | 130.7 | 94.5 | 136.4 | 144.3 | 94.5 | 158.6 | 167.7 |
| Maize | 145.2 | 197.3 | 135.9 | 146.9 | 186.6 | 127.0 | 150.1 | 193.7 | 129.1 | 156.1 | 198.8 | 127.3 | 163.4 | 254.2 | 155.6 |
| Ragi | 85.2 | 76.4 | 89.7 | 79.4 | 94.5 | 119.1 | 78.4 | 91.5 | 116.6 | 60.2 | 56.1 | 93.2 | 74.6 | 82.7 | 110.8 |
| Small Millets | 60.0 | 63.4 | 105.7 | 55.4 | 53.7 | 69.9 | 53.6 | 53.0 | 98.9 | 50.9 | 53.9 | 106.0 | 47.3 | 55.3 | 116.9 |
| Barley | 74.0 | 86.1 | 116.3 | 69.4 | 80.1 | 115.3 | 70.9 | 81.0 | 114.1 | 72.8 | 88.1 | 121.0 | 69.1 | 81.3 | 117.6 |
| Coarse Cereals | 114.2 | 154.0 | 134.9 | 107.6 | 136.1 | 126.5 | 107.7 | 138.5 | 128.5 | 106.4 | 138.5 | 130.1 | 106.4 | 166.3 | 156.2 |
| Cereals | 129.3 | 159.1 | 123.0 | 125.9 | 148.8 | 118.2 | 128.3 | 158.1 | 123.2 | 130.0 | 164.1 | 126.3 | 130.2 | 174.2 | 133.8 |
| Gram | 115.0 | 126.9 | 110.3 | 109.6 | 121.4 | 110.8 | 112.5 | 123.7 | 110.0 | 122.3 | 140.6 | 115.0 | 123.7 | 131.2 | 106.1 |
| *Arhar* | 98.4 | 98.7 | 100.4 | 98.5 | 98.3 | 99.9 | 100.2 | 114.7 | 114.5 | 99.7 | 97.0 | 97.3 | 104.9 | 129.5 | 123.4 |
| Other Pulses | 139.5 | 171.3 | 122.7 | 135.6 | 133.2 | 98.2 | 128.6 | 126.5 | 98.4 | 131.3 | 139.1 | 105.9 | 135.0 | 153.2 | 113.5 |
| Pulses | 123.5 | 134.2 | 108.7 | 119.9 | 119.8 | 100.0 | 117.7 | 122.6 | 104.1 | 122.1 | 130.3 | 106.7 | 125.3 | 138.7 | 110.6 |
| Food Grains | 128.2 | 155.1 | 121.0 | 124.7 | 144.2 | 115.6 | 126.3 | 152.5 | 120.8 | 128.5 | 158.8 | 123.6 | 129.2 | 168.6 | 130.4 |
| Groundnut | 132.4 | 192.5 | 145.4 | 146.8 | 160.4 | 109.3 | 148.9 | 189.3 | 127.1 | 124.1 | 115.2 | 92.8 | 141.8 | 221.8 | 156.4 |
| Sesamum | 74.1 | 120.7 | 163.0 | 80.3 | 104.1 | 129.6 | 75.1 | 99.0 | 131.9 | 74.2 | 95.5 | 128.7 | 76.7 | 121.2 | 158.0 |
| Rapeseed & Mustard | 85.6 | 117.3 | 137.1 | 115.3 | 141.6 | 122.8 | 114.7 | 151.6 | 132.2 | 107.0 | 138.7 | 129.6 | 90.6 | 108.2 | 119.4 |
| Sunflower | 161.9 | 37.8 | 75.5 | 174.6 | 155.9 | 89.3 | 189.1 | 189.1 | 100.0 | 175.0 | 161.3 | 92.2 | 152.1 | 183.3 | 120.5 |
| Soyabean | 175.3 | 223.7 | 127.6 | 202.5 | 196.7 | 97.2 | 206.1 | 236.7 | 114.8 | 222.7 | 253.2 | 113.7 | 237.5 | 285.7 | 120.3 |
| Nine Oilseeds | 112.5 | 154.0 | 136.9 | 130.9 | 150.2 | 114.8 | 132.5 | 171.6 | 129.5 | 126.1 | 147.1 | 116.7 | 126.2 | 174.6 | 138.3 |
| Cotton(Lint) | 101.1 | 129.3 | 127.9 | 116.9 | 154.7 | 132.3 | 115.5 | 174.2 | 150.9 | 121.7 | 213.1 | 175.2 | 125.4 | 243.0 | 193.8 |
| Jute | 110.9 | 218.9 | 197.4 | 101.1 | 200.7 | 198.5 | 99.3 | 212.9 | 214.5 | 103.6 | 220.3 | 212.7 | 106.6 | 217.4 | 204.0 |
| Mesta | 72.0 | 78.7 | 109.3 | 67.0 | 74.6 | 111.3 | 65.1 | 74.3 | 114.1 | 67.2 | 81.7 | 121.6 | 68.5 | 84.9 | 124.0 |
| Jute & Mesta | 102.5 | 199.1 | 194.2 | 93.7 | 182.9 | 195.1 | 91.9 | 193.3 | 210.4 | 95.7 | 200.7 | 209.7 | 98.4 | 199.0 | 202.3 |
| Fibres | 101.0 | 133.5 | 132.2 | 113.9 | 156.3 | 137.2 | 112.4 | 178.7 | 159.0 | 118.3 | 186.1 | 157.3 | 121.8 | 202.9 | 166.6 |
| Sugarcane | 109.0 | 98.6 | 90.4 | 101.4 | 99.9 | 98.6 | 116.3 | 118.5 | 101.9 | 142.6 | 149.8 | 105.1 | 139.6 | 143.5 | 102.8 |
| Potato | 123.5 | 141.1 | 114.3 | 126.3 | 144.6 | 114.5 | 134.2 | 146.3 | 109.0 | 142.0 | 135.2 | 95.2 | 142.0 | 135.2 | 95.2 |
| Tobacco | 89.1 | 93.2 | 104.6 | 88.4 | 93.1 | 105.4 | 89.9 | 93.6 | 104.2 | 88.8 | 88.1 | 99.1 | 88.8 | 88.1 | 99.1 |
| Non-Food Grains | 111.2 | 130.8 | 117.7 | 122.0 | 134.1 | 109.9 | 124.5 | 130.0 | 104.4 | 126.2 | 154.6 | 122.5 | 126.8 | 152.3 | 120.1 |
| All Crops | 122.6 | 143.1 | 116.7 | 123.5 | 139.2 | 112.7 | 125.4 | 146.7 | 117.0 | 127.4 | 167.2 | 131.2 | 128.1 | 168.6 | 131.6 |

* Based on Fourth Advance Estimates, 2007-08 as released on 09.07.2008

# Table 6.5

*Growth Rates of Area of Principal Crops in India from 1994-95 to 2007-08*
*(Base: T.E. 1993-94 = 100)*

(Per cent)

| Crop | 1994-95 | 1995-96 | 1996-97 | 1997-98 | 1998-99 | 1999-00 | 2000-01 | 2001-02 | 2002-03 | 2003-04 | 2004-05 | 2005-06 | 2006-07 | 2007-08* | Average Growth During 1994-95 to 2007-08 |
|---|---|---|---|---|---|---|---|---|---|---|---|---|---|---|---|
| 1 | 2 | 3 | 4 | 5 | 6 | 7 | 8 | 9 | 10 | 11 | 12 | 13 | 14 | 15 | 16 |
| Rice | 0.67 | 0.06 | 1.33 | 0.06 | 3.11 | 0.80 | -1.00 | 0.43 | -8.30 | 3.44 | -1.61 | 4.18 | 0.35 | -0.10 | **0.24** |
| Wheat | 2.23 | -2.65 | 3.50 | 3.10 | 3.10 | -0.14 | -6.39 | 2.39 | -4.36 | 5.55 | -0.80 | 0.38 | 5.71 | 0.57 | **0.87** |
| *Jowar* | -11.25 | -1.65 | 0.93 | -5.52 | -9.32 | 4.66 | -3.85 | -0.62 | -5.06 | 0.33 | -2.56 | -4.67 | -2.25 | -6.44 | **-3.38** |
| *Bajra* | 7.06 | -8.86 | 7.13 | -3.13 | -3.83 | -4.30 | 10.47 | -3.05 | -18.78 | 37.11 | -13.00 | 3.77 | -0.77 | 0.02 | **0.70** |
| Maize | 3.85 | -2.55 | 4.82 | 0.81 | -1.86 | 3.52 | 2.95 | -0.45 | 0.82 | 10.67 | 1.18 | 2.13 | 4.03 | 4.69 | **2.47** |
| Ragi | -6.33 | 0.55 | 0.55 | -7.13 | 6.13 | -7.04 | 7.63 | -6.39 | -14.07 | 17.77 | -6.82 | -1.21 | -23.24 | 24.00 | **-1.11** |
| Small millets | -5.05 | -7.21 | -3.70 | -4.47 | -2.28 | -5.60 | 0.94 | -7.98 | -8.39 | -0.77 | -7.62 | -3.30 | -5.08 | -6.95 | **-4.82** |
| Barley | 12.19 | -7.48 | -8.08 | 13.25 | -7.58 | -8.62 | 7.32 | -15.18 | 6.38 | -6.36 | -6.16 | 2.17 | 2.59 | -4.98 | **-1.47** |
| Coarse Cereals | 2.12 | 4.12 | 3.06 | 3.14 | 4.84 | 0.00 | 3.13 | -2.43 | -8.57 | 14.11 | -5.77 | 0.14 | -1.23 | 0.02 | **-0.86** |
| **Total Cereals** | **0.08** | **-2.07** | **2.43** | **-0.15** | **0.68** | **0.32** | **-1.26** | **0.07** | **-7.35** | **7.10** | **-2.67** | **1.95** | **1.32** | **0.12** | **0.04** |
| Gram | 18.61 | -5.61 | -3.79 | 10.47 | 11.98 | -27.43 | -15.64 | 23.74 | -7.95 | 19.33 | -4.73 | 2.70 | 8.67 | 1.15 | **2.25** |
| *Tur* | -6.17 | 3.99 | 1.97 | -4.37 | 2.40 | -0.36 | 5.99 | -8.39 | 0.94 | 4.67 | -0.08 | 1.77 | -0.53 | 5.26 | **0.51** |
| Other Pulses | 1.83 | -7.34 | 6.22 | -0.99 | -3.57 | -0.43 | -0.10 | 6.36 | -8.43 | 14.81 | -2.83 | -5.15 | 2.12 | 2.82 | **0.38** |
| **Total Pulses** | **5.62** | **-5.16** | **2.16** | **2.11** | **2.43** | **-10.15** | **-3.64** | **8.16** | **-6.87** | **14.45** | **-2.96** | **-1.77** | **3.71** | **2.65** | 0.77 |
| **Total Foodgrains** | **1.10** | **-2.64** | **2.39** | **0.23** | **1.00** | **-1.65** | **-1.67** | **1.43** | **-7.26** | **8.42** | **-2.73** | **1.24** | **1.76** | **0.60** | **0.16** |
| Groundnut | -5.71 | -4.09 | 0.90 | -6.67 | 4.35 | -7.15 | -4.50 | -4.89 | -4.85 | 0.87 | 10.91 | 1.44 | -16.64 | 14.21 | **-1.56** |
| Sesamum | -11.09 | -7.34 | 9.06 | -16.61 | -3.07 | -3.03 | 10.24 | -2.87 | -13.54 | 17.72 | 8.45 | -6.55 | -1.16 | 3.39 | **-1.17** |
| R & M | -3.63 | 8.06 | 0.00 | 7.56 | -7.50 | -7.47 | -25.72 | 13.32 | -10.43 | 19.46 | 34.79 | -0.55 | -6.69 | -15.33 | **0.42** |

*contd. ...*

*...contd. ...*

| 1 | 2 | 3 | 4 | 5 | 6 | 7 | 8 | 9 | 10 | 11 | 12 | 13 | 14 | 15 | 16 |
|---|---|---|---|---|---|---|---|---|---|---|---|---|---|---|---|
| Sunflower | -25.09 | 6.19 | -8.98 | -9.74 | 4.66 | -29.41 | -16.64 | 9.59 | 39.55 | 22.00 | 7.84 | 8.28 | -7.47 | 13.06 | **0.99** |
| Soyabean | -1.20 | 16.62 | 8.17 | 9.95 | 8.39 | -4.17 | 3.12 | -1.15 | -3.75 | 7.36 | 15.51 | 1.80 | 8.06 | 6.63 | **5.38** |
| Nine Oilseeds | -5.86 | 2.66 | 1.29 | -0.80 | 0.40 | -7.42 | -6.23 | -0.59 | -5.07 | 10.12 | 16.31 | 1.23 | -4.85 | 0.13 | **0.09** |
| **Total Oilseeds** | **-2.68** | **5.60** | **1.39** | **-1.37** | **1.60** | **-7.28** | **-4.69** | **1.61** | **-7.75** | **6.80** | **15.21** | **0.63** | **-2.31** | **0.83** | 0.54 |
| Cotton (Lint) | 7.60 | 14.69 | 1.00 | -2.80 | 5.34 | -6.77 | -2.01 | 7.00 | -16.01 | -0.94 | 15.65 | -1.25 | 5.39 | 3.05 | **2.14** |
| Jute | 6.28 | -0.21 | 21.70 | 1.02 | -6.40 | -0.20 | -2.21 | 5.46 | -0.98 | -1.79 | -8.85 | -1.82 | 4.36 | 2.91 | **1.38** |
| Mesta | -0.44 | -0.45 | 8.06 | -1.87 | -11.67 | 6.55 | 0.53 | -8.22 | -1.90 | -10.71 | -7.02 | -2.75 | 3.12 | 1.97 | **-1.77** |
| Jute & Mesta | 4.86 | -0.21 | 18.99 | 0.44 | -7.35 | 0.97 | -1.71 | 2.91 | -1.14 | -3.26 | -8.57 | -1.97 | 4.17 | 2.77 | **0.78** |
| **Total Fibres** | **7.14** | **13.13** | **2.56** | **-2.41** | **3.85** | **-5.96** | **-1.91** | **6.48** | **-14.46** | **-1.31** | **12.75** | **-1.32** | **5.25** | **3.01** | **1.92** |
| Sugarcane | 12.99 | 7.29 | 0.61 | -5.80 | 3.18 | 4.06 | 2.28 | 2.22 | 2.46 | -12.87 | -7.03 | 14.74 | 22.61 | -2.09 | **3.19** |
| Potato | 2.09 | 3.71 | 12.34 | -3.19 | 9.50 | 1.48 | -8.81 | -1.15 | 10.70 | -4.15 | 2.27 | 6.29 | 5.77 | 0.00 | **2.63** |
| Tobacco | 4.42 | 0.63 | 8.52 | 8.62 | 9.68 | -15.18 | -39.55 | 33.27 | -6.28 | 12.99 | -0.87 | 1.72 | -1.15 | 0.00 | **1.20** |
| **Non-Foodgrains** | **2.79** | **3.41** | **1.69** | **-1.16** | **1.07** | **-2.14** | **-2.74** | **0.54** | **-10.87** | **8.15** | **9.80** | **2.09** | **1.23** | **0.43** | **1.02** |
| **All Principal Crops** | **1.56** | **-0.65** | **2.11** | **-0.24** | **1.02** | **-1.79** | **-1.98** | **0.66** | **-7.83** | **8.34** | **0.75** | **1.50** | **1.60** | **0.55** | **0.40** |

* Based on Fourth Advance Estimates, 2007-08 as released on 09.07.2008

# Table 6.6

*Growth Rates of Production of Principal Crops in India from 1994-95 to 2007-08*
*(Base: T.E. 1993-94 = 100)*

(Per cent)

| Crop | 1994-95 | 1995-96 | 1996-97 | 1997-98 | 1998-99 | 1999-00 | 2000-01 | 2001-02 | 2002-03 | 2003-04 | 2004-05 | 2005-06 | 2006-07 | 2007-08* | Average Growth During 1994-95 to 2007-08 |
|---|---|---|---|---|---|---|---|---|---|---|---|---|---|---|---|
| 1 | 2 | 3 | 4 | 5 | 6 | 7 | 8 | 9 | 10 | 11 | 12 | 13 | 14 | 15 | 16 |
| Rice | 1.85 | -5.91 | 6.16 | 1.02 | 4.29 | 4.19 | -5.25 | 9.84 | -23.06 | 23.26 | -6.09 | 10.42 | 1.70 | 3.29 | 1.84 |
| Wheat | 9.91 | -5.60 | 11.69 | -4.32 | 7.45 | 7.13 | -8.76 | 4.43 | -9.63 | 9.73 | -4.87 | 1.05 | 9.30 | 3.42 | 2.21 |
| *Jowar* | -22.75 | 4.04 | 17.26 | -31.18 | 11.79 | 3.20 | -13.30 | 0.36 | -7.20 | -4.72 | 8.43 | 0.00 | -6.28 | 8.77 | -1.88 |
| *Bajra* | 43.92 | -24.83 | 46.10 | -2.83 | -9.01 | -16.87 | 16.90 | 22.56 | -43.04 | 156.61 | -34.52 | -3.12 | 9.63 | 16.24 | 12.70 |
| Maize | -7.44 | 7.26 | 12.99 | 0.42 | 3.06 | 3.25 | 4.64 | 9.27 | -15.26 | 34.37 | -5.42 | 3.80 | 2.63 | 27.91 | 5.82 |
| Ragi | -9.81 | 6.81 | -6.17 | -11.07 | 24.98 | -12.22 | 19.31 | -13.07 | -44.59 | 49.40 | 23.69 | -3.24 | -38.66 | 47.41 | 2.34 |
| Small millets | -13.00 | -2.34 | -6.51 | -11.48 | 4.02 | -7.84 | -5.06 | -1.74 | -20.36 | 22.75 | -15.30 | -1.26 | 1.70 | 2.59 | -3.84 |
| Barley | 31.46 | -12.49 | -3.19 | 14.85 | -8.43 | -5.90 | -1.13 | -0.43 | -1.20 | -7.80 | -6.97 | 1.12 | 8.79 | -7.67 | 0.07 |
| Coarse Cereals | -2.71 | -3.44 | 17.80 | -10.50 | 2.58 | -3.38 | 2.48 | 7.68 | -21.91 | 44.93 | -11.62 | 1.75 | 0.03 | 20.06 | 3.12 |
| Total Cereals | 3.68 | -5.51 | 9.37 | -2.29 | 5.10 | 4.24 | -5.57 | 7.83 | -18.71 | 21.04 | -6.47 | 6.25 | 3.82 | 6.14 | 2.07 |
| Gram | 29.23 | -22.62 | 11.76 | 10.20 | 10.90 | -24.74 | -24.67 | 41.96 | -22.59 | 34.95 | -4.33 | 1.94 | 13.60 | -6.69 | 3.49 |
| *Tur* | -20.39 | 7.80 | 14.98 | -30.37 | 46.41 | -0.52 | -16.61 | 0.58 | -3.27 | 7.80 | -0.41 | 16.66 | -15.48 | 33.57 | 2.91 |
| Other Pulses | 1.67 | -12.70 | 25.58 | -14.65 | 8.03 | 3.85 | -11.28 | 13.30 | 65.13 | 45.27 | -22.24 | -0.2 | 9.97 | 10.13 | 8.36 |
| Total Pulses | 8.93 | -14.60 | 16.91 | -6.57 | 15.18 | -11.54 | -18.20 | 22.02 | 12.34 | 33.10 | -10.73 | 2.28 | 6.29 | 6.45 | 4.42 |
| Total Foodgrains | 4.37 | -6.81 | 10.43 | -2.89 | 6.44 | 1.95 | -7.16 | 9.41 | -14.86 | 22.57 | -7.03 | 5.06 | 4.13 | 6.16 | 2.32 |
| Groundnut | 2.96 | -5.97 | 14.03 | -14.75 | 21.83 | -41.46 | 21.87 | 9.66 | -41.36 | 97.91 | -16.68 | 17.99 | -39.16 | 92.52 | 8.53 |
| Sesamum | 4.02 | -9.39 | 20.61 | -11.32 | -7.18 | -8.99 | 7.90 | 34.76 | -36.76 | 77.23 | -13.75 | -4.90 | -3.54 | 26.94 | 5.40 |
| R & M | 6.03 | 4.19 | 11.00 | -29.33 | 20.43 | 2.20 | -27.66 | 21.39 | -23.67 | 62.16 | 20.72 | 7.09 | -8.53 | -21.98 | 3.14 |

contd. ...

*...contd. ...*

| *1* | *2* | *3* | *4* | *5* | *6* | *7* | *8* | *9* | *10* | *11* | *12* | *13* | *14* | *15* | *16* |
|---|---|---|---|---|---|---|---|---|---|---|---|---|---|---|---|
| Sunflower | -9.66 | 3.25 | -0.61 | -29.17 | 6.64 | -26.51 | -6.81 | 5.12 | 28.42 | 6.62 | 312.43 | 21.26 | -14.70 | 13.65 | 1.79 |
| Soyabean | -17.16 | 29.60 | 5.49 | 20.22 | 10.52 | -0.86 | -25.50 | 13.02 | -21.94 | 67.98 | -12.07 | 20.32 | 6.98 | 12.83 | 7.82 |
| Nine Oilseeds | -0.22 | 2.62 | 10.49 | -14.11 | 16.10 | -16.92 | -10.09 | 12.28 | -28.53 | 69.84 | -2.47 | 14.24 | -14.27 | 18.68 | 4.12 |
| Total Oilseeds | 1.88 | 2.00 | 8.96 | -12.99 | 14.07 | -15.54 | -7.99 | 9.98 | -23.25 | 53.74 | -0.56 | 14.34 | -9.56 | 9.49 | 3.18 |
| Cotton (Lint) | 10.68 | 8.22 | 10.65 | -23.73 | 13.23 | -6.17 | -17.40 | 4.97 | -12.74 | 59.20 | 19.64 | 12.60 | 22.34 | 14.03 | 8.25 |
| Jute | 8.65 | -4.04 | 29.84 | -0.05 | -11.28 | 6.70 | -1.18 | 13.60 | -2.93 | -0.22 | -8.31 | 6.07 | 3.49 | -1.31 | 2.79 |
| Mesta | 0.99 | 4.88 | 3.31 | -9.61 | -7.82 | 15.95 | 9.69 | -11.69 | -8.47 | -8.03 | -5.21 | -0.33 | 9.86 | 3.99 | -0.18 |
| Jute & Mesta | 8.04 | -3.31 | 27.49 | -0.71 | -11.23 | 7.31 | -0.40 | 11.61 | -3.28 | -0.67 | -8.14 | 5.71 | 3.82 | -0.86 | 2.53 |
| Total Fibres | 10.37 | 7.14 | 11.94 | -21.63 | 10.37 | -4.91 | -15.62 | 5.76 | -11.49 | 50.61 | 17.08 | 5.39 | 4.12 | 9.07 | 3.79 |
| Sugarcane | 19.94 | 2.07 | -1.27 | 0.68 | 3.28 | 3.67 | -1.13 | 0.42 | -3.31 | -18.62 | 1.32 | 18.59 | 26.44 | -4.21 | 3.42 |
| Potato | 0.00 | 8.26 | 28.53 | -27.13 | 33.79 | 4.67 | -9.00 | 6.38 | -3.19 | -0.90 | 2.48 | 1.16 | -7.59 | 0.00 | 2.68 |
| Tobacco | 7.07 | -9.00 | 15.49 | 3.24 | 15.63 | -29.19 | -34.22 | 58.25 | -9.86 | 9.94 | -0.11 | 0.26 | -5.94 | 0.00 | 1.56 |
| Non-Foodgrains | 7.32 | 2.69 | 6.80 | -7.60 | 9.39 | -6.15 | -4.53 | 5.41 | -11.09 | 19.15 | 2.52 | 16.05 | 18.85 | -1.49 | 4.03 |
| All Principal Crops | 5.77 | -2.54 | 8.68 | -5.08 | 7.85 | -1.79 | -6.00 | 7.62 | -13.20 | 21.00 | -2.73 | 12.16 | 13.93 | 0.85 | 3.40 |

* Based on Fourth Advance Estimates, 2007-08 as released on 09.07.2008.

# Table 6.7

*Growth Rates of Yield of Principal Crops in India from 1994-95 to 2007-08*
*(Base: T.E. 1993-94 = 100)*

*(Per cent )*

| *Crop* | *1994-95* | *1995-96* | *1996-97* | *1997-98* | *1998-99* | *1999-00* | *2000-01* | *2001-02* | *2002-03* | *2003-04* | *2004-05* | *2005-06* | *2006-07* | *2007-08** | *Average Growth During 1994-95 to 2007-08* |
|---|---|---|---|---|---|---|---|---|---|---|---|---|---|---|---|
| *1* | *2* | *3* | *4* | *5* | *6* | *7* | *8* | *9* | *10* | *11* | *12* | *13* | *14* | *15* | *16* |
| Rice | 1.23 | -5.90 | 4.68 | 0.95 | 1.15 | 3.36 | -4.29 | 9.37 | -16.09 | 19.16 | -4.56 | 5.99 | 1.34 | 3.40 | 1.41 |
| Wheat | 7.46 | -2.96 | 7.91 | -7.24 | 4.22 | 7.27 | -2.53 | 1.99 | -5.51 | 3.95 | -4.11 | 0.66 | 3.40 | 2.84 | 1.24 |
| *Jowar* | -13.00 | 5.79 | 16.22 | -27.14 | 23.28 | -1.40 | -9.83 | 0.99 | -2.26 | -5.04 | 11.27 | 10.48 | -4.12 | 16.26 | 1.54 |
| *Bajra* | 34.51 | -17.60 | 36.56 | 0.31 | -5.39 | -13.13 | 5.81 | 26.41 | -29.87 | 87.15 | -24.72 | -6.64 | 10.47 | 16.22 | 8.58 |
| Maize | -10.82 | 10.06 | 7.82 | -0.44 | 5.02 | -0.26 | 1.64 | 9.77 | -15.95 | 21.41 | -6.53 | 1.64 | -1.34 | 22.18 | 3.16 |
| Ragi | -3.63 | 6.24 | -6.72 | -4.30 | 17.76 | -5.57 | 10.86 | -7.14 | -35.52 | 26.86 | 32.80 | -2.05 | -20.09 | 18.88 | 2.03 |
| Small millets | -8.29 | 5.13 | -2.87 | -7.28 | 6.44 | -2.37 | -5.95 | 6.79 | -13.07 | 23.70 | -8.30 | 2.11 | 7.14 | 10.25 | 0.96 |
| Barley | 17.37 | -5.43 | 5.28 | 1.41 | -0.92 | 2.97 | -7.87 | 17.39 | -7.13 | -1.54 | -0.86 | -1.03 | 6.05 | -2.83 | 1.63 |
| Coarse Cereals | -0.68 | 1.08 | 14.37 | -7.64 | 7.80 | -3.37 | -0.63 | 10.36 | -14.58 | 27.01 | -6.24 | 1.62 | 1.27 | 20.03 | 3.60 |
| Total Cereals | 3.58 | -3.45 | 6.69 | -2.12 | 4.39 | 3.91 | -4.36 | 7.75 | -12.26 | 13.02 | -3.91 | 4.25 | 2.47 | 6.01 | 1.86 |
| Gram | 8.91 | -18.00 | 16.18 | -0.27 | -0.96 | 3.70 | -10.71 | 14.72 | -15.91 | 13.09 | 0.41 | -0.75 | 4.54 | -7.75 | 0.51 |
| *Tur* | -15.15 | 3.61 | 12.75 | -27.12 | 42.98 | -0.16 | -21.32 | 9.79 | -4.16 | 3.00 | -0.49 | 14.64 | -15.03 | 26.90 | 2.16 |
| Other Pulses | -0.20 | -5.78 | 18.20 | -13.70 | 12.03 | 4.29 | -11.19 | 6.52 | 80.33 | 26.53 | -19.96 | 0.14 | 7.69 | 7.12 | 8.00 |
| Total Pulses | 3.15 | -10.01 | 14.51 | -8.51 | 12.45 | -1.55 | -15.12 | 12.82 | 20.63 | 16.30 | -8.01 | 4.12 | 2.48 | 3.70 | 3.35 |
| Total Foodgrains | 3.30 | -4.29 | 7.82 | -3.19 | 5.38 | 3.66 | -5.58 | 7.86 | -8.19 | 13.05 | -4.45 | 4.45 | 2.33 | 5.53 | 1.98 |
| Groundnut | 9.23 | -2.00 | 12.97 | -8.61 | 16.76 | -36.95 | 27.61 | 15.30 | -38.37 | 95.50 | -24.84 | 16.32 | -27.01 | 68.56 | 8.89 |
| Sesamum | 17.20 | -2.27 | 10.47 | 6.49 | -4.24 | -6.14 | -2.13 | 38.75 | -26.85 | 50.55 | -20.53 | 1.77 | -2.41 | 22.77 | 5.96 |
| R & M | 10.09 | -3.56 | 10.98 | -34.33 | 30.19 | 10.45 | -2.61 | 7.12 | -14.78 | 35.75 | -10.46 | 7.67 | -1.97 | -7.85 | 2.62 |

*contd. ...*

*...contd. ...*

| *1* | *2* | *3* | *4* | *5* | *6* | *7* | *8* | *9* | *10* | *11* | *12* | *13* | *14* | *15* | *16* |
|---|---|---|---|---|---|---|---|---|---|---|---|---|---|---|---|
| Sunflower | 20.71 | -2.72 | 9.13 | -21.58 | 1.89 | 4.10 | 11.79 | -4.08 | -7.98 | -12.60 | 18.27 | 11.98 | -7.81 | 30.72 | 3.70 |
| Soyabean | -16.11 | 11.09 | -2.40 | 9.28 | 1.96 | 3.45 | -27.75 | 14.33 | -18.90 | 56.47 | -23.86 | 18.19 | -1.00 | 5.81 | 2.18 |
| Nine Oilseeds | 6.01 | 0.00 | 9.00 | -13.38 | 15.64 | -10.26 | -4.12 | 12.95 | -24.72 | 54.23 | -16.16 | 12.85 | -9.90 | 18.52 | 3.62 |
| Total Oilseeds | 4.76 | -3.47 | 7.47 | -11.76 | 12.27 | -8.91 | -3.45 | 8.24 | -16.80 | 43.95 | -13.72 | 13.62 | -7.42 | 8.58 | 2.38 |
| Cotton (Lint) | 2.99 | -5.80 | 9.63 | -21.56 | 7.49 | 0.65 | -15.70 | -1.90 | 3.90 | 60.71 | 3.47 | 14.02 | 16.09 | 10.66 | 6.05 |
| Jute | 2.19 | -3.79 | 6.58 | -0.99 | -5.21 | 6.91 | 1.05 | 7.72 | -1.97 | 1.61 | 0.58 | 8.03 | -0.83 | -4.10 | 1.27 |
| Mesta | 1.48 | 5.35 | -4.43 | -7.92 | 4.36 | 8.82 | 9.11 | -3.78 | -6.70 | 3.01 | 1.91 | 2.49 | 6.54 | 1.98 | 1.59 |
| Jute & Mesta | 3.00 | -3.09 | 7.17 | -1.14 | -4.18 | 6.28 | 1.33 | 8.46 | -2.16 | 2.68 | 0.46 | 7.83 | -0.33 | -3.53 | 1.63 |
| Total Fibres | 3.07 | -5.23 | 9.04 | -19.72 | 6.28 | 1.12 | -13.99 | -0.68 | 3.46 | 52.62 | -18.48 | 6.80 | -1.08 | 5.88 | 2.08 |
| Sugarcane | 6.16 | -4.88 | -1.84 | 6.91 | 0.10 | -0.38 | -3.32 | -1.76 | -5.63 | -6.60 | 9.05 | 3.36 | 3.13 | -2.16 | 0.15 |
| Potato | -1.98 | 4.42 | 14.36 | -24.72 | 22.18 | 3.15 | -0.21 | 7.62 | -12.55 | 3.39 | 0.20 | -4.82 | -12.63 | 0.00 | -0.11 |
| Tobacco | 2.62 | -9.63 | 6.37 | -5.01 | 5.42 | -16.52 | 8.82 | 18.75 | -3.82 | -2.70 | 0.73 | -1.13 | -4.84 | 0.00 | -0.07 |
| Non-Foodgrains | 4.42 | -0.67 | 5.04 | -6.46 | 8.24 | -4.10 | -1.84 | 4.85 | -0.25 | 10.18 | -7.47 | 13.67 | 17.41 | -1.91 | 2.94 |
| All Principal Crops | 3.96 | -1.81 | 6.40 | -4.83 | 6.76 | 0.00 | -4.10 | 6.91 | -5.83 | 11.68 | 2.37 | 10.50 | 12.14 | 0.30 | 3.17 |

* Based on fourth advance estimates for 2007-08 except potato and tobacco for which final estimates for 2006-07 have been used.

## Table 6.8

*Index of Terms of Trade between Agricultural and Non-Agricultural Sectors from 1982-83 to 2006-07*

*(Base: Triennium Ending 1990-91=100)*

| Year | Index of Prices Received (IPR) | Index of Prices Paid (IPP) for | | | Combined Index Trade | Index of Terms of (ITT) |
|---|---|---|---|---|---|---|
| | | Final Consumption | Intermediate Consumption | Capital Formation | | |
| Weights | | 73.54 | 21.63 | 4.83 | 100.0 | |
| 1 | 2 | 3 | 4 | 5 | 6 | 7 |
| 1982-83 | 60.3 | 58.8 | 91.1 | 62.6 | 66.0 | 91.4 |
| 1983-84 | 64.2 | 64.2 | 91.0 | 67.4 | 70.1 | 91.6 |
| 1984-85 | 68.0 | 66.6 | 92.3 | 72.5 | 72.4 | 93.9 |
| 1985-86 | 70.4 | 69.5 | 94.3 | 76.4 | 75.2 | 93.6 |
| 1986-87 | 76.7 | 74.8 | 98.7 | 78.8 | 80.2 | 95.7 |
| 1987-88 | 86.0 | 84.6 | 102.3 | 82.5 | 88.3 | 97.4 |
| 1988-89 | 90.3 | 90.4 | 96.9 | 90.9 | 91.8 | 98.3 |
| 1989-90 | 97.5 | 97.6 | 99.2 | 100.6 | 98.1 | 99.4 |
| 1990-91 | 112.3 | 112.1 | 104.0 | 108.5 | 110.2 | 101.9 |
| 1991-92 | 130.8 | 124.9 | 119.4 | 127.2 | 123.8 | 105.6 |
| 1992-93 | 138.7 | 131.5 | 139.5 | 137.5 | 133.5 | 103.9 |
| 1993-94 | 151.4 | 143.9 | 152.9 | 147.3 | 146.1 | 103.6 |
| 1994-95 | 171.1 | 159.0 | 166.1 | 158.4 | 160.5 | 106.6 |
| 1995-96 | 182.9 | 173.4 | 174.2 | 176.1 | 173.7 | 105.3 |
| 1996-97 | 190.6 | 185.6 | 181.5 | 188.8 | 184.8 | 103.1 |
| 1997-98 | 205.9 | 195.7 | 192.0 | 196.7 | 194.9 | 105.6 |
| 1998-99 | 220.8 | 213.8 | 197.1 | 206.8 | 209.9 | 105.2 |
| 1999-2000 | 219.8 | 217.1 | 203.9 | 212.6 | 214.0 | 102.7 |
| 2000-01 | 225.0 | 220.5 | 230.4 | 227.0 | 223.0 | 100.9 |
| 2001-02 | 235.3 | 226.4 | 235.2 | 240.4 | 229.0 | 102.8 |
| 2002-03 | 247.9 | 234.9 | 252.7 | 245.2 | 239.3 | 103.6 |
| 2003-04 | 251.2 | 245.2 | 259.1 | 255.7 | 248.7 | 101.0 |
| 2004-05 | 258.2 | 252.3 | 264.5 | 305.6 | 257.5 | 100.3 |
| 2005-06 | 275.8 | 266.0 | 277.1 | 310.5 | 270.6 | 101.9 |
| 2006-07* | 291.1 | 283.4 | 282.7 | 327.8 | 285.4 | 102.0 |

*Note*: ITT is obtained by dividing IPR by the combined index, and multiplied by 100.

* Based on Third Advance Estimates, 2007-08 as releasec on 22-4-2008

# Table 7.1

*Share of Various Countries in World Production of Major Agricultural Commodities during 2006*

*(Production- '000 tonnes)*

| Country /Commodity | Production | % Share of the Country |
|---|---|---|
| **Rice/Paddy** | | |
| World | 634606 | **100.00** |
| China | 184070 | 29.01 |
| India | 136510 | 21.51 |
| Indonesia | 54400 | 8.57 |
| Bangladesh | 43729 | 6.89 |
| Viet Nam | 35827 | 5.65 |
| Thailand | 29269 | 4.61 |
| Myanmar | 25200 | 3.97 |
| Philipnes | 15326 | 2.42 |
| Brazil | 11505 | 1.81 |
| Japan | 10695 | 1.69 |
| USA | 8787 | 1.38 |
| Pakistan | 8137 | 1.28 |
| Korea Rep | 6305 | 0.99 |
| Egypt | 6500 | 1.02 |
| Nepal | 4209 | 0.66 |
| Cambodia | 6264 | 0.99 |
| Nigeria | 3924 | 0.62 |
| Iran | 3600 | 0.57 |
| Madagascar | 3485 | 0.55 |

| Country /Commodity | Production | % Share of the Country |
|---|---|---|
| **Maize** | | |
| World | 695228 | **100.00** |
| USA | 267598 | 38.49 |
| China | 145625 | 20.95 |
| Brazil | 42632 | 6.13 |
| Mexico | 21765 | 3.13 |
| France | 12902 | 1.86 |
| Argentina | 14446 | 2.08 |
| Romania | 8985 | 1.29 |
| India | 14710 | 2.12 |
| Italy | 9671 | 1.39 |
| Indonesia | 11611 | 1.67 |
| South Africa | 6935 | 1.00 |
| Ukraine | 6320 | 0.91 |
| Canada | 9268 | 1.33 |
| Hungary | 8441 | 1.21 |
| Egypt | 6838 | 0.98 |
| Serbia | 6017 | 0.87 |
| Philipnes | 6082 | 0.87 |
| Nigeria | 6404 | 0.92 |
| Spain | 3461 | 0.50 |
| Thailand | 3696 | 0.53 |

| *Country /Commodity* | *% Share of the* | |
|---|---|---|
| | *Production* | *Country* |
| **Wheat** | | |
| World | 605946 | **100.00** |
| China | 104470 | 17.24 |
| India | 69350 | 11.44 |
| USA | 57298 | 9.46 |
| Russia | 45006 | 7.43 |
| France | 35367 | 5.84 |
| Canada | 27277 | 4.50 |
| Germany | 22428 | 3.70 |
| Turkey | 20010 | 3.30 |
| Australia | 9819 | 1.62 |
| Pakistan | 21277 | 3.51 |
| Ukraine | 14000 | 2.31 |
| U.K. | 14735 | 2.43 |
| Argentina | 14000 | 2.31 |
| Iran Rep | 14500 | 2.39 |
| Kazakhstan | 13500 | 2.23 |
| Poland | 7060 | 1.17 |
| Italy | 7092 | 1.17 |
| Romania | 5526 | 0.91 |
| Egypt | 8308 | 1.37 |
| Spain | 5576 | 0.92 |

| *Country /Commodity* | *% Share of the* | |
|---|---|---|
| | *Production* | *Country* |
| **Groundnut In Shell** | | |
| World | 47768 | **100.00** |
| China | 14722 | 30.82 |
| India | 4680 | 10.43 |
| Nigeria | 3825 | 8.01 |
| USA | 1479 | 3.10 |
| Indonesia | 14700 | 30.77 |
| Sudan | 540 | 1.13 |
| Myanmar | 910 | 1.91 |
| Senegal | 460 | 0.96 |
| Viet Nam | 465 | 0.97 |
| Chad | 450 | 0.94 |
| Argentina | 496 | 1.04 |
| Ghana | 520 | 1.09 |
| Congo Dem Rep | 369 | 0.77 |
| Guinea | 294 | 0.62 |
| Burkina Faso | 221 | 0.46 |
| Cameroon | 160 | 0.33 |
| Brazil | 239 | 0.50 |
| Egypt | 180 | 0.38 |
| Mail | 172 | 0.36 |
| Malawi | 145 | 0.30 |

| Country /Commodity | % Share of the | |
|---|---|---|
| | Production | Country |
| **Tobacco Unmanufatureed** | | |
| World | 6719 | **100.00** |
| China | 2750 | 40.93 |
| Brazil | 905 | 13.47 |
| India | 550 | 8.19 |
| USA | 338 | 5.03 |
| Turkey | 140 | 2.08 |
| Indonesia | 141 | 2.10 |
| Greece | 126 | 1.88 |
| Argentina | 164 | 2.44 |
| Italy | 120 | 1.79 |
| Pakistan | 113 | 1.68 |
| Bulgaria | 42 | 0.63 |
| Malawi | 70 | 1.04 |
| Thailand | 70 | 1.04 |
| Korea Dem Rep | 35 | 0.52 |
| Zimbabwe | 65 | 0.97 |
| Japan | 47 | 0.70 |
| Philippines | 38 | 0.57 |
| Canada | 43 | 0.64 |
| Bangladesh | 39 | 0.58 |
| Korea Rep | 35 | 0.52 |

| Country /Commodity | % Share of the | |
|---|---|---|
| | Production | Country |
| **Sugar Cane** | | |
| World | 1392365 | **100.00** |
| Brazil | 455291 | 32.70 |
| India | 281170 | 20.19 |
| China | 100684 | 7.23 |
| Thailand | 47658 | 3.42 |
| Pakistan | 44666 | 3.21 |
| Mexico | 50597 | 3.63 |
| Colombia | 39849 | 2.86 |
| Australia | 38169 | 2.74 |
| Philppiness | 24345 | 1.75 |
| USA | 26835 | 1.93 |
| Indonesia | 30150 | 2.17 |
| Cuba | 11060 | 0.79 |
| Argentina | 18799 | 1.35 |
| South Africa | 20275 | 1.46 |
| Guatemala | 18721 | 1.34 |
| Egypt | 16317 | 1.17 |
| Vietnam | 15679 | 1.13 |
| Peru | 7600 | 0.55 |
| Venezuela | 8895 | 0.64 |
| Ecuador | 6834 | 0.49 |

*Source: www.Faostat.fao.org*

# Table 7.2

## *Area, Production and Yield of Principal Crops in Various Countries in 2006*

*Area ('000 Hectares)*
*Production ('000 Tonnes)*
*Yield (Kg./Hectare)*

| *Country* | *Area* | *Production* | *Yield* |
|---|---|---|---|
| *1* | 2 | 3 | 4 |
| **1. Paddy** | | | |
| World | 154324 | 634606 | 4112 |
| Bangladesh | 11200 | 43729 | 3904 |
| Brazil | 2975 | 11505 | 3868 |
| China | 29380 | 184070 | 6265 |
| Egypt | 613 | 6500 | 10598 |
| India | 43700 | 136510 | 3124 |
| Indonesia | 11400 | 54400 | 4772 |
| Japan | 1688 | 10695 | 6336 |
| Myanmar | 7200 | 25200 | 3500 |
| Pakistan | 2572 | 8137 | 3164 |
| Philippines | 4160 | 15327 | 3684 |
| Russian Federation | 156 | 686 | 4394 |
| Thailand | 10073 | 29269 | 2906 |
| USA | 1142 | 8787 | 7694 |
| Vietnam | 7324 | 35827 | 4891 |
| **2. Wheat** | | | |
| World | 216100 | 605946 | 2804 |
| Argentina | 5500 | 14000 | 2545 |
| Bangladesh | 481 | 772 | 1605 |
| Canada | 10534 | 27277 | 2589 |
| China | 23450 | 104470 | 4455 |
| Egypt | 1287 | 8308 | 6455 |
| France | 5247 | 35367 | 6740 |
| India | 26480 | 69350 | 2619 |
| Iran | 6000 | 14500 | 2417 |
| Italy | 1926 | 7092 | 3682 |
| Pakistan | 8448 | 21277 | 2519 |
| Romania | 1992 | 5526 | 2774 |
| Russian Federation | 23049 | 45006 | 1953 |
| Spain | 1958 | 5576 | 2848 |
| Syria | 1904 | 4669 | 2452 |
| Turkey | 9300 | 20010 | 2152 |
| U.K. | 1833 | 14735 | 8039 |
| USA | 20280 | 57298 | 2825 |
| Ukraine | 6395 | 14000 | 2189 |
| Australia | 11138 | 9819 | 882 |

*contd. ...*

*...contd. ...*

| 1 | 2 | 3 | 4 |
|---|---|---|---|
| **3. Maize** | | | |
| World | 144376 | 695228 | 4815 |
| Argentina | 2447 | 14446 | 5903 |
| Brazil | 12603 | 42632 | 3383 |
| Canada | 1093 | 9268 | 8480 |
| China | 27143 | 145625 | 5365 |
| France | 1503 | 12902 | 8586 |
| Egypt | 867 | 6838 | 7887 |
| Hungry | 1229 | 8441 | 6868 |
| India | 7590 | 14710 | 1938 |
| Indonesia | 3346 | 11611 | 3470 |
| Italy | 1108 | 9671 | 8729 |
| Mexico | 7338 | 21765 | 2966 |
| Pakistan | 1022 | 2971 | 2906 |
| Philippines | 2571 | 6082 | 2366 |
| Romania | 2513 | 8985 | 3575 |
| Russian Fed | 650 | 3811 | 5863 |
| Turkey | 28590 | 267598 | 9360 |
| USA | 1011 | 3669 | 3629 |
| **4. Groundnut (in shell)** | | | |
| World | 22232 | 47768 | 2149 |
| Argentina | 164 | 496 | 3031 |
| Brazil | 106 | 239 | 2251 |
| China | 4722 | 14722 | 3118 |
| India | 5800 | 4980 | 859 |
| Indonesia | 707 | 14700 | 20804 |
| Japan | 9 | 20 | 2326 |
| Myanmar | 650 | 910 | 1400 |
| Nigeria | 2224 | 3825 | 1720 |
| Senegal | 594 | 460 | 775 |
| Sudan | 935 | 540 | 578 |
| Thailand | 66 | 117 | 1773 |
| USA | 499 | 1479 | 2964 |
| Uganda | 230 | 154 | 670 |
| Vietnam | 249 | 465 | 1865 |

*contd. ...*

*...contd. ...*

| 1 | 2 | 3 | 4 |
|---|---|---|---|
| **5. Sugarcane** | | | |
| World | 20399 | 1392365 | 68257 |
| Argentina | 285 | 18799 | 66045 |
| Australia | 415 | 38169 | 91973 |
| Bangladesh | 157 | 6423 | 40862 |
| Brazil | 6153 | 455291 | 73996 |
| China | 1220 | 100684 | 82528 |
| Colombia | 426 | 39849 | 93601 |
| Cuba | 397 | 11060 | 27852 |
| Egypt | 135 | 16317 | 120887 |
| Guatemala | 190 | 18721 | 98534 |
| India | 4200 | 281170 | 66945 |
| Indonesia | 370 | 30150 | 81486 |
| Mauritius | 67 | 4749 | 71244 |
| Mexico | 668 | 50597 | 75710 |
| Thailand | 936 | 47658 | 50904 |
| Pakistan | 907 | 44666 | 49229 |
| Philippines | 392 | 24345 | 62061 |
| USA | 363 | 26835 | 73833 |
| **6. Tobacco Unmanufactured** | | | |
| World | 3897 | 6719 | 1724 |
| Argentina | 90 | 164 | 1817 |
| Bangladesh | 30 | 39 | 1285 |
| Brazil | 497 | 905 | 1821 |
| Bulgaria | 27 | 42 | 1533 |
| Canada | 16 | 43 | 2688 |
| China | 1377 | 2750 | 1997 |
| France | 7 | 19 | 2570 |
| Greece | 54 | 126 | 2308 |
| India | 370 | 550 | 1486 |
| Indonesia | 145 | 141 | 972 |
| Italy | 38 | 120 | 3158 |
| Japan | 19 | 47 | 2417 |
| Korea Republic | 14 | 35 | 2448 |
| Pakistan | 56 | 113 | 1996 |
| Philippines | 26 | 38 | 1458 |
| Poland | 17 | 38 | 2277 |
| Spain | 14 | 42 | 3111 |
| Thailand | 40 | 70 | 1750 |
| Turkey | 190 | 140 | 737 |
| USA | 135 | 338 | 2497 |
| Zimbabwe | 40 | 65 | 1625 |

*Source*: *www.faostat.fao.org*

# Table 7.3

## *India's Position in World Agriculture in 2006*

| *Item* | *India* | *World* | *India's Position* | | |
|---|---|---|---|---|---|
| | | | *% Share* | *Rank* | *Next to* |
| 1 | 2 | 3 | 4 | 5 | 6 |
| 1. **Area** ** (Million Hectares) | | | | | |
| Total Area | 329 | 13445 | 2.4 | Seventh | Russian Federation, Canada, U.S.A., China, Brazil, Australia |
| Land Area | 297 | 13015 | 2.3 | Seventh | Russian Federation, China, Canada, U.S.A., Brazil, Australia |
| Arable Land | 160 | 1423 | 11.2 | Second | U.S.A. |
| 2. **Population** (Million) | | | | | |
| Total | 1152 | 6593 | 17.5 | Second | China |
| Agriculture | 563 | 2606 | 21.6 | Second | China |
| 3. **Economically Active Population*** (Million) | | | | | |
| Total | 480 | 3141 | 15.3 | Second | China |
| Agriculture | 278 | 1353 | 20.5 | Second | China |
| 4. **Crop Production** (Million Tonnes) | | | | | |
| (A) **Total Cereals** | 239 | 2221 | 10.8 | Third | China, USA |
| Wheat | 69 | 606 | 11.4 | Second | China |
| Rice (Paddy) | 137 | 635 | 21.6 | Second | China |
| Total Pulses | 14 | 60 | 23.3 | First | |
| (B) Oilseeds | | | | | |
| Groundnut (in shell) | 5 | 48 | 10.4 | Third | China,Indonesia |
| Rapeseed | 8 | 49 | 16.3 | Third | China, Canada |
| 5. **Fruits & Vegetables** (Million Tonnes) | | | | | |
| Vegetables & Melons | 82 | 903 | 9.1 | Second | China |
| Fruits excluding Melons | 44 | 526 | 8.4 | Second | China |
| Potatoes | 24 | 315 | 7.6 | Third | China, Russian Federation |
| Onion (Dry) | 6 | 62 | 9.7 | Second | China |

*contd. ...*

*...contd. ...*

| *1* | *2* | *3* | *4* | *5* | *6* |
|---|---|---|---|---|---|
| **6. Commercial Crops** (Million Tonnes) | | | | | |
| Sugarcane | 281 | 1392 | 20.2 | Second | Brazil |
| Tea | 0.89 | 3.65 | 24.4 | Second | China |
| Coffee (green) | 0.27 | 7.84 | 3.4 | Sixth | Brazil, Vietnam Colombia, Indonesia |
| Jute & Jute like Fibres | 2.04 | 3.11 | 65.6 | First | |
| Cotton (Lint) | 3.56 | 2484 | 14.3 | Third | China, USA |
| Tobacco Leaves | 0.55 | 6.72 | 8.2 | Third | China, Brazil |
| **7. Livestock** (Million Heads) | | | | | |
| Cattle | 181 | 1383 | 13.1 | Second | Brazil |
| Buffaloes | 99 | 177 | 55.9 | First | |
| Camels | 0.63 | 22 | 2.9 | Sixth | Sudan, Ethiopia, Kenya, Chad, Pakistan |
| Sheep | 63 | 1102 | 5.7 | Third | China, Australia |
| Goats | 125 | 837 | 14.9 | Second | China |
| Chicken | 475 | 16888 | 2.8 | Fifth | China, USA, Indonesia, Brazil, |
| **8. Animal Products** | | | | | |
| Total Milk ('000 MT) | 95665 | 653789 | 14.6 | First | |
| Eggs Total ('000 MT) | 2604 | 66532 | 3.9 | Third | China, USA |
| Total Meat ('000 MT) | 6103 | 272884 | 2.2 | Fifth | China, USA, Brazil, Germany |
| **9. Implements** (Thousands numbers)** | | | | | |
| Tractors-in-use | 2528 | 27632 | 9.1 | Second | USA |

*Note*: * Figures relate to 2004
** Figures relate to 2005
*** Figures relate to 2003

*Source*: www.faostat.fao.org

# Table 7.4

## *Selected Indicators of Agriculture Development in India, Asia - Pacific Region and the World*

| Sl. No. | Indicator | Year | India | Asia Pacific Region | | | World |
|---|---|---|---|---|---|---|---|
| | | | | Developing Countries | Developed Countries | Total | |
| 1 | 2 | 3 | 4 | 5 | 6 | 7 | 8 |
| 1 | Agriculture Land ('000 hectares) | 1995 | 169750 | 503861 | 48566 | 552427 | 1521581 |
| | | 2005 | 169650 | 523500 | 57806 | 581306 | 1544621 |
| 2 | Agriculture Land as % of Total Land | 1995 | 57.1 | 19.1 | 5.8 | 16.0 | 11.7 |
| | | 2005 | 57.1 | 19.9 | 7.0 | 16.8 | 11.9 |
| 3 | Agriculture Population ('000) | 1995 | 523860 | 1858566 | 7776 | 1866342 | 2530256 |
| | | 2005 | 566140 | 1916130 | 4856 | 1920986 | 2610540 |
| 4 | Agriculture Population as % of Total Population | 1995 | 54.9 | 58.9 | 5.3 | 56.5 | 44.2 |
| | | 2005 | 49.9 | 53.2 | 3.2 | 51.2 | 40.1 |
| 5 | Ratio of Agriculture Land to Agriculture Population (Ha/Person) | 1995 | 0.32 | 0.27 | 6.25 | 6.52 | 0.60 |
| | | 2005 | 0.30 | 0.27 | 11.90 | 12.17 | 0.59 |
| 6 | Irrigated Land as % of Agricultural Land | 1995 | 31.2 | 32.8 | 11.2 | 30.9 | 18.2 |
| | | 2003 | 32.9 | 33.7 | 9.7 | 31.4 | 17.9 |
| 7 | Consumption Per hectare of Agriculture Land (Kg. Plant nutrient/hectare) | 1995 | 81.7 | 124.7 | 86.3 | 121.4 | 85.2 |
| | | 2005 | 121.4 | 166.6 | 76.9 | 157.7 | 103.1 |
| 8 | No. of Agricultural Tractors in Use | 1995 | 1354864 | 3593634 | 2514000 | 6107634 | 26231389 |
| | | 2003 | 2528122 | 5273643 | 2419000 | 7692643 | 27634185 |

*Note*: "Agricultural Land" as used in this publication refers to "Arable and Permanent Cropped Land", which excludes permanent meadows and pastures, fallow land resulting from shifting cultivation, and land under trees grown for food or timber. Double cropped area is counted once.

*Source*: Selected Indicators of Food and Agriculture Development in Asia-Pacific Region FAO Regional Office for Asia and the Pacific, Bangkok.

# Table 8.1

## *Minimum Support Prices of Various Agricultural Commodities (According to Crop Year)*

*(Rs. per quintal)*

| Sl. No. | Commodity | Variety | 2002-03 | Special Drought Relief Price | 2003-04 | 2004-05 | 2005-06 | 2006-07 | 2007-08 | 2008-09 |
|---|---|---|---|---|---|---|---|---|---|---|
| 1 | 2 | 3 | 4 | | 5 | 6 | 7 | 8 | 9 | 10 |
| | ***Kharif* Crops** | | | | | | | | | |
| 1 | Paddy | Common | 530 | 20 | 550 | 560 | 570 | 580^ | 645$$ | |
| | | **Grade 'A'** | 560 | 20 | 580 | 590 | 600 | 610^ | 675$$ | |
| 2 | *Jowar* | Hybrid | 485 | 5 | 505 | 515 | 525 | 540 | 600 | |
| | | Maldandi | - | - | - | - | - | 555 | 620 | |
| 3 | *Bajra* | | 485 | 10 | 505 | 515 | 525 | 540 | 600 | |
| 4 | Maize | | 485 | 5 | 505 | 525 | 540 | 540 | 620 | |
| 5 | Ragi | | 485 | 5 | 505 | 515 | 525 | 540 | 600 | |
| 6 | *Arhar (tur)* | | 1320 | 5 | 1360 | 1390 | 1400 | 1410 | 1550^^ | |
| 7 | *Moong* | | 1330 | 5 | 1370 | 1410 | 1520 | 1520 | 1700^^ | |
| 8 | *Urad* | | 1330 | 5 | 1370 | 1410 | 1520 | 1520 | 1700^^ | |
| 9 | Cotton | F-414/H-777/J34 | 1675 | 20 | 1725 | 1760 | 1760 | 1770* | 1800* | |
| | | H-4 | 1875 | 20 | 1925 | 1960 | 1980 | 1990** | 2030** | |
| 10 | Groundnut In Shell | | 1355 | 20 | 1400 | 1500 | 1520 | 1520 | 1550 | |
| 11 | Sunflower Seed | | 1195 | 15 | 1250 | 1340 | 1500 | 1500 | 1510 | |
| 12 | Soyabeen | Black | 795 | 10 | 840 | 900 | 900 | 900 | 910 | |
| | | Yellow | 885 | 10 | 930 | 1000 | 1010 | 1020 | 1050 | |
| 13 | Sesamum | | 1450 | 5 | 1485 | 1500 | 1550 | 1560 | 1580 | |
| 14 | Nigerseed | | 1120 | - | 1155 | 1180 | 1200 | 1220 | 1240 | |
| | ***Rabi* Crops** | | | | | | | | | |
| 15 | Wheat | | 620 | 10 | 630 | 640 | 650$ | 750$$ | 1000 | |
| 16 | Barley | | 500 | 5 | 525 | 540 | 550 | 565 | 650 | |
| 17 | Gram | | 1220 | 5 | 1400 | 1425 | 1435 | 1445 | 1600 | |
| 18 | *Masur* (Lentil) | | 1320 | 5 | 1500 | 1525 | 1535 | 1545 | 1700 | |
| 19 | Rapeseed/Mustard | | 1330 | 10 | 1600 | 1700 | 1715 | 1715 | 1800 | |
| 20 | Safflower | | 1300 | 5 | 1500 | 1550 | 1565 | 1565 | 1650 | |
| 21 | Toria | | 1295 | 10 | 1565 | 1665 | 1680 | 1680 | 1735 | |
| | **Other Crops** | | | | | | | | | |
| 22 | Copra | Milling | 3300 | - | 3320 | 3500 | 3570 | 3590 | 3620 | 3660 |
| | (Calender Year) | Ball | 3550 | - | 3570 | 3750 | 3820 | 3840 | 3870 | 3910 |
| 23 | Jute | | 850 | - | 860 | 890 | 910 | 1000 | 1055 | 1250 |
| 24 | Sugarcane@ | | 69.50 | - | 73.00 | 74.50 | 79.50 | 80.25 | 81.18 | 81.18 |
| 25 | Tobacco (vfc) | Black Soil (f2 Gr) | 28.00 | - | 31.00 | 32.00 | 32.00 | 32.00 | 32.00 | |
| | (Rs. per kg.) | Light Soil (L2 Gr) | 30.00 | - | 33.00 | 34.00 | 34.00 | 34.00 | 34.00 | |

@ Upto 2004-05 Statutory Minimum Price (SMP) linked to a basic recovery of 8.5% with proportionate premium for every 0.1% increase in recovery above that level. The SMP for 2002-2003 includes the one time drought relief of Rs. 5 per quintal recommended by CACP. From 2005-06 onwards SMP is linked to basic recovery of 9.0%.

$ An additional incentive bonus of Rs. 50 per quintal was payable over the Minimum Support Price (MSP).

$$ An additional incentive bonus of Rs. 100 per quintal is payable over the Minimum Support Price (MSP).

^ An additional incentive bonus of Rs. 40 per quintal is payable on procurement between 1.10.2006 to 31.03.2007. In case of Bihar and Kerala additional incentive bonus extended upto 31.5.2007 and in case of Andhra Pradesh, Chhattisgarh, Orissa, Tamil Nadu and West Bengal additional incentive bonus extended upto 30.9.2007.

* Medium Staple.

** Long Staple.

^^ A bonus of Rs. 40 per quintal is payble over & above the MSP.

# Table 8.2

*Central Issue Prices of Rice, Wheat and Coarse Grains*

*(Rs. per Quintal)*

| *Period* | *APL* | | *BPL* |
|---|---|---|---|
| | *Common* | *Grade "A"* | *Common/Grade"A"* |
| *1* | 2 | 3 | 4 |
| Rice | | | |
| 01.12.1997 to 28.01.1999 | 550@ | 700 | 350 |
| 29.01.1999 to 31.3.2000 | 700@ | 905 | 350 |
| 01.04.2000 to 24.7.2000 | 1135@ | 1180 | 590 |
| 25.07.2000 to 11.07.2001 | 1087@ | 1130 | 565 |
| 12.07.2001 to 31.03.2002 | 795@ | 830 | 565 |
| 01.04.2002 to 30.06.2002 | 695@ | 730 | 565 |
| 01.07.2002 to till date | 795@ | 830 | 565 |
| Wheat | | | |
| 01.06.1997 to 28.01.1999 | 450 | | 250 |
| 29.01.1999 to 31.03.1999 | 650 | | 250 |
| 01.04.1999 to 31.03.2000 | 682 | | 250 |
| 01.04.2000 to 24.07.2000 | 900 | | 450 |
| 25.07.2000 to 11.07.2001 | 830 | | 415 |
| 12.07.2001 to 31.03.2002 | 610 | | 415 |
| 01.04.2002 to 30.06.2002 | 510 | | 415 |
| 01.07.2002 to till date | 610 | | 415 |

| **Coarse Grains** | APL | BPL | AAY |
|---|---|---|---|
| 01.04.2005 to 26-11-2007 | 70% of Economic cost | 70% of Economic cost | Rs. 2.00 per kg. |
| 27-11-2007 to till date | Rs. 4.50 per kg. | Rs. 3.00 per kg. | Rs. 1.50 per kg. |

*Note*: @ Applicable only to J & K, H.P, North Eastern States, Sikkim and Uttarakhand.

Central issue price for AAY household fixed @ Rs.3 per kg. For rice and Rs.2 per kg for wheat since its inception in December 2000 and it was not been revised.

*Source*: Department of Food & Public Distribution.

# Table 8.3

*Cost Estimates of Principal Crops*

| *Crop/State* | *Year* | *Cost of Cultivation (Rs/Hect.)* | | *Cost of* | *Yield* |
|---|---|---|---|---|---|
| | | *A2+FL* | *C2* | *Production C2 (Rs/qtl.)* | *(Qtl./Hect.)* |
| *1* | *2* | *3* | *4* | *5* | *6* |
| **Paddy** | | | | | |
| Andhra Pradesh | 2003-04 | 18258.15 | 28532.78 | 488.83 | 53.72 |
| | 2004-05 | 18612.39 | 29056.11 | 503.73 | 53.65 |
| | 2005-06 | 18787.23 | 29256.97 | 540.96 | 50.21 |
| Assam | 2003-04 | 10561.12 | 14189.36 | 519.23 | 25.77 |
| | 2004-05 | 10937.62 | 14364.28 | 607.65 | 22.19 |
| | 2005-06 | 11176.62 | 15079.18 | 559.92 | 25.17 |
| Bihar | 2003-04 | 9078.61 | 13557.33 | 468.40 | 24.84 |
| | 2004-05 | 10118.01 | 14640.07 | 551.55 | 22.82 |
| | 2005-06 | 10969.84 | 15181.55 | 497.44 | 25.78 |
| Chhattisgarh | 2003-04 | 8969.88 | 13857.56 | 496.84 | 24.13 |
| | 2004-05 | 9295.87 | 14350.98 | 518.36 | 24.26 |
| | 2005-06 | 9831.59 | 15304.30 | 508.21 | 27.27 |
| Gujarat | 2005-06 | 14098.04 | 18854.53 | 478.35 | 33.20 |
| Haryana | 2003-04 | 20053.93 | 30462.71 | 692.74 | 43.13 |
| | 2004-05 | 21540.01 | 32038.25 | 704.69 | 44.70 |
| | 2005-06 | 18706.77 | 30543.61 | 618.45 | 48.72 |
| Himachal Pradesh | 2005-06 | 10240.81 | 15463.57 | 684.64 | 16.19 |
| Jharkhand | 2003-04 | 9819.85 | 13362.16 | 574.59 | 19.98 |
| | 2004-05 | 10143.64 | 13535.32 | 643.97 | 18.38 |
| | 2005-06 | 9723.39 | 12833.08 | 751.95 | 14.41 |
| Karnataka | 2003-04 | 23373.71 | 31359.68 | 605.09 | 45.77 |
| | 2004-05 | 23220.97 | 30555.70 | 569.46 | 46.94 |
| | 2005-06 | 20180.02 | 28350.97 | 518.54 | 49.11 |
| Kerala | 2003-04 | 21350.63 | 26333.83 | 728.26 | 29.77 |
| | 2004-05 | 21222.04 | 26644.52 | 707.13 | 31.40 |
| | 2005-06 | 22027.95 | 27866.20 | 671.66 | 35.37 |
| Madhya Pradesh | 2003-04 | 8159.52 | 11296.92 | 532.66 | 18.51 |
| | 2004-05 | 8724.89 | 11976.48 | 784.08 | 12.91 |
| | 2005-06 | 8884.01 | 12685.78 | 689.82 | 16.03 |
| Maharashtra | 2005-06 | 20388.07 | 25464.43 | 762.06 | 29.06 |
| Orissa | 2003-04 | 12851.79 | 17853.67 | 468.35 | 31.81 |
| | 2004-05 | 13100.24 | 18193.62 | 493.69 | 30.93 |
| | 2005-06 | 13252.67 | 18781.20 | 528.90 | 30.24 |
| Punjab | 2003-04 | 19541.08 | 28926.24 | 442.56 | 65.07 |
| | 2004-05 | 21004.06 | 31770.33 | 448.62 | 70.53 |
| | 2005-06 | 17247.31 | 30007.47 | 487.28 | 61.15 |
| Tamil Nadu | 2003-04 | 21449.14 | 29831.95 | 595.73 | 44.62 |
| | 2004-05 | 22104.95 | 30103.75 | 612.13 | 43.64 |
| | 2005-06 | 23305.15 | 32792.95 | 690.96 | 42.92 |
| Uttar Pradesh | 2003-04 | 12278.74 | 17812.29 | 440.56 | 38.07 |
| | 2004-05 | 13444.03 | 19520.32 | 570.70 | 31.95 |
| | 2005-06 | 13866.36 | 20557.42 | 559.19 | 34.37 |
| Uttarakhand | 2003-04 | 14062.68 | 18560.47 | 541.13 | 30.62 |
| | 2004-05 | 14257.74 | 18552.77 | 511.38 | 31.38 |
| | 2005-06 | 14516.05 | 21230.29 | 552.97 | 34.95 |
| West Bengal | 2003-04 | 18018.87 | 24109.80 | 545.77 | 36.75 |
| | 2004-05 | 17773.01 | 24403.45 | 578.11 | 35.69 |
| | 2005-06 | 18180.39 | 25162.96 | 581.02 | 37.18 |

*contd. ...*

...contd. ...

| Crop/State | Year | Cost of Cultivation (Rs/Hect.) | | Cost of Production C2 (Rs/qtl.) | Yield (Qtl./Hect.) |
|---|---|---|---|---|---|
| | | A2+FL | C2 | | |
| 1 | 2 | 3 | 4 | 5 | 6 |
| **Wheat** | | | | | |
| Bihar | 2003-04 | 10360.03 | 15456.82 | 616.07 | 21.90 |
| | 2004-05 | 10411.91 | 15456.25 | 591.94 | 22.70 |
| | 2005-06 | 10577.03 | 15010.70 | 691.20 | 18.78 |
| Chhattisgarh | 2003-04 | 8428.90 | 12170.15 | 808.63 | 12.43 |
| | 2004-05 | 7986.70 | 11443.44 | 907.96 | 10.61 |
| | 2005-06 | 8415.77 | 12480.33 | 914.95 | 10.78 |
| Gujarat | 2003-04 | 13103.75 | 17866.45 | 537.15 | 31.59 |
| | 2004-05 | 13397.64 | 18216.88 | 535.37 | 32.38 |
| | 2005-06 | 12874.20 | 18276.58 | 524.80 | 33.37 |
| Haryana | 2003-04 | 14862.56 | 23580.50 | 506.72 | 39.93 |
| | 2004-05 | 15285.93 | 24197.23 | 522.87 | 39.48 |
| | 2005-06 | 16598.35 | 27317.75 | 577.41 | 38.65 |
| Himachal Pradesh | 2003-04 | 8843.32 | 13957.49 | 818.16 | 12.57 |
| | 2004-05 | 8817.88 | 13603.61 | 678.84 | 14.96 |
| | 2005-06 | 9526.05 | 16363.25 | 720.55 | 15.80 |
| Jharkhand | 2003-04 | 9071.36 | 12875.28 | 844.07 | 13.42 |
| | 2004-05 | 10730.82 | 15007.35 | 946.56 | 13.60 |
| | 2005-06 | 12248.83 | 14609.04 | 947.19 | 12.62 |
| Madhya Pradesh | 2003-04 | 9290.76 | 14992.37 | 568.11 | 22.88 |
| | 2004-05 | 9274.54 | 14696.06 | 583.81 | 21.79 |
| | 2005-06 | 10124.17 | 16978.39 | 720.77 | 20.30 |
| Punjab | 2003-04 | 13713.78 | 22415.03 | 504.24 | 40.00 |
| | 2004-05 | 14908.76 | 24197.48 | 494.35 | 42.94 |
| | 2005-06 | 14558.15 | 26699.59 | 556.27 | 42.05 |
| Rajasthan | 2003-04 | 13558.54 | 19613.36 | 499.37 | 32.99 |
| | 2004-05 | 13367.56 | 19609.52 | 480.19 | 32.95 |
| | 2005-06 | 14770.73 | 22415.60 | 528.70 | 33.76 |
| Uttar Pradesh | 2003-04 | 12486.30 | 19241.29 | 482.78 | 34.01 |
| | 2004-05 | 13912.29 | 20812.56 | 597.81 | 29.29 |
| | 2005-06 | 14959.95 | 22821.62 | 654.22 | 28.59 |
| Uttarakhand | 2003-04 | 13602.23 | 18643.91 | 624.92 | 23.50 |
| | 2004-05 | 14723.74 | 20842.49 | 630.97 | 26.10 |
| | 2005-06 | 13804.46 | 20107.67 | 647.54 | 23.41 |
| West Bengal | 2005-06 | 16341.17 | 21981.49 | 987.41 | 19.90 |
| **Maize** | | | | | |
| Andhra Pradesh | 2003-04 | 14457.54 | 20026.54 | 521.82 | 35.17 |
| | 2004-05 | 12620.19 | 19077.26 | 581.89 | 30.39 |
| | 2005-06 | 12527.32 | 19924.87 | 578.27 | 31.37 |
| Bihar | 2003-04 | 10318.12 | 15843.62 | 406.18 | 34.94 |
| | 2004-05 | 9718.93 | 15321.52 | 359.03 | 37.03 |
| | 2005-06 | 11423.55 | 16694.47 | 382.62 | 38.05 |
| Chhattisgarh | 2003-04 | 4878.28 | 6842.12 | 579.55 | 10.57 |
| | 2004-05 | 5690.18 | 7750.67 | 635.28 | 10.06 |
| | 2005-06 | 3932.91 | 5712.22 | 571.22 | 8.67 |
| Gujarat | 2005-06 | 10969.77 | 13651.92 | 694.60 | 16.77 |
| Himachal Pradesh | 2003-04 | 7725.99 | 11425.26 | 487.97 | 16.17 |
| | 2004-05 | 7458.70 | 11095.92 | 510.90 | 14.69 |
| | 2005-06 | 9974.31 | 13951.25 | 736.27 | 12.47 |
| Karnataka | 2003-04 | 9909.98 | 13406.61 | 633.79 | 18.09 |
| | 2004-05 | 8517.74 | 12169.46 | 454.16 | 22.27 |

contd. ...

*...contd. ...*

| *Crop/State* | *Year* | *Cost of Cultivation (Rs/Hect.)* | | *Cost of Production C2 (Rs/qtl.)* | *Yield (Qtl./Hect.)* |
|---|---|---|---|---|---|
| | | *A2+FL* | *C2* | | |
| *1* | *2* | *3* | *4* | *5* | *6* |
| **Maize** | | | | | |
| Madhya Pradesh | 2003-04 | 6910.96 | 9297.54 | 641.88 | 12.08 |
| | 2004-05 | 6811.79 | 8741.79 | 726.67 | 9.46 |
| | 2005-06 | 9275.25 | 12677.31 | 743.56 | 13.65 |
| Rajasthan | 2003-04 | 11042.60 | 13972.44 | 555.46 | 19.11 |
| | 2004-05 | 9688.09 | 12191.31 | 715.04 | 12.35 |
| | 2005-06 | 10395.80 | 13427.77 | 916.28 | 10.15 |
| Tamil Nadu | 2005-06 | 13045.42 | 20171.54 | 498.42 | 37.99 |
| Uttar Pradesh | 2003-04 | 10289.92 | 13379.10 | 823.76 | 14.54 |
| | 2004-05 | 9203.51 | 13015.65 | 689.20 | 16.94 |
| | 2005-06 | 8774.50 | 12964.93 | 588.03 | 19.63 |
| Uttarakhand | 2003-04 | 10702.13 | 13244.93 | 889.70 | 13.07 |
| | 2004-05 | 7976.27 | 11452.43 | 887.25 | 12.06 |
| | 2005-06 | 9125.09 | 12814.10 | 808.44 | 14.19 |
| Jharkhand | 2003-04 | 7698.27 | 11256.52 | 640.52 | 15.72 |
| | 2004-05 | 7986.11 | 11426.41 | 655.81 | 15.46 |
| | 2005-06 | 7552.59 | 11175.11 | 756.39 | 13.85 |
| **Barley** | | | | | |
| Rajasthan | 2003-04 | 12071.66 | 17076.13 | 416.44 | 32.30 |
| | 2004-05 | 12118.61 | 18073.23 | 399.38 | 33.90 |
| | 2005-06 | 12978.93 | 19769.96 | 481.89 | 30.93 |
| Uttar Pradesh | 2003-04 | 9848.41 | 15879.72 | 468.49 | 28.51 |
| | 2004-05 | 11191.50 | 16886.09 | 576.70 | 24.36 |
| | 2005-06 | 14554.28 | 22450.14 | 614.10 | 30.03 |
| Uttarakhand | 2003-04 | 14884.97 | 18806.03 | 1126.69 | 11.50 |
| | 2004-05 | 15720.12 | 19247.75 | 977.09 | 14.29 |
| ***Jowar*** | | | | | |
| Andhra Pradesh | 2003-04 | 7994.48 | 11641.38 | 997.23 | 9.86 |
| | 2004-05 | 8178.30 | 11636.66 | 713.39 | 14.21 |
| | 2005-06 | 8581.98 | 12971.99 | 746.64 | 14.72 |
| Karnataka | 2003-04 | 4530.39 | 5718.71 | 1212.74 | 3.87 |
| | 2004-05 | 4336.01 | 6078.49 | 683.71 | 7.27 |
| | 2005-06 | 5195.64 | 7135.56 | 770.26 | 7.66 |
| Madhya Pradesh | 2003-04 | 7239.36 | 9315.36 | 729.52 | 9.79 |
| | 2004-05 | 5768.01 | 7848.16 | 674.84 | 9.10 |
| | 2005-06 | 5952.23 | 8159.33 | 757.02 | 8.52 |
| Maharashtra | 2003-04 | 9119.68 | 12530.39 | 579.54 | 12.80 |
| | 2004-05 | 7533.54 | 10306.06 | 613.62 | 10.44 |
| | 2005-06 | 11364.27 | 14846.56 | 676.07 | 12.90 |
| Rajasthan | 2003-04 | 5007.21 | 8525.26 | 600.69 | 6.30 |
| | 2004-05 | 4953.32 | 7287.81 | 524.36 | 7.14 |
| | 2005-06 | 5889.04 | 8549.13 | 683.50 | 4.14 |
| Tamil Nadu | 2003-04 | 7972.94 | 12105.23 | 700.41 | 11.84 |
| | 2004-05 | 7435.80 | 11157.44 | 735.30 | 8.93 |
| | 2005-06 | 6045.78 | 8222.52 | 1126.41 | 3.01 |
| ***Bajra*** | | | | | |
| Gujarat | 2003-04 | 9137.31 | 11201.55 | 509.00 | 15.67 |
| | 2004-05 | 9852.60 | 12305.52 | 558.48 | 16.17 |
| | 2005-06 | 10885.00 | 14010.83 | 579.99 | 17.57 |
| Haryana | 2003-04 | 5667.13 | 8190.98 | 569.76 | 11.22 |
| | 2004-05 | 6414.81 | 9325.70 | 675.02 | 10.88 |
| | 2005-06 | 8047.92 | 11462.76 | 782.21 | 10.61 |
| Karnataka | 2005-06 | 4380.74 | 5657.86 | 715.31 | 6.02 |

*contd. ...*

...contd. ...

| Crop/State | Year | Cost of Cultivation (Rs/Hect.) A2+FL | Cost of Cultivation (Rs/Hect.) C2 | Cost of Production C2 (Rs/qtl.) | Yield (Qtl./Hect.) |
|---|---|---|---|---|---|
| 1 | 2 | 3 | 4 | 5 | 6 |
| Maharashtra | 2003-04 | 9940.08 | 12186.62 | 739.03 | 9.71 |
| | 2004-05 | 8629.19 | 10859.95 | 632.69 | 14.08 |
| | 2005-06 | 9735.65 | 12966.62 | 681.82 | 14.70 |
| Rajasthan | 2003-04 | 4735.07 | 6597.15 | 328.30 | 12.11 |
| | 2004-05 | 4921.54 | 7166.10 | 478.84 | 8.88 |
| | 2005-06 | 4701.96 | 6902.64 | 596.07 | 5.92 |
| Uttar Pradesh | 2003-04 | 6839.42 | 10733.74 | 453.72 | 17.73 |
| | 2004-05 | 7097.08 | 11004.42 | 457.60 | 18.24 |
| | 2005-06 | 6636.69 | 11371.90 | 592.66 | 14.88 |
| **Ragi** | | | | | |
| Andhra Pradesh | 2003-04 | 8031.13 | 12866.70 | 1113.56 | 11.27 |
| Karnataka | 2003-04 | 8936.16 | 11480.27 | 731.93 | 11.00 |
| | 2004-05 | 9680.93 | 12108.17 | 713.77 | 11.69 |
| | 2005-06 | 12814.99 | 15459.41 | 759.13 | 14.69 |
| Maharashtra | 2005-06 | 16964.14 | 18990.74 | 1489.54 | 10.95 |
| Tamil Nadu | 2003-04 | 9805.62 | 14659.72 | 878.09 | 15.85 |
| | 2004-05 | 9268.03 | 13341.50 | 777.38 | 15.80 |
| | 2005-06 | 11595.73 | 16281.18 | 743.95 | 19.80 |
| **Gram** | | | | | |
| Andhra Pradesh | 2005-06 | 7706.43 | 13386.76 | 1267.54 | 10.39 |
| Bihar | 2003-04 | 4857.66 | 10279.46 | 1061.13 | 9.21 |
| | 2004-05 | 5011.46 | 9952.52 | 1097.08 | 8.71 |
| | 2005-06 | 6905.34 | 11574.91 | 1235.83 | 8.75 |
| Chhattisgarh | 2003-04 | 5083.15 | 8295.14 | 1070.23 | 7.37 |
| | 2004-05 | 5788.46 | 8522.40 | 1292.91 | 6.24 |
| | 2005-06 | 6678.50 | 14153.57 | 946.38 | 14.28 |
| Haryana | 2003-04 | 5849.45 | 10616.62 | 1597.86 | 6.23 |
| | 2004-05 | 5157.69 | 8640.21 | 1284.61 | 6.09 |
| Jharkhand | 2003-04 | 3317.58 | 5782.93 | 1049.68 | 5.25 |
| | 2004-05 | 5781.57 | 9179.00 | 1434.12 | 6.12 |
| | 2005-06 | 4767.83 | 6656.55 | 1137.93 | 5.59 |
| Madhya Pradesh | 2003-04 | 6473.60 | 11361.81 | 991.35 | 10.87 |
| | 2004-05 | 6463.65 | 11100.80 | 1004.76 | 10.42 |
| | 2005-06 | 7920.80 | 13978.33 | 1401.57 | 9.52 |
| Maharashtra | 2003-04 | 9082.80 | 14201.31 | 1307.17 | 10.16 |
| | 2004-05 | 8162.66 | 11143.03 | 1728.73 | 6.30 |
| | 2005-06 | 8611.16 | 12581.27 | 1552.46 | 7.92 |
| Rajasthan | 2003-04 | 5319.07 | 7368.06 | 1338.88 | 5.05 |
| | 2004-05 | 4866.33 | 7038.43 | 1094.82 | 5.87 |
| | 2005-06 | 7077.19 | 11877.99 | 1488.30 | 7.18 |
| Uttar Pradesh | 2003-04 | 6767.55 | 12039.27 | 1171.06 | 9.84 |
| | 2004-05 | 7075.28 | 12116.43 | 1114.71 | 10.38 |
| | 2005-06 | 8205.57 | 15092.59 | 1083.09 | 13.36 |
| Uttarakhand | 2003-04 | 5193.57 | 11955.12 | 1331.96 | 8.77 |
| | 2004-05 | 6198.34 | 11893.37 | 1330.37 | 8.54 |
| **Peas** | | | | | |
| Madhya Pradesh | 2005-06 | 7526.11 | 12105.21 | 1203.76 | 9.48 |
| Uttar Pradesh | 2005-06 | 10140.08 | 16894.20 | 1118.66 | 14.48 |
| **Lentil** | | | | | |
| Bihar | 2003-04 | 4771.75 | 10485.77 | 943.21 | 10.64 |
| | 2004-05 | 5026.61 | 9893.37 | 907.55 | 10.42 |
| | 2005-06 | 6075.72 | 10867.60 | 1134.62 | 8.98 |

contd. ...

...contd. ...

| Crop/State | Year | Cost of Cultivation(Rs/Hect.) | | Cost of Production C2 (Rs/qtl.) | Yield (Qtl./Hect.) |
|---|---|---|---|---|---|
| | | A2+FL | C2 | | |
| 1 | 2 | 3 | 4 | 5 | 6 |
| Jharkhand | 2003-04 | 2131.39 | 3958.18 | 1363.92 | 2.76 |
| | 2004-05 | 2764.14 | 5027.35 | 1520.30 | 3.14 |
| | 2005-06 | 4295.54 | 6536.45 | 1140.72 | 5.48 |
| Madhya Pradesh | 2003-04 | 5218.56 | 9782.08 | 962.66 | 9.71 |
| | 2004-05 | 5905.38 | 10942.51 | 958.30 | 10.91 |
| | 2005-06 | 5842.41 | 10998.14 | 1127.68 | 9.34 |
| Uttar Pradesh | 2003-04 | 6550.48 | 11351.73 | 1414.24 | 7.79 |
| | 2004-05 | 7003.50 | 11157.68 | 1745.60 | 6.11 |
| | 2005-06 | 7730.59 | 14012.82 | 1302.26 | 10.33 |
| ***Arhar*** | | | | | |
| Andhra Pradesh | 2003-04 | 6986.69 | 12201.96 | 1418.55 | 8.39 |
| | 2004-05 | 7320.57 | 12771.37 | 1313.49 | 9.37 |
| | 2005-06 | 8006.65 | 14136.92 | 1386.93 | 10.04 |
| Bihar | 2005-06 | 7221.51 | 14073.65 | 1436.98 | 9.15 |
| Gujarat | 2003-04 | 11015.59 | 14658.36 | 1411.55 | 9.52 |
| | 2004-05 | 10332.83 | 13759.07 | 1337.42 | 9.47 |
| | 2005-06 | 10129.20 | 13683.78 | 1442.19 | 8.72 |
| Karnataka | 2003-04 | 7762.89 | 10431.28 | 1994.27 | 5.07 |
| | 2004-05 | 8164.45 | 11175.22 | 1603.71 | 6.70 |
| | 2005-06 | 8191.81 | 11975.17 | 1378.82 | 8.33 |
| Madhya Pradesh | 2003-04 | 6505.48 | 10256.97 | 1383.81 | 6.79 |
| | 2004-05 | 6187.64 | 10433.68 | 1203.22 | 8.02 |
| | 2005-06 | 6167.05 | 10393.67 | 1392.28 | 6.95 |
| Maharashtra | 2003-04 | 9841.58 | 14486.38 | 1383.28 | 9.88 |
| | 2004-05 | 10000.11 | 14585.38 | 1253.81 | 10.98 |
| | 2005-06 | 13373.29 | 18460.07 | 1507.87 | 11.47 |
| Orissa | 2003-04 | 4794.09 | 7030.48 | 1707.77 | 3.89 |
| | 2004-05 | 5074.85 | 7479.86 | 1712.12 | 4.05 |
| | 2005-06 | 6021.80 | 9030.99 | 2583.43 | 3.37 |
| Tamil Nadu | 2005-06 | 11339.96 | 14640.19 | 2361.02 | 6.06 |
| Uttar Pradesh | 2003-04 | 5947.34 | 13862.40 | 1238.83 | 9.73 |
| | 2004-05 | 7298.69 | 16275.89 | 1295.31 | 10.89 |
| | 2005-06 | 7933.51 | 16403.49 | 1403.08 | 10.59 |
| ***Moong*** | | | | | |
| Andhra Pradesh | 2003-04 | 4438.91 | 6980.97 | 1339.20 | 5.12 |
| | 2004-05 | 4944.57 | 7541.83 | 1849.73 | 4.06 |
| | 2005-06 | 6438.31 | 11158.47 | 2294.22 | 4.83 |
| Karnataka | 2005-06 | 4010.11 | 5811.41 | 1516.24 | 3.53 |
| Maharashtra | 2003-04 | 6925.92 | 8892.61 | 1932.50 | 4.57 |
| | 2004-05 | 7409.42 | 9149.15 | 2737.83 | 3.30 |
| | 2005-06 | 9155.29 | 11387.29 | 2884.33 | 3.88 |
| Orissa | 2003-04 | 4553.61 | 6426.16 | 2132.11 | 2.86 |
| | 2004-05 | 4634.47 | 6533.24 | 1992.15 | 3.12 |
| | 2005-06 | 4753.94 | 7590.95 | 2141.61 | 3.39 |
| Rajasthan | 2003-04 | 4344.08 | 5971.87 | 1335.81 | 4.09 |
| | 2004-05 | 3924.68 | 5566.92 | 1672.78 | 2.90 |
| | 2005-06 | 4170.67 | 5957.83 | 2614.71 | 1.97 |
| ***Urad*** | | | | | |
| Andhra Pradesh | 2003-04 | 3962.74 | 5944.39 | 1268.16 | 4.51 |
| | 2004-05 | 4682.61 | 8029.85 | 1276.75 | 6.19 |
| | 2005-06 | 5813.79 | 13164.48 | 1385.68 | 9.49 |

contd. ...

...contd. ...

| Crop/State | Year | Cost of Cultivation(Rs/Hect.) A2+FL | C2 | Cost of Production C2 (Rs/qtl.) | Yield (Qtl./Hect.) |
|---|---|---|---|---|---|
| 1 | 2 | 3 | 4 | 5 | 6 |
| Chhattisgarh | 2003-04 | 4916.03 | 7163.36 | 1334.88 | 5.01 |
| | 2004-05 | 5292.09 | 7560.46 | 1561.10 | 4.55 |
| | 2005-06 | 4213.27 | 6377.58 | 1642.96 | 3.78 |
| Madhya Pradesh | 2003-04 | 5350.03 | 6732.69 | 2242.06 | 2.83 |
| | 2004-05 | 5359.68 | 7365.13 | 1602.78 | 4.39 |
| | 2005-06 | 5345.14 | 6961.40 | 3106.44 | 2.16 |
| Maharashtra | 2003-04 | 7499.35 | 9499.13 | 2011.49 | 4.60 |
| | 2004-05 | 7150.74 | 8705.72 | 3015.07 | 2.85 |
| | 2005-06 | 9448.00 | 11956.68 | 2331.37 | 5.09 |
| Orissa | 2003-04 | 4025.45 | 5638.59 | 1636.60 | 3.21 |
| | 2004-05 | 3817.78 | 5469.22 | 1554.42 | 3.30 |
| | 2005-06 | 4361.99 | 7149.74 | 1907.90 | 3.58 |
| Rajasthan | 2003-04 | 6169.38 | 8012.42 | 1574.24 | 4.31 |
| | 2004-05 | 5881.26 | 8290.36 | 1722.09 | 4.27 |
| | 2005-06 | 5256.01 | 7616.56 | 2434.11 | 2.20 |
| Tamil Nadu | 2003-04 | 5262.53 | 9027.27 | 1888.61 | 4.69 |
| | 2004-05 | 4793.53 | 7971.92 | 1681.69 | 4.62 |
| | 2005-06 | 5551.16 | 8871.55 | 2847.50 | 3.07 |
| Uttar Pradesh | 2003-04 | 3303.66 | 5233.54 | 1768.30 | 2.84 |
| | 2004-05 | 3683.83 | 5882.15 | 1476.59 | 3.81 |
| | 2005-06 | 3744.07 | 6448.03 | 2051.07 | 3.05 |
| Groundnut | | | | | |
| Andhra Pradesh | 2003-04 | 11726.03 | 17778.98 | 1679.61 | 9.64 |
| | 2004-05 | 11441.39 | 17318.32 | 1563.25 | 10.18 |
| | 2005-06 | 14138.10 | 19812.15 | 1989.42 | 9.24 |
| Gujarat | 2003-04 | 13480.06 | 19812.80 | 946.06 | 17.90 |
| | 2004-05 | 13152.49 | 17246.98 | 1533.51 | 9.48 |
| | 2005-06 | 14174.16 | 19597.61 | 1285.82 | 13.07 |
| Karnataka | 2003-04 | 9188.55 | 12044.71 | 2579.27 | 4.37 |
| | 2004-05 | 9590.46 | 12268.86 | 1899.28 | 6.06 |
| | 2005-06 | 9668.67 | 12982.60 | 1714.45 | 7.02 |
| Maharashtra | 2003-04 | 18943.10 | 24576.01 | 1709.29 | 13.32 |
| | 2004-05 | 18920.79 | 23993.43 | 2162.40 | 10.69 |
| | 2005-06 | 16568.71 | 21463.01 | 1855.77 | 10.53 |
| Tamil Nadu | 2003-04 | 17950.40 | 25139.42 | 1513.27 | 15.95 |
| | 2004-05 | 19332.46 | 25273.64 | 1376.31 | 17.61 |
| | 2005-06 | 15439.64 | 21583.31 | 1291.83 | 15.76 |
| Soyabean | | | | | |
| Chhattisgarh | 2003-04 | 9933.56 | 13027.22 | 1327.93 | 8.53 |
| | 2004-05 | 9245.47 | 12678.98 | 1105.28 | 10.53 |
| Madhya Pradesh | 2003-04 | 7624.03 | 12448.53 | 935.38 | 12.61 |
| | 2004-05 | 7898.56 | 12005.42 | 1078.15 | 10.58 |
| | 2005-06 | 7992.87 | 11990.71 | 994.97 | 11.28 |
| Maharashtra | 2003-04 | 10528.43 | 14631.94 | 973.62 | 14.68 |
| | 2004-05 | 12583.02 | 16447.79 | 1286.28 | 12.49 |
| | 2005-06 | 11547.32 | 14791.15 | 1227.04 | 11.70 |
| Rajasthan | 2003-04 | 7469.57 | 10498.09 | 921.64 | 10.74 |
| | 2004-05 | 7432.66 | 10268.42 | 698.03 | 13.24 |
| | 2005-06 | 8365.74 | 11102.03 | 902.06 | 11.37 |
| Sunflower | | | | | |
| Andhra Pradesh | 2003-04 | 7679.31 | 11182.69 | 1619.71 | 6.73 |
| | 2004-05 | 8183.03 | 11898.69 | 1729.03 | 6.81 |
| | 2005-06 | 7735.55 | 11559.34 | 1631.06 | 7.08 |

contd. ...

...contd. ...

| Crop/State | Year | Cost of Cultivation (Rs/Hect.) | | Cost of Production C2 (Rs/qtl.) | Yield (Qtl./Hect.) |
|---|---|---|---|---|---|
| | | A2+FL | C2 | | |
| 1 | 2 | 3 | 4 | 5 | 6 |
| Karnataka | 2003-04 | 6935.49 | 9100.29 | 2152.38 | 4.13 |
| | 2004-05 | 6177.10 | 8094.83 | 1801.88 | 4.40 |
| | 2005-06 | 6430.93 | 8519.56 | 1748.50 | 4.78 |
| Maharashtra | 2004-05 | 9973.86 | 12819.66 | 1540.26 | 8.31 |
| | 2005-06 | 7966.24 | 10446.67 | 1396.69 | 7.46 |
| **Rapeseed and Mustard** | | | | | |
| Assam | 2003-04 | 8408.17 | 10366.92 | 1965.49 | 5.27 |
| | 2004-05 | 8768.60 | 10697.67 | 1814.90 | 5.89 |
| | 2005-06 | 8581.05 | 10972.71 | 1721.62 | 6.37 |
| Gujarat | 2003-04 | 11085.33 | 17400.04 | 947.58 | 18.24 |
| | 2004-05 | 10802.73 | 15327.46 | 1038.51 | 14.53 |
| | 2005-06 | 10943.64 | 15812.13 | 932.63 | 16.79 |
| Haryana | 2003-04 | 9017.89 | 16933.05 | 1024.31 | 15.89 |
| | 2004-05 | 9393.36 | 15943.08 | 1326.11 | 11.52 |
| | 2005-06 | 11607.26 | 19810.23 | 1245.58 | 14.89 |
| Madhya Pradesh | 2003-04 | 8058.14 | 17114.21 | 986.22 | 17.08 |
| | 2004-05 | 6693.05 | 12907.32 | 1010.09 | 12.24 |
| | 2005-06 | 7614.00 | 13628.76 | 1138.56 | 11.37 |
| Punjab | 2003-04 | 7084.48 | 14136.39 | 1154.93 | 11.75 |
| | 2004-05 | 6558.07 | 11981.22 | 1270.95 | 8.80 |
| Rajasthan | 2003-04 | 7611.00 | 13585.61 | 863.36 | 15.49 |
| | 2004-05 | 7799.91 | 12766.04 | 917.39 | 13.58 |
| | 2005-06 | 8213.40 | 13506.41 | 1009.70 | 12.93 |
| Uttar Pradesh | 2003-04 | 8042.38 | 13899.68 | 1308.55 | 10.24 |
| | 2004-05 | 8581.50 | 14659.12 | 1237.80 | 11.37 |
| | 2005-06 | 9348.65 | 16604.48 | 1199.40 | 13.13 |
| West Bengal | 2003-04 | 10738.09 | 16240.63 | 1614.71 | 9.63 |
| | 2004-05 | 11384.58 | 16959.13 | 1408.15 | 11.70 |
| | 2005-06 | 12848.29 | 18184.70 | 1704.21 | 10.16 |
| Uttarakhand | 2003-04 | 7127.24 | 10898.03 | 1522.11 | 6.38 |
| | 2004-05 | 8798.15 | 11818.74 | 1964.46 | 5.19 |
| **Sesamum** | | | | | |
| Gujarat | 2003-04 | 6482.96 | 8867.68 | 1899.56 | 4.62 |
| | 2004-05 | 6782.21 | 8932.97 | 2895.16 | 3.07 |
| | 2005-06 | 8083.94 | 10868.23 | 2408.43 | 4.47 |
| Madhya Pradesh | 2003-04 | 4416.28 | 6670.53 | 2017.90 | 3.24 |
| | 2004-05 | 4743.28 | 7046.41 | 2507.90 | 2.74 |
| | 2005-06 | 3799.04 | 6188.24 | 3525.75 | 1.71 |
| Orissa | 2003-04 | 4819.46 | 6835.25 | 2760.85 | 2.37 |
| | 2004-05 | 4483.36 | 6687.16 | 2805.39 | 2.28 |
| | 2005-06 | 5062.10 | 7686.07 | 2164.61 | 3.41 |
| Rajasthan | 2003-04 | 4471.68 | 6902.07 | 1843.59 | 3.66 |
| | 2004-05 | 4459.23 | 6814.34 | 2299.32 | 2.92 |
| | 2005-06 | 4241.16 | 6249.90 | 2895.76 | 2.12 |
| Tamil Nadu | 2003-04 | 9214.67 | 15490.92 | 2654.94 | 5.80 |
| | 2004-05 | 7577.07 | 11543.17 | 2401.33 | 4.74 |
| | 2005-06 | 8101.40 | 12207.25 | 2639.60 | 4.57 |
| Uttar Pradesh | 2003-04 | 3182.50 | 5742.51 | 2154.86 | 2.41 |
| | 2004-05 | 3059.68 | 5257.30 | 1388.14 | 3.63 |
| West Bengal | 2005-06 | 8838.66 | 13635.56 | 1210.32 | 10.92 |
| **Nigerseed** | | | | | |
| Madhya Pradesh | 2005-06 | 3046.48 | 4725.67 | 1223.68 | 3.65 |

contd. ...

*...contd. ...*

| Crop/State | Year | Cost of Cultivation (Rs/Hect.) | | Cost of Production C2 (Rs/qtl.) | Yield (Qtl./Hect.) |
|---|---|---|---|---|---|
| | | A2+FL | C2 | | |
| 1 | 2 | 3 | 4 | 5 | 6 |
| Orissa | 2003-04 | 4140.05 | 5814.25 | 1705.11 | 3.37 |
| | 2004-05 | 4087.57 | 5620.15 | 1869.74 | 2.96 |
| | 2005-06 | 4942.53 | 5559.19 | 2123.05 | 2.60 |
| **Safflower** | | | | | |
| Karnataka | 2005-06 | 4342.28 | 6708.01 | 1065.45 | 6.11 |
| Maharashtra | 2003-04 | 5939.35 | 7576.60 | 2572.38 | 2.95 |
| | 2004-05 | 3184.03 | 5174.82 | 1030.69 | 5.02 |
| | 2005-06 | 5714.95 | 7822.56 | 1605.91 | 4.87 |
| **Cotton** | | | | | |
| Andhra Pradesh | 2003-04 | 20364.36 | 36202.37 | 1720.17 | 21.01 |
| | 2004-05 | 21183.39 | 32735.15 | 1810.16 | 18.05 |
| | 2005-06 | 19075.09 | 27625.34 | 2227.20 | 12.40 |
| Gujarat | 2003-04 | 16079.24 | 23395.87 | 1566.88 | 14.75 |
| | 2004-05 | 17570.35 | 23586.30 | 1501.24 | 15.45 |
| | 2005-06 | 22852.70 | 31437.07 | 1594.55 | 19.33 |
| Haryana | 2003-04 | 15331.62 | 24168.98 | 2008.12 | 11.76 |
| | 2004-05 | 17571.13 | 27288.54 | 1478.86 | 18.04 |
| | 2005-06 | 17708.31 | 26738.08 | 2166.09 | 11.90 |
| Karnataka | 2003-04 | 6326.50 | 9520.06 | 1910.61 | 4.87 |
| | 2004-05 | 8181.70 | 11631.02 | 1518.88 | 7.49 |
| | 2005-06 | 9084.33 | 12483.63 | 2071.34 | 5.84 |
| Madhya Pradesh | 2003-04 | 13297.24 | 18663.88 | 2580.92 | 6.98 |
| | 2004-05 | 16289.74 | 20714.67 | 2545.57 | 7.56 |
| | 2005-06 | 20620.42 | 33780.98 | 1959.11 | 16.32 |
| Maharashtra | 2003-04 | 15140.14 | 20989.96 | 2152.45 | 9.58 |
| | 2004-05 | 16225.56 | 21179.58 | 2112.60 | 9.72 |
| | 2005-06 | 16017.87 | 20793.82 | 2137.15 | 9.33 |
| Punjab | 2003-04 | 23206.29 | 33233.46 | 2140.28 | 15.08 |
| | 2004-05 | 23646.01 | 34070.90 | 1600.74 | 20.75 |
| | 2005-06 | 20209.59 | 33849.46 | 1606.12 | 20.37 |
| Rajasthan | 2003-04 | 11195.86 | 17410.09 | 1674.02 | 10.19 |
| | 2004-05 | 11380.07 | 17686.72 | 1335.65 | 12.96 |
| | 2005-06 | 11591.87 | 17594.12 | 1295.69 | 13.06 |
| Tamil Nadu | 2003-04 | 23682.40 | 34385.73 | 2490.93 | 13.70 |
| | 2004-05 | 20595.02 | 27559.95 | 2274.28 | 11.95 |
| | 2005-06 | 20530.55 | 28815.10 | 2708.65 | 10.55 |
| **Jute** | | | | | |
| Assam | 2003-04 | 11269.87 | 14040.98 | 768.87 | 16.53 |
| | 2004-05 | 12494.70 | 15285.88 | 974.93 | 14.18 |
| | 2005-06 | 15402.48 | 19089.76 | 980.82 | 18.15 |
| Orissa | 2003-04 | 12522.39 | 15868.33 | 1209.22 | 10.83 |
| | 2004-05 | 13165.06 | 17712.90 | 1043.01 | 15.07 |
| | 2005-06 | 17262.25 | 25012.03 | 1071.54 | 21.81 |
| West Bengal | 2003-04 | 18165.33 | 23642.34 | 901.40 | 23.07 |
| | 2004-05 | 19693.30 | 25622.92 | 954.40 | 23.83 |
| | 2005-06 | 18815.46 | 26807.76 | 1014.28 | 24.41 |
| **Sugarcane** | | | | | |
| Andhra Pradesh | 2003-04 | 24798.96 | 42322.98 | 65.24 | 645.75 |
| | 2004-05 | 31679.42 | 52665.27 | 78.65 | 662.96 |
| | 2005-06 | 39863.88 | 66958.06 | 96.08 | 687.49 |
| Haryana | 2003-04 | 28605.77 | 48552.32 | 72.85 | 638.91 |
| | 2004-05 | 20506.02 | 47566.20 | 69.91 | 649.33 |
| | 2005-06 | 26947.90 | 53357.29 | 90.48 | 580.07 |

*contd. ...*

*...contd. ...*

| *Crop/State* | *Year* | *Cost of Cultivation (Rs/Hect.)* A2+FL | C2 | *Cost of Production C2 (Rs/qtl.)* | *Yield (Qtl./Hect.)* |
|---|---|---|---|---|---|
| 1 | 2 | 3 | 4 | 5 | 6 |
| Karnataka | 2003-04 | 38091.87 | 59904.93 | 66.46 | 892.12 |
| | 2004-05 | 38873.46 | 59368.15 | 64.53 | 912.55 |
| | 2005-06 | 44682.70 | 71273.05 | 73.83 | 952.27 |
| Maharashtra | 2003-04 | 46149.40 | 60155.40 | 77.90 | 715.53 |
| | 2004-05 | 50382.80 | 71721.89 | 75.17 | 908.24 |
| | 2005-06 | 64943.79 | 84656.89 | 94.91 | 824.89 |
| Tamil Nadu | 2003-04 | 39028.21 | 56877.23 | 63.71 | 874.00 |
| | 2004-05 | 42167.96 | 62116.78 | 68.81 | 883.57 |
| | 2005-06 | 53350.99 | 74476.52 | 77.30 | 943.57 |
| Uttar Pradesh | 2003-04 | 15055.19 | 28997.12 | 55.64 | 490.36 |
| | 2004-05 | 18845.56 | 36068.21 | 64.61 | 529.63 |
| | 2005-06 | 22551.25 | 41445.02 | 68.86 | 568.93 |
| Uttarakhand | 2003-04 | 14065.19 | 25164.33 | 38.66 | 592.25 |
| | 2004-05 | 17035.94 | 29568.45 | 54.36 | 504.43 |
| | 2005-06 | 14289.21 | 30994.57 | 58.03 | 490.77 |
| **Potato** | | | | | |
| Bihar | 2003-04 | 29183.82 | 37154.01 | 244.64 | 151.87 |
| | 2004-05 | 24621.65 | 32684.14 | 215.18 | 151.89 |
| | 2005-06 | 30336.94 | 39609.63 | 215.37 | 183.92 |
| Himachal Pradesh | 2003-04 | 32212.70 | 42847.01 | 473.82 | 90.43 |
| | 2004-05 | 30549.80 | 39930.85 | 467.38 | 85.44 |
| | 2005-06 | 30317.46 | 41285.59 | 483.40 | 84.82 |
| Jharkhand | 2003-04 | 30427.95 | 33728.98 | 250.52 | 134.63 |
| | 2004-05 | 29668.17 | 33174.85 | 273.76 | 121.18 |
| | 2005-06 | 32653.00 | 37272.48 | 335.94 | 110.95 |
| Punjab | 2005-06 | 33762.47 | 52192.36 | 247.49 | 210.89 |
| Uttar Pradesh | 2003-04 | 28654.28 | 36343.52 | 236.76 | 153.51 |
| | 2004-05 | 34766.12 | 44175.87 | 226.90 | 194.70 |
| | 2005-06 | 38707.29 | 52261.42 | 241.75 | 216.18 |
| West Bengal | 2003-04 | 42105.99 | 59105.46 | 258.84 | 228.35 |
| | 2004-05 | 42812.80 | 56824.11 | 241.81 | 234.99 |
| | 2005-06 | 50670.10 | 72445.47 | 346.93 | 208.82 |
| **Onion** | | | | | |
| Andhra Pradesh | 2005-06 | 17691.76 | 31620.51 | 439.04 | 72.02 |
| Gujarat | 2005-06 | 33378.84 | 41280.15 | 171.34 | 237.17 |
| Karnataka | 2003-04 | 12110.19 | 21350.20 | 393.58 | 53.74 |
| | 2004-05 | 11415.05 | 16633.54 | 185.30 | 88.86 |
| Maharashtra | 2003-04 | 25258.46 | 45726.36 | 197.94 | 231.01 |
| | 2004-05 | 34824.10 | 43940.39 | 282.04 | 155.79 |
| | 2005-06 | 35535.55 | 49910.84 | 186.37 | 266.31 |

Cost A2+Fl=Cost A2+imputed value of Family Labour

Cost 2=Cost A1+rent paid for leased-in land.

Cost A1=All actual expenses in cash and kind incurred in production by owner.

Cost C2=Cost B2+imputed value of Family Labour.

Cost B2=Cost B1+rental vlaue of owned land (net of land revenue).

Cost B1=Cost A1+interest on value of owned capital assets (excluding land).

# Table 8.4

*Marketed Surplus Ratio (MSR) of Important Agricultural Commodities in Various States (Based on Comprehensive Scheme for Studying Cost of Cultivation of Principal Crops in India)*

| *Group of Crops* | *Crops* | *State* | *MSR* | |
|---|---|---|---|---|
| | | | *2004-05* | *2005-06* |
| *1* | *2* | *3* | *5* | *6* |
| **I. Foodgrains: Cereals** | 1. Rice | Andhra Pradesh | 83.06 | 197.66 |
| | | Assam | 41.26 | 22.43 |
| | | Bihar | 73.50 | 237.91 |
| | | Haryana | 98.22 | 96.82 |
| | | Karnataka | 84.41 | 253.71 |
| | | Kerala | 64.59 | 191.85 |
| | | Madhya Pradesh | 65.09 | 205.81 |
| | | Orissa | 65.43 | 158.06 |
| | | Punjab | 97.70 | 153.19 |
| | | Tamil Nadu | 78.44 | 121.42 |
| | | Uttar Pradesh | 46.51 | 219.76 |
| | | West Bengal | 60.39 | 176.17 |
| | **All-India Weighted Average*** | | **69.24** | **171.74** |
| | 2. Wheat | Bihar | 70.12 | 125.50 |
| | | Gujarat | 93.62 | 0.63 |
| | | Haryana | 74.64 | 151.06 |
| | | Himachal Pradesh | 29.49 | 425.12 |
| | | Madhya Pradesh | 74.54 | 98.90 |
| | | Punjab | 81.34 | 121.80 |
| | | Rajasthan | 27.20 | 191.46 |
| | | Uttar Pradesh | 50.21 | 101.63 |
| | **All-India Weighted Average*** | | **62.87** | **122.49** |
| | 3. Maize | Andhra Pradesh | 89.01 | 205.70 |
| | | Bihar | 67.40 | 157.76 |
| | | Himachal Pradesh | 44.30 | 208.99 |
| | | Karnataka | 93.47 | 133.99 |
| | | Madhya Pradesh | 66.82 | 90.15 |
| | | Rajasthan | 56.22 | 178.25 |
| | | Uttar Pradesh | 78.74 | 59.05 |
| | **All-India Weighted Average*** | | **72.78** | **145.77** |
| | 4. *Jowar* | Andhra Pradesh | 57.79 | 157.55 |
| | | Karnataka | 51.01 | 158.89 |
| | | Madhya Pradesh | 43.39 | 161.11 |
| | | Maharashtra | 54.10 | 151.91 |
| | | Rajasthan | 72.96 | 50.39 |
| | All-India Weighted Average* | | 54.01 | 150.63 |
| | 5. *Bajra* | Gujarat | 77.23 | 0.79 |
| | | Haryana | 82.07 | 115.58 |
| | | Maharashtra | 58.51 | 103.05 |
| | | Rajasthan | 66.27 | 119.35 |
| | | Uttar Pradesh | 72.36 | 69.93 |
| | **All-India Weighted Average*** | | **69.22** | **90.09** |

*contd. ...*

*...contd. ...*

| 1 | 2 | 3 | 5 | 6 |
|---|---|---|---|---|
| | 6. Barley | Rajasthan | 66.37 | - |
| | | Uttar Pradesh | 45.57 | 96.57 |
| | **All-India Weighted Average*** | | **55.08** | **47.93** |
| | 7. Ragi | Andhra Pradesh | 61.63 | 63.28 |
| | | Karnataka | 57.54 | 68.62 |
| | **All-India Weighted Average*** | | **57.86** | **68.36** |
| **II. Pulses** | 8. *Arhar* | Karnataka | 97.42 | 83.97 |
| | | Madhya Pradesh | 47.04 | 136.26 |
| | | Maharashtera | 90.42 | 125.80 |
| | | Orissa | 73.70 | 141.88 |
| | | Uttar Pradesh | 70.32 | 158.99 |
| | **All-India Weighted Average*** | | **78.35** | **128.55** |
| | 9. Gram | Bihar | 65.55 | 561.83 |
| | | Madhya Pradesh | 96.34 | 84.73 |
| | | Rajasthan | 85.26 | 77.78 |
| | | Uttar Pradesh | 96.25 | 118.83 |
| | **All-India Weighted Average*** | | **93.98** | **96.14** |
| | 10. *Urad* | Andhra Pradesh | 77.00 | 133.90 |
| | | Madhya Pradesh | **90.63** | 33.42 |
| | | Maharashtra | **88.74** | 416.44 |
| | | Orissa | 66.13 | 273.85 |
| | | Rajasthan | 58.62 | 17.28 |
| | | Tamil Nadu | 93.03 | 119.13 |
| | | Uttar Pradesh | 89.09 | 13.07 |
| | **All-India Weighted Average*** | | **80.43** | **157.63** |
| | 11. *Moong* | Andhra Pradesh | 61.20 | 423.88 |
| | | Maharashtra | 77.91 | 23.20 |
| | | Orissa | 51.16 | 206.33 |
| | | Rajasthan | 86.73 | 160.52 |
| | **All-India Weighted Average*** | | **82.04** | **158.43** |
| | 12. Lentil | Bihar | 73.44 | 86.07 |
| | | Madhya Pradesh | 91.92 | 103.68 |
| | | Uttar Pradesh | 86.01 | 173.75 |
| | **All-India Weighted Average*** | | **85.35** | **141.23** |
| **III. Oilseeds** | 13. Groundnut | Andhra Pradesh | 87.16 | 164.53 |
| | | Gujarat | **92.34** | 3.04 |
| | | Karnataka | 96.55 | 120.98 |
| | | Maharashtra | 54.69 | 233.52 |
| | | Tamil Nadu | 95.87 | 85.28 |
| | **All-India Weighted Average*** | | **90.03** | **98.38** |
| | 14. R & M | Gujarat | 99.46 | 1.09 |
| | | Haryana | 97.60 | 126.78 |
| | | Madhya Pradesh | 65.98 | 289.10 |
| | | Rajasthan | 97.50 | 108.72 |
| | | Uttar Pradesh | 69.55 | 218.11 |
| | | West Bengal | 67.18 | 142.70 |
| | **All-India Weighted Average*** | | **89.02** | **136.17** |

*contd. ...*

*...contd. ...*

| 1 | 2 | 3 | 5 | 6 |
|---|---|---|---|---|
| | 15. Soyabean | Madhya Pradesh | 92.47 | 100.78 |
| | | Maharashtra | 99.99 | 29.47 |
| | **All-India Weighted Average*** | | **94.88** | **76.85** |
| | 16. Sunflower | Andhra Pradesh | 100.00 | - |
| | | Karnataka | **98.40** | 129.85 |
| | | Maharashtra | 95.15 | 100.10 |
| | **All-India Weighted Average*** | | **98.58** | **89.52** |
| | 17. Sesamum | Gujarat | 98.07 | 12.15 |
| | | Madhya Pradesh | 84.08 | 56.49 |
| | | Orissa | 74.14 | 45.91 |
| | | Rajasthan | 81.14 | 87.38 |
| | **All-India Weighted Average*** | | **90.74** | **54.33** |
| | 18. Safflower | Karnataka | 94.36 | - |
| | | Maharashtra | **90.04** | - |
| | **All-India Weighted Average*** | | **91.22** | **98.22** |
| | 19. Nigerseed | Madhya Pradesh | **100.00** | **100.00** |
| | | Orissa | **94.23** | - |
| | **All-India Weighted Average*** | | **96.81** | - |
| **IV. Other Commercial Crops** | 20. Sugarcane | Andhra Pradesh | **94.17** | **98.06** |
| | | Haryana | 100.00 | 101.00 |
| | | Karnataka | 86.95 | 139.93 |
| | | Uttar Pradesh | **100.00** | **100.00** |
| | All-India Weighted Average* | | 98.08 | 103.49 |
| | 21. Cotton@ | Gujarat | 96.42 | 3.39 |
| | | Haryana | 89.06 | 64.83 |
| | | Karnataka | 95.40 | - |
| | | Madhya Pradesh | 100.00 | 176.48 |
| | | Rajasthan | 97.17 | 346.64 |
| | **All-India Weighted Average*** | | **95.08** | - |
| | 22. Jute# | West Bengal | 90.72 | 99.71 |
| | **All-India Weighted Average*** | | 90.72 | 76.80 |
| **V. Vegetables** | 23. Onion | Karnataka | 82.91 | 57.78 |
| | **All-India Weighted Average*** | | **0.00** | **0.00** |
| | 24. Potato | Bihar | 73.66 | 333.06 |
| | | Uttar Pradesh | 90.00 | **73.09** |
| | | West Bengal | 79.76 | 53.47 |
| | **All-India Weighted Average*** | | **84.26** | **80.74** |

* It is weighted average of states listed above.

Bold and underlined figures are estimated on the basis of average of preceding/succeeding years in cases where production/sale figure is not available in CS RT-610 or is inconsistent.

@ In case of cotton, production is in bales of 170 kgs. each.

# In case of jute, production is in bales of 180 kgs. each.

$ Adjusted sale with respect to opening stock & other receipts.

\- Not Reported

# Table 9.1

*State-wise Procurement of Rice and Wheat in Major Rice and Wheat Producing States from 1996-97 to 2008-09 (According to Marketing Year)*

*(Lakh Tonnes)*

| State | 1996-97 | 1997-98 | 1998-99 | 1999-2000 | 2000-01 | 2001-02 | 2002-03 | 2003-04 | 2004-05 | 2005-06 | 2006-07 | 2007-08 | 2008-09 |
|---|---|---|---|---|---|---|---|---|---|---|---|---|---|
| 1 | 2 | 3 | 4 | 5 | 6 | 7 | 8 | 9 | 10 | 11 | 12 | 13 | 14 |
| **Rice** (October-September)* | | | | | | | | | | | | | |
| Punjab | 42.49 | 60.59 | 44.04 | 68.15 | 69.64 | 72.82 | 79.39 | 86.62 | 91.06 | 88.55 | 78.29 | 78.61 | |
| Haryana | 12.05 | 12.69 | 3.00 | 9.87 | 14.81 | 14.84 | 13.25 | 13.34 | 16.62 | 20.54 | 17.77 | 15.72 | |
| Uttar Pradesh | 9.10 | 10.73 | 8.68 | 14.21 | 11.74 | 19.36 | 13.60 | 25.54 | 29.71 | 31.51 | 25.29 | 27.12 | |
| Andhra Pradesh | 45.25 | 38.55 | 51.19 | 54.98 | 71.74 | 64.25 | 26.35 | 42.30 | 39.06 | 49.71 | 53.28 | 61.29 | |
| Madhya Pradesh | 5.80 | 10.29 | 4.30 | 11.04 | 1.75 | 2.74 | 1.59 | 1.12 | 0.42 | 1.36 | 74 | 69 | |
| Orissa | 4.76 | 7.00 | 4.81 | 8.89 | 9.18 | 12.53 | 8.90 | 13.73 | 15.90 | 17.85 | 20.02 | 20.78 | |
| Tamil Nadu | 7.27 | 12.31 | 7.33 | 9.19 | 16.95 | 8.52 | 1.07 | 2.07 | 6.52 | 9.26 | 10.77 | 7.78 | |
| West Bengal | 1.59 | 2.03 | 1.41 | 3.51 | 4.34 | 0.48 | 1.26 | 9.25 | 9.44 | 12.75 | 6.42 | 8.11 | |
| Chhattisgarh | - | | - | - | 8.57 | 19.21 | 12.91 | 23.74 | 28.37 | 32.65 | 28.65 | 29.95 | |
| Uttaranchal | - | | - | - | 0.42 | 2.35 | 2.32 | 3.23 | 3.16 | 3.36 | 1.76 | 1.45 | |
| Others | 1.37 | 1.72 | 1.23 | 2.44 | 3.67 | 4.18 | 3.58 | 7.34 | 6.58 | 9.02 | 7.77 | 9.06 | |
| All-India | 129.68 | 155.91 | 125.99 | 182.28 | 212.81 | 221.28 | 164.22 | 228.28 | 246.84 | 276.56 | 251.06 | 260.56 | |
| **Wheat** (April-March)* | | | | | | | | | | | | | |
| Punjab | 56.42 | 59.61 | 61.46 | 78.37 | 94.23 | 105.60 | 98.80 | 89.38 | 92.40 | 90.10 | 69.46 | 67.81 | 99.37 |
| Uttar Pradesh | 2.61 | 6.17 | 21.41 | 12.62 | 15.45 | 24.46 | 21.10 | 12.13 | 17.41 | 5.60 | 0.49 | 5.46 | 28.87 |
| Haryana | 20.22 | 22.90 | 31.58 | 38.69 | 44.97 | 64.07 | 58.88 | 51.22 | 51.15 | 45.29 | 22.29 | 33.5 | 52.31 |
| Rajasthan | 2.29 | 3.20 | 6.67 | 6.37 | 5.39 | 6.76 | 4.61 | 2.59 | 2.79 | 1.59 | 0.2 | 3.83 | 9.32 |
| Chhattisgarh | - | - | - | - | - | - | - | Neg | Neg | - | - | - | - |
| Uttaranchal | - | - | - | - | - | 1.40 | 1.84 | 0.67 | 0.54 | 0.40 | Neg | 0.2 | 8.30 |
| Others | 0.04 | 1.09 | 5.40 | 5.44 | 3.51 | 4.01 | 5.31 | 2.02 | 3.66 | 4.87 | 0.00 | 6.6 | 31.55 |
| All-India | 81.58 | 92.97 | 126.52 | 141.44 | 163.55 | 206.30 | 190.54 | 158.01 | 167.95 | 147.85 | 92.26 | 111.28 | 222.25 |

* Position upto 23.6.2008.

*Source*: Department of Food and Public Distribution.

## Table 9.2

*State-wise Purchases of Cotton from 1998-99 to 2007-08 (October to September)*

*(Thousand Bales of 170 kgs. each)*

| State | 1998-99 | 1999-00 | 2000-01 | 2001-02 | 2002-03 | 2003-04 | 2004-05 | 2005-06 | 2006-07 | 2007-08 |
|---|---|---|---|---|---|---|---|---|---|---|
| 1 | 2 | 3 | 4 | 5 | 6 | 7 | 8 | 9 | 10 | 11 |
| Andhra Pradesh | 119 | 116 | 180 | 468 | 133 | 184 | 121 | 351 | 528 | 246 |
| Gujarat | 172 | 113 | 127 | 236 | 163 | 231 | 490 | 339 | 145 | 123 |
| Haryana | 5 | 19 | 34 | 10 | 3 | 11 | 166 | 17 | 15 | 19 |
| Karnataka | 18 | 24 | 15 | 90 | 18 | 9 | 147 | 64 | 12 | 16 |
| Madhya Pradesh | 35 | 44 | 62 | 63 | 39 | 26 | 154 | 150 | 251 | 38 |
| Punjab | 15 | 36 | 62 | 13 | 9 | 24 | 151 | 58 | 43 | 79 |
| Rajasthan | 63 | 124 | 117 | 68 | 26 | 84 | 281 | 72 | 60 | 70 |
| Tamil Nadu | 1 | 3 | 1 | 0 | 0 | 0 | 0 | 0 | 0 | 0 |
| Others | 2 | 4 | 5 | 19 | 6 | 24 | 24 | 6 | 17 | 11 |
| Maharashtra | 0 | 25 | 0 | 0 | 203 | 328 | 170 | 296 | 384 | 377 |
| **Total** | 430 | 508 | 603 | 967 | 600 | 921 | 2794 | 1352 | 1455 | 979 |

*Note*: The figures given in the table include purchases made under both price support operations (MSP based) and commercial operations undertaken by the Cotton Corporation of India Ltd.

*Source*; Cotton Corporation of India ltd Mumbai, Ministry of Textiles, Mumbai.

## Table 9.3

*State-wise Purchases of Raw Jute from 1998-99 to 2007-08*

*(Thousand Bales of 180 kgs. each)*

| *State* | *1998-99* | *1999-2000* | *2000-01* | *2001-02* | *2002-03* | *2003-04* | *2004-05* | *2005-06* | *2006-07* | *2007-08* |
|---|---|---|---|---|---|---|---|---|---|---|
| *1* | 2 | 3 | 4 | 5 | 6 | 7 | 8 | 9 | 10 | 11 |
| West Bengal | 48.70 | 91.80 | 276.90 | 159.00 | 880.63 | 741.86 | 312.22 | 104.34 | 309.55 | 502.63 |
| Bihar | 0.50 | 13.50 | 71.40 | 38.90 | 176.92 | 198.42 | 26.15 | 11.15 | 55.09 | 123.98 |
| Assam | 5.30 | 1.30 | 111.90 | 47.80 | 159.71 | 115.24 | 9.90 | 25.14 | 116.56 | 90.40 |
| Meghalaya | 0.00 | 0.00 | 1.40 | 0.30 | 7.12 | 5.42 | 0.59 | 0.00 | 0.00 | 0.00 |
| Orissa | 0.00 | 0.40 | 0.90 | 0.20 | 19.08 | 12.18 | 1.09 | 0.00 | 0.54 | 9.75 |
| Andhra Pradesh | 0.00 | 0.00 | 1.10 | 0.00 | 67.58 | 40.73 | 0.02 | 0.00 | 1.49 | 28.48 |
| Tripura | 0.00 | 0.00 | 0.00 | 0.00 | 2.87 | 3.99 | 2.27 | 0.02 | 0.57 | 0.68 |
| Total | 54.50 | 107.00 | 465.60 | 246.20 | 1313.91 | 1117.84 | 352.24 | 140.65 | 483.80 | 755.92 |

*Source*: Jute Corporation of India, Ministry of Textiles, Kolkata.

# Table 9.4

*Procurement of Oilseeds, Pulses and Cotton by NAFED under Price Support Scheme (PSS)*

| S. | Commodity | Year@ | Crop Season | MSP (Rs. per quintal) | Qty. Procured (Metric Tonnes) | Value (Rs. Lakh) |
|---|---|---|---|---|---|---|
| 1 | 2 | 3 | 4 | 5 | 6 | 7 |
| 1. | Soyabean | 1999-2000 | *Kharif*-1999 | 845 (Yellow)<br>755 (Black) | 494938 | 45699.94 |
| | | 2000-2001 | *Kharif*-2000 | 865 (Yellow)<br>775 (Black) | 54745 | 5206.25 |
| | | 2005-2006 | *Kharif*-2005 | 1010 | 893 | 98.23 |
| | | 2006-2007 | *Kharif*-2006 | 1020 | 7 | 0.78 |
| 2. | Groundnut | 2000-2001 | *Rabi*-2000 | 1155 | 151 | 20.00 |
| | | 2000-2001 | *Kharif*-2000 | 1220 | 28707 | 4035.88 |
| | | 2001-2002 | *Rabi*-2001 | 1220 | 425 | 61.71 |
| | | 2001-2002 | *Kharif*-2001 | 1340 | 163493 | 23571.96 |
| | | 2004-2005 | *Kharif*-2004 | 1500 | 418 | 71.18 |
| | | 2005-2006 | *Kharif*-2005 | 1520 | 3463 | 587.17 |
| | | 2006-2007 | *Rabi*-2006 | 1520 | 116 | 19.37 |
| 3. | Safflower Seed | 2000-2001 | *Rabi*-2000 | 1100 | 6534 | 823.43 |
| | | 2001-2002 | *Rabi*-2001 | 1200 | 3201 | 426.69 |
| | | 2002-2003 | *Rabi*-2002 | 1300 | 2020 | 293.46 |
| | | 2004-2005 | *Rabi*-2004 | 1500 | 63 | 10.31 |
| | | 2005-2006 | *Rabi*-2005 | 1550 | 31805 | 5422.76 |
| | | 2006-2007 | *Rabi*-2006 | 1565 | 50779 | 8733.98 |
| | | 2007-2008 | *Rabi*-2007 | 1565 | 121 | 20.87 |
| 4. | Mustard Seed | 2000-2001 | *Rabi*-2000 | 1100 | 247956 | 29194.18 |
| | | 2001-2002 | *Rabi*-2001 | 1200 | 329524 | 42486.44 |
| | | 2002-2003 | *Rabi*-2002 | 1300 | 467629 | 65192.23 |
| | | 2004-2005 | *Rabi*-2004 | 1600 | 18580 | 3180.30 |
| | | 2005-2006 | *Rabi*-2005 | 1700 | 2084114 | 380469.24 |
| | | 2006-2007 | *Rabi*-2006 | 1715 | 2205000 | 421400.00 |
| | | 2007-2008 | *Rabi*-2007 | 1715 | 22091.00 | 424146.00 |
| 5. | Sunflower Seed | 1999-2000 | *Rabi*-1999 | 1060 | 3689 | 432.17 |
| | | 1999-2000 | *Kharif*-1999 | 1155 | 44261 | 6036.69 |
| | | 2000-2001 | *Kharif*-2000 | 1170 | 19502 | 2583.91 |
| | | 2001-2002 | *Rabi*-2001 | 1170 | 25 | 3.32 |
| | | 2002-2003 | *Rabi*-2002 | 1185 | 28 | 3.81 |
| | | 2004-2005 | *Rabi*-2004 | 1250 | 2316 | 310.89 |
| | | 2005-2006 | *Rabi*-2005 | 1340 | 33 | 4.42 |
| | | 2005-2006 | *Kharif*-2005 | 1500 | 3121 | 514.96 |
| | | 2006-2007 | *Rabi*-2006 | 1500 | 5242 | 864.93 |

*contd. ...*

*...contd. ...*

| *1* | *2* | *3* | *4* | *5* | *6* | *7* |
|---|---|---|---|---|---|---|
| 6 | Toria | 2005-2006 | *Rabi*-2005 | 1665 | 89 | 16.07 |
| 7 | Copra | 2000-2001 | 2000 Season | 3550 (Milling)<br>3500 (Ball) | 235491 | 78842.81 |
| | | | | 3500 (Ball) | | |
| | | 2001-2002 | 2001 Season | 3300 (Milling) | 50300 | 18305.22 |
| | | | | 3550 (Ball) | | |
| | | 2002-2003 | 2002 Season | 3300 (Milling) | 7468 | 2518.50 |
| | | | | 3550 (Ball) | | |
| | | 2003-2004 | 2003 Season | 2840 | 787 | 223.51 |
| | | (Copra Milling) | | (A & N Islands) | | |
| | | 2005-2006 | 2005 Season | 3570 | 3750 | 1498.06 |
| | | 2006-2007 | 2006 Season | 3590 | 8787 | 3414.16 |
| | | (Copra Milling) | | | | |
| | | 2007-2008 | 2007 Season | 3620 | 9522 | 3117.27 |
| | | (Copra Ball) | | | | |
| | | 2007-2008 | 2007 Season | 3870 | 19159 | 8041.03 |
| | | (Copra Milling) | | | | |
| 8 | *Tur/Arhar* | 2000-2001 | *Kharif*-2000 | 1200 | 98 | 12.32 |
| | | 2001-2002 | *Kharif*-2001 | 1320 | 3776 | 525.69 |
| | | 2002-2003 | *Kharif*-2002 | 1320+5* | 50 | 6.68 |
| 9 | *Urad* | 2002-2003 | *Kharif*-2002 | 1330+5* | 17148 | 2462.11 |
| | | 2003-2004 | *Kharif*-2003 | 1370 | 152357 | 22487.15 |
| | | 2004-2005 | *Kharif*-2004 | 1410 | 529 | 80.66 |
| 10 | *Moong* | 2003-2004 | *Kharif*-2003 | 1370 | 2488 | 383.65 |
| 11 | *Gram* | 2004-2005 | *Rabi*-2004 | 1400 | 281088 | 42698.68 |
| | | 2005-2006 | *Rabi*-2005 | 1425 | 107797 | 16668.26 |
| 12 | Cotton | 2005-2006 | *Rabi*-2005 | 1640 to 1960 | 18415 | 3356.96 |

@ Financial Year

* Special Drought Relief

*Source*: Cooperation Division, Ministry of Agriculture, New Delhi.

# Table 9.5

*Procurement of Onion by NAFED under Market Intervention Scheme*

| *Year* | *Price (Rs. Per Quintal)* | *Quantity Procured (Tonnes)* | *Value (Rs. Lakh)* | *Major State(s) of Procurement* |
|---|---|---|---|---|
| *1* | 2 | 3 | 4 | 5 |
| 1996-97 | 300 | 60 | 1.98 | Karnataka |
| 1999-2000 | 250 | 65000 | 1625.00 | Maharashtra |
| 2000-01 | Nil | Nil | Nil | - |
| 2001-02 | Nil | Nil | Nil | - |
| 2002-03 | Nil | Nil | Nil | - |
| 2003-04 | Nil | Nil | Nil | - |
| 2004-05 | 280 | 5000 | 0.73 | Rajasthan |
| 2005-06 | Nil | Nil | Nil | |
| 2006-07 | Nil | Nil | Nil | |

*Source*: NAFED.

# Table 10.1

*Net Availability of Foodgrains (Per Day) in India from 1951 to 2007*

*(Grams per day)*

| *Years* | *Rice* | *Wheat* | *Other Cereals* | *Cereals* | *Gram* | *Pulses* | *Foodgrains* |
|---|---|---|---|---|---|---|---|
| *1* | 2 | 3 | 4 | 5 | 6 | 7 | 8 |
| 1951 | 158.9 | 65.7 | 109.6 | 334.2 | 22.5 | 60.7 | 394.9 |
| 1956 | 187.7 | 61.5 | 111.2 | 360.4 | 29.0 | 70.3 | 430.7 |
| 1961 | 201.1 | 79.1 | 119.5 | 399.7 | 30.2 | 69.0 | 468.7 |
| 1966 | 161.9 | 95.4 | 102.6 | 359.9 | 18.3 | 48.2 | 408.1 |
| 1971 | 192.6 | 103.6 | 121.4 | 417.6 | 20.0 | 51.2 | 468.8 |
| 1976 | 187.2 | 79.5 | 107.4 | 373.8 | 20.2 | 50.5 | 424.3 |
| 1981 | 197.8 | 129.6 | 89.9 | 417.3 | 13.4 | 37.5 | 454.8 |
| 1985 | 188.8 | 138.6 | 87.9 | 415.3 | 12.9 | 38.1 | 453.4 |
| 1990 | 212.1 | 132.6 | 86.8 | 431.5 | 10.7 | 41.1 | 472.6 |
| 1991 | 221.7 | 166.8 | 80.0 | 468.5 | 13.4 | 41.6 | 510.1 |
| 1992 | 217.0 | 158.6 | 58.9 | 434.5 | 10.1 | 34.3 | 468.8 |
| 1993 | 201.1 | 140.2 | 86.6 | 427.9 | 10.7 | 36.2 | 464.1 |
| 1994 | 207.4 | 159.5 | 67.1 | 434.0 | 11.8 | 37.2 | 471.2 |
| 1995 | 220.0 | 172.7 | 64.9 | 457.6 | 14.9 | 37.8 | 495.5 |
| 1996 | 204.4 | 176.0 | 62.0 | 442.5 | 11.3 | 32.7 | 475.2 |
| 1997 | 214.0 | 179.1 | 72.9 | 466.0 | 12.4 | 37.1 | 503.1 |
| 1998 | 200.3 | 151.5 | 62.4 | 414.2 | 13.4 | 32.8 | 447.0 |
| 1999 | 203.4 | 162.3 | 63.4 | 429.2 | 14.6 | 36.5 | 465.7 |
| 2000 | 203.7 | 160.0 | 59.0 | 422.7 | 10.8 | 31.8 | 454.4 |
| 2001 | 190.5 | 135.8 | 56.2 | 386.2 | 8.0 | 30.0 | 416.2 |
| 2002 | 228.7 | 166.6 | 63.4 | 458.7 | 10.7 | 35.4 | 494.1 |
| 2003 | 181.4 | 180.4 | 46.7 | 408.5 | 8.5 | 29.1 | 437.6 |
| 2004 | 195.4 | 162.2 | 69.3 | 426.9 | 11.2 | 35.8 | 462.7 |
| 2005 | 177.3 | 154.3 | 59.4 | 390.9 | 10.6 | 31.5 | 422.4 |
| 2006 | 198.0 | 154.3 | 60.5 | 412.8 | 10.7 | 32.5 | 445.3 |
| 2007(P) | 196.8 | 156.1 | 58.9 | 409.9 | 11.9 | 29.4 | 439.3 |

*Notes:* 1. The net availability of foodgrains is estimated to be Gross Production (-) seed, feed & wastage, (-) exports (+) imports, (+/-) change in stocks.

2. The net availability of foodgrains divided by the population estimates for a particular year indicate per capita availability of foodgrains in terms of kg/year. Net availability, thus worked out further divided by the number of days in a year I.e , 365 days gives us net availability of foodgrains in terms of grams/day.

3. Figures in respect of per capita net availability given abcve are not strictly representative of actual level of consumption in the country especially as they do not take in to account any change in stocks in possession of traders, producers and consumers.

4. For calculation of per capita net availability the figures of net imports from 1981 to 1994 are based on imports and exports on Government of India account only. Net imports from 1995 ownwards are the total exports and imports (on Government as well as private accounts)

5. Cereals includes rice, wheat and other cereals

6. Pulses includes all *kharif* and *rabi* pulses

7. Foodgrains includes rice, wheat, other cereals and all pulses.

# Table 10.2

*Net Availability of Foodgrains (per Annum) in India from 1951 to 2007*

*(Grams per day)*

| *Years* | *Rice* | *Wheat* | *Other cereals* | *Cereals* | *Gram* | *Pulses* | *Foodgrains* |
|---|---|---|---|---|---|---|---|
| *1* | *2* | *3* | *4* | *5* | *6* | *7* | *8* |
| 1951 | 58.0 | 24.0 | 40.0 | 122.0 | 8.2 | 22.1 | 144.1 |
| 1956 | 68.7 | 22.5 | 40.7 | 131.9 | 10.6 | 25.7 | 157.6 |
| 1961 | 73.4 | 28.9 | 43.6 | 145.9 | 11.0 | 25.2 | 171.1 |
| 1966 | 59.1 | 34.8 | 37.5 | 131.4 | 6.7 | 17.6 | 149.0 |
| 1971 | 70.3 | 37.8 | 44.3 | 152.4 | 7.3 | 18.7 | 171.1 |
| 1976 | 68.5 | 29.1 | 39.2 | 13.8 | 7.4 | 18.5 | 155.3 |
| 1981 | 72.2 | 47.3 | 32.8 | 152.3 | 4.9 | 13.7 | 166.0 |
| 1985 | 68.9 | 50.6 | 32.1 | 151.6 | 4.7 | 13.9 | 165.5 |
| 1990 | 77.4 | 48.4 | 31.7 | 157.5 | 3.9 | 15.0 | 172.5 |
| 1991 | 80.9 | 60.0 | 29.2 | 171.0 | 4.9 | 15.2 | 186.2 |
| 1992 | 79.2 | 57.9 | 21.5 | 158.6 | 3.7 | 12.5 | 171.1 |
| 1993 | 73.4 | 51.2 | 31.6 | 156.2 | 3.9 | 13.2 | 169.4 |
| 1994 | 75.7 | 58.2 | 24.5 | 158.4 | 4.3 | 13.6 | 172.0 |
| 1995 | 80.3 | 63.0 | 23.7 | 167.0 | 5.4 | 13.8 | 180.8 |
| 1996 | 74.6 | 64.3 | 22.6 | 161.5 | 4.1 | 12.0 | 173.5 |
| 1997 | 78.1 | 65.4 | 26.6 | 170.1 | 4.5 | 13.5 | 183.6 |
| 1998 | 73.1 | 55.3 | 22.8 | 151.2 | 4.9 | 12.0 | 163.2 |
| 1999 | 74.2 | 59.2 | 23.1 | 156.7 | 5.3 | 13.3 | 170.0 |
| 2000 | 74.3 | 58.4 | 21.5 | 154.3 | 3.9 | 11.6 | 165.9 |
| 2001 | 69.5 | 49.6 | 20.5 | 141.0 | 2.9 | 10.9 | 151.9 |
| 2002 | 83.5 | 60.8 | 23.1 | 167.4 | 3.9 | 12.9 | 180.4 |
| 2003 | 66.2 | 65.8 | 17.1 | 149.1 | 3.1 | 10.6 | 159.7 |
| 2004 | 71.3 | 59.2 | 25.3 | 155.8 | 4.1 | 13.1 | 168.9 |
| 2005 | 64.7 | 56.3 | 21.7 | 142.7 | 3.9 | 11.5 | 154.2 |
| 2006 | 72.3 | 56.3 | 22.1 | 150.7 | 3.9 | 11.8 | 162.5 |
| 2007(P) | 71.8 | 57 | 20.8 | 149.6 | 4.3 | 10.7 | 160.4 |

*Notes*:
1. The net availability of foodgrains is estimated to be Gross Production (-) seed, feed & wastage, (-) exports (+) imports, (+/-) change in stocks.
2. The net availability of foodgrains divided by the population estimates for a particular year indicate per capita availability of foodgrains in terms of kg/year. Net availability, thus worked out further divided by the number of days in a year I.e., 365 days gives us net availability of foodgrains in terms of grams/day.
3. Figures in respect of per capita net availability given above are not strictly representative of actual level of consumption in the country especially as they do not take in to account any change in stocks in possession of traders, producers and consumers.
4. For calculation of per capita net availability the figures of net imports from 1981 to 1994 are based on imports and exports on Government of India account only. Net imports from 1995 ownwards are the total exports and imports (on Government as well as private accounts)
5. Cereals includes rice, wheat and other cereals
6. Pulses includes all *kharif* and *rabi* pulses
7. Foodgrains includes rice, wheat, other cereals and all pulses.

# Table 10.3

## *Per Capita Availability of Certain Important Articles of Consumption*

| *Year* | *Edible# Oils@ (kgs.)* | *Vanaspati (kgs.) #* | *Sugar (Nov-Oct) (kgs.) +* | *Cloth (metres)* | | | *Tea* (gms) ##* | *Coffee* (gms)* |
|---|---|---|---|---|---|---|---|---|
| | | | | *Cotton @@* | *Man-made* | *Total* | | |
| *1* | *2* | *3* | *4* | *5* | *6* | *7* | *8* | *9* |
| 1960-61 | 3.2 | 0.8 | 4.8 | 13.8 | 1.2 | 15.0 | 296.0 | 80.0 |
| 1965-66 | 2.7 | 0.8 | 5.7 | 14.7 | 1.7 | 16.4 | 346.0 | 72.0 |
| 1971-72 | 3.0 | 1.1 | 6.8 | 12.4 | 2.2 | 14.6 | 426.0 | 65.0 |
| 1972-73 | 2.4 | 1.0 | 6.2 | 13.2 | 2.0 | 15.2 | 458.0 | 69.0 |
| 1973-74 | 3.4 | 0.8 | 6.1 | 12.0 | 1.9 | 13.9 | 492.0 | 64.0 |
| 1974-75 | 3.3 | 0.6 | 5.8 | 12.9 | 1.7 | 14.6 | 471.0 | 62.0 |
| 1975-76 | 3.5 | 0.8 | 6.1 | 12.6 | 2.0 | 14.6 | 446.0 | 62.0 |
| 1976-77 | 3.2 | 0.9 | 6.0 | 11.4 | 2.4 | 13.8 | 450.0 | 71.0 |
| 1977-78 | 3.8 | 0.9 | 7.2 | 9.5 | 4.0 | 13.5 | 516.0 | 73.0 |
| 1978-79 | 3.8 | 1.0 | 9.6 | 10.2 | 4.8 | 15.0 | 599.0 | 77.0 |
| 1979-80 | 3.7 | 1.0 | 7.8 | 10.1 | 4.6 | 14.7 | 521.0 | 73.0 |
| 1980-81 | 3.8 | 1.2 | 7.3 | 12.9 | 4.4 | 17.3 | 511.0 | 74.0 |
| 1981-82 | 5.1 | 1.3 | 8.2 | 12.2 | 4.9 | 17.1 | 466.0 | 79.0 |
| 1982-83 | 4.5 | 1.3 | 9.0 | 11.8 | 4.3 | 16.1 | 525.0 | 82.0 |
| 1983-84 | 5.8 | 1.2 | 10.5 | 12.6 | 4.7 | 17.3 | 519.0 | 78.0 |
| 1984-85 | 5.5 | 1.3 | 10.7 | 12.6 | 4.6 | 17.2 | 576.0 | 72.0 |
| 1985-86 | 5.0 | 1.3 | 11.1 | 15.4 | 6.1 | 21.5 | 589.0 | 71.0 |
| 1986-87 | 5.0 | 1.2 | 11.4 | 15.2 | 6.6 | 21.8 | 545.0 | 76.0 |
| 1987-88 | 5.8 | 1.2 | 11.7 | 14.0 | 7.0 | 21.0 | 592.0 | 72.0 |
| 1988-89 | 5.3 | 1.2 | 12.1 | 15.0 | 8.0 | 23.0 | 612.0 | 79.0 |
| 1989-90 | 5.3 | 1.1 | 12.3 | 14.6 | 8.1 | 22.7 | 571.0 | 65.0 |
| 1990-91 | 5.5 | 1.0 | 12.7 | 15.1 | 9.0 | 24.1 | 612.0 | 59.0 |
| 1991-92 | 5.4 | 1.0 | 13.0 | 13.7 | 9.2 | 22.9 | 655.0 | 64.0 |
| 1992-93 | 5.8 | 1.0 | 13.7 | 15.6 | 8.9 | 24.5 | 649.0 | 60.0 |
| 1993-94 | 6.1 | 1.0 | 12.5 | 15.9 | 10.3 | 26.2 | 667.0 | 56.0 |
| 1994-95 | 6.3 | 1.0 | 13.2 | 15.2 | 10.8 | 26.0 | 664.0 | 55.0 |
| 1995-96 | 7.0 | 1.0 | 14.1 | 16.3 | 11.7 | 28.0 | 646.0 | 55.0 |
| 1996-97 | 8.0 | 1.0 | 14.6 | 16.2 | 13.1 | 29.3 | 657.0 | 58.0 |
| 1997-98 | 6.2 | 1.0 | 14.5 | 15.9 | 15.0 | 30.9 | 635.0 | 58.0 |
| 1998-99 | 8.5 | 1.3 | 14.9 | 13.1 | 15.1 | 28.2 | 684.0 | 65.0 |
| 1999-2000 | 9.0 | 1.4 | 15.6 | 14.2 | 16.4 | 30.6 | 642.0 | 55.0 |
| 2000-01 | 8.2 | 1.3 | 15.8 | 14.2 | 16.5 | 30.7 | 631.0 | 58.0 |
| 2001-02 | 8.8 | 1.4 | 16.0 | 14.8 | 17.2 | 32.0 | 650.0 | 67.0 |
| 2002-03 | 7.2 | 1.4 | 16.3 | 14.4 | 17.0 | 31.4 | 596.0 | 67.0 |
| 2003-04 | NA | NA | NA | 13.4 | 17.6 | 31.0 | 662.0 | 70.0 |
| 2004-05 | NA | NA | NA | 14.1 | 19.4 | 33.5 | 663.0 | 75.0 |
| 2005-06 | NA | NA | NA | 16.4 | 19.7 | 36.1 | 687.0 | 80.0 |
| 2006-07(P) | NA | NA | NA | 18.3 | 21.9 | 40.2 | 687.0 | 85.0 |

@ Includes Groundnut Oil, Sesamum Oil, Rapeseed Oil, Coconut Oil, Nigerseed Oil, Soyabean Oil and Sunflower oil but excludes oil used for manufacture of vanaspati.

@@ Figures for blended/mixed fabrics were not separately available prior to 1969. These have been included under man made fibre fabrics after 1969.

* Figures upto 1971-72 relate to coffee season and thereafter on calendar year basis. The figures for 1972-73 correspond to 1973 and so on.

# Relates to financial year.

## Relates to calendar year.

+ Relates to actual releases for domestic consumption.

(P) Provisional NA- Not Available

*Note:* 1. From 1967-68 sugar season is October-September.

*Source*: *Economic Survey 2006-07*.

## Table 10.4(a)

*Share of Monthly Per Capita Expenditure on Cereals and Non-Cereals in Total Food Items*

(Per cent)

| Year | Rural Areas | | Urban Areas | |
|---|---|---|---|---|
| | Cereals | Non-Cereals | Cereals | Non-Cereals |
| 1 | 2 | 3 | 4 | 5 |
| 1987-88 | 40.99 | 59.01 | 26.46 | 73.54 |
| 1993-94 | 38.30 | 61.70 | 25.69 | 74.31 |
| 1999-2000 | 37.31 | 62.69 | 25.70 | 74.30 |
| 2000-01 | 35.58 | 64.42 | 25.13 | 74.87 |
| 2001-02 | 34.55 | 65.45 | 24.30 | 75.70 |
| 2002 (July-Dec.) | 33.10 | 66.90 | 23.42 | 76.58 |
| 2003 (Jan-Dec) | 33.41 | 66.59 | 23.76 | 76.24 |
| 2004 (Jan-June) | 33.55 | 66.45 | 23.92 | 76.08 |
| 2004-05 (July-June) | 32.72 | 67.28 | 23.65 | 76.35 |
| 2005-06 (July-June) | 31.91 | 68.09 | 23.47 | 76.53 |

*Source*: National Sample Survey Organisation, 62nd round of NSS (July-June,2005-06).

## Table 10.4(b)

*Monthly Per Capita Expenditure*

(in Kg.)

| Crop | 55th Round (1999-2000) | | 61st Round (2004-05) | |
|---|---|---|---|---|
| | Rural | Urban | Rural | Urban |
| 1 | 2 | 3 | 4 | 5 |
| Rice | 6.75 | 5.21 | 6.55 | 4.85 |
| Wheat | 4.49 | 4.76 | 4.29 | 4.65 |
| Coarse Cereals | 1.48 | 0.45 | 1.28 | 0.44 |
| Cereals | 12.72 | 10.42 | 12.12 | 9.94 |
| Pulses | 0.84 | 1.00 | 0.67 | 0.78 |

*Source*: 55th and 61st Round of National Sample Survey Organisation.

## Table 11.1 (a)

*Stock of Cereals held by Central and State Agencies (as on 1st January)*

*(Million Tonnes)*

| *Year* | *Rice* | *Wheat* | *Coarse Cereals* | *Total* |
|---|---|---|---|---|
| *1* | 2 | 3 | 4 | 5 |
| 1991 | 9.63 | 9.38 | 0.12 | 19.13 |
| 1992 | 9.29 | 5.43 | 0.01 | 14.73 |
| 1993 | 9.48 | 3.47 | 0.18 | 13.13 |
| 1994 | 11.95 | 11.10 | 0.47 | 23.52 |
| 1995 | 17.42 | 12.88 | - | 30.30 |
| 1996 | 15.41 | 13.15 | - | 28.56 |
| 1997 | 12.94 | 7.08 | - | 20.02 |
| 1998 | 11.49 | 6.76 | - | 18.25 |
| 1999 | 11.68 | 12.70 | - | 24.38 |
| 2000 | 14.18 | 17.17 | - | 31.35 |
| 2001 | 20.70 | 25.04 | 0.03 | 45.77 |
| 2002 | 25.62 | 32.41 | 0.08 | 58.11 |
| 2003 | 19.37 | 28.83 | - | 48.20 |
| 2004 | 11.73 | 12.69 | 0.60 | 25.02 |
| 2005 | 12.76 | 8.93 | 0.60 | 21.70 |
| 2006 | 12.64 | 6.19 | 0.43 | 19.26 |
| 2007 | 11.98 | 5.43 | 0.09 | 17.49 |
| 2008 | 11.47 | 7.71 | 0.00 | 19.18 |

## Table 11.1 (b)

*Buffer Norms*

| *Date* | *Rice* | | *Wheat* | | *Total* | |
|---|---|---|---|---|---|---|
| | *Old* | *New* | *Old* | *New* | *Old* | *New* |
| *1* | 2 | 3 | 4 | 5 | 6 | 7 |
| 1-Apr | 11.80 | 12.20 | 4.00 | 4.00 | 15.80 | 16.20 |
| 1-Jul | 10.00 | 9.80 | 14.30 | 17.10 | 24.30 | 26.90 |
| 1-Oct | 6.50 | 5.20 | 11.60 | 11.00 | 18.10 | 16.20 |
| 1-Jan | 8.40 | 11.80 | 8.40 | 8.20 | 16.80 | 20.00 |

*Note*: New buffer norm are effective from April 2005.

*Source*: Department of Food & Public Distribution.

## Table 11.2

*Consumption and Stocks of Raw Jute From 1982-83 to 2007-08*

(Lakh Bales of 180 kgs. each)

| Year (July-June) | Consumption | Stocks (End of Season) |
|---|---|---|
| 1 | 2 | 3 |
| 1982-83 | 75.60 | 18.98 |
| 1983-84 | 63.83 | 20.91 |
| 1984-85 | 78.33 | 7.50 |
| 1985-86 | 87.68 | 42.42 |
| 1986-87 | 82.00 | 43.00 |
| 1987-88 | 74.00 | 34.00 |
| 1988-89 | 84.50 | 20.50 |
| 1989-90 | 86.00 | 9.50 |
| 1990-91 | 87.40 | 16.10 |
| 1991-92 | 79.50 | 32.00 |
| 1992-93 | 85.00 | 25.00 |
| 1993-94 | 87.50 | 9.50 |
| 1994-95 | 85.50 | 9.00 |
| 1995-96 | 87.00 | 5.00 |
| 1996-97 | 91.10 | 18.50 |
| 1997-98 | 130.00 | 27.00 |
| 1998-99 | 97.00 | 22.00 |
| 1999-2000 | 97.50 | 10.50 |
| 2000-01 | 97.50 | 7.00 |
| 2001-02 | 100.00 | 16.00 |
| 2002-03 | 101.00 | 34.00 |
| 2003-04 | 96.00 | 33.00 |
| 2004-05 | 98.00 | 15.00 |
| 2005-06 | 98.00 | 5.00 |
| 2006-07 | 82.00 | 25.00 |
| 2007-08 | 102.00 | 26.00 |

*Source*: Jute Corporation of India, Kolkata.

# Table 12.1 (a)

## *Agriculture Subsidies in India during 1993-94 to 2006-07*

*(Rs. Crore)*

| Item | (At current prices) At 1993-94 Series | | | | | | | | At Revised 1999-00 Series | | | | | | |
|---|---|---|---|---|---|---|---|---|---|---|---|---|---|---|---|
| | *1993-94* | *1994-95* | *1995-96* | *1996-97* | *1997-98* | *1998-99* | *1999-00* | *1999-00* | *2000-01* | *2001-02* | *2002-03* | *2003-04* | *2004-05* | *2005-06* | *2006-07* |
| 1 | 2 | 3 | 4 | 5 | 6 | 7 | 8 | 9 | 10 | 11 | 12 | 13 | 14 | 15 | 16 |
| 1. Food subsidy | NA | NA | NA | NA | NA | NA | NA | 9434 | 12060 | 17499 | 24176 | 25181 | 25798 | 23077 | 24014 |
| 2. Fertiliser (Total) | 4562 | 5769 | 6735 | 7578 | 9918 | 11596 | 13244 | 13244 | 13800 | 12595 | 11015 | 11847 | 15879 | 18460 | |
| 3. Electricity** | 2400 | 2338 | 1977 | 8356 | 4937 | 3819 | 4276 | 6033 | 8919 | 10410 | 8521 | 14544 | 17852 | 20301 | |
| 4. Irrigation ## | 5872 | 6772 | 7931 | 9221 | 10318 | 11827 | 11487 | 11196 | 13259 | 13009 | 12794 | 10921 | 12508 | 14625 | |
| 5. Other subsidies given to marginal farmers and Farmers' Cooperative societies in the form of seeds, development of oil seeds, pulses, Cotton, Rice, Maize and crop insurance | 1235 | 1246 | 1034 | 895 | 983 | 1182 | 1810 | 3118 | 2733 | 3234 | 3173 | 4132 | 3598 | 6501 | |
| Total | 14069 | 16125 | 17677 | 26050 | 26156 | 28424 | 30817 | 43025 | 50771 | 56747 | 59679 | 66625 | 75635 | 82967 | 24014 |

*Notes*: 1. Fertilisers and other subsidies given to marginal farmers Expenditure Budget 2007-08, Central Government.

2. Electricity & Irrigation: Central Statistical Organisation.

3. The estimates based on Revised Series Base 1999-00 are Provisional.

* Provisional

** Includes all subsidies to Electricity Boards and Corporations. Separate estimates of Electricity subsidy accountable exclusively to agricultural sector are not available

##: The rates for supply of water to farmers are kept low as a matter of policy, resulting in losses to the Government irrigation system. The excess of operating costs over the gross revenue is treated as imputed irrigation subsidy.

*Source*: Central Statistical Organisation, New Delhi.

# Table 12.1 (b)

## *Fertiliser Subsidies in India during 1993-94 to 2005-06*

| *Item* | *(At current prices) At 1993-94 Series* | | | | | | | | *At Revised 1999-00 Series* | | | | | |
|---|---|---|---|---|---|---|---|---|---|---|---|---|---|---|
| | *1993-94* | *1994-95* | *1995-96* | *1996-97* | *1997-98* | *1998-99* | *1999-00* | *1999-00* | *2000-01* | *2001-02* | *2002-03* | *2003-04* | *2004-05* | *2005-06* |
| *1* | *2* | *3* | *4* | *5* | *6* | *7* | *8* | *9* | *10* | *11* | *12* | *13* | *14* | *15* |
| Indigenous fertiliser | 3800 | 4075 | 4300 | 4743 | 6600 | 7473 | 8670 | 8670 | 9480 | 8044 | 7790 | 8521 | 10243 | 10653 |
| Imported fertiliser | 762 | 1166 | 1935 | 1163 | 722 | 333 | 74 | 74 | 1 | 47 | 0 | 0 | 494 | 1211 |
| Sale of decontrolled fertilser with concession to farmers | 0 | 528 | 500 | 1672 | 2596 | 3790 | 4500 | 4500 | 4319 | 4504 | 3225 | 3326 | 5142 | 6596 |
| Fertiliser (Total) | 4562 | 5769 | 6735 | 7578 | 9918 | 11596 | 13244 | 13244 | 13800 | 12595 | 11015 | 11847 | 15879 | 18460 |

*Note*: The estimates based on Revised series Base 1999 are provisional.

*Source*: Fertilisers Expenditure Budget 2007-08, Vol. I of Central Government.

## Table 12.2

*Share of Agricultural Subsidies (PSE) in Selected OECD Countries*

*(US $ million)*

| Country | 1986 - 1988* | | 2003-2005* | | 2003 | | 2004 | | 2005(P) | |
|---|---|---|---|---|---|---|---|---|---|---|
| | PSE | %PSE | PSE | %PSE | PSE | %PSE | PSE | %PSE | PSE | %PSE |
| 1 | 2 | 3 | 4 | 5 | 6 | 7 | 8 | 9 | 10 | 11 |
| Canada | 6066 | 36 | 5884 | 18 | 5941 | 25 | 5695 | 21 | 6015 | 21 |
| EU | 90924 | 41 | 107563 | 34 | 119149 | 36 | 136144 | 33 | 133785 | 32 |
| USA | 36958 | 22 | 40479 | 16 | 35929 | 15 | 42869 | 16 | 42669 | 16 |
| JAPAN | 49579 | 64 | 48324 | 58 | 48171 | 59 | 49368 | 58 | 47435 | 56 |
| Australia | 1320 | 8 | 1404 | 5 | 1339 | 5 | 1421 | 5 | 1453 | 5 |
| OECD | 242474 | 37 | 272738 | 30 | 258798 | 30 | 279572 | 29 | 279845 | 29 |

*Note*: P = provisional, * average of three years, PSE (Producer Support Estimates): an indicator of the annual money value of gross transfers from consumers and taxpayers to agricultural producers, measured at the farm-gate level, arising from policy measures that support agriculture, regardless of their nature, objectives or impacts on farm production or income. %PSE: is support expressed as a percentage of gross farm receipts, which shows the amount of support to farmers, irrespective of the sectoral structure of a given country.

*Source*: OECD 2006 PSE/CSE data base 2006.

## Table 13.1

*India's Imports and Exports of Agricultural Commodities vis-à-vis Total National Imports/Exports during 1990-91 to 2007-08*

*(Value in Rupees Crore)*

| Year | Agriculture Imports | Total National Imports | %age Agriculture Imports to Total National Imports | Agriculture Exports | Total National Exports | % age Agriculture Exports to Total National Exports |
|---|---|---|---|---|---|---|
| 1 | 2 | 3 | 4 | 5 | 6 | 7 |
| 1990-91 | 1205.86 | 43170.82 | 2.79 | 6012.76 | 32527.28 | 18.49 |
| 1991-92 | 1478.27 | 47850.84 | 3.09 | 7838.04 | 44041.81 | 17.80 |
| 1992-93 | 2876.25 | 63374.52 | 4.54 | 9040.30 | 53688.26 | 16.84 |
| 1993-94 | 2327.33 | 73101.01 | 3.18 | 12586.55 | 69748.85 | 18.05 |
| 1994-95 | 5937.21 | 89970.70 | 6.60 | 13222.76 | 82673.40 | 15.99 |
| 1995-96 | 5890.10 | 122678.14 | 4.80 | 20397.74 | 106353.35 | 19.18 |
| 1996-97 | 6612.60 | 138919.88 | 4.76 | 24161.29 | 118817.32 | 20.33 |
| 1997-98 | 8784.19 | 154176.29 | 5.70 | 24832.45 | 130100.64 | 19.09 |
| 1998-99 | 14566.48 | 178331.69 | 8.17 | 25510.64 | 139751.77 | 18.25 |
| 1999-00 | 16066.73 | 215528.53 | 7.45 | 25313.66 | 159095.20 | 15.91 |
| 2000-01 | 12086.23 | 228306.64 | 5.29 | 28657.37 | 201356.45 | 14.23 |
| 2001-02 | 16256.61 | 245199.72 | 6.63 | 29728.61 | 209017.97 | 14.22 |
| 2002-03 | 17608.83 | 297205.87 | 5.92 | 34653.94 | 255137.28 | 13.58 |
| 2003-04 | 21972.68 | 359107.66 | 6.12 | 37266.52 | 293366.75 | 12.70 |
| 2004-05 | 22811.84 | 501064.54 | 4.55 | 41602.65 | 375339.53 | 11.08 |
| 2005-06 | 21499.22 | 660408.90 | 3.26 | 49216.96 | 456417.86 | 10.78 |
| 2006-07 | 29637.86 | 840506.31 | 3.53 | 62411.42 | 571779.28 | 10.92 |
| 2007-08(P) | 29777.01 | 964849.76 | 3.09 | 77769.71 | 640172.14 | 12.15 |

*Note*: P- Provisional

*Source* : DGCI&S, Ministry of Commerce, Kolkata.

## Table 13.2

*India's Imports of Agricultural Products during 2001-02 to 2007-08*

(Quantity: '000 tonnes)
(Value: Rs. Crore)

| Commodity | April 2001-March 02 | | April 2002-March 03 | | April 2003-March 04 | |
|---|---|---|---|---|---|---|
| | Quantity | Value | Quantity | Value | Quantity | Value |
| 1 | 2 | 3 | 4 | 5 | 6 | 7 |
| Pulses | 2217.82 | 3160.16 | 1992.29 | 2737.05 | 1723.33 | 2284.87 |
| Wheat | 1.35 | 0.84 | - | - | 0.46 | 0.25 |
| Rice | 0.06 | 0.07 | 0.87 | 1.09 | 0.54 | 0.27 |
| Other Cereals | 4.76 | 3.58 | 1.12 | 0.67 | 1.53 | 1.87 |
| Cereal Preparation | 35.13 | 82.34 | 51.74 | 116.79 | 32.51 | 86.85 |
| Milk & Cream | 1.17 | 8.39 | 1.14 | 9.55 | 11.55 | 89.56 |
| Cashew Nuts | 161.79 | 431.14 | 402.98 | 1236.24 | 442.59 | 1371.80 |
| Fruits & Nuts Excluding Cashew Nuts | - | 756.76 | - | 641.76 | - | 802.27 |
| Spices | 86.29 | 503.87 | 121.28 | 586.44 | 124.17 | 564.42 |
| Sugar | 25.58 | 32.60 | 41.43 | 32.83 | 74.40 | 62.70 |
| Oil Seeds | - | 1.34 | - | 11.49 | - | 13.89 |
| Vegetable Oils Fixed (Edible) | 4321.83 | 6464.97 | 4365.03 | 8779.64 | 5290.30 | 11683.24 |
| Vegetable & Animal fats | 1.21 | 10.37 | 1.45 | 11.61 | 1.76 | 12.68 |
| Cotton (Raw & Waste) | 387.04 | 2053.62 | 233.85 | 1237.61 | 252.90 | 1570.03 |
| Jute (Raw) | 79.69 | 95.68 | 143.23 | 134.77 | 111.97 | 93.92 |
| Tea | 10.31 | 67.01 | 23.64 | 125.30 | 10.77 | 64.60 |
| Wood & Wood Products | - | 2583.87 | - | 1945.99 | - | 3269.46 |
| Total Agricultural Imports | 16256.61 | 17608.83 | 21972.68 | | | |
| Total National Imports | 245199.72 | 297205.87 | 359107.66 | | | |
| % Share of Agricultural Import in National Imports | 6.63 | 5.92 | 6.12 | | | |

*contd. ...*

*...contd. ...*

| *Commodity* | *April 2004-March 2005* | | *April 2005-March 2006* | | *April 2006-March 2007* | | *April 2007-March 2008(P)* | |
|---|---|---|---|---|---|---|---|---|
| | *Quantity* | *Value* | *Quantity* | *Value* | *Quantity* | *Value* | *Quantity* | *Value* |
| *1* | *8* | *9* | *10* | *11* | *12* | *13* | *14* | *15* |
| Pulses | 1339.45 | 1777.58 | 1695.95 | 2476.25 | 2270.97 | 3891.91 | 2791.10 | 5278.02 |
| Wheat | - | - | - | - | 6079.56 | 5850.49 | 1793.21 | 2657.51 |
| Rice | - | - | 0.26 | 0.34 | 0.16 | 0.41 | 0.15 | 0.42 |
| Other Cereals | 6.62 | 6.56 | 27.88 | 30.09 | 7.96 | 11.73 | 10.35 | 19.27 |
| Cereal Preparation | 45.46 | 112.16 | 42.40 | 129.23 | 38.22 | 132.92 | 43.84 | 161.58 |
| Milk & Cream | 2.21 | 12.87 | 1.63 | 14.20 | 3.09 | 28.90 | 2.06 | 29.59 |
| Cashew Nuts | 479.71 | 1804.96 | 543.94 | 2089.46 | 586.49 | 1820.75 | 591.61 | 1714.24 |
| Fruits & Nuts Excluding Cashew Nuts | - | 1100.83 | - | 1390.32 | - | 1913.11 | - | 1857.84 |
| Spices | 107.22 | 593.83 | 108.93 | 687.81 | 118.51 | 738.90 | 140.45 | 938.22 |
| Sugar | 932.74 | 976.18 | 558.77 | 651.59 | 1.05 | 3.48 | 0.51 | 2.29 |
| Oil Seeds | - | 28.41 | - | 47.03 | - | 104.47 | - | 147.88 |
| Vegetable Oils Fixed (Edible) | 4751.19 | 11076.89 | 4288.10 | 8960.99 | 4269.38 | 9539.90 | 4902.81 | 10298.68 |
| Vegetable & Animal fats | 1.42 | 12.78 | 1.47 | 14.24 | 1.11 | 11.19 | 1.10 | 12.48 |
| Cotton (Raw & Waste) | 192.18 | 1135.56 | 98.75 | 703.66 | 81.48 | 663.07 | 136.34 | 911.89 |
| Jute (Raw) | 20.74 | 31.79 | 61.34 | 93.16 | 83.05 | 115.21 | 147.36 | 161.34 |
| Tea | 31.76 | 146.92 | 18.75 | 108.14 | 23.29 | 127.06 | 19.84 | 131.66 |
| Wood & Wood Products | - | **3994.52** | - | **4102.71** | - | 4684.35 | - | 5454.10 |
| **Total Agricultural Imports** | - | **22811.84** | - | **21499.22** | - | **29637.86** | - | **29777.01** |
| **Total National Imports** | - | **501064.54** | - | **660408.90** | - | **840506.31** | - | **964849.76** |
| **% Share of Agricultural Import in** National Imports | - | **4.55** | - | **3.26** | - | **3.53** | - | **3.09** |

P Provisional

*Source*: Director General of Commercial Intelligence & Statistics, Ministry of Commerce, Kolkata.

# Table 13.3

*India's Exports of Agricultural Products during 2001-02 to 2007-08*

(Quantity: '000 tonnes)
(Value: Rs. Crore)

| Commodity | April 2001 March 02 | | April 2002 March 03 | | April 2003 March 2004 | |
|---|---|---|---|---|---|---|
| | Quantity | Value | Quantity | Value | Quantity | Value |
| 1 | 2 | 3 | 4 | 5 | 6 | 7 |
| Pulses | 161.64 | 369.13 | 148.08 | 345.02 | 153.88 | 328.60 |
| Rice *Basmati* | 667.07 | 1842.77 | 708.79 | 2058.47 | 771.49 | 1993.05 |
| Rice (Other than *Basmati*) | 1541.49 | 1331.37 | 4259.08 | 3772.77 | 2640.57 | 2174.94 |
| Wheat | 2649.38 | 1330.21 | 3671.25 | 1759.87 | 4093.08 | 2391.15 |
| Other Cereals | 144.73 | 115.92 | 106.08 | 91.06 | 604.23 | 397.55 |
| Tea | 180.10 | 1719.22 | 182.86 | 1652.07 | 177.77 | 1637.35 |
| Coffee | 176.26 | 1094.92 | 184.87 | 993.98 | 188.44 | 1085.92 |
| Tobacco Unmanufactured | 84.48 | 582.05 | 100.47 | 733.52 | 120.64 | 801.41 |
| Tobacco Manufactured | 13.46 | 225.66 | - | 289.37 | - | 295.06 |
| Poultry & Dairy Products | - | 207.96 | - | 176.45 | - | 161.56 |
| Floriculture Products | - | 127.43 | - | 180.77 | - | 250.47 |
| Spices | 239.29 | 1496.97 | 277.02 | 1655.49 | 267.47 | 1544.18 |
| Cashewnut Shell Liquid | 1.99 | 5.25 | 6.14 | 8.56 | 4.33 | 5.02 |
| Cashew | 98.20 | 1788.68 | 129.43 | 2052.94 | 99.68 | 1699.82 |
| Sesamum Seed | 218.97 | 562.23 | 118.31 | 372.89 | 189.11 | 708.90 |
| Nigerseed | 22.22 | 47.85 | 36.13 | 77.99 | 17.89 | 45.41 |
| Groundnut | 112.81 | 250.94 | 67.89 | 178.30 | 176.93 | 179.11 |
| Guargum Meal | 117.89 | 403.09 | 111.94 | 486.64 | 119.33 | 120.56 |
| Oil Meals | 2781.72 | 2262.93 | 1776.13 | 1487.35 | 3172.31 | 3249.89 |
| Castor Oil | 213.68 | 625.94 | 177.69 | 609.81 | 152.36 | 656.06 |
| Shellac | 5.70 | 72.99 | 5.72 | 89.85 | 10.50 | 179.74 |
| Sugar | 1456.45 | 1728.29 | 1662.37 | 1769.49 | 1200.60 | 1216.59 |
| Molasses | 221.12 | 53.56 | 207.86 | 45.05 | 98.62 | 19.38 |
| Fruits/Vegetable Seeds | 4.93 | 62.19 | 8.92 | 97.96 | 5.18 | 53.61 |
| Fresh Fruits | - | 417.15 | - | 447.32 | - | 784.03 |

*contd. ...*

*contd. ...*

| 1 | 2 | 3 | 4 | 5 | 6 | 7 |
|---|---|---|---|---|---|---|
| Fresh Vegetables | - | 575.20 | - | 642.78 | - | 953.93 |
| Processed Vegetables | - | 198.79 | - | 256.73 | - | 291.15 |
| Processed Fruit Juices | - | 512.55 | - | 574.13 | - | 343.66 |
| Miscellaneous Processed Items | - | 724.18 | - | 910.08 | - | 1058.69 |
| Meat & Preparations | - | 1193.28 | - | 1377.19 | - | 1714.41 |
| Marine Products | 468.03 | 5898.34 | 527.87 | 6928.05 | 409.49 | 6105.63 |
| Cotton Raw including Waste | 8.23 | 42.69 | 11.75 | 50.28 | 179.61 | 942.37 |
| Jute Hessian | 51.28 | 186.22 | - | 349.31 | - | 410.11 |
| Poultry Products | - | 148.29 | - | 182.07 | - | 253.59 |
| Paper/Wood products | - | 1524.37 | - | 1950.33 | - | 2362.58 |
| Total Agricultural Exports | 29728.61 | - | 34653.94 | - | 37266.52 | |
| Total National Exports | 209017.97 | 255137.28 | 293366.75 | | | |
| % Share of Agricultural Exports in National Exports | 14.22 | 13.58 | 12.70 | | | |

| Commodity | April 2004 March 05 | | April 2005 March 06 | | April 2006 March 07 | | April 2007 March 08(P) | |
|---|---|---|---|---|---|---|---|---|
| | *Quantity* | *Value* | *Quantity* | *Value* | *Quantity* | *Value* | *Quantity* | *Value* |
| 1 | 8 | 9 | 10 | 11 | 12 | 13 | 14 | 15 |
| Pulses | 271.18 | 602.57 | 447.44 | 1115.21 | 250.70 | 773.34 | 163.67 | 526.95 |
| Rice *Basmati* | 1163.00 | 2823.90 | 1166.57 | 3043.10 | 1045.73 | 2792.81 | 1181.66 | 4334.77 |
| Rice (Other than *Basmati*) | 3615.10 | 3945.02 | 2921.60 | 3178.17 | 3702.22 | 4243.10 | 5314.18 | 7396.23 |
| Wheat | 2009.35 | 1459.82 | 746.18 | 557.53 | 46.64 | 35.35 | 0.23 | 0.23 |
| Other Cereals | 1178.03 | 793.82 | 567.22 | 453.82 | 730.38 | 599.25 | 3227.79 | 2978.86 |
| Tea | 183.40 | 1840.30 | 162.86 | 1730.73 | 185.63 | 1969.51 | 198.55 | 2022.32 |
| Coffee | 167.55 | 1069.08 | 177.68 | 1588.69 | 213.65 | 1969.00 | 178.12 | 1868.02 |
| Tobacco Unmanufactured | 135.74 | 940.07 | 142.70 | 1021.32 | 158.25 | 1251.28 | 173.51 | 1430.49 |
| Tobacco Manufactured | - | 314.54 | - | 309.34 | - | 433.89 | | 498.16 |
| Poultry & Dairy Products | - | 458.79 | - | 794.61 | - | 497.09 | | 921.18 |

*contd. ...*

| 1 | 8 | 9 | 10 | 11 | 12 | 13 | 14 | 15 |
|---|---|---|---|---|---|---|---|---|
| Floriculture Products | - | 222.92 | - | 301.45 | - | 652.70 | | 338.01 |
| Spices | 364.53 | 1883.18 | 400.24 | 2115.98 | 482.80 | 3157.90 | 619.61 | 4176.07 |
| Cashewnut Shell Liquid | 5.33 | 11.94 | 5.94 | 8.67 | 8.09 | 15.34 | 14.75 | 25.02 |
| Cashew | 118.11 | 2477.18 | 125.10 | 2584.70 | 122.78 | 2291.18 | 111.28 | 2209.73 |
| Sesamum Seed | 168.28 | 708.95 | 199.81 | 746.60 | 233.34 | 939.58 | 314.11 | 1624.01 |
| Nigerseed | 26.14 | 64.74 | 28.42 | 60.25 | 30.02 | 66.87 | 21.68 | 90.03 |
| Groundnut | 177.15 | 547.02 | 190.06 | 513.69 | 251.43 | 79.86 | 268.14 | 1041.72 |
| Guargum Meal | 131.31 | 689.48 | 186.73 | 1094.23 | 189.33 | 1125.79 | 207.81 | 1110.55 |
| Oil Meals | 3603.38 | 3177.60 | 5976.00 | 4875.01 | 6437.43 | 5504.32 | 6690.84 | 7953.79 |
| Castor Oil | 271.69 | 1077.98 | 254.72 | 939.74 | 294.87 | 1090.11 | 281.99 | 1274.58 |
| Shellac | 8.54 | 164.87 | 9.30 | 159.98 | 7.51 | 147.20 | 7.98 | 123.55 |
| Sugar | 108.69 | 149.53 | 321.20 | 569.10 | 1643.40 | 3127.47 | 4641.14 | 5404.18 |
| Molasses | 8.16 | 5.52 | 72.94 | 28.81 | 326.87 | 133.37 | 863.46 | 250.08 |
| Fruits/Vegetable Seeds | 6.74 | 66.04 | 7.52 | 92.96 | 8.10 | 121.59 | 10.30 | 142.66 |
| Fresh Fruits | - | 862.26 | - | 1120.69 | - | 1413.98 | | 1446.99 |
| Fresh Vegetables | - | 862.99 | - | 919.81 | - | 1546.53 | | 1473.31 |
| Processed Vegetables | - | 362.46 | - | 494.48 | - | 650.23 | | 605.54 |
| Processed Fruit Juices | - | 369.16 | - | 599.91 | - | 711.40 | | 769.20 |
| Miscellaneous Processed Items | - | 908.03 | - | 989.53 | - | 1125.05 | | 1337.16 |
| Meat & Preparations | - | 1905.27 | - | 2750.17 | - | 3314.03 | | 3734.48 |
| Marine Products | 483.52 | 6469.22 | 554.20 | 7035.91 | 611.55 | 8001.04 | 487.50 | 6854.68 |
| Cotton Raw including Waste | 86.64 | 422.58 | 614.80 | 2904.35 | 1162.22 | 6107.81 | 1417.46 | 7999.69 |
| Jute Hessian | - | 427.20 | - | 490.90 | - | 375.81 | | 463.76 |
| Poultry Products | - | 281.96 | - | 313.37 | - | 313.82 | | 406.68 |
| Paper/Wood products | - | 3236.66 | - | 3759.15 | - | 4915.22 | | 4937.03 |
| **Total Agricultural Exports** | - | **41602.65** | - | **49216.96** | - | 62411.42 | | 77769.71 |
| **Total National Exports** | | **375339.53** | - | **456417.86** | - | 571779.28 | | 640172.14 |
| % Share of Agricultural Exports in National Exports | | 11.08 | | 10.78 | | 10.92 | | 12.15 |

P- Provisional

*Source*: Director General of Commercial Intelligence & Statistics, Ministry of Commerce, Kolkata.

## Table 13.4

*Tariffs and Bound Rates on Major Agricultural Commodities/Groups*

| S. No. | Item Description | Basic Customs Duty (%) (As on 01.04.2008) | Bound Duty (%) (As on 01.08.2007) |
|---|---|---|---|
| 1 | 2 | 3 | 4 |
| | **I. Cereals and Pulses** | | |
| | Pulses other than peas (pisum sativam) | Free | 100 |
| 2 | Wheat | Free | 100 |
| 3 | Maize (Corn) seed | Free | 70 |
| 4 | Rice in the husk | 80 | 80 |
| 5 | Husked (brown) rice; broken rice | 80 | 80 |
| 6 | Semi-milled or wholly milled rice whether or not polished | Free | 70 |
| 7 | Millet, *Jowar* | 50 | 70 |
| 8 | Sorghum | 80 | 80 |
| 9 | Buck wheat and canary seed | Free | 100 |
| 10 | Other cereals (rye, barley etc.) | Free | 100 |
| | **II. Cereals Products** | | |
| 1 | *Atta* | Free | 150 |
| 2 | *Maida* | 30 | 150 |
| 3 | *Sooji* | 30 | 150 |
| 4 | Wheat and potato starch | 30 | 35 |
| 5 | Flour, meal and powder of dried leguminous vegetables including sago, tamarind and mango | 30 | 150 |
| 6 | Other starches | 30 | 100 |
| 7 | Roasted malt | 30 | 100 |
| 8 | Unroasted malt | 30 | 40 |
| | **III. Dairy Products** | | |
| 1 | Fresh milk and cream | 30 | 100 |
| 2 | Butter and melted butter (ghee) | 40 | 40 |
| 3 | Cheese | 30 | 40 |
| 4 | Milk powder | 60 | 60 |
| 5 | Yoghurt | 30 | 150 |
| | **IV. Plantation Crops** | | |
| 1 | Tea | 100 | 150 |
| 2 | Coffee | 100 | 100 |
| 3 | Coconut | 70 | 100 |
| 4 | Copra | 70 | 100 |
| 5 | Cassia and cinnamon | 30 | 100 |
| 6 | Cloves | 35 | 100 |
| 7 | Other Spices | 30/70 | 150/100 |

contd. ...

*...contd. ...*

| S. No. | Item Description | Basic Customs Duty (%) (As on 01.04.2008) | Bound Duty (%) (As on 01.08.2007) |
|---|---|---|---|
| 1 | 2 | 3 | 4 |
| | **V. Meat & Poultry** | | |
| 1 | Chicken sausages | 100 | 150 |
| 2 | Chicken leg (processed) | 100 | 150 |
| 3 | Meat of poultry, not cut in pieces, fresh or chilled | 30 | 100 |
| 4 | Raw hams, pig fat; meat of bovine animals | 30 | 100 |
| 5 | Other meat and offal | 30 | 100 |
| 6 | Processed hams | 30 | 55 |
| 7 | Fish | 30 | Unbound |
| | **VI. Sugar** | 60 | 150 |
| | **VII. Horticulture** | | |
| 1 | Apples | 50 | 50 |
| 2 | Grapefruit | 25 | 25 |
| 3 | Strawberries, dried apricots etc. | 30 | 100 |
| 4 | Pears and quinces | 30 | 35 |
| 5 | Oranges; lemons and limes; fresh grapes | 30 | 40/100 |
| 6 | Fresh pomegranates, *litchi*, tamarind (fresh), custard apple | 15 | 100 |
| 7 | Other fruits except Sapota (Black berries etc.) | 30 | 100/150 |
| 8 | Garlic | 100 | 100 |
| 9 | Onions | 5 | 100 |
| 10 | Mushrooms | 30 | 100 |
| 11 | Potato | 30 | 150 |
| 12 | Sweet Potato | 30 | 150 |
| 13 | Frozen vegetables-peas, beans, spinach, sweet corn etc. | 30 | 55 |
| 14 | Other edible roots and tubers with high starch or insulin content, fresh or chilled (cassava) | 30 | 100 |
| 15 | All other vegetables | 30 | 100 |
| 16 | Arecanut | 100 | 100 |
| 17 | Hop cones (ground, powdered or inpellets) | 30 | 75 |
| 18 | Hop cones (other than ground, powdered or inpellets) | 30 | 100 |
| 19 | Betel leaves | 30 | 100 |
| 20 | Planting materials of oilseeds | 5 | 10 |
| 21 | Oilseeds, misc. grains, seeds of fruits, industrial or medicinal plants (other than the kind used for sowing and hop cones) | 30 | 100 |
| 22 | Basil, hyasop, rosemary, sage, savory, comboge fruit rind | 15 | 100 |

*contd. ...*

*...contd. ...*

| *S. No.* | *Item Description* | *Basic Customs Duty (%) (As on 01.04.2008)* | *Bound Duty (%) (As on 01.08.2007)* |
|---|---|---|---|
| *1* | 2 | 3 | 4 |
| 23 | Apricot, peach or plum stones and kernels | 30 | 100 |
| 24 | Seeds of a kind used for sowing (other than vegetable seeds) | 15 | 100 |
| 25 | Vegetable seeds | 5 | 10 |
| 26 | Oilcake and oilcake meal solvent extracted (defatted) variety of coconut/copra; oilcake and oilcake meal expeller variety of coconut/ copra | 15 | 100 |
| | **VIII. Edible Oils (Crude)** | | |
| 1 | Soyabean Oil | Free | 45 |
| 2 | Palm Oil (for manufacture of vanaspati) | Free | 300 |
| 3 | Groundnut Oil | Free | 300 |
| 4 | Sunflower/Safflower | Free | 300 |
| 5 | Coconut Oil | Free | 300 |
| 6 | Rapeseed Oil | Free | 75 |
| 7 | Colza or Mustard Oil | Free | 75 |
| 8 | Castor Oil/Tung Oil | Free | 100 |
| 9 | Other Oils | Free | 300 |
| 10 | Safflower | Free | 300 |
| | **IX. Edible Oils (Refined)** | | |
| 1 | Soyabean Oil | 7.5 | 45 |
| 2 | RBD Palmolein | 7.5 | 300 |
| 3 | Palm Oil | 7.5 | 300 |
| 4 | Groundnut Oil | 7.5 | 300 |
| 5 | Sunflower | 7.5 | 300 |
| 6 | Coconut Oil Edible grade | 7.5 | 300 |
| 7 | Coconut Oil Other | 7.5 | 300 |
| 8 | Rapeseed Oil | 7.5 | 75 |
| 9 | Colza or Mustard Oil | 7.5 | 75 |
| 10 | Castor Oil/Tung Oil | 7.5 | 100 |
| 11 | Safflower | 7.5 | 300 |
| 12 | Other Oils edible grade | 7.5 | 300 |
| 13 | Other Oils other than edible grade | 65 | 100/300 |

*Note*: 1 Duty on pulses was reduced to Zero on 8.6.2006 and is valid upto 31.3.2009.

2 With effect from 30.3.2007,duly on wheat was reduced to zero for imports by all.

3 With effect from 1.3.2007,All edible oils have been exempied from additional duty of customs of 4%.

4 The zero import duty on rice (semi-milled or wholly milled) and that on Atta is available for imports till 31.3.2009

5 The zero duty on maize is available for a tariff rate quota (TRQ) of 5 lakh metric tonnes.

*Source*: Department of Revenue, Ministry of Finance.

# Table 13.5

*Trend in Wholesale Price Index of Foodgrains from 1997-98 to 2007-08*

*(Base year-1993-94=100)*

| Commodity | Weight | 1997-98 | 1998-99 | 1999-00 | 2000-01 | 2001-02 | 2002-03 | 2003-04 | 2004-05 | 2005-06 | 2006-07 | 2007-08 | Average % Variation in WPI from 1997-98 to 2002-03 | 2003-04 to 2007-08 |
|---|---|---|---|---|---|---|---|---|---|---|---|---|---|---|
| 1 | 2 | 3 | 4 | 5 | 6 | 7 | 8 | 9 | 10 | 11 | 12 | 13 | 14 | 15 |
| (I) All Commodities | 100.00 | 132.8 | 140.7 | 145.3 | 155.7 | 161.3 | 166.8 | 175.9 | 187.3 | 195.6 | 206.2 | 215.7 | 4.87 | 5.04 |
| (a) Primary Articles | 22.03 | 139.4 | 156.2 | 158.0 | 162.5 | 168.4 | 174.0 | 181.5 | 188.1 | 193.6 | 208.7 | 224.4 | 4.47 | 4.40 |
| (b) Fuel, Power & Lubricants | 14.23 | 143.8 | 148.5 | 162.0 | 208.1 | 226.7 | 239.2 | 254.5 | 280.2 | 306.8 | 210.5 | 222.0 | 12.70 | 0.02 |
| (c) Manufactured Products | 63.75 | 128.0 | 133.6 | 137.2 | 141.7 | 144.3 | 148.1 | 156.5 | 166.3 | 171.4 | | | 3.02 | -16.47 |
| (II) Foodgrains | 5.01 | 139.3 | 152.0 | 176.4 | 173.8 | 172.4 | 174.3 | 176.3 | 177.5 | 187.0 | 206.0 | 215.5 | 4.83 | 3.69 |
| (a) Cereals | 4.41 | 138.4 | 150.9 | 177.8 | 173.0 | 170.1 | 173.5 | 176.3 | 177.9 | 185.8 | 199.4 | 211.7 | 4.89 | 3.26 |
| 1 Rice | 2.45 | 134.3 | 146.2 | 171.3 | 167.5 | 167.0 | 166.0 | 168.8 | 168.2 | 174.5 | 179.6 | 191.7 | 5.61 | 1.48 |
| 2 Wheat | 1.38 | 138.1 | 151.8 | 174.7 | 176.6 | 175.3 | 175.7 | 181.4 | 184.1 | 191.5 | 216.5 | 225.7 | 5.19 | 4.41 |
| 3 *Jowar* | 0.22 | 172.3 | 191.1 | 240.6 | 199.7 | 182.4 | 218.7 | 218.8 | 233.4 | 244.5 | 272.0 | 309.4 | 1.85 | 8.52 |
| 4 *Bajra* | 0.11 | 143.4 | 156.6 | 201.2 | 184.0 | 151.3 | 190.3 | 178.6 | 183.0 | 210.8 | 227.6 | 235.7 | 1.76 | 9.05 |
| 5 Maize | 0.19 | 148.5 | 154.4 | 193.6 | 178.9 | 171.5 | 189.5 | 181.2 | 187.9 | 205.0 | 224.9 | 235.7 | 1.96 | 5.72 |
| 6 Barley | 0.03 | 166.3 | 151.7 | 199.2 | 174.3 | 160.2 | 193.8 | 179.5 | 188.6 | 205.3 | 220.8 | 228.8 | 3.31 | 7.01 |
| 7 Ragi | 0.03 | 138.4 | 152.4 | 178.7 | 182.1 | 158.0 | 174.3 | 197.0 | 184.5 | 186.8 | 205.6 | 227.0 | 2.17 | 5.66 |
| (b) Pulses | 0.60 | 145.9 | 160.1 | 166.1 | 179.6 | 189.2 | 180.6 | 176.6 | 174.4 | 194.9 | 254.2 | 243.1 | 4.68 | 6.83 |
| 8 Gram | 0.22 | 124.6 | 110.9 | 115.4 | 139.2 | 170.3 | 149.7 | 142.5 | 137.1 | 157.0 | 208.7 | 189.9 | 12.98 | 5.35 |
| 9 *Arhar* | 0.13 | 136.2 | 186.2 | 176.3 | 150.3 | 142.6 | 157.5 | 172.8 | 179.3 | 170.8 | 182.0 | 207.4 | -1.74 | 5.15 |
| 10 *Moong* | 0.11 | 164.7 | 181.2 | 180.6 | 186.9 | 205.6 | 208.0 | 195.9 | 187.4 | 219.2 | 303.0 | 267.3 | 4.38 | 9.24 |
| 11 *Masur* | 0.04 | 176.6 | 202.6 | 216.0 | 206.8 | 203.9 | 214.1 | 233.4 | 234.7 | 242.8 | 254.0 | 319.0 | 0.14 | 4.53 |
| 12 *Urad* | 0.10 | 175.3 | 197.3 | 234.0 | 295.7 | 273.9 | 239.8 | 217.4 | 216.3 | 270.4 | 403.8 | 336.8 | 6.57 | 10.41 |
| (III) Sugar, *Khandsari* & *Gur* | 3.93 | 134.4 | 153.5 | 156.0 | 153.2 | 146.1 | 134.6 | 139.3 | 163.3 | 178.8 | 179.8 | 155.2 | 4.49 | 4.58 |
| 13 Sugar | 3.62 | 133.8 | 135.4 | 137.4 | 142.6 | 135.4 | 123.5 | 124.4 | 148.1 | 164.0 | 164.9 | 142.7 | 3.25 | 4.46 |
| 14 *Khandsari* | 0.17 | 133.5 | 144.7 | 137.3 | 137.3 | 145.3 | 133.2 | 130.3 | 164.1 | 197.8 | 203.5 | 169.0 | 3.01 | 7.77 |
| 15 *Gur* | 0.06 | 126.1 | 130.3 | 129.9 | 129.1 | 137.8 | 128.2 | 122.0 | 172.9 | 192.5 | 185.3 | 170.3 | 7.54 | 7.50 |
| Food Articles | 15.40 | 141.4 | 159.4 | 165.5 | 170.5 | 176.1 | 179.2 | 181.5 | 186.3 | 195.3 | 210.5 | 222.0 | 5.17 | 3.66 |
| Non-Food Articles | 6.14 | 137.5 | 151.8 | 143.0 | 146.5 | 152.9 | 165.4 | 186.3 | 187.6 | 179.1 | 188.2 | 211.9 | 2.78 | 4.41 |

*Source*: Office of the Economic Adviser, Ministry of Commerce & Industry.

# Table 13.6

*Trend in Wholesale Price Index of Commercial Crops from 1997-98 to 2007-08*

(Base Year: 1993-94 = 100)

| Commodity | Weight | Wholesale Price Index | | | | | | | | | | | Average % Variation in WPI from | |
|---|---|---|---|---|---|---|---|---|---|---|---|---|---|---|
| | | 1997-98 | 1998-99 | 1999-00 | 2000-01 | 2001-02 | 2002-03 | 2003-04 | 2004-05 | 2005-06 | 2006-07 | 2007-08 | 1997-98 to 2002-03 | 2003.-04 to 2007-08 |
| 1 | 2 | 3 | 4 | 5 | 6 | 7 | 8 | 9 | 10 | 11 | 12 | 13 | 14 | 15 |
| **Oilseeds** | 2.666 | 128.3 | 148.5 | 133.4 | 129.3 | 137.6 | 160.2 | 177.8 | 180.8 | 167.1 | 175.7 | 217.7 | 1.43 | 5.32 |
| of which | | | | | | | | | | | | | | |
| Groundnut Seed | 1.029 | 133.9 | 150.8 | 139.6 | 139.4 | 143.4 | 169.3 | 180.6 | 182.0 | 171.9 | 190.0 | 239.4 | 1.42 | 6.09 |
| Rape/Mustard Seed | 0.581 | 116.9 | 163.6 | 138.7 | 118.0 | 126.7 | 143.3 | 184.3 | 172.3 | 164.7 | 172.1 | 203.7 | 3.02 | 7.07 |
| Cotton Seed | 0.045 | 135.9 | 159.8 | 153.6 | 159.2 | 159.1 | 156.2 | 181.6 | 166.8 | 154.8 | 167.8 | 181.4 | 5.72 | 1.51 |
| Copra | 0.029 | 142.6 | 129.4 | 141.5 | 95.1 | 89.6 | 130.5 | 163.3 | 193.5 | 159.8 | 138.7 | 128.0 | -6.16 | 11.76 |
| Gingelly Seed | 0.132 | 115.4 | 157.2 | 154.4 | 132.9 | 134.5 | 157.8 | 169.5 | 177.3 | 169.3 | 187.2 | 205.2 | 2.07 | 7.01 |
| Niger Seed | 0.025 | 124.3 | 145.0 | 177.6 | 183.0 | 156.8 | 189.6 | 223.4 | 214.1 | 184.1 | 219.6 | 390.0 | 4.86 | 7.72 |
| Safflower/Kardi | 0.056 | 108.0 | 142.8 | 129.7 | 112.6 | 128.9 | 150.5 | 188.8 | 171.4 | 146.6 | 143.8 | 155.9 | 4.56 | 3.25 |
| Sunflower | 0.204 | 128.2 | 144.5 | 117.4 | 118.1 | 164.6 | 192.5 | 194.3 | 197.8 | 188.7 | 197.7 | 253.3 | 6.82 | 3.96 |
| Soyabean | 0.446 | 138.8 | 124.3 | 107.5 | 121.1 | 130.0 | 153.4 | 158.4 | 187.8 | 157.5 | 144.9 | 186.9 | -1.62 | 3.63 |
| **Edible Oils** | 2.755 | 113.5 | 139.1 | 122.1 | 103.3 | 112.9 | 138.0 | 157.9 | 156.4 | 146.0 | 154.6 | 175.1 | 0.57 | 7.00 |
| of which | | | | | | | | | | | | | | |
| Gingelly Oil | 0.055 | 129.0 | 157.8 | 149.6 | 122.8 | 138.2 | 170.9 | 193.8 | 193.7 | 174.7 | 185.5 | 224.7 | 1.43 | 6.65 |
| Rapeseed/Mustard Oil | 0.486 | 115.5 | 179.6 | 140.1 | 112.1 | 121.1 | 147.3 | 193.9 | 177.1 | 159.8 | 165.7 | 194.0 | 3.32 | 7.74 |
| Coconut Oil | 0.172 | 140.8 | 131.1 | 142.5 | 98.7 | 95.0 | 121.6 | 156.2 | 178.8 | 150.7 | 141.7 | 139.2 | -4.41 | 9.89 |
| Groundnut Oil | 0.174 | 123.6 | 150.2 | 137.6 | 129.1 | 135.8 | 166.6 | 181.2 | 173.8 | 168.2 | 189.8 | 231.8 | 1.94 | 7.36 |
| Cotton Seed Oil | 0.275 | 101.7 | 128.8 | 114.4 | 96.3 | 111.8 | 126.5 | 126.6 | 127.5 | 112.4 | 124.8 | 159.8 | 2.41 | 2.57 |
| Sunflower Oil | 0.135 | 120.9 | 147.6 | 118.0 | 106.7 | 136.1 | 157.1 | 161.6 | 163.0 | 157.6 | 166.7 | 189.7 | 3.82 | 4.40 |
| Soyabean Oil | 0.178 | 100.5 | 134.8 | 113.0 | 86.7 | 86.9 | 122.7 | 135.8 | 138.1 | 137.4 | 132.4 | 142.9 | -1.80 | 9.91 |

contd. ...

*...contd. ...*

| 1 | 2 | 3 | 4 | 5 | 6 | 7 | 8 | 9 | 10 | 11 | 12 | 13 | 14 | 15 |
|---|---|---|---|---|---|---|---|---|---|---|---|---|---|---|
| **Fruits & Vegetables** | 2.917 | 142.8 | 185.4 | 154.5 | 160.0 | 188.9 | 190.20 | 195.60 | 204.20 | 218.90 | 228.2 | 238.2 | 6.14 | 3.89 |
| **Fruits** of which | 1.458 | 146.3 | 169.3 | 166.8 | 180.7 | 206.1 | 217.50 | 231.10 | 237.30 | 246.60 | 258.3 | 248.4 | 7.11 | 4.64 |
| Banana | 0.282 | 149.4 | 147.3 | 130.2 | 134.7 | 154.0 | 166.60 | 178.80 | 190.30 | 213.20 | 214.4 | 227.3 | 0.28 | 6.91 |
| Apple | 0.436 | 156.2 | 188.7 | 251.4 | 391.6 | 574.1 | 542.1 | 539.70 | 493.00 | 470.60 | 570.5 | 521.8 | 31.37 | -5.67 |
| Orange | 0.054 | 289.3 | 535.9 | 505.7 | 513.5 | 530.9 | 535.70 | 555.30 | 573.20 | 711.40 | 739.9 | 504.9 | 17.13 | 7.30 |
| **Vegetables** of which | 1.459 | 139.5 | 201.4 | 142.1 | 139.4 | 171.7 | 162.90 | 160.10 | 171.10 | 191.30 | 198.1 | 224.0 | 5.85 | 3.10 |
| Potato | 0.256 | 84.4 | 226.2 | 93.9 | 84.6 | 198.5 | 190.00 | 111.40 | 169.80 | 199.60 | 235.0 | 257.1 | 37.22 | 8.30 |
| Onion | 0.094 | 136.7 | 261.8 | 132.1 | 107.1 | 123.9 | 130.40 | 169.80 | 137.00 | 169.70 | 160.8 | 233.3 | 14.52 | 6.96 |
| **Condiment & Spices** of which | 0.662 | 176.9 | 220.2 | 226.4 | 202.5 | 188.1 | 194.00 | 205.00 | 188.10 | 176.60 | 228.2 | 239.6 | ·1.96 | 4.73 |
| Black pepper | 0.023 | 421.1 | 511.2 | 525.9 | 501.4 | 246.6 | 246.20 | 220.40 | 192.10 | 185.60 | 260.3 | 374.2 | 11.12 | 2.47 |
| Chillies (Dry) | 0.189 | 231.3 | 342.8 | 305.8 | 254.4 | 249.9 | 253.90 | 294.30 | 225.90 | 213.40 | 345.1 | 347.2 | -0.51 | 10.09 |
| Turmeric | 0.077 | 186.0 | 234.3 | 191.8 | 134.6 | 119.4 | 181.20 | 230.40 | 224.10 | 201.70 | 190.4 | 174.7 | 2.47 | 12.24 |
| Raw Cotton | 1.357 | 155.4 | 166.9 | 147.3 | 157.3 | 148.7 | 142.1 | 180.60 | 165.60 | 144.30 | 151.7 | 178.1 | 2.73 | 1.24 |
| Raw Jute | 0.109 | 101.8 | 107.9 | 112.9 | 149.8 | 182.5 | 153.60 | 137.10 | 159.90 | 198.80 | 203.7 | 190.6 | 3.82 | 3.43 |

*Source:* Office of the Economic Adviser, Ministry of Commerce & Industry.

## Table 14.1

*Distribution of Agricultural Land by Use in India for 1950-51, 1960-61, 1970-71, 1980-81 and 1990-91 to 2005-06*

*(Million Hectares)*

| Sl. No. | Classification | 1950-51 | 1960-61 | 1970-71 | 1980-81 | 1990-91 | 1991-92 | 1992-93 | 1993-94 | 1994-95 | 1995-96 | 1996-97 |
|---|---|---|---|---|---|---|---|---|---|---|---|---|
| 1 | 2 | 3 | 4 | 5 | 6 | 7 | 8 | 9 | 10 | 11 | 12 | 13 |
| I. | Geographical Area | 328.73 | 328.73 | 328.73 | 328.73 | 328.73 | 328.73 | 328.73 | 328.73 | 328.73 | 328.73 | 328.73 |
| II. | Reporting Area for Land Utilisation Statistics (1 to 5) | 284.32 | 298.46 | 303.76 | 304.16 | 304.86 | 304.90 | 304.85 | 304.86 | 304.83 | 304.88 | 304.62 |
| | 1. Forest | 40.48 | 54.05 | 63.92 | 67.47 | 67.81 | 67.87 | 67.98 | 68.31 | 68.60 | 68.82 | 69.10 |
| | % | 14.2 | 18.1 | 21.0 | 22.2 | 22.2 | 22.3 | 22.3 | 22.4 | 22.5 | 22.6 | 22.7 |
| | 2. Not Available for Cultivation (A+B) | | | | | | | | | | | |
| | A. Area Under Non-Agril. Uses | 9.36 | 14.84 | 16.48 | 19.66 | 21.09 | 21.47 | 21.87 | 22.21 | 22.56 | 22.36 | 22.55 |
| | % | 3.3 | 5.0 | 5.4 | 6.5 | 6.9 | 7.0 | 7.2 | 7.3 | 7.4 | 7.3 | 7.4 |
| | B. Barren & Un-culturable Land | 38.16 | 35.91 | 28.16 | 19.96 | 19.39 | 19.27 | 19.04 | 18.71 | 18.46 | 19.01 | 17.96 |
| | % | 13.4 | 12.0 | 9.3 | 6.6 | 6.4 | 6.3 | 6.2 | 6.1 | 6.1 | 6.2 | 5.9 |
| | 3. Other Uncultivated Land Excluding Fallow Land (A+B+C) | | | | | | | | | | | |
| | A. Permanent Pastures & Other Grazing Lands | 6.68 | 13.97 | 13.26 | 11.97 | 11.40 | 11.30 | 11.07 | 10.96 | 11.03 | 11.06 | 10.88 |
| | % | 2.3 | 4.7 | 4.4 | 3.9 | 3.7 | 3.7 | 3.6 | 3.6 | 3.6 | 3.6 | 3.6 |
| | B. Land Under Misc. Tree Crops & Groves not included in Net Area Sown | 19.83 | 4.46 | 4.30 | 3.61 | 3.82 | 3.76 | 3.76 | 3.71 | 3.73 | 3.48 | 3.65 |
| | % | 7.0 | 1.5 | 1.4 | 1.2 | 1.3 | 1.2 | 1.2 | 1.2 | 1.2 | 1.1 | 1.2 |
| | C. Culturable Waste Land | 22.94 | 19.21 | 17.50 | 16.74 | 15.00 | 14.99 | 14.57 | 14.41 | 14.26 | 14.10 | 14.02 |
| | % | 8.1 | 6.4 | 5.8 | 5.5 | 4.9 | 4.9 | 4.8 | 4.7 | 4.7 | 4.6 | 4.6 |

contd. ...

...contd. ...

| 1 | 2 | 3 | 4 | 5 | 6 | 7 | 8 | 9 | 10 | 11 | 12 | 13 |
|---|---|---|---|---|---|---|---|---|---|---|---|---|
| | 4. Fallow Lands (A+B) | | | | | | | | | | | |
| | A. Fallow Lands Other Than Current Fallows | 17.45 | 11.18 | 8.76 | 9.92 | 9.66 | 9.94 | 9.68 | 9.83 | 9.97 | 10.02 | 10.19 |
| | % | 6.1 | 3.7 | 2.9 | 3.3 | 3.2 | 3.3 | 3.2 | 3.2 | 3.3 | 3.3 | 3.3 |
| | B. Current Fallows | 10.68 | 11.64 | 11.12 | 14.83 | 13.70 | 14.67 | 14.16 | 14.38 | 13.25 | 13.83 | 13.32 |
| | % | 3.8 | 3.9 | 3.7 | 4.9 | 4.5 | 4.8 | 4.6 | 4.7 | 4.3 | 4.5 | 4.4 |
| | 5. Net Area Sown (6-7) | 118.75 | 133.20 | 140.27 | 140.00 | 143.00 | 141.63 | 142.72 | 142.34 | 142.96 | 142.20 | 142.93 |
| | % | 41.8 | 44.6 | 46.2 | 46.0 | 46.9 | 46.5 | 46.8 | 46.7 | 46.9 | 46.6 | 46.9 |
| | 6. Total Cropped Area (Gross Cropped Area) | 131.89 | 152.77 | 165.79 | 172.63 | 185.74 | 182.24 | 185.70 | 186.58 | 188.05 | 187.47 | 189.50 |
| | 7. Area Sown More than Once | 13.15 | 19.57 | 25.52 | 32.63 | 42.74 | 40.61 | 42.98 | 44.25 | 45.09 | 45.27 | 46.57 |
| | 8. Cropping Intensity* | 111.1 | 114.7 | 118.2 | 123.3 | 129.9 | 128.7 | 130.1 | 131.1 | 131.5 | 131.8 | 132.6 |
| III. | Net Irrigated Area | 20.85 | 24.66 | 31.10 | 38.72 | 48.02 | 49.87 | 50.29 | 51.34 | 53.00 | 53.40 | 55.11 |
| IV. | Gross Irrigated Area | 22.56 | 27.98 | 38.20 | 49.78 | 63.20 | 65.68 | 66.76 | 68.26 | 70.65 | 71.35 | 76.03 |

contd. ...

...contd. ...

| Sl. No. | Classification | 1997-98 | 1998-99 | 1999-2000 | 2000-01 | 2001-02 (P) | 2002-03 (P) | 2003-04 (P) | 2004-05 (P) | 2005-06 (P) |
|---|---|---|---|---|---|---|---|---|---|---|
| 1 | 2 | 14 | 15 | 16 | 17 | 18 | 19 | 20 | 21 | 22 |
| I. | Geographical Area | 328.73 | 328.73 | 328.73 | 328.73 | 328.73 | 328.73 | 328.73 | 328.73 | 328.73 |
| II. | Reporting Area for Land Utilisation Statistics (1 to 5) | 304.61 | 304.98 | 304.95 | 305.17 | 305.11 | 305.34 | 305.40 | 305.31 | 305.27 |
| | 1. Forest | 69.25 | 69.21 | 69.16 | 69.53 | 69.57 | 69.70 | 69.73 | 69.72 | 69.79 |
| | % | 22.7 | 22.7 | 22.7 | 22.8 | 22.8 | 22.8 | 22.8 | 22.8 | 22.9 |
| | 2. Not Available for Cultivation (A+B) | | | | | | | | | |
| | A. Area Under Non-Agril. Uses | 23.12 | 23.30 | 23.59 | 23.86 | 24.06 | 24.27 | 24.66 | 24.76 | 25.03 |
| | % | 7.6 | 7.6 | 7.7 | 7.8 | 7.9 | 7.9 | 8.1 | 8.1 | 8.2 |
| | B. Barren & Un-Culturable Land | 17.46 | 17.54 | 17.51 | 17.60 | 17.72 | 17.81 | 17.58 | 17.58 | 17.48 |
| | % | 5.7 | 5.8 | 5.7 | 5.8 | 5.8 | 5.8 | 5.8 | 5.8 | 5.7 |
| | 3. Other Uncultivated Land Excluding Fallow Land (A+B+C) | | | | | | | | | |
| | A. Permanent Pastures & Other Grazing Lands | 10.84 | 10.89 | 10.80 | 10.66 | 10.58 | 10.50 | 10.44 | 10.42 | 10.42 |
| | % | 3.6 | 3.6 | 3.5 | 3.5 | 3.5 | 3.4 | 3.4 | 3.4 | 3.4 |
| | B. Land Under Misc. Tree Crops & Groves not included in Net Area Sown | 3.71 | 3.70 | 3.74 | 3.43 | 3.38 | 3.37 | 3.40 | 3.38 | 3.38 |
| | % | 1.2 | 1.2 | 1.2 | 1.1 | 1.1 | 1.1 | 1.1 | 1.1 | 1.1 |
| | C. Culturable Waste Land | 13.93 | 13.87 | 13.74 | 13.61 | 13.40 | 13.53 | 13.13 | 13.16 | 13.12 |
| | % | 4.6 | 4.5 | 4.5 | 4.5 | 4.4 | 4.4 | 4.3 | 4.3 | 4.3 |

contd. ...

...contd. ...

| 1 | 2 | 14 | 15 | 16 | 17 | 18 | 19 | 20 | 21 | 22 |
|---|---|---|---|---|---|---|---|---|---|---|
| | 4. Fallow Lands (A+B) | | | | | | | | | |
| | A. Fallow Lands Other Than Current Fallows | 10.08 | 10.11 | 10.30 | 10.31 | 10.32 | 11.78 | 11.21 | 10.71 | 10.50 |
| | % | 3.3 | 3.3 | 3.4 | 3.4 | 3.4 | 3.9 | 3.7 | 3.5 | 3.4 |
| | B. Current Fallows | 14.28 | 13.59 | 15.05 | 14.77 | 14.64 | 21.69 | 14.28 | 14.43 | 13.67 |
| | % | 4.7 | 4.5 | 4.9 | 4.8 | 4.8 | 7.1 | 4.7 | 4.7 | 4.5 |
| | 5. Net Area Sown (6-7) | 141.95 | 142.76 | 141.06 | 141.40 | 141.45 | 132.69 | 140.97 | 141.14 | 141.89 |
| | % | 46.6 | 46.8 | 46.3 | 46.3 | 46.4 | 43.5 | 46.2 | 46.2 | 46.5 |
| | 6. Total Cropped Area (Gross Cropped Area) | 190.02 | 191.69 | 188.43 | 185.37 | 189.71 | 175.62 | 190.20 | 190.42 | 192.80 |
| | 7. Area Sown More than Once | 48.07 | 48.92 | 47.37 | 43.97 | 48.26 | 42.93 | 49.23 | 49.28 | 50.90 |
| | 8. Cropping Intensity* | 133.9 | 134.3 | 133.6 | 131.1 | 134.1 | 132.4 | 134.9 | 134.9 | 135.9 |
| III. Net Irrigated Area | | 55.17 | 57.41 | 57.11 | 55.08 | 56.67 | 53.78 | 56.62 | 58.87 | 60.20 |
| IV. Gross Irrigated Area | | 75.42 | 78.37 | 78.79 | 76.57 | 78.73 | 73.55 | 78.00 | 80.00 | 82.63 |

P Provisional

* Cropping Intensity is percentage of the gross cropped area to the net area sown.

*Note*: The decline in net area sown in 2002-03 reflects the impact of the severe drought of 2002-03 on agriculture operations.

## Table 14.2

*Selected Categories of Land Use from 1950-51 to 2005-06*

(Million Hectares)

| Year | Net Area Sown | Total Cropped Area | Area Sown more than Once (3-2) | Net Irrigated Area | Gross Irrigated Area | Area Irrigated more than Once (6-5) |
|---|---|---|---|---|---|---|
| 1 | 2 | 3 | 4 | 5 | 6 | 7 |
| 1950-51 | 118.75 | 131.89 | 13.15 | 20.85 | 22.56 | 1.71 |
| 1951-52 | 119.40 | 133.23 | 13.83 | 21.05 | 23.18 | 2.13 |
| 1952-53 | 123.44 | 137.68 | 14.23 | 21.12 | 23.31 | 2.18 |
| 1953-54 | 126.81 | 142.48 | 15.67 | 21.87 | 24.36 | 2.49 |
| 1954-55 | 127.85 | 144.09 | 16.24 | 22.09 | 24.95 | 2.86 |
| 1955-56 | 129.16 | 147.31 | 18.16 | 22.76 | 25.64 | 2.88 |
| 1956-57 | 130.85 | 149.49 | 18.64 | 22.53 | 25.71 | 3.17 |
| 1957-58 | 129.08 | 145.83 | 16.75 | 23.16 | 26.63 | 3.47 |
| 1958-59 | 131.83 | 151.63 | 19.80 | 23.40 | 26.95 | 3.55 |
| 1959-60 | 132.94 | 152.82 | 19.89 | 24.04 | 27.45 | 3.42 |
| 1960-61 | 133.20 | 152.77 | 19.57 | 24.66 | 27.98 | 3.32 |
| 1961-62 | 135.40 | 156.21 | 20.81 | 24.88 | 28.46 | 3.58 |
| 1962-63 | 136.34 | 156.76 | 20.42 | 25.67 | 29.45 | 3.79 |
| 1963-64 | 136.48 | 156.96 | 20.48 | 25.89 | 29.71 | 3.82 |
| 1964-65 | 138.12 | 159.23 | 21.11 | 26.60 | 30.71 | 4.11 |
| 1965-66 | 136.20 | 155.28 | 19.08 | 26.34 | 30.90 | 4.56 |
| 1966-67 | 137.23 | 157.36 | 20.12 | 26.91 | 32.68 | 5.78 |
| 1967-68 | 139.88 | 163.74 | 23.86 | 27.19 | 33.21 | 6.01 |
| 1968-69 | 137.31 | 159.53 | 22.22 | 29.01 | 35.48 | 6.47 |
| 1969-70 | 138.77 | 162.27 | 23.49 | 30.20 | 36.97 | 6.78 |
| 1970-71 | 140.27 | 165.79 | 25.52 | 31.10 | 38.20 | 7.09 |
| 1971-72 | 139.72 | 165.19 | 25.47 | 31.55 | 38.43 | 6.88 |
| 1972-73 | 137.14 | 162.15 | 25.01 | 31.83 | 39.06 | 7.22 |
| 1973-74 | 142.42 | 169.87 | 27.46 | 32.55 | 40.28 | 7.74 |
| 1974-75 | 137.79 | 164.19 | 26.40 | 33.71 | 41.74 | 8.03 |
| 1975-76 | 141.65 | 171.30 | 29.64 | 34.59 | 43.36 | 8.77 |
| 1976-77 | 139.48 | 167.33 | 27.86 | 35.15 | 43.55 | 8.40 |
| 1977-78 | 141.95 | 172.23 | 30.28 | 36.55 | 46.08 | 9.53 |
| 1978-79 | 142.98 | 174.80 | 31.82 | 38.06 | 48.31 | 10.25 |
| 1979-80 | 138.90 | 169.59 | 30.69 | 38.52 | 49.21 | 10.69 |
| 1980-81 | 140.00 | 172.63 | 32.63 | 38.72 | 49.78 | 11.06 |
| 1981-82 | 141.93 | 176.75 | 34.82 | 40.50 | 51.41 | 10.91 |

contd. ...

*...contd. ...*

| *1* | *2* | *3* | *4* | *5* | *6* | *7* |
|---|---|---|---|---|---|---|
| 1982-83 | 140.22 | 172.75 | 32.53 | 40.69 | 51.83 | 11.14 |
| 1983-84 | 142.84 | 179.56 | 36.72 | 41.95 | 53.82 | 11.88 |
| 1984-85 | 140.89 | 176.33 | 35.44 | 42.15 | 54.53 | 12.38 |
| 1985-86 | 140.90 | 178.46 | 37.56 | 41.87 | 54.28 | 12.42 |
| 1986-87 | 139.58 | 176.41 | 36.83 | 42.57 | 55.76 | 13.19 |
| 1987-88 | 134.09 | 170.74 | 36.65 | 42.89 | 56.04 | 13.14 |
| 1988-89 | 141.89 | 182.28 | 40.39 | 46.15 | 61.13 | 14.98 |
| 1989-90 | 142.34 | 182.27 | 39.93 | 46.70 | 61.85 | 15.15 |
| 1990-91 | 143.00 | 185.74 | 42.74 | 48.02 | 63.20 | 15.18 |
| 1991-92 | 141.63 | 182.24 | 40.61 | 49.87 | 65.68 | 15.81 |
| 1992-93 | 142.72 | 185.70 | 42.98 | 50.29 | 66.76 | 16.47 |
| 1993-94 | 142.34 | 186.58 | 44.25 | 51.34 | 68.26 | 16.92 |
| 1994-95 | 142.96 | 188.05 | 45.09 | 53.00 | 70.65 | 17.65 |
| 1995-96 | 142.20 | 187.47 | 45.27 | 53.40 | 71.35 | 17.95 |
| 1996-97 | 142.93 | 189.50 | 46.57 | 55.11 | 76.03 | 20.91 |
| 1997-98 | 141.95 | 190.02 | 48.07 | 55.17 | 75.42 | 20.25 |
| 1998-99 | 142.76 | 191.69 | 48.92 | 57.41 | 78.37 | 20.96 |
| 1999-00 | 141.06 | 188.43 | 47.37 | 57.11 | 78.79 | 21.69 |
| 2000-01 | 141.40 | 185.37 | 43.97 | 55.08 | 76.57 | 21.50 |
| 2001-02 (p) | 141.45 | 189.71 | 48.26 | 56.67 | 78.73 | 22.06 |
| 2002-03 (p) | 132.69 | 175.62 | 42.93 | 53.78 | 73.55 | 19.77 |
| 2003-04 (p) | 140.97 | 190.20 | 49.23 | 56.62 | 78.00 | 21.38 |
| 2004-05 (p) | 141.14 | 190.42 | 49.28 | 58.87 | 80.00 | 21.13 |
| 2005-06 (p) | 141.89 | 192.80 | 50.90 | 60.20 | 82.63 | 22.43 |

P : Provisional

# Table 14.3

*Percentage Distribution of Gross Cropped Area across Different Crops during 2004-05 and 2005-06*

| *Crop* | *2004-05* | *2005-06* |
|---|---|---|
| *1* | 2 | 3 |
| Rice | 22.11 | 22.71 |
| Wheat | 13.88 | 13.79 |
| *Jowar* | 4.76 | 4.49 |
| *Bajra* | 4.86 | 5.01 |
| Maize | 3.96 | 4.00 |
| Ragi | 0.85 | 0.85 |
| Barley | 0.33 | 0.33 |
| Other Cereals & Millets | 0.58 | 0.54 |
| Coarse Cereals | 15.34 | 15.22 |
| Total Cereals | 51.33 | 51.72 |
| Gram | 3.52 | 3.59 |
| *Tur* | 1.84 | 1.84 |
| Other Pulses | 7.07 | 6.15 |
| Total Pulses | 12.43 | 11.58 |
| Total Foodgrains | 63.76 | 63.31 |
| Sugarcane | 2.10 | 2.41 |
| Condiments & Spices | 1.61 | 1.46 |
| Total Fruits | 2.05 | 2.02 |
| Potatoes | 0.72 | 0.75 |
| Onions | 0.32 | 0.34 |
| Total Vegetables | 2.89 | 2.82 |
| Groundnut | 3.52 | 3.55 |
| Sesamum | 1.13 | 1.05 |
| Rapeseed & Mustard | 3.70 | 3.81 |
| Linseed | 0.23 | 0.22 |
| Other Oil Seeds | 7.28 | 7.41 |
| Total Oil Seeds | 15.85 | 16.04 |
| Cotton | 4.38 | 4.47 |
| Jute | 0.41 | 0.40 |
| Mesta | 0.07 | 0.07 |
| Total Fibres | 4.90 | 4.97 |
| Tobacco | 0.19 | 0.19 |
| Other Crops | 6.65 | 6.78 |
| Gross Cropped Area | 100.00 | 100.00 |

## Table 14.4

*State-wise Percent Coverage of Irrigated Area under Principal Crops during 2005-06 (Provisional)*

| State/ Union Territory | Rice | Jowar | Bajra | Maize | Wheat | Barley | Total Cereals | Gram | Tur or Arhar | Total Pulses | Total Food-grains | Sugar-cane | Ground nut | R & M | Soya-bean | Sun-flower | Total Oil-Seeds | Cotton | Tob-acco | Total Area under All Crops |
|---|---|---|---|---|---|---|---|---|---|---|---|---|---|---|---|---|---|---|---|---|
| (1) | (2) | (3) | (4) | (5) | (6) | (7) | (8) | (9) | (10) | (11) | (12) | (13) | (14) | (15) | (16) | (17) | (18) | (19) | (20) | (21) |
| Andhra Pradesh | 96.6 | 7.9 | 22.1 | 34.3 | 77.2 | - | 77.6 | 1.1 | 0.4 | 1.1 | 58.6 | 93.5 | 15.0 | 12.4 | 2.6 | 31.7 | 16.5 | 20.1 | 24.5 | 44.9 |
| Arunachal Pradesh** | 36.5 | - | - | - | - | - | 23.3 | - | - | - | 22.4 | - | - | - | - | - | - | - | - | 16.7 |
| Assam** | 4.6 | - | - | - | 4.7 | - | 4.5 | - | - | - | 4.3 | 16.7 | - | 1.9 | - | - | 2.5 | - | - | 4.3 |
| Bihar** | 51.1 | 7.0 | 0.5 | 57.3 | 90.7 | 29.3 | 64.9 | 17.7 | 1.2 | 3.6 | 59.3 | 25.7 | 9.1 | 38.3 | - | 86.5 | 37.1 | - | 81.1 | 57.0 |
| Chhattisgarh | 30.3 | 0.3 | - | 4.0 | 62.9 | 1.1 | 28.8 | 16.2 | - | 4.7 | 24.3 | 90.7 | 14.9 | 7.3 | 0.6 | 96.8 | 5.5 | 75.4 | 89.7 | 23.9 |
| Goa | 32.7 | - | - | - | - | - | 32.4 | - | - | - | 26.8 | 100.0 | 89.4 | - | - | - | 12.8 | - | - | 22.2 |
| Gujarat** | 57.5 | 7.6 | 16.9 | 6.7 | 87.5 | 100.0 | 43.6 | 28.7 | 9.2 | 10.2 | 37.1 | 100.0 | 7.5 | 98.0 | - | 55.6 | 24.1 | 44.7 | 89.0 | 38.0 |
| Haryana | 99.9 | 75.4 | 29.8 | 17.0 | 99.0 | 92.6 | 87.7 | 18.3 | 84.1 | 31.9 | 85.2 | 99.2 | 95.4 | 74.2 | - | 90.9 | 74.3 | 99.5 | 100.0 | 83.7 |
| Himachal Pradesh** | 61.0 | - | 9.6 | 8.0 | 19.5 | 14.5 | 19.0 | 6.7 | - | 13.8 | 18.8 | 49.3 | 12.4 | 12.1 | 27.3 | - | 18.4 | 15.9 | - | 18.8 |
| Jammu & Kashmir** | 90.9 | 99.7 | 0.5 | 8.0 | 28.2 | 3.7 | 39.7 | 8.1 | - | 14.8 | 38.9 | 71.4 | 0.0 | 76.7 | - | - | 68.4 | 1.7 | - | 41.6 |
| Jharkhand** | 5.6 | - | - | 1.8 | 86.0 | - | 7.9 | 3.6 | - | 2.1 | 7.0 | 84.4 | 0.0 | 18.0 | - | 2.5 | 2.6 | - | 13.3 | 10.0 |
| Karnataka | 75.8 | 8.7 | 9.1 | 41.7 | 51.6 | - | 33.3 | 16.3 | 3.3 | 5.4 | 26.0 | 100.0 | 20.2 | 14.8 | 10.9 | 17.1 | 21.1 | 14.7 | 3.3 | 27.9 |
| Kerala | 57.9 | - | - | - | - | - | 57.3 | - | - | - | 55.2 | 84.5 | - | - | - | - | 17.7 | - | - | 15.4 |
| Madhya Pradesh | 13.6 | 0.1 | 0.1 | 1.9 | 78.3 | 44.4 | 43.2 | 47.5 | 1.2 | 33.5 | 39.7 | 99.8 | 9.0 | 50.2 | 0.5 | 60.7 | 7.7 | 36.4 | 73.6 | 30.0 |
| Maharashtra | 28.7 | 10.5 | 6.3 | 9.7 | 66.5 | - | 18.1 | 24.0 | 1.7 | 11.3 | 16.7 | 100.0 | 36.2 | 6.3 | 0.5 | 28.8 | 7.7 | 4.8 | 21.5 | 18.8 |
| Manipur | 30.7 | - | - | - | - | - | 30.2 | - | - | - | 28.8 | - | - | - | - | - | - | - | - | 23.0 |

contd. ...

...contd. ...

| (1) | (2) | (3) | (4) | (5) | (6) | (7) | (8) | (9) | (10) | (11) | (12) | (13) | (14) | (15) | (16) | (17) | (18) | (19) | (20) | (21) |
|---|---|---|---|---|---|---|---|---|---|---|---|---|---|---|---|---|---|---|---|---|
| Meghalaya | 39.0 | - | - | - | 39.7 | - | 32.9 | - | - | - | 32.0 | - | - | 40.0 | - | - | 29.3 | - | - | 24.9 |
| Mizoram | 29.0 | - | - | - | - | - | 24.0 | - | - | - | 21.8 | - | - | 0.0 | - | - | 42.0 | - | - | 19.4 |
| Nagaland | 36.2 | - | - | - | 100.0 | - | 29.3 | - | - | - | 25.7 | - | - | 100.0 | - | 100.0 | 51.4 | - | - | 27.4 |
| Orissa** | 42.6 | - | - | 11.8 | 100.0 | - | 40.0 | - | - | 8.3 | 32.0 | 100.0 | 37.5 | 8.9 | - | 95.8 | 14.3 | - | 10.9 | 30.8 |
| Punjab** | 99.3 | - | 92.7 | 60.4 | 98.0 | 96.1 | 97.6 | 81.0 | 95.1 | 86.9 | 97.6 | 96.5 | 47.9 | 92.4 | - | 98.7 | 87.4 | 99.8 | - | 96.6 |
| Rajasthan | 40.3 | 0.5 | 6.4 | 4.7 | 99.0 | 96.8 | 30.0 | 33.6 | 3.9 | 12.0 | 25.0 | 96.5 | 70.6 | 82.5 | 11.4 | 72.3 | 63.6 | 94.8 | 81.0 | 36.0 |
| Sikkim** | 62.5 | - | - | - | - | - | 13.9 | - | - | - | 12.7 | - | - | - | - | - | - | - | - | 8.2 |
| Tamil Nadu | 93.0 | 9.4 | 8.4 | 29.5 | - | - | 72.3 | 1.5 | 3.2 | 5.5 | 61.7 | 100.0 | 30.7 | 0.6 | 68.1 | 68.0 | 49.4 | 40.1 | 99.9 | 56.3 |
| Tripura** | 41.5 | - | - | - | - | - | 41.0 | - | - | - | 39.5 | - | - | - | - | - | - | - | - | 35.0 |
| Uttarakhand** | 64.9 | - | - | 1.6 | 52.3 | 2.2 | 41.7 | 62.4 | 0.2 | 6.2 | 39.6 | 97.2 | 6.0 | 51.4 | 1.7 | - | 27.6 | 20.0 | - | 44.2 |
| Uttar Pradesh** | 73.0 | 0.9 | 4.5 | 28.4 | 97.5 | 67.2 | 79.5 | 17.9 | 14.3 | 23.4 | 71.6 | 91.0 | 1.0 | 75.6 | 0.5 | 98.0 | 57.0 | 95.1 | 100.0 | 73.1 |
| West Bengal** | 49.8 | - | - | 7.0 | 65.7 | - | 50.1 | 12.5 | - | 7.6 | 48.6 | 20.0 | - | 72.6 | - | - | 58.8 | - | 100.0 | 57.5 |
| All India | 56.0 | 9.0 | 8.9 | 21.1 | 89.6 | 68.3 | 52.3 | 31.1 | 4.2 | 15.0 | 45.5 | 92.5 | 19.6 | 72.1 | 1.7 | 24.9 | 28.0 | 36.1 | 40.9 | 42.9 |

** The figures related to irrigated area (Part-II) are either estimated based on the data for the the latest available year received from the State/ UT or are estimated/taken from Agriculture Census.

# Table 14.5

*State-wise Distribution of Agricultural Land by Type of Use from 2003-04 to 2005-06 (Provisional)*

*(Thousand Hectares)*

| State/ Union- Territory/ Year | Geographical Area | Reporting Area for Land Utilisation Statistics (col. 4+7+11+14+15) | Forest | Land not Available for Cultivation | Other uncultivated land excluding Fallow Land: Permanent Pastures & Other Grazing Lands | Land under Misc. Tree Crops & Groves not incl. in Net Area Sown | Culturable Wasteland | Total | Fallow: Fallow Lands other than Current Fallows | Current Fallows | Land Total | Net Area Sown | Total Cropped Area | Cropping Intensity | Agri. Land/ Cultivable land/ Culturable land |
|---|---|---|---|---|---|---|---|---|---|---|---|---|---|---|---|
| 1 | 2 | 3 | 4 | 5 | 6 | 7 | 8 | 9 | 10 | 11 | 12 | 13 | 14 | 15 | 16 |
| Andhra Pradesh | | | | | | | | | | | | | | | |
| 2003-04 | 27507 | 27440 | 6199 | 4775 | 676 | 277 | 701 | 1654 | 1658 | 3036 | 4693 | 10118 | 12366 | 122.2 | 15789 |
| 2004-05 | 27507 | 27336 | 6199 | 4691 | 676 | 278 | 694 | 1648 | 1651 | 2819 | 4469 | 10327 | 12519 | 121.2 | 15769 |
| 2005-06 | 27507 | 27440 | 6199 | 4793 | 676 | 278 | 692 | 1646 | 1623 | 2434 | 4057 | 10745 | 13362 | 124.4 | 15772 |
| Arunachal Pradesh* | | | | | | | | | | | | | | | |
| 2003-04 | 8374 | 5547 | 5154 | 32 | | 46 | 28 | 74 | 64 | 23 | 87 | 200 | 259 | 129.7 | 361 |
| 2004-05 | 8374 | 5547 | 5154 | 32 | | 46 | 28 | 74 | 64 | 23 | 87 | 200 | 260 | 130.2 | 361 |
| 2005-06 | 8374 | 5547 | 5154 | 32 | | 46 | 28 | 74 | 64 | 23 | 87 | 200 | 267 | 133.4 | 361 |
| Assam | | | | | | | | | | | | | | | |
| 2003-04* | 7844 | 7850 | 1933 | 2533 | 160 | 209 | 77 | 445 | 66 | 99 | 164 | 2774 | 3958 | 142.7 | 3224 |
| 2004-05* | 7844 | 7850 | 1933 | 2533 | 160 | 209 | 77 | 445 | 66 | 99 | 164 | 2774 | 3756 | 135.4 | 3224 |
| 2005-06* | 7844 | 7850 | 1933 | 2533 | 160 | 209 | 77 | 445 | 66 | 99 | 164 | 2774 | 3731 | 134.5 | 3224 |
| Bihar | | | | | | | | | | | | | | | |
| 2003-04 | 9416 | 9360 | 622 | 2081 | 18 | 238 | 46 | 302 | 130 | 513 | 643 | 5712 | 7882 | 138.0 | 6640 |
| 2004-05 | 9416 | 9360 | 622 | 2082 | 17 | 239 | 46 | 303 | 133 | 648 | 781 | 5572 | 7399 | 132.8 | 6639 |
| 2005-06* | 9416 | 9360 | 622 | 2082 | 17 | 239 | 46 | 303 | 133 | 648 | 781 | 5572 | 7405 | 132.9 | 6639 |
| Chhattisgarh | | | | | | | | | | | | | | | |
| 2003-04 | 13519 | 13790 | 6300 | 1039 | 848 | 1 | 344 | 1192 | 232 | 248 | 480 | 4779 | 5707 | 119.4 | 5604 |
| 2004-05 | 13519 | 13790 | 6300 | 1045 | 850 | 1 | 337 | 1188 | 250 | 238 | 487 | 4770 | 5716 | 119.8 | 5595 |
| 2005-06 | 13519 | 13790 | 6353 | 992 | 855 | 2 | 339 | 1196 | 235 | 250 | 485 | 4764 | 5746 | 120.6 | 5590 |

contd. ...

...contd. ...

| 1 | 2 | 3 | 4 | 5 | 6 | 7 | 8 | 9 | 10 | 11 | 12 | 13 | 14 | 15 | 16 |
|---|---|---|---|---|---|---|---|---|---|---|---|---|---|---|---|
| Goa | | | | | | | | | | | | | | | |
| 2003-04 | 370 | 361 | 125 | 37 | 1 | 1 | 55 | 57 | | | | 141 | 169 | 119.2 | 197 |
| 2004-05 | 370 | 361 | 125 | 37 | 1 | 1 | 55 | 57 | | 8 | 8 | 134 | 169 | 126.6 | 197 |
| 2005-06 | 370 | 361 | 125 | 37 | 1 | 1 | 53 | 55 | | 7 | 7 | 137 | 170 | 124.5 | 197 |
| Gujarat | | | | | | | | | | | | | | | |
| 2003-04 | 19602 | 18868 | 1854 | 3753 | 850 | 4 | 1977 | 2831 | 11 | 568 | 579 | 9852 | 11421 | 115.9 | 12412 |
| 2004-05* | 19602 | 18868 | 1854 | 3753 | 850 | 4 | 1977 | 2831 | 11 | 568 | 579 | 9852 | 11009 | 111.7 | 12412 |
| 2005-06* | 19602 | 18868 | 1854 | 3753 | 850 | 4 | 1977 | 2831 | 11 | 568 | 579 | 9852 | 11304 | 114.7 | 12412 |
| Haryana | | | | | | | | | | | | | | | |
| 2003-04 | 4421 | 4374 | 45 | 532 | 25 | 6 | 36 | 67 | 4 | 192 | 196 | 3534 | 6388 | 180.8 | 3771 |
| 2004-05 | 4421 | 4374 | 44 | 524 | 25 | 6 | 35 | 65 | 11 | 201 | 213 | 3527 | 6425 | 182.2 | 3780 |
| 2005-06 | 4421 | 4372 | 44 | 518 | 25 | 12 | 31 | 68 | 6 | 169 | 175 | 3566 | 6504 | 182.4 | 3784 |
| Himachal Pradesh | | | | | | | | | | | | | | | |
| 2003-04 | 5567 | 4544 | 1099 | 1126 | 1515 | 62 | 128 | 1706 | 17 | 56 | 73 | 541 | 956 | 176.8 | 804 |
| 2004-05* | 5567 | 4544 | 1099 | 1126 | 1515 | 62 | 128 | 1706 | 17 | 56 | 73 | 541 | 984 | 182.0 | 804 |
| 2005-06* | 5567 | 4544 | 1099 | 1126 | 1515 | 62 | 128 | 1706 | 17 | 56 | 73 | 541 | 940 | 173.9 | 804 |
| Jammu & Kashmir | | | | | | | | | | | | | | | |
| 2003-04 | 22224 | 3781 | 2023 | 582 | 125 | 72 | 142 | 339 | 12 | 78 | 90 | 747 | 1102 | 147.6 | 1050 |
| 2004-05 | 22224 | 3781 | 2023 | 582 | 125 | 71 | 141 | 338 | 12 | 73 | 85 | 752 | 1102 | 146.4 | 1050 |
| 2005-06* | 22224 | 3781 | 2023 | 582 | 125 | 71 | 141 | 338 | 12 | 73 | 85 | 752 | 1090 | 144.9 | 1050 |
| Jharkhand* | | | | | | | | | | | | | | | |
| 2003-04 | 7972 | 7970 | 2333 | 1366 | 88 | 113 | 274 | 476 | 783 | 1244 | 2027 | 1769 | 2164 | 122.3 | 4184 |
| 2004-05 | 7972 | 7970 | 2333 | 1366 | 88 | 113 | 274 | 476 | 783 | 1244 | 2027 | 1769 | 2068 | 116.9 | 4184 |
| 2005-06 | 7972 | 7970 | 2333 | 1366 | 88 | 113 | 274 | 476 | 783 | 1244 | 2027 | 1769 | 2116 | 119.6 | 4184 |
| Karnataka | | | | | | | | | | | | | | | |
| 2003-04 | 19179 | 19050 | 3071 | 2124 | 947 | 301 | 419 | 1668 | 487 | 1854 | 2341 | 9847 | 11450 | 116.3 | 12908 |
| 2004-05 | 19179 | 19050 | 3070 | 2128 | 945 | 297 | 420 | 1663 | 443 | 1247 | 1689 | 10499 | 12807 | 122.0 | 12906 |
| 2005-06 | 19179 | 19050 | 3072 | 2137 | 936 | 292 | 419 | 1647 | 452 | 1233 | 1684 | 10509 | 13027 | 124.0 | 12905 |
| Kerala | | | | | | | | | | | | | | | |
| 2003-04 | 3886 | 3885 | 1082 | 425 | 0 | 11 | 67 | 78 | 41 | 69 | 110 | 2190 | 2954 | 134.9 | 2378 |
| 2004-05 | 3886 | 3885 | 1082 | 459 | 0 | 10 | 70 | 81 | 41 | 69 | 110 | 2155 | 2996 | 139.0 | 2345 |
| 2005-06 | 3886 | 3886 | 1082 | 481 | 0 | 10 | 66 | 76 | 45 | 70 | 115 | 2132 | 2986 | 140.0 | 2323 |

contd. ...

...contd. ...

| 1 | 2 | 3 | 4 | 5 | 6 | 7 | 8 | 9 | 10 | 11 | 12 | 13 | 14 | 15 | 16 |
|---|---|---|---|---|---|---|---|---|---|---|---|---|---|---|---|
| Madhya Pradesh | | | | | | | | | | | | | | | |
| 2003-04 | 30825 | 30756 | 8683 | 3350 | 1360 | 19 | 1177 | 2557 | 621 | 599 | 1220 | 14945 | 19788 | 132.4 | 17362 |
| 2004-05 | 30825 | 30756 | 8688 | 3364 | 1341 | 20 | 1175 | 2536 | 596 | 596 | 1192 | 14975 | 20203 | 134.9 | 17362 |
| 2005-06 | 30825 | 30756 | 8692 | 3389 | 1339 | 19 | 1161 | 2519 | 595 | 591 | 1185 | 14971 | 19608 | 131.0 | 17337 |
| Maharashtra | | | | | | | | | | | | | | | |
| 2003-04 | 30771 | 30758 | 5214 | 3115 | 1249 | 251 | 917 | 2418 | 1216 | 1364 | 2580 | 17432 | 22190 | 127.3 | 21180 |
| 2004-05 | 30771 | 30758 | 5213 | 3119 | 1251 | 249 | 918 | 2417 | 1204 | 1316 | 2520 | 17490 | 22368 | 127.9 | 21176 |
| 2005-06 | 30771 | 30758 | 5212 | 3127 | 1252 | 249 | 914 | 2415 | 1204 | 1327 | 2531 | 17473 | 22556 | 129.1 | 21167 |
| Manipur | | | | | | | | | | | | | | | |
| 2003-04* | 2233 | 1945 | 1693* | 27 | 1 | 6 | 1 | 8 | 0 | 0 | 0 | 217 | 217 | 100.0 | 224 |
| 2004-05* | 2233 | 1966 | 1693* | 27 | 1 | 6 | 1 | 8 | 0 | 0 | 0 | 238 | 238 | 100.0 | 245 |
| 2005-06* | 2233 | 1950 | 1693* | 27 | 1 | 6 | 1 | 8 | 0 | 0 | 0 | 223 | 223 | 100.0 | 230 |
| Meghalaya | | | | | | | | | | | | | | | |
| 2003-04 | 2243 | 2227 | 951 | 215 | | 161 | 444 | 605 | 167 | 63 | 229 | 227 | 272 | 120.2 | 1061 |
| 2004-05 | 2243 | 2227 | 942 | 228 | | 157 | 452 | 609 | 166 | 63 | 230 | 219 | 265 | 121.0 | 1057 |
| 2005-06 | 2243 | 2227 | 942 | 227 | | 158 | 453 | 611 | 169 | 66 | 235 | 212 | 258 | 121.9 | 1058 |
| Mizoram | | | | | | | | | | | | | | | |
| 2003-04 | 2108 | 2085 | 1594 | 134 | 6 | 15 | 6 | 27 | 194 | 38 | 233 | 98 | 98 | 100.0 | 352 |
| 2004-05 | 2108 | 2075 | 1594 | 134 | 5 | 10 | 5 | 21 | 181 | 50 | 231 | 96 | 96 | 100.0 | 342 |
| 2005-06 | 2108 | 1951 | 1594 | 134 | 5 | 10 | 5 | 21 | 82 | 24 | 106 | 97 | 97 | 100.0 | 218 |
| Nagaland | | | | | | | | | | | | | | | |
| 2003-04 | 1658 | 1583 | 863 | 75 | | 122 | 58 | 179 | 76 | 84 | 160 | 305 | 370 | 121.2 | 645 |
| 2004-05 | 1658 | 1584 | 863 | 76 | | 121 | 57 | 178 | 75 | 82 | 157 | 309 | 380 | 123.1 | 645 |
| 2005-06 | 1658 | 1582 | 863 | 75 | | 120 | 56 | 177 | 76 | 82 | 158 | 309 | 387 | 125.2 | 644 |
| Orissa | | | | | | | | | | | | | | | |
| 2003-04 | 15571 | 15571 | 5813 | 1842 | 443 | 482 | 392 | 1317 | 434 | 370 | 804 | 5795 | 8637 | 149.0 | 7473 |
| 2004-05 | 15571 | 15571 | 5813 | 1842 | 443 | 482 | 392 | 1317 | 434 | 426 | 860 | 5739 | 8718 | 151.9 | 7473 |
| 2005-06* | 15571 | 15571 | 5813 | 1842 | 443 | 482 | 392 | 1317 | 434 | 426 | 860 | 5739 | 8716 | 151.9 | 7473 |
| Punjab | | | | | | | | | | | | | | | |
| 2003-04 | 5036 | 5033 | 308 | 451 | 4 | 4 | 9 | 17 | 0 | 13 | 14 | 4243 | 7985 | 188.2 | 4270 |
| 2004-05* | 5036 | 5033 | 308 | 451 | 4 | 4 | 9 | 17 | 0 | 13 | 14 | 4243 | 8069 | 190.2 | 4270 |
| 2005-06* | 5036 | 5033 | 308 | 451 | 4 | 4 | 9 | 17 | 0 | 13 | 14 | 4243 | 8085 | 190.6 | 4270 |

contd. ...

*...contd. ...*

| 1 | 2 | 3 | 4 | 5 | 6 | 7 | 8 | 9 | 10 | 11 | 12 | 13 | 14 | 15 | 16 |
|---|---|---|---|---|---|---|---|---|---|---|---|---|---|---|---|
| Rajasthan | | | | | | | | | | | | | | | |
| 2003-04 | 34224 | 34266 | 2661 | 4259 | 1708 | 14 | 4547 | 6269 | 2407 | 1275 | 3683 | 17394 | 21664 | 124.5 | 25638 |
| 2004-05 | 34224 | 34266 | 2661 | 4267 | 1709 | 14 | 4603 | 6325 | 2162 | 2302 | 4464 | 16549 | 21062 | 127.3 | 25630 |
| 2005-06 | 34224 | 34266 | 2675 | 4262 | 1708 | 21 | 4590 | 6319 | 2264 | 1910 | 4174 | 16836 | 21699 | 128.9 | 22739 |
| Sikkim* | | | | | | | | | | | | | | | |
| 2003-04 | 710 | 726 | 319 | 250 | 4 | 5 | 2 | 12 | 30 | 5 | 35 | 110 | 121 | 109.8 | 153 |
| 2004-05 | 710 | 728 | 319 | 250 | 4 | 5 | 2 | 12 | 30 | 5 | 35 | 112 | 123 | 109.8 | 155 |
| 2005-06 | 710 | 728 | 319 | 250 | 4 | 5 | 2 | 12 | 30 | 5 | 35 | 112 | 123 | 109.8 | 155 |
| Tamil Nadu | | | | | | | | | | | | | | | |
| 2003-04 | 13006 | 13027 | 2122 | 2623 | 113 | 283 | 379 | 776 | 1863 | 954 | 2817 | 4689 | 5316 | 113.4 | 8168 |
| 2004-05 | 13006 | 13027 | 2122 | 2634 | 114 | 290 | 374 | 778 | 1704 | 692 | 2396 | 5097 | 5889 | 115.5 | 8157 |
| 2005-06 | 13006 | 13027 | 2111 | 2642 | 110 | 274 | 369 | 753 | 1518 | 759 | 2277 | 5244 | 6033 | 115.0 | 8164 |
| Tripura* | | | | | | | | | | | | | | | |
| 2003-04 | 1049 | 1049 | 606 | 134 | | 27 | 1 | 28 | 1 | 1 | 2 | 280 | 291 | 104.0 | 310 |
| 2004-05 | 1049 | 1049 | 606 | 134 | | 27 | 1 | 28 | 1 | 1 | 2 | 280 | 297 | 106.0 | 310 |
| 2005-06 | 1049 | 1049 | 606 | 134 | | 27 | 1 | 28 | 1 | 1 | 2 | 280 | 299 | 106.6 | 310 |
| Uttarakhand | | | | | | | | | | | | | | | |
| 2003-04 | 5348 | 5668 | 3465 | 463 | 229 | 251 | 386 | 866 | 71 | 43 | 114 | 761 | 1222 | 160.6 | 1511 |
| 2004-05 | 5348 | 5670 | 3465 | 464 | 229 | 249 | 386 | 864 | 68 | 42 | 110 | 767 | 1235 | 161.0 | 1512 |
| 2005-06* | 5348 | 5670 | 3465 | 464 | 229 | 249 | 386 | 864 | 68 | 42 | 110 | 767 | 1266 | 165.1 | 1512 |
| Uttar Pradesh | | | | | | | | | | | | | | | |
| 2003-04 | 24093 | 24201 | 1686 | 3143 | 66 | 359 | 468 | 892 | 594 | 1137 | 1731 | 16750 | 25425 | 151.8 | 19307 |
| 2004-05 | 24093 | 24201 | 1688 | 3178 | 64 | 344 | 454 | 862 | 574 | 1217 | 1791 | 16683 | 24578 | 147.3 | 19272 |
| 2005-06* | 24093 | 24201 | 1688 | 3178 | 64 | 344 | 454 | 862 | 574 | 1217 | 1791 | 16683 | 25105 | 150.5 | 19272 |
| West Bengal | | | | | | | | | | | | | | | |
| 2003-04 | 8875 | 8688 | 1171 | 1636 | 5 | 58 | 34 | 97 | 22 | 333 | 355 | 5428 | 9661 | 178.0 | 5876 |
| 2004-05 | 8875 | 8687 | 1175 | 1700 | 5 | 59 | 36 | 99 | 25 | 314 | 339 | 5374 | 9523 | 177.2 | 5808 |
| 2005-06 | 8875 | 8683 | 1175 | 1753 | 6 | 63 | 43 | 111 | 30 | 319 | 349 | 5295 | 9533 | 180.0 | 5749 |
| A&N Island* | | | | | | | | | | | | | | | |
| 2003-04 | 825 | 740 | 717 | 1 | | 2 | 1 | 3 | 2 | 1 | 3 | 16 | 46 | 286.1 | 22 |
| 2004-05 | 825 | 740 | 717 | 1 | | 2 | 1 | 3 | 2 | 1 | 3 | 16 | 47 | 291.6 | 22 |
| 2005-06 | 825 | 740 | 717 | 1 | | 2 | 1 | 3 | 2 | 1 | 3 | 16 | 46 | 285.4 | 22 |

*contd. ...*

*...contd. ...*

| *1* | *2* | *3* | *4* | *5* | *6* | *7* | *8* | *9* | *10* | *11* | *12* | *13* | *14* | *15* | *16* |
|---|---|---|---|---|---|---|---|---|---|---|---|---|---|---|---|
| Chandigarh | | | | | | | | | | | | | | | |
| 2003-04 | 11 | 7 | 0 | 5 | | 0 | | 0 | 0 | 0 | 0 | 1 | 2 | 147.5 | 2 |
| 2004-05 | 11 | 7 | 0 | 5 | | 0 | | 0 | 0 | 0 | 0 | 1 | 2 | 150.0 | 2 |
| 2005-06 | 11 | 7 | 0 | 5 | | 0 | | 0 | 0 | 0 | 0 | 1 | 2 | 151.5 | 2 |
| D & N Haveli | | | | | | | | | | | | | | | |
| 2003-04 | 49 | 49 | 20 | 4 | 1 | | 0 | 1 | 1 | 1 | 2 | 22 | 30 | 133.2 | 24 |
| 2004-05 | 49 | 49 | 20 | 4 | 1 | | 0 | 1 | 1 | 1 | 1 | 23 | 29 | 126.8 | 24 |
| 2005-06* | 49 | 49 | 20 | 4 | 1 | | 0 | 1 | 1 | 1 | 1 | 23 | 28 | 121.0 | 24 |
| Daman & Diu* | | | | | | | | | | | | | | | |
| 2003-04 | 11 | 3 | | | | | | | | | | 3 | 3 | 103.3 | 3 |
| 2004-05 | 11 | 3 | | | | | | | | | | 3 | 3 | 100.0 | 3 |
| 2005-06 | 11 | 3 | | | | | | | | | | 3 | 3 | 103.3 | 3 |
| Delhi | | | | | | | | | | | | | | | |
| 2003-04 | 148 | 147 | 1 | 89 | 0 | 1 | 10 | 11 | 7 | 12 | 19 | 27 | 44 | 165.0 | 57 |
| 2004-05 | 148 | 147 | 1 | 88 | 0 | 1 | 10 | 11 | 7 | 12 | 19 | 28 | 45 | 165.1 | 58 |
| 2005-06 | 148 | 147 | 1 | 88 | 0 | 1 | 10 | 11 | 8 | 12 | 19 | 28 | 44 | 159.2 | 58 |
| Lakshadweep* | | | | | | | | | | | | | | | |
| 2003-04 | 3 | 3 | | | | | | | | | | 3 | 3 | 100.4 | 3 |
| 2004-05 | 3 | 3 | | | | | | | | | | 3 | 3 | 100.4 | 3 |
| 2005-06 | 3 | 3 | | | | | | | | | | 3 | 3 | 100.4 | 3 |
| Pondicherry | | | | | | | | | | | | | | | |
| 2003-04 | 48 | 49 | | 17 | | 1 | 4 | 5 | 2 | 3 | 6 | 21 | 37 | 181.1 | 31 |
| 2004-05 | 48 | 49 | | 17 | 0 | 1 | 4 | 5 | 3 | 3 | 5 | 21 | 39 | 184.5 | 31 |
| 2005-06 | 48 | 49 | | 18 | | 1 | 4 | 5 | 3 | 2 | 5 | 21 | 36 | 175.7 | 31 |
| All India | | | | | | | | | | | | | | | |
| 2003-04 | 328726 | 305399 | 69725 | 42236 | 10444 | 3403 | 13130 | 26976 | 11214 | 14277 | 25491 | 140971 | 190200 | 134.9 | 182994 |
| 2004-05 | 328726 | 305313 | 69725 | 42341 | 10423 | 3378 | 13164 | 26965 | 10715 | 14428 | 25143 | 141139 | 190422 | 134.9 | 182824 |
| 2005-06 | 328726 | 305269 | 69785 | 42503 | 10415 | 3376 | 13123 | 26914 | 10504 | 13672 | 24176 | 141891 | 192796 | 135.9 | 179684 |

* Estimated

Primary Source: State Governments.

# Table 15.1(a)

*Performance of South West Monsoon during 1989 to 2007*

*(1 June - 30 September)*

| Year | Number of Meteorological Sub-Divisions @ | | Percentage of Districts with Normal/ Excess Rainfall | Actual Rainfall as % of Normal Rainfall (All India) |
|---|---|---|---|---|
| | Excess/ Normal Rainfall | Deficient/ Scanty Rainfall | | |
| 1 | 2 | 3 | 4 | 5 |
| 1989 | 29 | 6 | 72 | 101 |
| 1990 | 32 | 3 | 88 | 119 |
| 1991 | 27 | 8 | 68 | 91 |
| 1992 | 32 | 3 | 65 | 93 |
| 1993 | 31 | 4 | 78 | 100 |
| 1994 | 25 | 10 | 77 | 110 |
| 1995 | 33 | 2 | 79 | 100 |
| 1996 | 32 | 3 | 82 | 103 |
| 1997 | 32 | 3 | 81 | 102 |
| 1998 | 33 | 2 | 83 | 106 |
| 1999 | 28 | 7 | 67 | 96 |
| 2000 | 28 | 7 | 66 | 92 |
| 2001 | 30 | 5 | 68 | 92 |
| 2002 | 15 | 21 | 39 | 81 |
| 2003 | 33 | 3 | 75 | 102 |
| 2004 | 23 | 13 | 56 | 87 |
| 2005 | 32 | 4 | 72 | 99 |
| 2006 | 26 | 10 | 59 | 99 |
| 2007 | 30 | 6 | 72 | 105 |

@ Total number of Meteorological sub-divisions was 35 upto 2001. From 2002 onwards, the no. of meteorological sub-divisions is 36.

Excess: + 20 per cent or more of Long Period Average Rainfall

Normal: Between + 19 per cent and -19 per cent of Long Period Average Rainfall

Deficient: Between -20 per cent and -59 per cent of Long Period Average Rainfall

Scanty: Between -60 per cent and -99 per cent of Long Period Average Rainfall

## Table 15.1(b)

*All India rainfall distribution from 1992-93 to 2007-08*

*(In Millimeters)*

| Year | Monsoon Season (June-September) | | | Post-Monsoon (October-December) | | | Winter Season (January-February) | | | Pre-monsoon Season (March-May) | | | Over all rainfall (June-May) | | |
|---|---|---|---|---|---|---|---|---|---|---|---|---|---|---|---|
| | Actual | Normal | %Departure | Actual | Normal | %Departure | Actual | Normal | %Departure | Actual | Normal | %Departure | Actual | Normal | %Departure |
| 1 | 2 | 3 | 4 | 5 | 6 | 7 | 8 | 9 | 10 | 11 | 12 | 13 | 14 | 15 | 16 |
| 1992-93 | 830.7 | 899.2 | -7.6 | 106.5 | 114.1 | -6.7 | 37.9 | 41.0 | -7.6 | 116.5 | 121.3 | -4.0 | 1091.6 | 1175.6 | -7.1 |
| 1993-94 | 902.1 | 908.9 | -0.7 | 131.6 | 119.6 | 10.0 | 44.5 | 40.8 | 9.1 | 106.1 | 123.3 | -13.9 | 1184.3 | 1192.6 | -0.7 |
| 1994-95 | 999.2 | 906.8 | 10.2 | 121.5 | 119.6 | 1.6 | 53.1 | 41.1 | 29.2 | 123.5 | 123.2 | 0.2 | 1297.3 | 1190.7 | 9.0 |
| 1995-96 | 904.5 | 904.7 | 0.0 | 117.8 | 119.9 | -1.8 | 37.4 | 40.8 | -8.3 | 94.9 | 123.9 | -23.4 | 1154.6 | 1189.3 | -2.9 |
| 1996-97 | 927.6 | 905.7 | 2.4 | 128.0 | 120.8 | 6.0 | 21.0 | 40.6 | -48.3 | 118.9 | 123.2 | -3.5 | 1195.5 | 1190.3 | 0.4 |
| 1997-98 | 927.4 | 908.6 | 2.1 | 187.7 | 119.5 | 57.1 | 44.1 | 41.9 | 5.3 | 132.3 | 128.3 | 3.1 | 1291.5 | 1198.3 | 7.8 |
| 1998-99 | 945.2 | 903.6 | 4.6 | 178.8 | 121.8 | 46.8 | 28.4 | 42.8 | -33.6 | 123.1 | 130.6 | -5.7 | 1275.5 | 1198.8 | 6.4 |
| 1999-2000 | 866.9 | 903.2 | -4.0 | 144.7 | 121.8 | 18.8 | 43.1 | 42.5 | 1.4 | 128.8 | 129.5 | -0.5 | 1183.5 | 1197.0 | -1.1 |
| 2000-01 | 833.7 | 902.3 | -7.6 | 64.1 | 121.7 | -47.3 | 16.2 | 42.2 | -61.6 | 129.7 | 129.3 | 0.3 | 1043.7 | 1195.5 | -12.7 |
| 2001-02 | 826.0 | 901.1 | -8.3 | 137.7 | 121.7 | 13.1 | 35.0 | 41.2 | -15.0 | 121.5 | 132.0 | -8.0 | 1120.2 | 1196.0 | -6.3 |
| 2002-03 | 737.1 | 911.7 | -19.2 | 83.4 | 123.7 | -32.6 | 53.2 | 38.3 | 38.9 | 107.7 | 131.7 | -18.2 | 981.4 | 1205.4 | -18.6 |
| 2003-04 | 947.3 | 902.7 | 4.9 | 134.6 | 125.0 | 7.7 | 34.5 | 39.2 | -12.0 | 161.6 | 129.6 | 24.7 | 1278.0 | 1196.5 | 6.8 |
| 2004-05 | 779.6 | 893.3 | -12.7 | 111.8 | 125.7 | -11.1 | 69.8 | 43.8 | 59.0 | 124.7 | 134.5 | -7.0 | 891.4 | 1019.0 | -12.5 |
| 2005-06 | 879.3 | 892.5 | -1.0 | 138.4 | 125.8 | 10.0 | 27.8 | 43.9 | -37.0 | 139.9 | 134.6 | -4.0 | 1017.7 | 1018.3 | -0.1 |
| 2006-07 | 886.6 | 892.2 | -0.6 | 99.3 | 125.9 | -21.1 | 34.3 | 43.8 | -21.7 | 112.8 | 133.6 | -15.6 | 1133.0 | 1195.5 | -5.2 |
| 2007-08* | 936.9 | 892.2 | 5.0 | 85.4 | 125.9 | -32.2 | 42.6 | 43.2 | -1.4 | 76.9 | 82 | -6.2 | 1141.8 | 1143.3 | -0.1 |

* Pre-Monsoon Season till 07th May, 2008.

# Table 15.2

*Status of Important Reservoirs as on 31.03.2008*

*(Capacity in Billion Cubic Meter)*

| Sl. No. | Reservoir/States | Districts/Area Benefitted | Live Capacity at FRL (BCM) | Capacity as % of live Capacity at FRL | |
|---|---|---|---|---|---|
| | | | | As on 31st March 2008 | As on 31st March 2007 |
| 1 | 2 | 3 | 4 | 5 | 6 |
| **Andhra Pradesh** | | | | | |
| 1 | Srisailam | Kurnool, Cuddapah | 8.288 | 58 | 31 |
| 2 | Nagarjuna Sagar | Guntur, Krishna, Nalgonda, Khammam, Prakasham | 6.841 | 21 | 30 |
| 3 | Sriramsagar | Nizamabad, Karimnagar, Warangal, Adilabad | 2.300 | -18 | 8 |
| 4 | Somasila | Nellore. | 1.994 | 73 | 19 |
| 5 | Lower Manair | Karimnagar, Warangal | 0.621 | 42 | 45 |
| Sub-Total (Andhra Pradesh) | | | 20.044 | 212 | 133 |
| **Chhattisgarh** | | | | | |
| 6 | Minimata Bango | Korba, Janjgir-Champa, Raigarh | 3.046 | 37 | 44 |
| 7 | Mahanadi | Dhamtari | 0.767 | 50 | 67 |
| Sub-Total (Chhattisgarh) | | | 3.813 | 87 | 111 |
| **Gujarat** | | | | | |
| 8 | Ukai | Surat, Bahruch, Valsad, Navsari | 6.615 | 49 | 55 |
| 9 | Sabarmati (Dharoi) | Sabarkantha, Mehsana, Patana, Gandhinagar | 0.735 | 42 | 53 |
| 10 | Kadana | Godhra (Panchmahal), Kheda, Anand | 1.472 | 49 | 61 |
| 11 | Shetrunji | Bhavnagar | 0.300 | 44 | 21 |
| 12 | Bhadar | Rajkot, Junagarh | 0.188 | 29 | 26 |
| 13 | Damanganga | Valsad, Dadra & Nagar Haveli, Daman | 0.502 | 1 | 60 |

...contd...

...contd...

| 1 | 2 | 3 | 4 | 5 | 6 |
|---|---|---|---|---|---|
| 14 | Dantiwada | Banaskantha, Paten | 0.399 | 12 | 30 |
| 15 | Panam | Panchmahals, Baroda | 0.697 | 71 | 70 |
| Sub-Total (Gujarat) | | | 10.908 | 297 | 376 |
| **Himachal Pradesh** | | | | | |
| 16 | Gobind Sagar (Bhakra) | Ludhiana, Patiala, Sangrur, Kapurthala, Jullandur, Hoshiarpur, Bhatinda, Ferozpur, Ganganagar | 6.229 | 14 | 41 |
| 17 | Pong Dam | Amritsar, Bhatinda, Ferozpur, Ganganagar | 6.157 | 12 | 52 |
| Sub-Total (Himachal Pradesh) | | | 12.386 | 26 | 93 |
| **Jharkhand** | | | | | |
| 18 | Tenughat | Nil | 0.821 | 33 | 37 |
| 19 | Maithon | Bankura, Burdwan | 0.471 | 82 | 70 |
| 20 | Panchet Hill | Howrah, Hoogli | 0.184 | 76 | 78 |
| 21 | Konar | Hazaribagh, Giridih | 0.176 | 2 | 46 |
| 22 | Tilaiya | Hazaribagh, Nawada | 0.142 | 20 | 35 |
| Sub-Total (Jharkhand) | | | 1.794 | 213 | 411 |
| **Karnataka** | | | | | |
| 23 | Krishnaraja Sagara | Mysore, Mandya | 1.163 | 66 | 27 |
| 24 | Tungabhadra | Bellary, Raichur, Anantpur, Kurnool | 3.276 | 19 | 4 |
| 25 | Ghataprabha | Belgaum, Bijapur | 1.391 | 9 | 10 |
| 26 | Bhadra | Chikmagalore, Shimoga, Devangere | 1.785 | 57 | 58 |
| 27 | Linganamakki | Hydel Project | 4.294 | 41 | 45 |
| 28 | Narayanpur | Bijapur, Bagalkot, Gulbarga, Raichur | 0.863 | 67 | 72 |
| 29 | Malaprabha (Renuka) | Belgaum, Dharwad, Bagalkot, Gadag | 0.972 | 12 | 14 |
| 30 | Kabini | Mysore, Chamarajanagar | 0.275 | 47 | 25 |

...contd...

...contd...

| 1 | 2 | 3 | 4 | 5 | 6 |
|---|---|---|---|---|---|
| 31 | Hemavathy | Hassan, Mandya, Tumkur, Mysore, Coorg | 0.927 | 30 | 37 |
| 32 | Harangi | Hassan, Mysore, Coorg | 0.220 | 24 | 23 |
| 33 | Supa | Hydel Project | 4.120 | 41 | 57 |
| 34 | Vanivilas Sagar | Chitradurga | 0.802 | 8 | 14 |
| 35 | Almatti | Bagalkot, Bijapur | 3.105 | 94 | 10 |
| Sub-Total (Karnataka) | | | 23.193 | 515 | 396 |
| **Kerala** | | | | | |
| 36 | Kallada (Parappar) | Kollam, Pathanamthitta, Allappuzha | 0.507 | 48 | 61 |
| 37 | Idamalayar | Ernakulam | 1.018 | 45 | 44 |
| 38 | Idukki | Idukki, Kottayam, Ernakulam | 1.460 | 42 | 45 |
| 39 | Kakki | Hydel project | 0.447 | 54 | 57 |
| 40 | Periyar | Theni, Dindigul, Madurai, Sivagangai | 0.173 | 69 | 54 |
| Sub-Total (Kerala) | | | 3.605 | 258 | 261 |
| **Madhya Pradesh** | | | | | |
| 41 | Gandhi Sagar | Bundi, Tonk, Sawai Madhopur, Kota | 6.827 | 16 | 0 |
| 42 | Tawa | Hoshangabad, Harda | 1.944 | 16 | 23 |
| 43 | Bargi | Jabalpur, Naresimhapur | 3.180 | 23 | 30 |
| Sub-Total (Madhya Pradesh) | | | 11.951 | 55 | 53 |
| **Maharashtra** | | | | | |
| 44 | Jayakwadi (Paithon) | Pune, Parbhani | 2.171 | 50 | 63 |
| 45 | Koyana | Hydel Project | 2.652 | 62 | 48 |
| 46 | Bhima (Ujjani) | Solapur, Pune, Ahmednagar | 1.517 | 33 | 53 |
| 47 | Isapur | N.A. | 0.965 | 47 | 24 |
| 48 | Mula | Ahmednagar | 0.609 | 42 | 30 |
| 49 | Yeldari | N.A. | 0.809 | 6 | 26 |

*...contd...*

| 1 | 2 | 3 | 4 | 5 | 6 |
|---|---|---|---|---|---|
| 50 | Girna | Jalgaon | 0.524 | 27 | 27 |
| 51 | Khadakvasla | Pune | 0.056 | 9 | 20 |
| 52 | Upper Vaitarna | Hydel Project | 0.331 | 47 | 60 |
| 53 | Upper Tapi | Jalgaon, Nashik | 0.255 | 50 | 49 |
| Sub-Total (Maharashtra) | | | 9.889 | 373 | 400 |
| **Orissa** | | | | | |
| 54 | Hirakud | Sambalpur, Balangir | 5.378 | 44 | 54 |
| 55 | Balimela | Hydel Project | 2.676 | 32 | 22 |
| 56 | Salanadi | Balasore, Bhadrak | 0.558 | 49 | 32 |
| 57 | Rengali | Dhenkanal, Cuttack, Keonjhar, Ballasore | 3.432 | 17 | 39 |
| 58 | Machkund (Jalaput) | Hydel Project | 0.893 | 85 | 55 |
| 59 | Upper Kolab | Koraput | 0.935 | 29 | 42 |
| 60 | Upper Indravati | Hydel Project | 1.456 | 28 | 49 |
| Sub-Total (Orissa) | | | 15.328 | 284 | 293 |
| **Punjab** | | | | | |
| 61 | Thein | Gurdaspur,Amritsar,Kathua | 2.344 | 21 | 21 |
| Sub-Total (Punjab) | | | 2.344 | 21 | 21 |
| **Rajasthan** | | | | | |
| 62 | Mahi Bajaj Sagar | Banswara, Dungapur | 1.711 | 40 | 38 |
| 63 | Jhakam | Udaipur, Chittorgarh | 0.132 | 4 | 8 |
| 64 | Rana Pratap Sagar | Bundi, Tonk, Sawai Madhopur, Kota | 1.436 | 78 | 10 |
| Sub-Total (Rajasthan) | | | 3.279 | 122 | 56 |
| **Tamil Nadu** | | | | | |
| 65 | Lower Bhawani | Erode, Karur | 0.792 | 90 | 84 |
| 66 | Mettur (Stanley) | Triuvarur, Salem, Tiruchirapali, Erode, Namakkal, Nagai, Karaikkal, Puthukottai | 2.647 | 84 | 89 |

*...contd...*

*...contd...*

| 1 | 2 | 3 | 4 | 5 | 6 |
|---|---|---|---|---|---|
| 67 | Vaigai | Madurai, Dindugal, Sivagangai | 0.172 | 100 | 51 |
| 68 | Parambikulam | Nil | 0.380 | 62 | 57 |
| 69 | Aliyar | Coimbatore, Palaghat | 0.095 | 75 | 51 |
| 70 | Sholayar | Coimbatore | 0.143 | 5 | 6 |
| Sub-Total (Tamil Nadu) | | | 4.229 | 416 | 338 |
| **Tripura** | | | | | |
| 71 | Gumti | | 0.312 | 62 | 9 |
| Sub-Total (Tripura) | | | 0.312 | 62 | 9 |
| **Uttar Pradesh** | | | | | |
| 72 | Matatila | Jhansi, Datia, Orai, Hamirpur | 0.707 | 40 | 38 |
| 73 | Rihand | Hydel project | 5.649 | 16 | 15 |
| Sub-Total (Uttar Pradesh) | | | 6.356 | 56 | 53 |
| **Uttrakhand** | | | | | |
| 74 | Ramganga | Bijnor, Moradabad | 2.196 | 40 | 25 |
| Sub-Total (Uttranchal) | | | 2.196 | 40 | 25 |
| **West Bengal** | | | | | |
| 75 | Mayurakshi | Birbhum, Burdwan, Murshidabad | 0.480 | 36 | 22 |
| 76 | Kangsabati | Bankura, Hoogli, Midnapur | 0.914 | 19 | 4 |
| Sub-Total (West Bengal) | | | 1.394 | 55 | 26 |
| Total (All India) | | | 133.021 | 3092 | 3055 |

*Source:* Central Water Commission, New Delhi.

## Table 15.3

*Production and Use of Agricultural Inputs in India from 1991-92 to 2007-08*

| Programme | Unit | 1991-92 | 1992-93 | 1993-94 | 1994-95 | 1995-96 | 1996-97 | 1997-98 |
|---|---|---|---|---|---|---|---|---|
| 1 2 | 3 | 4 | 5 | 6 | 7 | 8 | 9 | |
| **1. Seeds** | | | | | | | | |
| (i) Production of Breeder Seeds | Thousand Qtls. | 34.90 | 36.00 | 37.00 | 40.11 | 43.36 | 46.03 | 46.13 |
| (ii) Production of Foundation Seeds | Lakh Qtls. | 3.75 | 3.93 | 4.06 | 4.73 | 4.76 | 5.76 | 6.84 |
| (iii) Distribution of Certified/Quality Seeds | Lakh Qtls. | 57.50 | 60.33 | 62.20 | 65.86 | 69.90 | 73.27 | 78.79 |
| **2. Consumption of Chemical Fertilisers** | | | | | | | | |
| Nitrogenous (N) | Lakh Tonnes | 80.46 | 84.26 | 87.88 | 95.07 | 98.23 | 103.02 | 109.02 |
| Phosphatic(P) | Lakh Tonnes | 33.21 | 28.43 | 26.69 | 29.32 | 28.98 | 29.77 | 39.14 |
| Potassic(K) | Lakh Tonnes | 13.61 | 8.84 | 9.09 | 11.25 | 11.56 | 10.29 | 13.72 |
| Total (N+P+K) | Lakh Tonnes | 127.28 | 121.53 | 123.66 | 135.64 | 138.77 | 143.08 | 161.88 |
| Per Hectare ** | Kg. | 69.84 | 65.48 | 66.27 | 72.13 | 74.02 | 75.47 | 84.94 |
| **3. Consumption of Pesticides** | | | | | | | | |
| (Technical Grade Material) | Thousand Tonnes | 72.13 | 70.79 | 63.65 | 61.36 | 61.26 | 56.11 | 52.24 |
| 4. Area Covered Under Soil Conservation | | | | | | | | |
| (Cumulative) | Million Hactares | 35.70 | 36.51 | 37.30 | 38.20 | 39.30 | 39.4@ | 37.37@ |

...contd...

...contd...

| Programme | Unit | 1999-2000 | 2000-2001 | 2001-2002 | 2002-2003 | 2003-2004 | 2004-2005 | 2005-2006 | 2006-2007 | 2007-2008* |
|---|---|---|---|---|---|---|---|---|---|---|
| 1 | 2 | 11 | 12 | 13 | 14 | 15 | 16 | 17 | 18 | 19 |
| **1. Seeds** | | | | | | | | | | |
| (i) Production of Breeder Seeds | Thousand Qtls. | 51.13 | 42.69 | 45.54 | 48.42 | 61.82 | 66.46 | 68.64 | 73.83 | 80.08 |
| (ii) Production of Foundation Seeds | Lakh Qtls. | 4.66 | 5.91 | 5.44 | 6.14 | 6.50 | 6.90 | 7.40 | 7.96 | 8.50 |
| (iii) Distribution of Certified/Quality Seeds | Lakh Qtls. | 87.98 | 86.27 | 91.80 | 98.03 | 108.59 | 120.26 | 126.75 | 155.01 | 162.00 |
| **2. Consumption of Chemical Fertilisers** | | | | | | | | | | |
| Nitrogenous (N) | Lakh Tonnes | 115.92 | 109.20 | 113.10 | 104.74 | 110.77 | 117.13 | 127.23 | 137.73 | |
| Phosphatic(P) | Lakh Tonnes | 47.99 | 42.15 | 43.82 | 40.19 | 41.24 | 46.24 | 52.04 | 55.43 | |
| Potassic(K) | Lakh Tonnes | 16.78 | 15.67 | 16.67 | 16.01 | 15.98 | 20.61 | 24.13 | 23.35 | |
| Total (N+P+K) | Lakh Tonnes | 180.69 | 167.02 | 173.60 | 160.94 | 167.99 | 183.98 | 203.40 | 216.51 | |
| Per Hectare ** | Kg. | 94.94 | 89.63 | 91.13 | 91.45 | 88.05 | 94.52 | 104.50 | 112.69 | |
| **3. Consumption of Pesticides** | | | | | | | | | | |
| (Technical Grade Material) | Thousand Tonnes | 46.20 | 43.58 | 47.02 | 48.30 | 41.00 | 40.67 | 39.77 | 37.95 | NA |
| **4. Area Covered Under Soil Conservation** | | | | | | | | | | |
| (Cumulative) | Million Hactares | 39.44@ | 39.47@ | 39.49@ | 39.62@ | 39.64@ | 39.80@ | NA | NA | NA |

E = Estimated

* Targets inrespect of seeds.

NA: Not Available.

** Based on Gross Cropped Area.

@ Excluding state sector soil conservation programme

*Sources:* 1) Department of Fertilizers, New Delhi.

## Table 15.4

*All-India Consumption of Fertilisers in Terms of Nutrients (N, P & K) from 1951-52 to 2006-07*

*('000 Tonnes)*

| *Year* | *N* | *P* | *K* | *Total* |
|---|---|---|---|---|
| *1* | *2* | *3* | *4* | *5* |
| 1951-52 | 58.7 | 6.9 | - | 65.6 |
| 1952-53 | 57.8 | 4.6 | 3.3 | 65.7 |
| 1953-54 | 89.3 | 8.3 | 7.5 | 105.1 |
| 1954-55 | 94.8 | 15.0 | 11.1 | 120.9 |
| 1955-56 | 107.5 | 13.0 | 10.3 | 130.8 |
| 1956-57 | 123.1 | 15.9 | 14.8 | 153.8 |
| 1957-58 | 149.0 | 21.9 | 12.8 | 183.7 |
| 1958-59 | 172.0 | 29.5 | 22.4 | 223.9 |
| 1959-60 | 229.3 | 53.9 | 21.3 | 304.5 |
| 1960-61 | 210.0 | 53.1 | 29.0 | 292.1 |
| 1961-62 | 249.8 | 60.5 | 28.0 | 338.3 |
| 1962-63 | 333.0 | 82.8 | 36.4 | 452.2 |
| 1963-64 | 376.1 | 116.5 | 50.6 | 543.2 |
| 1964-65 | 555.2 | 148.7 | 69.3 | 773.2 |
| 1965-66 | 574.8 | 132.5 | 77.3 | 784.6 |
| 1966-67 | 737.8 | 248.6 | 114.2 | 1100.6 |
| 1967-68 | 1034.6 | 334.8 | 169.6 | 1539.0 |
| 1968-69 | 1208.6 | 382.1 | 170.0 | 1760.7 |
| 1969-70 | 1356.0 | 416.0 | 210.0 | 1982.0 |
| 1970-71 | 1487.0 | 462.0 | 228.0 | 2177.0 |
| 1971-72 | 1798.0 | 558.2 | 300.6 | 2656.8 |
| 1972-73 | 1839.0 | 581.3 | 347.6 | 2767.9 |
| 1973-74 | 1829.0 | 649.7 | 359.8 | 2838.5 |
| 1974-75 | 1765.7 | 471.5 | 336.1 | 2573.3 |
| 1975-76 | 2148.6 | 466.8 | 278.3 | 2893.7 |
| 1976-77 | 2456.9 | 634.9 | 319.2 | 3411.0 |
| 1977-78 | 2913.0 | 866.6 | 506.2 | 4285.8 |
| 1978-79 | 3419.5 | 1105.9 | 591.5 | 5116.9 |
| 1979-80 | 3498.1 | 1150.9 | 606.4 | 5255.4 |
| 1980-81 | 3678.1 | 1213.6 | 623.9 | 5515.6 |
| 1981-82 | 4068.6 | 1322.3 | 673.2 | 6064.1 |
| 1982-83 | 4224.3 | 1437.0 | 727.0 | 21651.0 |

*...contd...*

*...contd...*

| 1 | 2 | 3 | 4 | 5 |
|---|---|---|---|---|
| 1983-84 | 5204.4 | 1730.3 | 775.4 | 7710.1 |
| 1984-85 | 5486.1 | 1886.4 | 838.5 | 8211.0 |
| 1985-86 | 5660.8 | 2005.2 | 808.1 | 8474.1 |
| 1986-87 | 5716.0 | 2078.9 | 850.0 | 8644.9 |
| 1987-88 | 5716.8 | 2187.0 | 880.5 | 8784.3 |
| 1988-89 | 7251.0 | 2720.7 | 1068.3 | 11040.0 |
| 1989-90 | 7386.0 | 3014.2 | 1168.0 | 11568.2 |
| 1990-91 | 7997.2 | 3221.0 | 1328.0 | 12546.2 |
| 1991-92 | 8046.3 | 3321.2 | 1360.5 | 12728.0 |
| 1992-93 | 8426.8 | 2843.8 | 883.9 | 12154.5 |
| 1993-94 | 8788.3 | 2669.3 | 908.4 | 12366.0 |
| 1994-95 | 9507.1 | 2931.7 | 1124.7 | 13563.5 |
| 1995-96 | 9822.8 | 2897.5 | 1155.8 | 13876.1 |
| **1996-97** | 10301.8 | 2976.8 | 1029.6 | 14308.1 |
| **1997-98** | 10901.8 | 3913.6 | 1372.5 | 16187.9 |
| **1998-99** | 11353.8 | 4112.2 | 1331.5 | 16797.5 |
| 1999-2000 | 11592.7 | 4798.3 | 1678.7 | 18069.7 |
| 2000-01 | 10920.2 | 4214.6 | 1567.5 | 16702.3 |
| 2001-02 | 11310.2 | 4382.4 | 1667.1 | 17359.7 |
| 2002.03 | 10474.1 | 4018.8 | 1601.2 | 16094.1 |
| 2003-04 | 11077.0 | 4124.3 | 1597.9 | 16799.1 |
| 2004-05 | 11713.9 | 4623.8 | 2060.6 | 18398.3 |
| 2005-06 | 12723.3 | 5203.7 | 2413.3 | 20340.3 |
| 2006-07 | 13772.9 | 5543.3 | 2334.8 | 21651.0 |

*Note:* Upto 1982-83, figures relate to Feb.-Jan. and from 1983-84 onwards, these relate to April-March.

# Table 15.5

*Consumption, Production and Import of Fertilisers from 1981-82 to 2006-07*

(Lakh Tonnes)

| Year | Consumption | | | | Production | | | | Imports | | | | C.I.F. Value of Imports (Rs. Crore) |
|---|---|---|---|---|---|---|---|---|---|---|---|---|---|
| | N | P | K | TOTAL | N | P | K | TOTAL | N | P | K | TOTAL | |
| 1 | 2 | 3 | 4 | 5 | 6 | 7 | 8 | 9 | 10 | 11 | 12 | 13 | 14 |
| 1981-82 | 40.69 | 13.22 | 6.73 | 60.64 | 31.44 | 9.49 | - | 40.93 | 10.54 | 3.43 | 6.44 | 20.41 | 716.62 |
| 1982-83 | 42.24 | 14.37 | 7.27 | 63.88 | 34.24 | 9.80 | - | 44.04 | 4.25 | 0.63 | 6.44 | 11.32 | 273.53 |
| 1983-84 | 52.05 | 17.30 | 7.75 | 77.10 | 34.85 | 10.48 | - | 45.33 | 6.56 | 1.43 | 5.56 | 13.55 | 365.05 |
| 1984-85 | 54.87 | 18.86 | 8.38 | 82.11 | 39.17 | 12.63 | - | 51.80 | 20.08 | 7.45 | 8.71 | 36.24 | 1500.00 |
| 1985-86 | 56.61 | 20.05 | 8.08 | 84.74 | 43.28 | 14.28 | - | 57.56 | 16.80 | 8.16 | 9.03 | 33.99 | 1405.00 |
| 1986-87 | 57.16 | 20.79 | 8.50 | 86.45 | 54.10 | 16.60 | - | 70.70 | 11.03 | 2.55 | 9.52 | 23.10 | 651.00 |
| 1987-88 | 57.17 | 21.87 | 8.80 | 87.84 | 54.66 | 16.65 | - | 71.31 | 1.75 | - | 8.09 | 9.84 | 223.77 |
| 1988-89 | 72.51 | 27.21 | 10.68 | 110.40 | 67.12 | 22.52 | - | 89.64 | 2.19 | 4.07 | 9.82 | 16.08 | 644.52 |
| 1989-90 | 73.86 | 30.14 | 11.68 | 115.68 | 67.47 | 17.96 | - | 85.43 | 5.23 | 13.11 | 12.80 | 31.14 | 1538.77 |
| 1990-91 | 79.97 | 32.21 | 13.28 | 125.46 | 69.93 | 20.52 | - | 90.45 | 4.14 | 10.16 | 13.28 | 27.58 | 1335.82 |
| 1991-92 | 80.46 | 33.21 | 13.61 | 127.28 | 73.01 | 25.62 | - | 98.63 | 5.66 | 9.67 | 12.36 | 27.69 | 1934.19 |
| 1992-93 | 84.26 | 28.43 | 8.84 | 121.53 | 74.30 | 23.06 | - | 97.36 | 11.37 | 6.89 | 10.82 | 29.08 | 2216.00 |
| 1993-94 | 87.88 | 26.69 | 9.09 | 123.66 | 72.31 | 18.16 | - | 90.47 | 15.88 | 7.22 | 8.57 | 31.67 | 1300.18 |
| 1994-95 | 95.07 | 29.32 | 11.25 | 135.64 | 79.45 | 24.93 | - | 104.38 | 14.76 | 3.80 | 11.09 | 29.65 | 1785.11 |
| 1995-96 | 98.23 | 28.98 | 11.56 | 138.77 | 87.77 | 25.58 | - | 113.35 | 19.93 | 6.47 | 13.15 | 39.55 | 2840.13 |
| 1996-97 | 103.01 | 29.77 | 10.30 | 143.08 | 85.99 | 25.56 | - | 111.55 | 11.67 | 4.26 | 6.13 | 22.06 | 1701.74 |
| 1997-98 | 109.01 | 39.14 | 13.73 | 161.88 | 100.86 | 29.76 | - | 130.62 | 13.62 | 6.72 | 11.40 | 31.74 | 1296.57 |
| 1998-99 | 113.54 | 41.12 | 13.32 | 167.98 | 104.80 | 31.44 | - | 136.24 | 6.35 | 9.68 | 15.42 | 31.45 | 240.00 |

...contd...

...contd...

| 1 | 2 | 3 | 4 | 5 | 6 | 7 | 8 | 9 | 10 | 11 | 12 | 13 | 14 |
|---|---|---|---|---|---|---|---|---|---|---|---|---|---|
| 1999-2000 | 115.92 | 47.99 | 16.78 | 180.69 | 108.90 | 33.99 | - | 142.89 | 8.33 | 15.03 | 17.34 | 40.75 | 197.16 |
| 2000-01 | 109.20 | 42.15 | 15.67 | 167.02 | 109.61 | 37.43 | - | 147.04 | 1.54 | 3.96 | 15.41 | 20.91 | * |
| 2001-02 | 113.10 | 43.82 | 16.67 | 173.59 | 107.68 | 38.60 | - | 146.28 | 2.69 | 4.29 | 17.01 | 23.99 | 113.09 |
| 2002-03 | 104.74 | 40.19 | 16.01 | 160.94 | 105.61 | 39.04 | - | 144.65 | 0.67 | 1.70 | 15.20 | 17.57 | * |
| 2003-04 | 110.77 | 41.24 | 15.98 | 167.99 | 106.34 | 36.32 | - | 142.66 | 1.32 | 3.38 | 15.48 | 20.18 | * |
| 2004-05 | 117.14 | 46.24 | 20.61 | 183.99 | 113.39 | 40.64 | - | 154.03 | 4.11 | 2.96 | 20.45 | 27.52 | 684.73 |
| 2005-06 | 127.23 | 52.04 | 24.13 | 203.40 | 113.54 | 42.21 | - | 155.75 | 13.85 | 11.21 | 27.47 | 52.53 | 1747.34 |
| 2006-07 | 137.73 | 55.43 | 23.35 | 216.51 | 115.78 | 45.18 | | 160.96 | 26.82 | 13.07 | 20.69 | 60.58 | 4647.18 |

* There was no import of Urea in 2000-01, 2002-03 and 2003-04 in Government account.

*Note:* 1. Figures relate to imports made on Govt. Account only.

2. The imports made after decanalisation of phosphatic fertilisers (w.e.f. 17.9.1992) and potassic fertilisers (w.e.f. 17.6.1993) include the quantities imported by private parties also.

*Source:* Department of Fertilisers.

## State-wise Consumption of Fertilisers in 2004-05 and 2005-06, 2006-07

(Tonnes)

| Sl. No. | State/Zone | 2004-05 | | | | 2005-06 | | | | 2006-07 | | | |
|---|---|---|---|---|---|---|---|---|---|---|---|---|---|
| | | N | P | K | Total | N | P | K | Total | N | P | K | Total |
| 1 | 2 | 3 | 4 | 5 | 6 | 7 | 8 | 9 | 10 | 11 | 12 | 13 | 14 |
| **South Zone** | | | | | | | | | | | | | |
| 1 | Andhra Pradesh | 115653 | 53898 | 29236 | 198787 | 152209 | 69010 | 34036 | 255255 | 146608 | 68593 | 33205 | 248406 |
| 2 | Karnataka | 65598 | 36398 | 27327 | 129323 | 75444 | 43514 | 33534 | 152492 | 75616 | 43836 | 29131 | 148583 |
| 3 | Kerala | 8825 | 4179 | 7149 | 20153 | 8249 | 4424 | 7573 | 20246 | 8866 | 4561 | 7465 | 20892 |
| 4 | Tamil Nadu | 48273 | 21132 | 25798 | 95203 | 55924 | 25612 | 28386 | 109922 | 58498 | 26963 | 27035 | 112496 |
| 5 | Pondicherry | 2288 | 1009 | 907 | 4204 | 2229 | 1090 | 939 | 4258 | 2518 | 1140 | 756 | 4414 |
| 6 | A&N Islands | 27 | 17 | 7 | 51 | 36 | 14 | 9 | 59 | 29 | 28 | 4 | 61 |
| 7 | Lakshadweep | 0 | 0 | 0 | 0 | 0 | 0 | 0 | 0 | 0 | 0 | 0 | 0 |
| Sub-Total (South Zone) | | 240664 | 116633 | 90424 | 447721 | 294091 | 143664 | 104477 | 542232 | 292135 | 145121 | 97596 | 534852 |
| **West Zone** | | | | | | | | | | | | | |
| 8 | Gujarat | 75400 | 29626 | 9622 | 114648 | 83473 | 32846 | 11673 | 127992 | 92757 | 36113 | 12009 | 140879 |
| 9 | Madhya Pradesh | 61772 | 39325 | 5530 | 106627 | 55993 | 32211 | 5881 | 94085 | 73006 | 40982 | 6523 | 120511 |
| 10 | Chhattisgarh | 23036 | 9687 | 3504 | 36227 | 23209 | 10440 | 3783 | 37432 | 27222 | 11600 | 4801 | 43623 |
| 11 | Maharashtra | 95404 | 52674 | 25843 | 173921 | 106152 | 57761 | 32925 | 196838 | 120945 | 67750 | 37202 | 225897 |
| 12 | Rajasthan | 54185 | 20314 | 1633 | 76132 | 61981 | 24239 | 1986 | 88206 | 66489 | 25825 | 1299 | 93613 |
| 13 | Goa | 255 | 148 | 173 | 576 | 248 | 140 | 164 | 552 | 285 | 143 | 167 | 595 |
| 14 | Daman & Diu | 1 | 0 | 0 | 1 | 0 | 0 | 0 | 0 | 42 | 6 | 1 | 49 |
| 15 | D&N Haveli | 67 | 35 | 4 | 106 | 69 | 38 | 6 | 113 | 72 | 47 | 6 | 125 |
| Sub-Total (West Zone) | | 310120 | 151809 | 46309 | 508238 | 331125 | 157675 | 56418 | 545218 | 380818 | 182466 | 62008 | 625292 |
| **North Zone** | | | | | | | | | | | | | |
| 16 | Haryana | 78940 | 24103 | 1959 | 105002 | 84743 | 25257 | 2867 | 112867 | 86264 | 24411 | 1822 | 112497 |

...contd...

...contd...

| 1 | 2 | 3 | 4 | 5 | 6 | 7 | 8 | 9 | 10 | 11 | 12 | 13 | 14 |
|---|---|---|---|---|---|---|---|---|---|---|---|---|---|
| 17 | Punjab | 120223 | 31719 | 4322 | 156264 | 125459 | 36946 | 6310 | 168715 | 129900 | 35361 | 3859 | 169120 |
| 18 | Uttar Pradesh | 238933 | 73938 | 18149 | 331020 | 243867 | 82375 | 20184 | 346426 | 271704 | 84744 | 16160 | 372608 |
| 19 | Uttaranchal | 8499 | 2124 | 764 | 11387 | 8832 | 2369 | 867 | 12068 | 10851 | 2561 | 902 | 14314 |
| 20 | Himachal Pradesh | 3069 | 853 | 703 | 4625 | 3037 | 974 | 786 | 4797 | 3079 | 1022 | 797 | 4898 |
| 21 | Jammu & Kashmir | 5046 | 2160 | 317 | 7523 | 6178 | 2585 | 463 | 9226 | 5549 | 2373 | 681 | 8603 |
| 22 | Delhi | 54 | 2 | 0 | 56 | 45 | 0 | 0 | 45 | 82.00 | 21.00 | 1.00 | 104 |
| 23 | Chandigarh | 1 | 0 | 0 | 1 | 0 | 0 | 0 | 0 | 0 | 0 | 0 | 0 |
| Sub-Total (North Zone) | | 454765 | 134899 | 26214 | 615878 | 472161 | 150506 | 31477 | 654144 | 507429 | 150493 | 24222 | 682144 |
| **East Zone** | | | | | | | | | | | | | |
| 24 | Bihar @ | 61880 | 7090 | 4198 | 73168 | 68827 | 13196 | 9873 | 91896 | 80937 | 17907 | 8325 | 107169 |
| 25 | Jharkhand | 7616 | 4296 | 437 | 12349 | 8621 | 4106 | 718 | 13445 | 9220 | 4200 | 431 | 13851 |
| 26 | Orissa | 22354 | 7799 | 5377 | 35530 | 24321 | 9105 | 6062 | 39488 | 25654 | 9277 | 5357 | 40288 |
| 27 | West Bengal | 63094 | 33962 | 29089 | 126145 | 61139 | 35778 | 27051 | 123968 | 67843 | 38625 | 30047 | 136515 |
| Sub-Total (East Zone) | | 154944 | 53147 | 39101 | 247192 | 162908 | 62185 | 43704 | 268797 | 183654 | 70009 | 44160 | 297823 |
| **North East (NE) Zone** | | | | | | | | | | | | | |
| 28 | Assam | 7794 | 5092 | 3704 | 16590 | 9351 | 5604 | 4857 | 19812 | 10261 | 5108 | 4999 | 20368 |
| 29 | Tripura | 933 | 258 | 153 | 1344 | 993 | 290 | 234 | 1517 | 1019 | 385 | 250 | 1654 |
| 30 | Manipur | 1705 | 240 | 95 | 2040 | 1209 | 138 | 73 | 1420 | 1418 | 359 | 130 | 1907 |
| 31 | Meghalaya | 285 | 190 | 18 | 493 | 299 | 170 | 22 | 491 | 295 | 194 | 24 | 513 |
| 32 | Nagaland | 31 | 22 | 7 | 60 | 33 | 22 | 7 | 62 | 32 | 24 | 8 | 64 |
| 33 | Arunachal Pradesh | 45 | 20 | 11 | 76 | 44 | 20 | 11 | 75 | 45 | 19 | 9 | 73 |
| 34 | Mizoram | 61 | 51 | 30 | 142 | 97 | 83 | 48 | 228 | 160 | 138 | 72 | 370 |
| 35 | Sikkim | 44 | 18 | 0 | 62 | 21 | 11 | 3 | 35 | 21 | 11 | 33 | 35 |
| Sub-Total (NE Zone) | | 10898 | 5891 | 4018 | 20807 | 12047 | 6338 | 5255 | 23640 | 13251 | 6238 | 5495 | 24984 |
| All India | | 1171391 | 462379 | 206066 | 1839836 | 1272332 | 520368 | 241331 | 2034031 | 1377287 | 554327 | 233481 | 2165095 |

*Source:* INM Division, Ministry of Agriculture.

# Table 15.6(b)

*State-wise Estimated Consumption of Fertiliser per Hectare during, 2004-05, 2005-06 and 2006-07*

(Kgs./Hectare)

| Sl. No. | State/Zone | 2004-05 | | | | 2005-06 | | | | 2006-07 | | | |
|---|---|---|---|---|---|---|---|---|---|---|---|---|---|
| | | N | P | K | Total | N | P | K | Total | N | P | K | Total |
| 1 | 2 | 3 | 4 | 5 | 6 | 7 | 8 | 9 | 10 | 11 | 12 | 13 | 14 |
| **South Zone** | | | | | | | | | | | | | |
| 1 | Andhra Pradesh | 92.25 | 42.99 | 23.32 | 158.57 | 121.41 | 55.05 | 27.15 | 203.61 | 130.88 | 61.23 | 29.64 | 221.75 |
| 2 | Karnataka | 50.48 | 28.01 | 21.03 | 99.51 | 58.05 | 33.48 | 25.80 | 117.34 | 58.05 | 33.65 | 22.36 | 114.06 |
| 3 | Kerala | 24.85 | 11.77 | 20.13 | 56.74 | 23.22 | 12.45 | 21.32 | 57.00 | 29.69 | 15.28 | 25.00 | 69.97 |
| 4 | Tamil Nadu | 80.66 | 35.31 | 43.11 | 159.07 | 93.44 | 42.80 | 47.43 | 183.67 | 96.97 | 44.69 | 44.81 | 186.47 |
| 5 | Pondicherry | 591.21 | 260.72 | 234.37 | 1086.30 | 575.97 | 281.65 | 242.64 | 1100.26 | 662.63 | 300.00 | 198.95 | 1161.58 |
| 6 | A&N Islands | 5.78 | 3.64 | 1.50 | 10.92 | 7.71 | 3.00 | 1.93 | 12.63 | 6.30 | 6.09 | 0.87 | 13.26 |
| 7 | Lakshadweep | 0.00 | 0.00 | 0.00 | 0.00 | 0.00 | 0.00 | 0.00 | 0.00 | 0.00 | 0.00 | 0.00 | 0.00 |
| Average (South Zone) | | 68.45 | 33.17 | 25.72 | 127.35 | 83.65 | 40.86 | 29.72 | 154.23 | 87.64 | 43.54 | 29.28 | 160.46 |
| **West Zone** | | | | | | | | | | | | | |
| 8 | Gujarat | 65.43 | 25.71 | 8.35 | 99.49 | 72.44 | 28.50 | 10.13 | 111.07 | 74.54 | 29.02 | 9.65 | 113.21 |
| 9 | Madhya Pradesh | 30.95 | 19.70 | 2.77 | 53.42 | 28.05 | 16.14 | 2.95 | 47.13 | 37.98 | 21.32 | 3.39 | 62.69 |
| 10 | Chhattisgarh | 41.45 | 17.43 | 6.31 | 65.19 | 41.76 | 18.79 | 6.81 | 67.36 | 42.59 | 18.15 | 7.51 | 68.25 |
| 11 | Maharashtra | 40.97 | 22.62 | 11.10 | 74.68 | 45.58 | 24.80 | 14.14 | 84.52 | 53.62 | 30.04 | 16.49 | 100.15 |
| 12 | Rajasthan | 22.29 | 8.36 | 0.67 | 31.33 | 25.50 | 9.97 | 0.82 | 36.29 | 31.00 | 12.04 | 0.61 | 43.65 |
| 13 | Goa | 15.09 | 8.76 | 10.24 | 34.08 | 14.67 | 8.28 | 9.70 | 32.66 | 16.76 | 8.41 | 9.82 | 34.99 |
| 14 | Daman & Diu | 3.33 | 0.00 | 0.00 | 3.33 | 0.00 | 0.00 | 0.00 | 0.00 | 139.67 | 20.33 | 3.33 | 163.33 |
| 15 | D&N Haveli | 26.07 | 13.62 | 1.56 | 41.25 | 26.85 | 14.79 | 2.33 | 43.97 | 25.71 | 16.79 | 2.14 | 44.64 |
| Average (West Zone) | | 36.56 | 17.90 | 5.46 | 59.91 | 39.03 | 18.59 | 6.65 | 64.27 | 46.29 | 22.18 | 7.54 | 76.01 |
| 16 | Haryana | 116.60 | 35.60 | 2.89 | 155.10 | 125.18 | 37.31 | 4.23 | 166.72 | 135.85 | 38.44 | 2.87 | 177.16 |
| 17 | Punjab | 149.69 | 39.49 | 5.38 | 194.56 | 156.20 | 46.00 | 7.86 | 210.06 | 160.67 | 43.74 | 4.77 | 209.18 |

...contd...

*...contd...*

| 1 | 2 | 3 | 4 | 5 | 6 | 7 | 8 | 9 | 10 | 11 | 12 | 13 | 14 |
|---|---|---|---|---|---|---|---|---|---|---|---|---|---|
| 18 | Uttar Pradesh | 96.82 | 29.96 | 7.35 | 134.13 | 98.82 | 33.38 | 8.18 | 140.37 | 111.84 | 34.88 | 6.65 | 153.37 |
| 19 | Uttaranchal | 66.37 | 16.59 | 5.97 | 88.93 | 68.97 | 18.50 | 6.77 | 94.24 | 89.56 | 21.14 | 7.44 | 118.14 |
| 20 | Himachal Pradesh | 31.19 | 8.67 | 7.14 | 47.00 | 30.86 | 9.90 | 7.99 | 48.75 | 32.76 | 10.87 | 8.48 | 52.11 |
| 21 | Jammu & Kashmir | 44.47 | 19.04 | 2.79 | 66.30 | 54.45 | 22.78 | 4.08 | 81.31 | 48.22 | 20.62 | 5.92 | 74.76 |
| 22 | Delhi | 12.62 | 0.47 | 0.00 | 13.08 | 10.51 | 0.00 | 0.00 | 10.51 | 18.64 | 4.77 | 0.00 | 23.41 |
| 23 | Chandigarh | 9.09 | 0.00 | 0.00 | 9.09 | 0.00 | 0.00 | 0.00 | 0.00 | 0.00 | 0.00 | 0.00 | 0.00 |
| Average (North Zone) | | 105.95 | 31.43 | 6.11 | 143.48 | 110.00 | 35.06 | 7.33 | 152.40 | 120.60 | 35.77 | 5.76 | 162.13 |
| **East Zone** | | | | | | | | | | | | | |
| 24 | Bihar | 84.38 | 9.67 | 5.72 | 99.78 | 93.86 | 17.99 | 13.46 | 125.32 | 101.85 | 22.53 | 10.48 | 134.86 |
| 25 | Jharkhand | 38.30 | 21.60 | 2.20 | 62.10 | 43.35 | 20.65 | 3.61 | 67.61 | 34.94 | 15.92 | 1.63 | 52.49 |
| 26 | Orissa | 32.46 | 11.32 | 7.81 | 51.59 | 35.31 | 13.22 | 8.80 | 57.33 | 29.73 | 10.75 | 6.21 | 46.69 |
| 27 | West Bengal | 64.89 | 34.93 | 29.92 | 129.73 | 62.88 | 36.80 | 27.82 | 127.50 | 71.17 | 40.52 | 31.52 | 143.21 |
| Average (East Zone) | | 59.75 | 20.49 | 15.08 | 95.32 | 62.82 | 23.98 | 16.85 | 103.65 | 63.89 | 24.35 | 15.36 | 103.60 |
| **North-East (NE) Zone** | | | | | | | | | | | | | |
| 28 | Assam | 19.38 | 12.66 | 9.21 | 41.25 | 23.25 | 13.93 | 12.08 | 49.26 | 26.95 | 13.41 | 13.13 | 53.49 |
| 29 | Tripura | 24.11 | 6.67 | 3.95 | 34.74 | 25.67 | 7.50 | 6.05 | 39.21 | 21.23 | 8.02 | 5.21 | 34.46 |
| 30 | Manipur | 71.85 | 10.11 | 4.00 | 85.97 | 50.95 | 5.82 | 3.08 | 59.84 | 50.46 | 12.78 | 4.63 | 67.87 |
| 31 | Meghalaya | 10.44 | 6.96 | 0.66 | 18.05 | 10.95 | 6.22 | 0.81 | 17.98 | 11.43 | 7.52 | 0.01 | 18.96 |
| 32 | Nagaland | 0.75 | 0.53 | 0.17 | 1.46 | 0.80 | 0.53 | 0.17 | 1.50 | 0.01 | 0.01 | 0.00 | 0.02 |
| 33 | Arunachal Pradesh | 1.77 | 0.79 | 0.43 | 2.98 | 1.73 | 0.79 | 0.43 | 2.94 | 1.69 | 0.01 | 0.00 | 1.70 |
| 34 | Mizoram | 6.81 | 5.69 | 3.35 | 15.85 | 10.83 | 9.26 | 5.36 | 25.45 | 16.49 | 14.23 | 7.42 | 38.14 |
| 35 | Sikkim | 3.56 | 1.46 | 0.00 | 5.01 | 1.70 | 0.89 | 0.24 | 2.83 | 1.71 | 0.01 | 0.00 | 1.72 |
| Average (NE Zone) | | 18.79 | 10.16 | 6.93 | 35.88 | 20.77 | 10.93 | 9.06 | 40.76 | 23.24 | 10.94 | 9.64 | 43.82 |
| **All India** (Average) | | 60.18 | 23.76 | 10.59 | 94.52 | 65.37 | 26.73 | 12.40 | 104.50 | 71.69 | 28.85 | 12.15 | 112.69 |

*Source:* INM Division, Ministry of Agriculture.

## Table 15.7

*Fertiliser Consumption per Hectare of Agricultural Land in Selected Countries in 2003-04 and 2004-05*

*(Kg./Hectare)*

| *Country* | *2003-04* | | | | *2004-05* | | | |
|---|---|---|---|---|---|---|---|---|
| | *N* | *P* | *K* | *Total* | *N* | *P* | *K* | *Total* |
| *1* | *6* | *7* | *8* | *9* | *10* | *11* | *12* | *13* |
| **Africa** | | | | | | | | |
| Egypt | 468.9 | 52.3 | 13.9 | 535.1 | 476.2 | 67.6 | 11.3 | 555.1 |
| Morocco | 6.8 | 7.2 | 2.4 | 16.4 | 8.7 | 8.4 | 2.3 | 19.4 |
| South Africa | 5.2 | 1.7 | 1.1 | 8.0 | 4.1 | 2.9 | 1.4 | 8.4 |
| **North & Central America** | | | | | | | | |
| Canada | 24.4 | 10.2 | 5.2 | 39.8 | 22.8 | 9.0 | 4.9 | 36.7 |
| Mexico | 11.2 | 4.5 | 1.7 | 17.4 | 10.9 | 4.6 | 2.3 | 17.8 |
| USA | 28.3 | 10.5 | 12.0 | 50.8 | 27.0 | 10.1 | 11.3 | 48.4 |
| **South America** | | | | | | | | |
| Brazil | 8.9 | 11.8 | 14.8 | 35.5 | 8.3 | 13.0 | 16.1 | 37.4 |
| Chile | 14.5 | 12.2 | 5.8 | 32.5 | 22.3 | 11.5 | 8.0 | 41.8 |
| **Asia** | | | | | | | | |
| Bangladesh | 121.7 | 35.4 | 18.0 | 175.1 | 128.6 | 37.4 | 18.5 | 184.5 |
| China, Main | 51.6 | 17.6 | 8.9 | 78.1 | 51.9 | 18.3 | 10.5 | 80.7 |
| **India** | 61.5 | 22.9 | 8.9 | 93.3 | 65.0 | 25.7 | 11.4 | 102.1 |
| Japan | 119.9 | 133.0 | 112.9 | 365.8 | 116.0 | 137.4 | 109.5 | 362.9 |
| Jordan | 56.1 | 105.4 | - | 161.5 | 20.8 | 37.5 | 37.8 | 96.1 |
| Korea Rep. | 188.7 | 75.4 | 92.5 | 356.6 | 208.3 | 82.9 | 103.5 | 394.7 |
| Nepal | 4.2 | 4.2 | 0.3 | 8.7 | 3.6 | 2.7 | 0.3 | 6.6 |
| Pakistan | 91.0 | 28.9 | 2.2 | 122.1 | 92.6 | 26.6 | | 119.2 |
| Sri Lanka | 64.0 | 16.0 | 26.0 | 106.0 | 69.3 | 14.8 | 25.8 | 109.9 |
| **Europe** | | | | | | | | |
| Austria | 54.1 | 18.8 | 22.5 | 95.4 | 59.5 | 18.7 | 18.5 | 96.7 |
| Belarus | 36.1 | 7.9 | 53.8 | 97.8 | 47.7 | 12.1 | 55.6 | 115.4 |
| Denmark | 82.1 | 5.9 | 28.3 | 116.3 | 80.3 | 13.0 | 36.7 | 130.0 |
| France | 80.0 | 25.1 | 33.0 | 138.1 | 80.1 | 26.0 | 33.1 | 139.2 |
| Germany | 107.5 | 24.1 | 28.6 | 160.2 | 104.5 | 21.1 | 28.1 | 153.7 |
| Netherlands | 150.7 | 28.0 | 104.3 | 283.0 | 142.4 | 29.6 | 77.4 | 249.4 |
| Norway | 100.2 | 12.1 | 44.5 | 156.8 | 101.4 | 12.4 | 44.7 | 158.5 |
| Poland | 38.7 | 19.3 | 28.1 | 86.1 | 53.4 | 24.8 | 37.9 | 116.1 |
| Russian Federation | 2.4 | 1.4 | 2.2 | 6.0 | 3.6 | 1.5 | 3.1 | 8.2 |
| Spain | 38.8 | 20.2 | 16.6 | 75.6 | 34.9 | 18.9 | 17.3 | 71.1 |
| Sweden | 67.3 | 15.2 | 16.9 | 99.4 | 53.9 | 19.4 | 17.2 | 90.5 |
| United Kingdom | 65.0 | 15.9 | 22.0 | 102.9 | 66.8 | 16.0 | 22.3 | 105.1 |
| Ukraine | 8.0 | 1.0 | 1.8 | 10.8 | 8.5 | 0.5 | 2.5 | 11.5 |
| **Oceania** | | | | | | | | |
| Australia | 2.1 | 2.3 | 0.5 | 4.9 | 2.4 | 2.5 | 0.6 | 5.5 |
| New Zealand | 20.1 | 25.6 | 9.1 | 54.8 | 22.5 | 28.3 | 9.2 | 60.0 |
| World | 17.5 | 7.0 | 5.2 | 29.7 | 18.2 | 7.5 | 5.6 | 31.3 |

*Source:* Fertiliser Association of India, New Delhi.

# Table 15.8

*Mineral Fertilisers Consumption in Countries of Asia-Pacific Region*

*('000 Tonnes of plant nutrient)*

| S.No. | Country | 1995 | 2002 | 2003 | 2004 | 2005 | Annual Growth Rate % 1995-2005 |
|---|---|---|---|---|---|---|---|
| 1 | 2 | 3 | 4 | 5 | 6 | 7 | 8 |
| **Developing Countries** | | | | | | | |
| 1. | Cambodia | 9.8 | 21.6 | 14.3 | 19.9 | 0.0 | 0.0 |
| 2. | Indonesia | 2529.2 | 3165.1 | 2884.1 | 3673.4 | 3442.8 | 3.1 |
| 4. | Malaysia | 1092.0 | 1189.9 | 1324.4 | 1767.2 | 1517.8 | 3.3 |
| 5. | Myanmar | 180.0 | 16.1 | 22.0 | 7.7 | 2.1 | -35.9 |
| 6. | Philippines | 598.4 | 722.3 | 938.4 | 948.8 | 812.7 | 3.1 |
| 7. | Thailand | 1507.0 | 1700.8 | 2267.5 | 2008.7 | 1724.0 | 1.4 |
| 8. | Viet Nam | 1223.7 | 2012.8 | 2252.7 | 2307.5 | 1984.9 | 5.0 |
| 9. | Bangladesh | 1195.0 | 1556.8 | 1578.5 | 1663.9 | 1757.3 | 3.9 |
| 11. | India | 13876.2 | 16114.9 | 16092.8 | 17903.6 | 20588.3 | 4.0 |
| 12. | Maldives | 0.0 | 0.0 | 0.0 | 0.0 | 0.1 | - |
| 13. | Nepal | 93.7 | 55.6 | 37.0 | 27.7 | 22.7 | -13.2 |
| 14. | Pakistan | 2515.1 | 3043.0 | 3324.5 | 3221.6 | 4069.3 | 4.9 |
| 15. | Sri Lanka | 206.0 | 285.1 | 249.4 | 258.8 | 281.3 | 3.2 |
| 16. | China | 35580.2 | 44186.1 | 43356.7 | 44768.4 | 48866.2 | 3.2 |
| 17. | DPR Korea | 104.2 | - | - | - | - | - |
| 18. | Iran (Islamic Rep.of) | 1017.2 | 1134.8 | 1134.8 | 405.7 | 1260.1 | 2.2 |
| 19. | Mongolia | 2.0 | 4.4 | 3.7 | 5.5 | 4.2 | 7.8 |
| 20. | Rep. of Korea | 979.0 | 689.9 | 678.2 | 747.2 | 722.4 | -3.0 |
| 21. | Cook Islands | - | - | - | - | - | |
| 22. | Fiji Islands | 18.0 | 12.2 | 3.3 | 7.7 | 6.7 | -9.5 |
| 23. | Papua New Guinea | - | - | - | - | - | |
| 24. | Samoa | 0.0 | 0.0 | 0.0 | 0.0 | 0.1 | |
| 25. | Solomon Islands | - | - | - | - | - | |
| 26. | Tonga | - | - | - | - | - | |
| 27. | Vanuatu | - | - | - | - | - | |
| | Sub-Total | 62849.9 | 76044.2 | 76314.7 | 79940.3 | 87235.4 | 3.3 |
| **Developed Countries** | | | | | | | |
| 28. | Australia | 1867.4 | 2265.2 | 2173.1 | 2431.5 | 2215.3 | 1.7 |
| 29. | Japan | 1641.1 | 1734.0 | 1732.3 | 1711.2 | 1176.5 | -3.3 |
| 30. | New Zealand | 682.6 | 874.1 | 945.8 | 1033.6 | 1053.9 | 4.4 |
| | Sub-Total | 4191.1 | 4873.4 | 4851.2 | 5176.3 | 4445.8 | 0.6 |
| Asia-Pacific Total | | 67041.0 | 80917.6 | 81165.9 | 85116.6 | 91681.1 | 3.2 |
| Rest of World | | 62640.4 | 64222.4 | 70631.6 | 70971.5 | 67498.8 | 0.7 |
| World | | 129681.4 | 145140.1 | 151797.5 | 156088.1 | 159179.9 | 2.1 |

*Source:* FAO Regional Office for Asia and the Pacific, Bangkok.

# Table 15.9

*Economics of Application of Fertilisers on Paddy and Wheat in India from 1995-96 to 2006-07*

(Quantity in kgs.)

| Particulars | 1995-96 | | 1996-97 | | 1997-98 | 1998-99 | 1999-00 | | 2000-01 | 2001-02 | | 2002-03 | 2003-04 | 2004-05 | 2005-06 | 2006-07 |
|---|---|---|---|---|---|---|---|---|---|---|---|---|---|---|---|---|
| | Kharif | Rabi | Kharif | Rabi | | | Prior to 29.02 2000 | w.e.f. 29.02 2000 | | Prior to 28.02 2002 | w.e.f. 28.02 2002 | | | | | |
| 1 | 2 | 3 | 4 | 5 | 6 | 7 | 8 | 9 | 10 | 11 | 12 | 13 | 14 | 15 | 16 | 17 |
| **I. Paddy** | | | | | | 198** | | | | | | | | | | |
| Quantity of Paddy required to Buy | | | | | | | | | | | | | | | | |
| a. 1 kg of N | 2.01 | 2.01 | 1.90 | 1.90 | 1.92 | 1.81 | 1.78 | 2.04 | 1.96 | 1.89 | 1.98 | 1.98 | 1.91 | 1.88 | 1.84 | 1.81 |
| b. 1 kg of $P_2O_5$ | | | | | | | | | | | | | | | | |
| - as SSP | 3.97 to 4.91 | 4.51 to 5.06 | 3.67 to 5.07 | 4.28 to 5.07 | 3.77 to 4.52 | 3.55 to 4.26 | 3.19 to 3.83 | 3.19 to 3.83 | 3.06 to 4.29 | 2.95 to 4.13 | 2.95 to 4.13 | 3.07 to 4.13 | 2.95 to 4.35 | 2.90 to 4.18 | 2.85 to 4.63 | 2.91 to 4.55 |
| - as DAP | 4.71 to 5.13 | 5.03 to 5.40 | 3.59 to 4.26 | 3.93 to 4.46 | 3.60 | 3.39 | 2.99 | 3.15 | 3.03 | 2.91 | 3.06 | 3.06 | 2.95 | 2.90 | 2,85 | 2.8 |
| - as complex | 5.46 to 8.66 | 5.59 to 8.36 | 3.56 to 7.88 | 4.41 to 8.09 | 3.80 to 6.80 | 3.59 to 6.42 | 3,16 to 5.61 | 3.33 to 5.73 | 3.20 to 5.51 | 3.08 to 5.30 | 3.32 to 5.61 | 3.32 to 5.61 | 3.19 to 5.40 | 3.14 to 5.31 | 3.08 to 5.21 | 3.03 to 5.12 |
| c. 1 kg of $K_2O$ | 1.68 1.35 to 2.10 | 1.94 1.33 to 2.22 | 1.63 1.30 to 1.89 | 1.74 1.28 to 1.97 | 1.49 | 1.40 | 1.26 | 1.45 | 1.39 | 1.34 | 1.40 | 1.40 | 1.35 | 1.33 | 1.30 | 1.28 |

...contd...

...contd...

| 1 | 2 | 3 | 4 | 5 | 6 | 7 | 8 | 9 | 10 | 11 | 12 | 13 | 14 | 15 | 16 | 17 |
|---|---|---|---|---|---|---|---|---|---|---|---|---|---|---|---|---|
| **II. Wheat** | | | | | | 1.71** | | | | | | | | | | |
| Quantity of Wheat required to Buy | 1.90 | 1.90 | 1.52 | 1.52 | 1.56 | 1.45 | 1.50 | 1.72 | 1.64 | 1.61 | 1.69 | 1.69 | 1.67 | 1.64 | 1.62 | 1.40 |
| a. 1 kg of N | | | | | | | | | | | | | | | | |
| b. 1 kg of $P_2O_5$ | | | | | | | | | | | | | | | | |
| - as SSP | 3.76 to 4.65 | 4.28 to 4.79 | 2.93 to 4.05 | 3.42 to 4.05 | 3.06 to 3.68 | 2.84 to 3.41 | 2.69 to 3.23 | 2.69 to 3.23 | 2.56 to 3.59 | 2.52 to 3.53 | 2.52 to 3.53 | 2.62 to 3.53 | 2.58 to 3.80 | 2.54 to 3.65 | 2.50 to 4.06 | 2.25 to 3.52 |
| - as DAP | 4.46 to 4.86 | 4.77 to 5.12 | 2.87 to 3.41 | 3.14 to 3.57 | 2.93 | 2.71 | 2.52 | 2.66 | 2.53 | 2.49 | 2.62 | 2.62 | 2.57 | 2.53 | 2.50 | 2.16 |
| - as complex | 5.18 to 8.21 | 5.29 to 7.92 | 2.85 to 6.30 | 3.53 to 6.47 | 3.09 to 5.54 | 2.87 to 5.13 | 2.67 to 4.74 | 2.81 to 4.84 | 2.67 to 4.60 | 2.63 to 4.53 | 2.83 to 4.79 | 2.83 to 4.79 | 2.79 to 4.72 | 2.75 to 4.64 | 2.70 to 4.57 | 2.34 to 3.96 |
| c. 1 kg of $K_2O$ | 1.59 to 1.99 | 1.84 to 2.11 | 1.30 to 1.51 | 1.39 to 1.58 | 1.21 | 1.12 | 1.06 | 1.22 | 1.16 | 1.14 | 1.20 | 1.20 | 1.18 | 1.16 | 1.14 | 0.99 |

** With effect from 29-1-1999

*Source:* The Fertiliser Association of India.

## Table 15.10

*Crop-wise Distribution of Certified/Quality Seeds from 1991-92 to 2006-2007*

*(Lakh Quintals)*

| *Crops* | *1991-1992* | *1992-1993* | *1993-1994* | *1994-1995* | *1995-1996* | *1996-1997* | *1997-1998* | *1998-1999* | *1999-2000* | *2000-2001* | *2001-2002* | *2002-2003* | *2003-2004* | *2004-2005* | *2005-2006* | *2006-2007* |
|---|---|---|---|---|---|---|---|---|---|---|---|---|---|---|---|---|
| *1* | *2* | *3* | *4* | *5* | *6* | *7* | *8* | *9* | *10* | *11* | *12* | *13* | *14* | *15* | *16* | *17* |
| **Cereals** | | | | | | | | | | | | | | | | |
| Wheat | 14.04 | 15.15 | 18.23 | 20.21 | 22.41 | 23.22 | 24.42 | 26.14 | 29.52 | 27.04 | 32.59 | 33.00 | 35.97 | 41.31 | 44.21 | 54.55 |
| Paddy | 14.47 | 14.28 | 13.58 | 14.63 | 15.37 | 16.57 | 20.22 | 24.05 | 23.71 | 24.41 | 25.58 | 25.79 | 26.51 | 30.98 | 32.41 | 43.51 |
| Maize | 1.50 | 1.50 | 1.35 | 1.35 | 1.55 | 1.88 | 2.36 | 2.26 | 2.72 | 3.11 | 2.75 | 3.42 | 3.38 | 4.13 | 4.64 | 5.74 |
| *Jowar* | 3.46 | 3.91 | 3.69 | 3.25 | 2.73 | 2.73 | 2.83 | 2.57 | 2.67 | 2.39 | 2.20 | 2.15 | 2.37 | 2.31 | 2.28 | 2.32 |
| *Bajra* | 1.68 | 1.69 | 1.68 | 1.61 | 1.66 | 1.71 | 1.63 | 1.87 | 2.00 | 1.94 | 1.80 | 1.79 | 1.76 | 1.76 | 2.17 | 2.16 |
| Ragi | 0.15 | 0.14 | 0.15 | 0.24 | 0.23 | 0.19 | 0.19 | 0.19 | 0.20 | 0.19 | 0.14 | 0.24 | 0.21 | 0.25 | 0.22 | 0.21 |
| Barley | 0.05 | 0.05 | 0.06 | 0.06 | 0.08 | 0.13 | 0.13 | 0.19 | 0.32 | 0.39 | 0.50 | 0.58 | 0.62 | 0.67 | 0.80 | 1.08 |
| Sub-Total (Cereals) | 35.35 | 36.72 | 38.74 | 41.35 | 44.03 | 46.43 | 51.78 | 57.27 | 61.14 | 59.47 | 65.56 | 66.97 | 70.82 | 81.41 | 86.73 | 109.87 |
| **Pulses** | | | | | | | | | | | | | | | | |
| Gram | 1.03 | 1.02 | 1.11 | 1.14 | 1.22 | 1.44 | 1.06 | 1.10 | 1.19 | 1.09 | 1.53 | 2.64 | 3.90 | 3.62 | 3.45 | 5.08 |
| Lentil | 0.08 | 0.07 | 0.08 | 0.05 | 0.05 | 0.08 | 0.11 | 0.13 | 0.07 | 0.13 | 0.29 | 0.26 | 0.35 | 0.52 | 0.45 | 0.54 |
| Peas | 0.31 | 0.27 | 0.23 | 0.27 | 0.27 | 0.25 | 0.30 | 0.25 | 0.29 | 0.26 | 0.31 | 0.37 | 0.60 | 0.61 | 0.82 | 0.93 |
| *Urad* | 0.65 | 0.79 | 0.79 | 0.75 | 0.69 | 1.07 | 0.96 | 1.02 | 0.86 | 0.88 | 0.97 | 1.43 | 1.32 | 1.01 | 0.92 | 0.8 |
| *Moong* | 0.62 | 0.63 | 0.77 | 0.68 | 0.63 | 0.65 | 0.69 | 0.86 | 0.80 | 0.78 | 0.83 | 0.81 | 1.21 | 0.76 | 0.77 | 0.23 |
| *Arhar* | 0.49 | 0.48 | 0.49 | 0.56 | 0.57 | 0.58 | 0.59 | 0.64 | 0.56 | 0.58 | 0.64 | 0.64 | 0.64 | 0.72 | 0.77 | 0.85 |
| Cowpea | 0.08 | 0.09 | - | 0.15 | 0.15 | 0.12 | - | - | 0.06 | 0.06 | 0.05 | 0.33 | 0.05 | 0.08 | 0.11 | 0.09 |
| Others | 0.03 | 0.05 | 0.15 | - | - | - | 0.18 | 0.06 | 0.04 | 0.07 | 0.07 | 0.12 | 0.10 | 0.08 | 0.08 | 0.11 |
| Sub-Total (Pulses) | 3.29 | 3.40 | 3.62 | 3.60 | 3.58 | 4.19 | 3.89 | 4.06 | 3.87 | 3.85 | 4.69 | 6.60 | 8.17 | 7.40 | 7.37 | 9.63 |
| **Oilseeds** | | | | | | | | | | | | | | | | |
| Groundnut | 6.72 | 7.00 | 6.73 | 7.01 | 6.85 | 7.27 | 6.81 | 6.20 | 6.46 | 5.40 | 5.25 | 5.42 | 7.00 | 7.18 | 6.96 | 9.89 |
| Rapeseed & Mustard | 0.77 | 0.75 | 0.79 | 0.75 | 0.86 | 1.02 | 1.05 | 1.13 | 0.94 | 0.69 | 0.86 | 0.82 | 0.91 | 1.31 | 1.24 | 1.36 |
| *Til* | 0.08 | 0.09 | 0.14 | 0.12 | 0.11 | 0.10 | 0.15 | 0.12 | 0.19 | 0.10 | 0.16 | 0.13 | 0.14 | 0.12 | 0.14 | 0.16 |

...contd...

...contd...

| 1 | 2 | 3 | 4 | 5 | 6 | 7 | 8 | 9 | 10 | 11 | 12 | 13 | 14 | 15 | 16 | 17 |
|---|---|---|---|---|---|---|---|---|---|---|---|---|---|---|---|---|
| Sunflower | 0.55 | 0.63 | 0.73 | 0.68 | 0.82 | 0.58 | 0.75 | 0.62 | 0.51 | 0.42 | 0.48 | 0.61 | 0.56 | 0.80 | 0.90 | 0.89 |
| Soyabean | 1.23 | 1.91 | 2.57 | 2.98 | 3.57 | 3.08 | 3.68 | 5.41 | 4.45 | 5.46 | 4.99 | 6.88 | 10.20 | 13.36 | 14.29 | 14.05 |
| Linseed | 0.01 | 0.01 | 0.01 | 0.01 | 0.01 | 0.02 | 0.01 | 0.01 | 0.02 | 0.01 | 0.02 | 0.01 | 0.01 | 0.01 | 0.01 | 0.02 |
| Castorseed | 0.21 | 0.22 | 0.22 | 0.22 | 0.26 | 0.30 | 0.25 | 0.29 | 0.30 | 0.39 | 0.26 | 0.37 | 0.39 | 0.41 | 0.50 | 0.36 |
| Safflower | 0.08 | 0.13 | 0.19 | 0.23 | 0.15 | 0.16 | 0.16 | 0.05 | 0.09 | 0.06 | 0.07 | 0.07 | 0.04 | 0.08 | 0.08 | 0.07 |
| Others | 0.01 | 0.01 | - | 0.01 | 0.01 | - | 0.01 | - | 0.02 | 0.01 | 0.01 | 0.04 | 0.14 | 0.15 | 0.23 | 0.20 |
| Sub-Total (Oilseeds) | 9.66 | 10.75 | 11.38 | 12.01 | 12.64 | 12.53 | 12.87 | 13.83 | 12.98 | 12.54 | 12.10 | 13.36 | 19.39 | 23.42 | 24.35 | 27.00 |
| **Fibres** | | | | | | | | | | | | | | | | |
| Cotton | 1.77 | 1.87 | 1.81 | 1.96 | 2.34 | 2.93 | 2.87 | 2.61 | 2.66 | 2.61 | 2.62 | 2.31 | 2.45 | 2.56 | 2.69 | 2.45 |
| Jute | 0.26 | 0.22 | 0.20 | 0.24 | 0.24 | 0.25 | 0.34 | 0.31 | 0.26 | 0.27 | 0.21 | 0.19 | 0.20 | 0.19 | 0.19 | 0.26 |
| Mesta/Others | | | | | | | | | 0.01 | 0.03 | 0.06 | 0.24 | 0.13 | 0.01 | 0.01 | 0.34 |
| Sub-Total (Fibers) | 2.03 | 2.09 | 2.01 | 2.20 | 2.58 | 3.18 | 3.21 | 2.92 | 2.93 | 2.91 | 2.89 | 2.74 | 2.78 | 2.76 | 2.89 | 3.05 |
| **Other Miscellaneous** | | | | | | | | | | | | | | | | |
| Potato | 6.90 | 7.10 | 6.17 | 6.62 | 6.85 | 6.69 | 6.83 | 6.86 | 6.89 | 7.23 | 6.33 | 7.16 | 7.01 | 5.05 | 5.08 | 5.12 |
| Others | 0.27 | 0.27 | 0.28 | 0.08 | 0.24 | 0.25 | 0.21 | 0.03 | 0.17 | 0.27 | 0.23 | 0.21 | 0.23 | 0.22 | 0.33 | 0.34 |
| Sub-Total (Other Miscellaneous) | 7.17 | 7.37 | 6.45 | 6.70 | 7.09 | 6.94 | 7.04 | 6.89 | 7.06 | 7.50 | 6.56 | 7.37 | 7.24 | 5.27 | 5.41 | 5.46 |
| Grand Total | 57.50 | 60.33 | 62.20 | 65.86 | 69.92 | 73.27 | 78.79 | 84.97 | 87.98 | 86.27 | 91.80 | 97.04 | 108.40 | 120.26 | 126.75 | 155.01 |

*Source:* Seeds Division, Ministry of Agriculture, New Delhi.

## Table 15.11(a)

*Consumption of Electricity for Agricultural Purposes from 1982-83 to 2006-07*

| Year | Consumption for Agricultural Purposes (GWh) | Total Consumption (GWh) | % Share of Agricultural Consumption to Total Consumption |
|---|---|---|---|
| 1 | 2 | 3 | 4 |
| 1982-83 | 17817 | 95589 | 18.64 |
| 1983-84 | 18234 | 102344 | 17.82 |
| 1984-85 | 20960 | 114068 | 18.38 |
| 1985-86 | 23422 | 122999 | 19.04 |
| 1986-87 | 29444 | 135952 | 21.66 |
| 1987-88 | 35267 | 145613 | 24.22 |
| 1988-89 | 38878 | 160196 | 24.27 |
| 1989-90 | 44056 | 175419 | 25.11 |
| 1990-91 | 50321 | 190357 | 26.44 |
| 1991-92 | 58557 | 207645 | 28.20 |
| 1992-93 | 63328 | 220674 | 28.70 |
| 1993-94 | 70699 | 238569 | 29.63 |
| 1994-95 | 79301 | 259630 | 30.54 |
| 1995-96 | 85732 | 277029 | 30.95 |
| 1996-97 | 84019 | 280206 | 29.98 |
| 1997-98 | 91242 | 296749 | 30.75 |
| 1998-99 | 97195 | 309734 | 31.38 |
| 1999-00 | 90934 | 312841 | 29.07 |
| 2000-01 | 84729 | 316600 | 26.76 |
| 2001-02 | 81673 | 322459 | 25.33 |
| 2002-03 | 84486 | 339598 | 24.88 |
| 2003-04 | 87089 | 360937 | 24.13 |
| 2004-05 | 88555 | 386134 | 22.93 |
| 2005-06 | 90292 | 411887 | 21.92 |
| 2006-07* | 99023 | 455748 | 21.73 |

*Source:* Central Electricity Authority, New Delhi.

* Provisional

# Table 15.11(b)

## *State-wise Consumption of Electricity for Agriculture Purpose in 2006-07**

| Region | State/UTs | Consumption for Agricultural Purpose (GWh) | Total Consumption (GWh) | % Share of Consumption for Agricultural Purpose |
|---|---|---|---|---|
| 1 | 2 | 3 | 4 | 5 |
| **Northern** | | | | |
| | Haryana | 6871.92 | 16642.97 | 41.29 |
| | Himachal Pradesh | 26.40 | 4300.46 | 0.61 |
| | Jammu & Kashmir | 201.91 | 4030.85 | 5.01 |
| | Punjab | 8229.49 | 26386.91 | 31.19 |
| | Rajasthan | 6658.21 | 20033.94 | 33.23 |
| | Uttar Pradesh | 6965.33 | 34543.73 | 20.16 |
| | Uttrakhand | 359.00 | 3885.92 | 9.24 |
| | Chandigarh | 1.53 | 1064.34 | 0.14 |
| | Delhi | 52.36 | 14677.14 | 0.36 |
| | Sub-Total | 29366.15 | 125566.26 | 23.39 |
| **Western** | | | | |
| | Gujarat | 11262.02 | 40772.65 | 27.62 |
| | Madhya Pradesh | 7028.24 | 20099.72 | 34.97 |
| | Chhattisgarh | 1408.11 | 9443.90 | 14.91 |
| | Maharashtra | 9748.80 | 62267.24 | 15.66 |
| | Goa | 15.55 | 2324.37 | 0.67 |
| | Daman & Diu | 3.05 | 1166.23 | 0.26 |
| | D & N Haveli | 7.67 | 2718.00 | 0.28 |
| | Sub-Total | 29473.44 | 138792.11 | 21.24 |
| **Southern** | | | | |
| | Andhra Pradesh | 16327.92 | 45872.25 | 35.59 |
| | Karnataka | 11208.08 | 30690.56 | 35.52 |
| | Kerala | 221.24 | 11310.22 | 1.98 |
| | Tamil Nadu | 10358.00 | 49934.00 | 20.74 |
| | Lakshadweep | 101.05 | 2024.99 | 4.99 |
| | Pondicherry | 0.00 | 24.17 | 0.00 |
| | Sub-Total | 38216.29 | 139856.19 | 27.33 |
| **Eastern** | | | | |
| | Bihar | 787.73 | 3825.87 | 20.59 |
| | Jharkhand | 68.41 | 10945.68 | 0.62 |
| | Orissa | 144.36 | 9483.74 | 1..52 |
| | West Bengal | 940.81 | 22524.12 | 4.18 |

*...contd...*

...contd...

| 1 | 2 | 3 | 4 | 5 |
|---|---|---|---|---|
| | A & N Islands | 0.00 | 136.63 | 0.00 |
| | Sikkim | 0.00 | 211.93 | 0.00 |
| | Sub-Total | 1941.31 | 47127.97 | 4.12 |
| **North Eastern** | | | | |
| | Assam | 16.78 | 25.75.35 | 0.65 |
| | Manipur | 0.09 | 217.38 | 0.04 |
| | Meghalaya | 0.42 | 778.48 | 0.06 |
| | Nagaland | 0.00 | 166.49 | 0.00 |
| | Tripura | 8.91 | 394.59 | 2.26 |
| | Arunachal Pradesh | 0.00 | 138.50 | 0.00 |
| | Mizoram | 0.00 | 145.15 | 0.00 |
| | Sub-Total | 26.20 | 4405.94 | 0.59 |
| | Total (All India) | 99023.39 | 451395.78 | 21.73 |

GWh: Giga Watthour * Provisional

*Source:* Central Electricity Authority, New Delhi.

# Table 15.12

## *Flow of Institutional Credit to Agriculture Sector from 1998-99 to 2007-08*

*(Rs. Crore)*

| S. No. | Institution | 1998-1999 | 1999-2000 | 2000-2001 | 2001-2002 | 2002-2003 | 2003-2004 | 2004-2005 | 2005-2006 | 2006-2007 | 2007-2008q |
|---|---|---|---|---|---|---|---|---|---|---|---|
| 1 | 2 | 3 | 4 | 5 | 6 | 7 | 8 | 9 | 10 | 11 | 12 |
| 1 | **A. Production (Short Term) Credit** | | | | | | | | | | |
| | Co-operative Banks | 12514 | 14771 | 16528 | 18787 | 19668 | 22640 | 27157 | 34930 | 38622 | |
| | Regional Rural Banks | 1710 | 2423 | 3245 | 3777 | 4775 | 6088 | 10010 | 12712 | 16631 | |
| | Commercial Banks* | 9622 | 11697 | 13486 | 17904 | 21104 | 26192 | 36793 | 57640 | 76378 | |
| | Other Agencies | 59 | 74 | 55 | 41 | 39 | 57 | 104 | 68 | - | |
| | Sub-Total (A) | 23905 | 28965 | 33314 | 40509 | 45586 | 54977 | 74064 | 105350 | 131631 | |
| 2 | **B. MT/LT Total** | | | | | | | | | | |
| | Co-operative Banks | 3356 | 3489 | 4190 | 4737 | 3968 | 4235 | 4074 | 4474 | 3858 | |
| | Regional Rural Banks | 750 | 749 | 974 | 1077 | 1295 | 1493 | 2394 | 2511 | 3804 | |
| | Commercial Banks* | 8821 | 13036 | 14321 | 15683 | 18670 | 26249 | 44688 | 67837 | 64004 | |
| | Other Agencies | 30 | 29 | 28 | 39 | 41 | 27 | 89 | 314 | - | |
| | Sub-Total (B) | 12957 | 17303 | 19513 | 21536 | 23974 | 32004 | 51245 | 75136 | 71666 | |
| 3 | **B. MT/LT Credit** | | | | | | | | | | |
| | Co-operative Banks | 15870 | 18260 | 20718 | 23524 | 23636 | 26875 | 31231 | 39404 | 42480 | 33070 |
| | Regional Rural Banks | 2460 | 3172 | 4219 | 4854 | 6070 | 7581 | 12404 | 15223 | 20435 | 15925 |
| | Commercial Banks* | 18443 | 24733 | 27807 | 33587 | 39774 | 52441 | 81481 | 125477 | 140382 | 88765 |
| | Other Agencies | 87 | 103 | 83 | 80 | 80 | 84 | 193 | 382 | - | 0 |
| | Grand Total (A+B) | 36860 | 46268 | 52827 | 62045 | 69560 | 86981 | 125309 | 180486 | 203297 | 137760 |

q Up to 30-11-2007

* commercial Banks includes Public Sector and Private Sector Banks

*Source:* Credit Division, Ministry of Agriculture, New Delhi.

# Table 15.13

*State-wise Number of Kisan Credit Cards issued upto 31 March 2007*

| *Sl. No.* | *State/Uts* | *No** | *Cooperative Banks(CB)* | *No** | *Regional Rural Banks(RRB)* | *Commercial Banks* | *Total* |
|---|---|---|---|---|---|---|---|
| *1* | *2* | *3* | *4* | *5* | *6* | *7* | *8* |
| 1 | Andhra Pradesh** | 22 | 3594385 | 5 | 1409665 | 5954397 | 10958447 |
| 2 | Assam | 1 | 5947 | 2 | 66642 | 224510 | 297099 |
| 3 | Arunachal Pradesh | 1 | 980 | 1 | 1401 | 11351 | 13732 |
| 4 | Bihar** | 22 | 775438 | 5 | 374506 | 869915 | 2019859 |
| 5 | Gujarat | 18 | 1161928 | 3 | 198070 | 956321 | 2316319 |
| 6 | Goa ^ | 1 | 3729 | 0 | 0 | 7907 | 11636 |
| 7 | Haryana | 19 | 1194877 | 2 | 259284 | 583773 | 2037934 |
| 8 | Himachal Pradesh | 3 | 54896 | 2 | 23809 | 147532 | 226237 |
| 9 | Jammu & Kashmir | 4 | 48872 | 3 | 9878 | 6600 | 65350 |
| 10 | Karnataka | 21 | 1349972 | 6 | 780084 | 1704873 | 3834929 |
| 11 | Kerala | 14 | 1030315 | 2 | 327710 | 1051697 | 2409722 |
| 12 | Madhya Pradesh | 38 | 2819980 | 10 | 346143 | 1054064 | 4220187 |
| 13 | Maharashtra | 30 | 4413984 | 7 | 215134 | 1675513 | 6304631 |
| 14 | Meghalaya # | 1 | 5975 | 1 | 12218 | 28434 | 46627 |
| 15 | Mizoram # | 1 | 2104 | 1 | 5100 | 8456 | 15660 |
| 16 | Manipur # | 1 | 13475 | 1 | 1037 | 17334 | 31846 |
| 17 | Nagaland # | 1 | 1735 | 1 | 1037 | 13617 | 16389 |
| 18 | Orissa | 17 | 2674014 | 6 | 382358 | 701391 | 3757763 |
| 19 | Punjab | 19 | 848851 | 3 | 84799 | 961568 | 1895218 |
| 20 | Rajasthan | 28 | 2764330 | 6 | 317801 | 1008041 | 4090172 |
| 21 | Sikkim #^ | 1 | 2548 | 0 | | 3857 | 6405 |
| 22 | Tamil Nadu | 22 | 1697972 | 2 | 177544 | 2525556 | 4401072 |
| 23 | Tripura # | 1 | 3090 | 1 | 20158 | 25328 | 48576 |
| 24 | Uttar Pradesh | 51 | 5680065 | 16 | 2684949 | 4305333 | 12670347 |
| 25 | West Bengal | 20 | 1205415 | 9 | 203217 | 898083 | 2306715 |
| 26 | A & N Islands #^ | 1 | 2801 | 0 | 0 | 1093 | 3894 |
| 27 | Chandigarh ^ | 0 | 0 | 0 | 0 | 935 | 935 |
| 28 | Daman & Diu @# | 0 | 0 | 0 | 0 | 93 | 93 |
| 29 | New Delhi #^ | 1 | 1986 | 0 | 0 | 3785 | 5771 |
| 30 | D & N Haveli @^ | 0 | 0 | 0 | 0 | 28 | 28 |
| 31 | Lakshdweep @^ | 0 | 0 | 0 | 0 | 463 | 463 |
| 32 | Pondicherry ^ | 1 | 6916 | 0 | 0 | 29473 | 36389 |
| 33 | Jharkhand | 8 | 278892 | 2 | 174634 | 260946 | 714472 |
| 34 | Chhattisgarh | 7 | 783643 | 3 | 183761 | 170632 | 1138036 |
| 35 | Uttrakhand | 10 | 280219 | 2 | 23350 | 168864 | 472433 |
| | Breakup not available for CBs (1998-99) | | | | | 188005 | 188005 |
| | Total | 385 | 32709334 | 102 | 8284289 | 25569768 | 66563391 |

\# SCB functions as CFA., @ No Cooperative Banks in these UTs. ^ No RRB in these States/UTs.

* No of Banks Implementing the scheme ** Coopraltive data are under reconcilation

*Source:* Credit Division, Ministry of Agriculture, New Delhi.

# Table 15.14

## *National Agricultural Insurance Scheme: Trend in Area, Sum Insured, Premium and Claims*

| Sl. No. | Season | Farmers Covered(No) | Area (In Hec) | Sum Insured (Rs. Crore) | Premium (Rs. Crore) | Claims Reported (Rs. Crore) |
|---|---|---|---|---|---|---|
| 1 | 2 | 3 | 4 | 5 | 6 | 7 |
| 1 | *Rabi* 1999-00 | 579940 | 780569.36 | 35640.71 | 542.48 | 769.26 |
| 2 | *Kharif* 2000 | 8409374 | 13219828.68 | 690338.38 | 20673.55 | 122248.15 |
| 3 | *Rabi* 2000-01 | 2091733 | 3111423.25 | 160268.46 | 2778.76 | 5948.63 |
| 4 | *Kharif* 2001 | 8696587 | 12887710.38 | 750246.11 | 26161.82 | 49353.55 |
| 5 | *Rabi* 2001-02 | 1955431 | 3145872.65 | 149751.11 | 3014.79 | 6465.80 |
| 6 | *Kharif* 2002 | 9768711 | 15532348.53 | 943169.37 | 32546.68 | 182431.26 |
| 7 | *Rabi* 2002-03 | 2326811 | 4037824.35 | 183754.52 | 3850.43 | 18854.83 |
| 8 | *Kharif* 2003 | 7970830 | 12355513.83 | 811412.55 | 28333.19 | 65202.67 |
| 9 | *Rabi* 2003-04 | 4421287 | 6468662.75 | 304949.21 | 6405.87 | 49692.25 |
| 10 | *Kharif* 2004 | 12687046 | 24273241.97 | 1317049.22 | 45893.97 | 103787.25 |
| 11 | *Rabi* 2004-05 | 3531045 | 5343243.62 | 377420.53 | 7584.93 | 16058.60 |
| 12 | *Kharif* 2005 | 12674080 | 20530776.70 | 1351725.43 | 44986.19 | 105976.06 |
| 13 | *Rabi* 2005-06 | 4048524 | 7218420.95 | 507166.12 | 10482.40 | 33830.20 |
| 14 | *Kharif* 2006 | 12934035 | 19674159.96 | 1475911.74 | 46727.43 | 126800.13 |
| 15 | *Rabi* 2006-07@ | 4967280 | 7758975.31 | 615358.12 | 14209.12 | 549.29 |
| | Total | 97073461 | 156211397.40 | 9718111.85 | 294271.77 | 985692.44 |

'@ As on 15 Feb 2007

*Source:* Credit Division, Ministry of Agriculture,New Delhi.

# Table 16.1

*Distribution of Operational Holdings—All India*

*No. of Holdings: ('000 Number)*
*Area Operated: ('000 Hectares)*
*Average Size: (Hectares)*

| *Category of Holdings* | *No of Operational Holdings* | | *Area Operated* | | *Average Size of Operational Holdings* | |
|---|---|---|---|---|---|---|
| | *1995-96* | *2000-01** | *1995-96* | *2000-01** | *1995-96* | *2000-01** |
| *1* | 2 | *3* | *4* | *5* | *6* | *7* |
| Marginal (Less than 1 hectare) | 71179 (61.6) | 75408 (62.3) | 28121 (17.2) | 29814 (18.7) | 0.40 | 0.40 |
| Small (1.0 to 2.0 hectares) | 21643 (18.7) | 22695 (19.0) | 30722 (18.8) | 32139 (20.2) | 1.42 | 1.42 |
| Semi-Medium (2.0 to 4.0 hectares) | 14261 (12.3) | 14021 (11.8) | 38953 (23.8) | 38193 (24.0.) | 2.73 | 2.72 |
| Medium (4.0 to 10.0 hectares) | 7092 (6.1) | 6577 (5.5) | 41398 (25.3) | 38217 (24.0) | 5.84 | 5.81 |
| Large (10.0 hectares and above) | 1404 (1.2) | 1230 (1.0) | 24163 (14.8) | 21072 (13.2) | 17.21 | 17.12 |
| All Holdings | 115580 (100.0) | 119231 (100.0) | 163357 (100.0) | 159436 (100.0) | 1.41 | 1.33 |

*Note*: Figures in parentheses indicate the percentage of respective column total.
* Excluding Jharkhand

*Source*: Agricultural Census Division, Ministry of Agriculture, New Delhi.

# Table 16.2(a)

*State-wise Number of Operational Holdings by Major Size-Groups, 2000-01**

*('000 Number)*

| *State/UT* | *Marginal* | *Small* | *Semi-Medium* | *Medium* | *Large* | *All Holdings* |
|---|---|---|---|---|---|---|
| *1* | *2* | *3* | *4* | *5* | *6* | *7* |
| Andhra Pradesh | 7023 | 2518 | 1424 | 501 | 66 | 11532 |
| Arunachal Pradesh | 15 | 20 | 36 | 30 | 6 | 107 |
| Assam | 1699 | 561 | 352 | 96 | 5 | 2712 |
| Bihar | 9743 | 1069 | 589 | 164 | 9 | 11574 |
| Chhattisgarh | 1747 | 716 | 508 | 245 | 39 | 3255 |
| Goa | 52 | 7 | 3 | 1 | 1 | 64 |
| Gujarat | 1298 | 1257 | 1043 | 576 | 65 | 4239 |
| Haryana | 704 | 294 | 278 | 202 | 50 | 1528 |
| Himachal Pradesh | 615 | 174 | 90 | 31 | 4 | 914 |
| Jammu & Kashmir | 1175 | 179 | 76 | 13 | 1 | 1443 |
| Jharkhand ## | 0 | 0 | 0 | 0 | 0 | 0 |
| Karnataka | 3252 | 1909 | 1259 | 569 | 89 | 7079 |
| Kerala | 6335 | 227 | 76 | 16 | 3 | 6657 |
| Madhya Pradesh | 2838 | 1951 | 1488 | 917 | 166 | 7360 |
| Maharashtra | 5306 | 3606 | 2274 | 865 | 87 | 12138 |
| Manipur | 75 | 49 | 22 | 3 | 0 | 149 |
| Meghalaya | 115 | 57 | 35 | 7 | 0 | 214 |
| Mizoram | 34 | 28 | 13 | 1 | 0 | 76 |
| Nagaland | 5 | 9 | 28 | 68 | 34 | 144 |
| Orissa | 2295 | 1114 | 501 | 145 | 13 | 4067 |
| Punjab | 123 | 173 | 328 | 301 | 72 | 997 |
| Rajasthan | 1849 | 1210 | 1200 | 1100 | 460 | 5819 |
| Sikkim | 36 | 15 | 10 | 3 | 1 | 67 |
| Tamil Nadu | 5846 | 1226 | 571 | 193 | 23 | 7859 |
| Tripura | 406 | 55 | 17 | 2 | 0 | 479 |
| Uttrakhand | 628 | 158 | 78 | 24 | 1 | 891 |
| Uttar Pradesh | 16659 | 3087 | 1427 | 463 | 32 | 21668 |
| West Bengal | 5462 | 1009 | 283 | 35 | 1 | 6790 |
| A & N Islands | 4 | 3 | 3 | 2 | 0 | 11 |
| Chandigarh | 1 | 0 | 0 | 0 | 0 | 1 |
| Dadar & Nagar Haveli | 7 | 4 | 2 | 1 | 0 | 14 |
| Daman & Diu | 5 | 1 | 0 | 0 | 0 | 6 |
| Delhi | 16 | 6 | 4 | 2 | 0 | 28 |
| Lakshadweep | 10 | 0 | 0 | 0 | 0 | 10 |
| Pondicherry | 31 | 4 | 2 | 1 | 0 | 38 |
| Total | 75409 | 22696 | 14020 | 6577 | 1228 | 119930 |

Neg: Negligible

* includes institutional holdings also.

## Data for the year 2000-01 not collected.

*Note*: The sum of States/UTs may not exactly tally with all-India total due to rounding off.

*Source*: Agricultural Census Division, Ministry of Agriculture, New Delhi.

# Table 16.2 (b)

*State-wise Area Operated by Major Size-Groups, 2000-01*

*('000 Hectares)*

| *State/UT* | *Marginal* | *Small* | *Semi-Medium* | *Medium* | *Large* | *All Holdings* |
|---|---|---|---|---|---|---|
| *1* | *2* | *3* | *4* | *5* | *6* | *7* |
| Andhra Pradesh | 3104 | 3565 | 3795 | 2855 | 1080 | 14400 |
| Arunachal Pradesh | 7 | 27 | 97 | 171 | 92 | 394 |
| Assam | 663 | 731 | 958 | 499 | 264 | 3114 |
| Bihar | 2907 | 1296 | 1544 | 861 | 140 | 6747 |
| Chhattisgarh | 776 | 1018 | 1375 | 1410 | 643 | 5223 |
| Goa | 17 | 8 | 8 | 8 | 12 | 54 |
| Gujarat | 692 | 1835 | 2902 | 3343 | 1104 | 9877 |
| Haryana | 317 | 422 | 781 | 1210 | 821 | 3550 |
| Himachal Pradesh | 252 | 245 | 243 | 176 | 63 | 979 |
| Jammu & Kashmir | 429 | 250 | 202 | 69 | 12 | 962 |
| Jharkhand ## | 0 | 0 | 0 | 0 | 0 | 0 |
| Karnataka | 1492 | 2742 | 3429 | 3317 | 1327 | 12307 |
| Kerala | 883 | 300 | 191 | 85 | 112 | 1569 |
| Madhya Pradesh | 1398 | 2828 | 4122 | 5448 | 2576 | 16372 |
| Maharashtra | 2649 | 5127 | 6109 | 4880 | 1338 | 20103 |
| Manipur | 40 | 63 | 55 | 14 | 0 | 172 |
| Meghalaya | 63 | 83 | 91 | 36 | 5 | 278 |
| Mizoram | 22 | 36 | 29 | 6 | 1 | 93 |
| Nagaland | 2 | 11 | 72 | 420 | 543 | 1047 |
| Orissa | 1155 | 1544 | 1344 | 818 | 220 | 5081 |
| Punjab | 78 | 242 | 876 | 1731 | 1096 | 4022 |
| Rajasthan | 894 | 1744 | 3416 | 6811 | 8386 | 21251 |
| Sikkim | 15 | 22 | 27 | 25 | 15 | 104 |
| Tamil Nadu | 2159 | 1712 | 1551 | 1094 | 455 | 6972 |
| Tripura | 126 | 75 | 43 | 9 | 16 | 270 |
| Uttrakhand | 243 | 221 | 212 | 132 | 36 | 843 |
| Uttar Pradesh | 6648 | 4366 | 3906 | 2580 | 484 | 17983 |
| West Bengal | 2759 | 1607 | 784 | 178 | 219 | 5547 |
| A & N Islands | 1 | 4 | 8 | 7 | 2 | 23 |
| Chandigarh | 0 | 0 | 0 | 1 | 0 | 2 |
| Dadar & Nagar Haveli | 4 | 5 | 5 | 4 | 2 | 21 |
| Daman & Diu | 2 | 1 | 1 | 0 | 0 | 4 |
| Delhi | 7 | 8 | 11 | 13 | 4 | 43 |
| Lakshadweep | 2 | 0 | 0 | 0 | 0 | 3 |
| Pondicherry | 9 | 6 | 5 | 4 | 2 | 26 |
| Total | 29815 | 32144 | 38192 | 38215 | 21070 | 159436 |

Neg: Negligible

## Data for the year 2000-01 not collected.

*Source*: Website of Agricultural Census Division, Ministry of Agriculture, New Delhi.

## Table 16.3

*State-wise Average Size of Operational Holdings by Major Size Groups, 2000-01**

(Hectares)

| State/UT | Marginal | Small | Semi-Medium | Medium | Large | All Holdings |
|---|---|---|---|---|---|---|
| 1 | 2 | 3 | 4 | 5 | 6 | 7 |
| Andhra Pradesh | 0.44 | 1.42 | 2.67 | 5.70 | 16.34 | 1.25 |
| Arunachal Pradesh | 0.50 | 1.32 | 2.66 | 5.77 | 16.13 | 3.69 |
| Assam | 0.39 | 1.30 | 2.73 | 5.22 | 53.02 | 1.15 |
| Bihar | 0.30 | 1.21 | 2.62 | 5.24 | 15.50 | 0.58 |
| Chhattisgarh | 0.44 | 1.42 | 2.70 | 5.76 | 16.49 | 1.60 |
| Goa | 0.32 | 1.26 | 2.56 | 5.64 | 23.77 | 0.84 |
| Gujarat | 0.53 | 1.46 | 2.78 | 5.80 | 16.91 | 2.33 |
| Haryana | 0.45 | 1.43 | 2.81 | 5.99 | 16.48 | 2.32 |
| Himachal Pradesh | 0.41 | 1.40 | 2.71 | 5.69 | 15.91 | 1.07 |
| Jammu & Kashmir | 0.37 | 1.40 | 2.66 | 5.39 | 21.13 | 0.67 |
| Karnataka | 0.46 | 1.44 | 2.72 | 5.83 | 14.83 | 1.74 |
| Kerala | 0.14 | 1.32 | 2.52 | 5.29 | 40.93 | 0.24 |
| Madhya Pradesh | 0.49 | 1.45 | 2.77 | 5.94 | 15.50 | 2.22 |
| Maharashtra | 0.50 | 1.42 | 2.69 | 5.64 | 15.38 | 1.66 |
| Manipur | 0.53 | 1.29 | 2.47 | 4.86 | 11.38 | 1.15 |
| Meghalaya | 0.55 | 1.45 | 2.58 | 5.41 | 13.12 | 1.30 |
| Mizoram | 0.64 | 1.28 | 2.33 | 4.78 | 13.14 | 1.24 |
| Nagaland | 0.52 | 1.19 | 2.55 | 6.20 | 15.83 | 7.28 |
| Orissa | 0.50 | 1.39 | 2.69 | 5.63 | 16.48 | 1.25 |
| Punjab | 0.63 | 1.40 | 2.67 | 5.75 | 15.14 | 4.03 |
| Rajasthan | 0.48 | 1.44 | 2.85 | 6.19 | 18.21 | 3.65 |
| Sikkim | 0.42 | 1.40 | 2.74 | 5.79 | 20.67 | 1.57 |
| Tamil Nadu | 0.37 | 1.40 | 2.72 | 5.68 | 19.48 | 0.89 |
| Tripura | 0.31 | 1.37 | 2.55 | 5.16 | 78.77 | 0.56 |
| Uttrakhand | 0.39 | 1.39 | 2.71 | 5.47 | 25.07 | 0.95 |
| Uttar Pradesh | 0.40 | 1.41 | 2.74 | 5.57 | 15.07 | 0.83 |
| West Bengal | 0.51 | 1.59 | 2.77 | 5.12 | 278.95 | 0.82 |
| A & N Islands | 0.39 | 1.38 | 2.53 | 4.31 | 46.79 | 2.00 |
| Chandigarh | 0.39 | 1.42 | 2.79 | 5.92 | 12.00 | 1.44 |
| Dadar & Nagar Haveli | 0.52 | 1.32 | 2.75 | 5.78 | 15.95 | 1.48 |
| Daman & Diu | 0.29 | 1.37 | 2.63 | 5.86 | 20.25 | 0.59 |
| Delhi | 0.42 | 1.38 | 2.86 | 5.77 | 15.27 | 1.52 |
| Lakshadweep | 0.19 | 1.27 | 2.56 | 5.47 | 22.33 | 0.27 |
| Pondicherry | 0.29 | 1.42 | 2.74 | 5.68 | 19.50 | 0.70 |
| All-India* | 0.24 | 1.42 | 2.39 | 4.42 | 13.16 | 1.33 |

*Note*: 1. * No data available for Jharkhand

2. Includes institutional holdings also.

*Source*: Agricultural Census Division, Ministry of Agriculture, New Delhi.

## Table 16.4

*Distribution of Number of Holdings and Net Area Sown According to Size Class of Holdings and Irrigation Status for all Size Groups, 2000-01*

*Number: ('000)*
*Area: ('000 Hectares)*

| Size-Group | Wholly Irrigated | | Partly Irrigated | | | Wholly Unirrigated | | Total | |
|---|---|---|---|---|---|---|---|---|---|
| | Number | Area | Number | Area | Net Area | Number | Area | Number | Area |
| 1 | 2 | 3 | 4 | 5 | 6 | 7 | 8 | 9 | 10 |
| Marginal | 27768 | 11175 | 5498 | 2389 | 1805 | 28077 | 10702 | 61343 | 24266 |
| Small | 6593 | 8876 | 3067 | 4154 | 2612 | 11252 | 14824 | 20912 | 27854 |
| Semi-medium | 3534 | 9024 | 2514 | 6315 | 3937 | 7020 | 17269 | 13068 | 32608 |
| Medium | 1538 | 8154 | 1537 | 7870 | 4686 | 3188 | 15996 | 6263 | 32020 |
| Large | 230 | 3370 | 301 | 3758 | 2351 | 659 | 8802 | 1190 | 15930 |
| All Sizes | 39663 | 40599 0 | 12917 | 24486 | 15391 0 | 50196 | 67593 0 | 102776 | 132678 |

*Source*: Agricultural Census Division, Ministry of Agriculture, New Delhi.

## Table 16.5

*Distribution of Irrigated Area According to Source of Irrigation by Major Size-Classes of Operational Holdings for Major Size Groups, 2000-01*

*('000 Hectares)*

| Size Class | Canals | Tanks | Wells | Tubewells | Others | Total |
|---|---|---|---|---|---|---|
| 1 | 2 | 3 | 4 | 5 | 6 | 7 |
| Marginal | 3405 | 855 | 1296 | 5419 | 1409 | 12384 |
| Small | 2929 | 587 | 1971 | 4335 | 1069 | 10891 |
| Semi-medium | 3219 | 463 | 2500 | 4740 | 1010 | 11932 |
| Medium | 3447 | 276 | 2511 | 4502 | 811 | 11547 |
| Large | 1578 | 77 | 952 | 1715 | 550 | 4872 |
| Total | 14578 | 2258 | 9231 | 20711 | 4849 | 51627 |

*Source*: Agricultural Census Division, Ministry of Agriculture, New Delhi.

## Table 16.6

*Distribution of Irrigated and Unirrigated Area Under Major Crops in India According to Size-Class of Operational Holdings, 2000-01*

('000 Hectares)

| Crops | Irrigated | | | | | | Unirrigated | | | | | |
|---|---|---|---|---|---|---|---|---|---|---|---|---|
| | Marginal | Small | Semi-Medium | Medium | Large | Total | Marginal | Small | Semi-medium | Medium | Large | Total |
| 1 | 2 | 3 | 5 | 6 | 7 | 8 | 9 | 10 | 11 | 12 | 13 | 14 |
| Paddy | 7547 | 5413 | 5013 | 4069 | 1563 | 23606 | 5515 | 2955 | 4682 | 3316 | 1131 | 17599 |
| Wheat | 5117 | 3952 | 4670 | 5107 | 2336 | 21181 | 575 | 404 | 678 | 166 | 266 | 2089 |
| Total Cereals | 13294 | 10148 | 10667 | 10125 | 4274 | 48507 | 10006 | 11125 | 12385 | 10813 | 5178 | 49507 |
| Pulses | 304 | 431 | 592 | 792 | 358 | 2477 | 2043 | 2917 | 3557 | 3740 | 1884 | 14141 |
| Sugarcane | 906 | 1063 | 1124 | 821 | 189 | 4103 | 106 | 93 | 92 | 59 | 13 | 363 |
| Oilseeds | 757 | 852 | 1110 | 1304 | 589 | 4612 | 1941 | 3386 | 4477 | 4822 | 1907 | 16533 |
| Fibres | 368 | 429 | 594 | 937 | 583 | 2911 | 600 | 1789 | 2150 | 1935 | 542 | 7016 |
| All Crops | 17260 | 14223 | 15539 | 15482 | 6733 | 69237 | 16503 | 20769 | 24497 | 23859 | 12256 | 97884 |

*Source*: Agricultural Census Division, Ministry of Agriculture, New Delhi

# Table 16.7

*Usage of Chemical Fertilisers, Farm Yard Manure and Pesticides for All Crops in India by Major Size-Class of Holdings and by Irrigation Status in 2000-01*

(Area: '000 Hectares)
(Fertilisers/Manure/Pesticides: Tonnes)

| Sl. No. | Size Groups (Hectares) | | Gross Cropped Area | Area Treated with Fertilisers | Total Consumption of Nutrients | | | | Area Treated with | | Quantity of Farm Yard Manure Used |
|---|---|---|---|---|---|---|---|---|---|---|---|
| | | | | | N | P | K | N+P+K (Total) | Farm Yard Manure | Pesticides | |
| 1 | 2 | | 3 | 4 | 5 | 6 | 7 | 8 | 9 | 10 | 11 |
| 1. | Marginal | I | 15033 | 12469 | 1446168 | 609914 | 245641 | 2301723 | 3506 | 3452 | 53774877 |
| | (Below 1.00) | UI | 14049 | 6163 | 439267 | 194822 | 84224 | 718312 | 3825 | 1569 | 45314074 |
| | | T | 29081 | 18632 | 1885435 | 804736 | 329865 | 3020036 | 7331 | 5021 | 99088951 |
| 2. | Small | I | 13287 | 10803 | 1054890 | 475001 | 187559 | 1717450 | 3236 | 3139 | 28374017 |
| | (1.0-1.99) | UI | 15881 | 7478 | 400449 | 218674 | 72620 | 691743 | 4450 | 1621 | 26404379 |
| | | T | 29168 | 18281 | 455338 | 693675 | 260179 | 1409192 | 7687 | 4760 | 54778396 |
| 3. | Semi-Medium | I | 15651 | 12674 | 233807 | 538578 | 159280 | 931665 | 3489 | 3257 | 24421182 |
| | (2.0-3.99) | UI | 20323 | 9196 | 448708 | 259398 | 69412 | 777517 | 4993 | 1810 | 22040713 |
| | | T | 35974 | 21870 | 682515 | 797976 | 228691 | 1709183 | 8483 | 5067 | 46461895 |
| 4. | Medium | I | 16310 | 13232 | 308691 | 541595 | 103205 | 953491 | 3261 | 3274 | 17457565 |
| | (4.0-9.99) | UI | 22143 | 8855 | 382287 | 238152 | 45988 | 666427 | 4600 | 1511 | 15710932 |
| | | T | 38454 | 22086 | 690978 | 779747 | 149193 | 1619918 | 7862 | 4785 | 33168497 |
| 5. | Large | I | 6943 | 5704 | 571634 | 231822 | 20332 | 823788 | 1137 | 1627 | 4553291 |
| | (10.0 and above) | UI | 13356 | 3435 | 120988 | 81592 | 11847 | 214427 | 2127 | 709 | 5090335 |
| | | T | 20298 | 9139 | 692622 | 313414 | 32179 | 1038215 | 3264 | 2335 | 9643625 |
| | All Groups | I | 67223 | 54882 | 5615190 | 2396910 | 716017 | 8728116 | 14630 | 14748 | 128580931 |
| | | UI | 85752 | 35126 | 1791698 | 992638 | 284090 | 3068427 | 19996 | 7220 | 114560432 |
| | | T | 152976 | 90008 | 7406888 | 3389548 | 1000107 | 11796543 | 34626 | 21968 | 243141364 |

*Note*: The sum of I (Irrigated) and UI (Unirrigated) area may not exactly tally with T (Total) area due to rounding off.

*Source*: Agricultral Census Division, Ministry of Agricultre, New Delhi.

# Table 16.8

## *All India Estimates of Institutional Credit Taken for Agriculural Purposes from Various Sources by Major Size-Class of Holdings in 2000-01*

(Number in '000)
(Amount in Rs. Lakh)

| Sl. No. | Size-Class (Hectares) | Total Number of Operational Holdings | Estimated Number of Operational Holdings that took Institutional Credit | No. of Operational Holdings that took Credit from | | | | Amount of Institutional Credit Taken | | | |
|---|---|---|---|---|---|---|---|---|---|---|---|
| | | | | PACS | PLDB/ SLDB | CBB | RRBB | PACS | PLDB/ SLDB | CBB | RRBB |
| 1 | 2 | 3 | 4 | 5 | 6 | 7 | 8 | 9 | 10 | 11 | 12 |
| 1. | Marginal (Below 1.00) | 53857 | 5118 | 3582 | 370 | 781 | 413 | 165483 | 33867 | 60570 | 21653 |
| 2. | Small (1.00-1.99) | 16790 | 2961 | 1855 | 184 | 633 | 288 | 118270 | 26537 | 63364 | 19039 |
| 3. | Semi-Medium (2.00-3.99) | 11051 | 2199 | 1309 | 141 | 523 | 225 | 114579 | 34036 | 77399 | 21429 |
| 4. | Medium (4.00-9.99) | 5751 | 1328 | 772 | 97 | 192 | 266 | 95631 | 42420 | 50260 | 46096 |
| 5. | Large (10.00 and above) | 1245 | 287 | 173 | 32 | 33 | 48 | 29778 | 21469 | 11760 | 13259 |
| | All Classes | 88694 | 11893 | 7691 | 824 | 2162 | 1240 | 523741 | 158329 | 263353 | 121476 |

*Note:* The sum of individual size-classes may not exacity tai.y with ' All Sizes' due to rounding off.
*Source:* Agricultural Census Division, Ministry of Agriculture, New Delhi.

# Table 16.9

## *All India Estimates of Institutional Credit Availed for Agriculture by terms/ Components of Credit by Major Size-Class of Holdings in 2000-01*

(Rs. Lakh)

| Sl. No. | Size-Class (Hectares) | Short Term Loan | | | | Medium Term Loan | Long Term Loan | Total (6+7+8) |
|---|---|---|---|---|---|---|---|---|
| | | Value of Quantity of Fertiliser Given | Value of Quantity of Other Inputs | Given in Cash | Total | | | |
| 1 | 2 | 3 | 4 | 5 | 6 | 7 | 8 | 9 |
| 1. | Marginal (Below 1.00) | 43321 | 7468 | 152384 | 203173 | 453382 | 33061 | 689618 |
| 2. | Small (1.0-1.99) | 43232 | 6786 | 117986 | 168004 | 28249 | 30958 | 227031 |
| 3. | Semi-medium (2.0-3.99) | 51359 | 7624 | 112270 | 171253 | 27050 | 49140 | 247443 |
| 4. | Medium (4.0-9.99) | 51745 | 6506 | 87297 | 145548 | 29415 | 59445 | 234408 |
| 5. | Large (10.0 and above) | 17725 | 2097 | 19959 | 39781 | 9137 | 27348 | 76268 |
| | All Groups | 207382 | 30481 | 489896 | 727759 | 547233 | 199952 | 1474944 |

*Source:* Agricultural Census Division, Ministry of Agriculture, New Delhi.

# Table 17.1

*State-wise Estimated Number of Rural Households, Total and Indebted Farmer Households in India*

| State | Estimated Number of Rural Households ('00) | Estimated Number of Farmer Households ('00) | Estimated Number of Indebted Farmer Households ('00) | % of Farmer Households Indebted |
|---|---|---|---|---|
| 1 | 2 | 3 | 4 | 5 |
| Andhra Pradesh | 142512 | 60339 | 49493 | 82 |
| Arunachal Pradesh | 15412 | 1227 | 72 | 5.9 |
| Assam | 41525 | 25040 | 4536 | 18.1 |
| Bihar | 116853 | 70804 | 23383 | 33 |
| Chhattisgarh | 36316 | 27598 | 11092 | 40.2 |
| Gujarat | 63015 | 37845 | 19644 | 51.9 |
| Haryana | 31474 | 19445 | 10330 | 53.1 |
| Himachal Pradesh | 11928 | 9061 | 3030 | 33.4 |
| Jammu & Kashmir | 10418 | 9432 | 3003 | 31.8 |
| Jharkhand | 36930 | 28238 | 5893 | 20.9 |
| Karnataka | 69906 | 40413 | 24897 | 61.6 |
| Kerala | 49942 | 21946 | 14126 | 64.4 |
| Madhya Pradesh | 93898 | 63206 | 32110 | 50.8 |
| Maharashtra | 118177 | 65817 | 36098 | 54.8 |
| Manipur | 2685 | 2146 | 533 | 24.8 |
| Meghalaya | 3401 | 2543 | 103 | 4.1 |
| Mizoram | 942 | 780 | 184 | 23.6 |
| Nagaland | 973 | 805 | 294 | 36.5 |
| Orissa | 66199 | 42341 | 20250 | 47.8 |
| Punjab | 29847 | 18442 | 12069 | 65.4 |
| Rajasthan | 70172 | 53080 | 27828 | 52.4 |
| Sikkim | 812 | 531 | 174 | 38.8 |
| Tamil Nadu | 110182 | 38880 | 28954 | 74.5 |
| Tripura | 5977 | 2333 | 1148 | 49.2 |
| Uttar Pradesh | 221499 | 171575 | 69199 | 40.3 |
| Uttrakhand | 11959 | 8962 | 644 | 7.2 |
| West Bengal | 121667 | 69226 | 34696 | 50.1 |
| UTs | 2325 | 732 | 372 | 50.8 |
| All India | 1478988 | 893504 | 434242 | 48.6 |

*Source:* Report No. 498(59/33/1), Situation Assessment Survey of Farmers: Indebtedness of Farmer Households, National Sample Survey 59th Round (January-December 2003)

## Table 17.2

*Indebtedness of Farmer Households (all-India) in Different Size Classes of Land Possessed*

| Land Possessed (ha) | Estimated Number of Farmer Households (lakh) | % to Total in Each Class | Estimated No. of Indebted Farmer Households (lakh) | % to Total in Each Class | % of Indebted Farmer Households to Total | Average Outstanding Loan Amount (Rs.) |
|---|---|---|---|---|---|---|
| 1 | 2 | 3 | 4 | 5 | 6 | 7 |
| <0.01 | 12.59 | 1.40 | 5.71 | 1.3 | 45.3 | 6121 |
| 0.01-0.40 | 292.87 | 32.80 | 130.11 | 30.0 | 44.4 | 6545 |
| 0.41-1.00 | 283.61 | 31.70 | 129.21 | 29.8 | 45.6 | 8623 |
| 1.01-2.00 | 160.60 | 18.00 | 81.92 | 18.8 | 51.0 | 13762 |
| 2.01-4.00 | 93.50 | 10.50 | 54.41 | 12.5 | 58.2 | 23456 |
| 4.01-10.00 | 42.58 | 4.80 | 27.73 | 6.4 | 65.1 | 42532 |
| >10.00 | 7.75 | 0.80 | 5.15 | 1.2 | 66.4 | 76232 |
| All | 893.50 | 100.00 | 434.24 | 100.0 | 48.6 | 12585 |

*Source:* Report No. 498(59/33/1), Situation Assessment Survey of Farmers: Indebtedness of Farmer Households, National Sample Survey 59th Round (January-December 2003)

## Table 17.3

*Incidence of Indebtedness in Major States*

| State | Estimated Number of Indebted Farmer Households | % Share in Estimated Farmer Households | % Share in Total Foodgrains | |
|---|---|---|---|---|
| | | | Area | Production |
| 1 | 2 | 3 | 4 | 5 |
| 1. Uttar Pradesh | 69199 | 15.9 | 16.3 | 20.7 |
| 2. Maharashtra | 36098 | 8.3 | 10.9 | 5.8 |
| 3. Madhya Pradesh | 32110 | 7.4 | 9.6 | 6.7 |
| 4. Rajasthan | 27828 | 6.4 | 9.5 | 5.5 |
| 5. Karnataka | 24897 | 5.7 | 6.1 | 4.6 |
| 6. Andhra Pradesh | 49493 | 11.4 | 5.9 | 7.0 |
| 7. Bihar | 23383 | 5.4 | 5.8 | 5.9 |
| 8. West Bengal | 34696 | 8.0 | 5.5 | 7.5 |
| 9. Punjab | 12069 | 2.8 | 5.1 | 12.2 |
| 10. Orissa | 20250 | 4.7 | 4.4 | 2.8 |
| Sub-total | 330023 | 76.0 | 79.1 | 78.7 |
| All India | 434242 | 100.0 | 100.0 | 100.0 |

*Source:* Report No. 498(59/33/1), Situation Assessment Survey of Farmers: Indebtedness of Farmer Households, National Sample Survey 59th Round (January-December 2003)

## Table 17.4

*Incidence of Indebtedness based on Size of Land Possessed*

| State | % of Marginal Indebted Farmer Households (Upto 1.0 ha of Land) | % of Small Indebted Farmer Households (1.01 to 2.00 ha) | % of Semi-Medium Indebted Farmer Households (2.01 to 4.00 ha) | % of Medium Indebted Farmer Households (4.01 to 10.00 ha) | % of Large Indebted Farmer Households (>10.00 ha) |
|---|---|---|---|---|---|
| 1 | 2 | 3 | 4 | 5 | 6 |
| 1. Uttar Pradesh | 71.3 | 17.4 | 7.8 | 3.4 | 0.3 |
| 2. Maharashtra | 36.0 | 26.2 | 23.3 | 12.2 | 2.4 |
| 3. Madhya Pradesh | 33.0 | 27.1 | 23.1 | 13.0 | 3.9 |
| 4. Rajasthan | 43.9 | 19.8 | 17.8 | 14.1 | 4.5 |
| 5. Karnataka | 50.7 | 22.8 | 15.9 | 9.3 | 1.2 |
| 6. Andhra Pradesh | 55.7 | 21.8 | 15.1 | 6.6 | 0.7 |
| 7. Bihar | 86.9 | 9.2 | 2.8 | 0.7 | 0.6 |
| 8. West Bengal | 88.7 | 8.5 | 2.4 | 0.4 | 0.0 |
| 9. Punjab | 53.3 | 15.8 | 17.0 | 11.8 | 2.2 |
| 10. Orissa | 70.3 | 20.6 | 7.3 | 1.7 | 0.0 |
| All India | 61.0 | 18.9 | 12.5 | 6.4 | 1.2 |

*Source:* Report No. 498(59/33/1), Situation Assessment Survey of Farmers: Indebtedness of Farmer Households, National Sample Survey 59th Round (January-December 2003)

# Table 18.1

*Ceilings on Land Holdings*

*(In Acres)*

| *State* | *Irrigated Land with Two Crops* | *Irrigated Land with One Crop* | *Dry Land* |
|---|---|---|---|
| 1 | 2 | 3 | 4 |
| As Recommended in 1972 National Guidelines | 10-18 | 27 | 54 |
| Proposed in Agenda Notes 1985 of RMC | 12 | 18 | 30 |
| Andhra Pradesh | 10 to 18 | 15 to 27 | 35 to 54 |
| Assam | 17 | 17 | 17 |
| Bihar | 15 to 18 | 25 | 30 to 45 |
| Gujarat | 10 to 18 | 15 to 27 | 20 to 54 |
| Haryana | 18 | 27 | 54 |
| Himachal Pradesh | 10 | 15 | 30 to 70 |
| Jammu & Kashmir | 9 to 12.5 | 9 to 12.5 | 15 to 23 (in Ladakh 19) |
| Karnataka | 10 to 20 | 25 to 30 | 54 |
| Kerala | 12 to 15 | 12 to 15 | 12 to 15 |
| Madhya Pradesh | 18 | 27 | 54 |
| Maharashtra | 18 | 27 | 54 |
| Manipur | 12 | 12 | 15 |
| Orissa | 10 | 15 | 30 to 45 |
| Punjab | 17 | 27 | 51 |
| Rajasthan | 18 | 27 | 54 to 175 |
| Tamil Nadu | 12 | 30 | 60 |
| Sikkim | 12.5 | 12.5 | 50 |
| Tripura | 10 | 10 | 30 |
| Uttar Pradesh | 18 | 27 | 45 |
| West Bengal | 12 | 12 | 17 |

*RMC:* Regional Minister's Conference

*Note:* 1. The actual limits for lands in Karnataka and Uttar Pradesh are higher due to classification of land.

2. The actual ceiling limits in Himachal Pradesh and Rajasthan are higher due to hilly terrain and desert lands.

3. 1 Acre = 0.404686 Hectare.

*Source:* Ministry of Rural Development, New Delhi.

## Table 18.2

*Daily Rates of Minimum Wages for Agricultural Workers Fixed by Different State/ UT Administrations under the Minimum Wages Act, 1948[@]*

*As on: 08.05.2008*

| *S.No.* | *Appropriate Governments* | *Minimum Wages for Unskilled Agricultural Workers (in Rupees per day)* |
|---|---|---|
| 1 | 2 | 3 |
| | Central Sphere | 117.65 - 131.22 |
| | States/UTs | |
| 1 | Andhra Pradesh | 64.00 |
| 2 | Arunachal Pradesh | 55.00 |
| 3 | Assam | 76.35 |
| 4 | Bihar | 66.00 |
| 5 | Chhattisgarh | 69.00 |
| 6 | Goa | 110.00 |
| 7 | Gujarat | 50.00 |
| 8 | Haryana | 135.00 |
| 9 | Himachal Pradesh | 75.00 |
| 10 | Jharkhand | 50.06 |
| 11 | Jammu & Kashmir | 66.00 |
| 12 | Karnataka | 71.28 |
| 13 | Kerala | 72.00 for Light Work<br>125.00 for Hard Work |
| 14 | Madhya Pradesh | 61.37 |
| 15 | Maharashtra | Zone-I 72.00<br>zone-II 70.00<br>Zone-III 68.00<br>Zone-IV 66.00 |
| 16 | Manipur | 72.40 |
| 17 | Meghalaya | 70.00 |
| 18 | Mizoram | 103.00 |
| 19 | Nagaland | 66.00 |
| 20 | Orissa | 70.00 |
| 21 | Punjab | 100.70(Without meal)<br>91.00(With meal) |
| 22 | Rajasthan | 73.00 |
| 23 | Tamil Nadu | 70.00 |
| 24 | Tripura | 85.00 |
| 25 | Uttar Pradesh | 100.00 |
| 26 | Uttarakhand | 73.00 |
| 27 | West Bengal | 71.13 with meal<br>74.33 without meal |
| 28 | Andaman & Nicobar | 130.00 (Andaman)<br>139.00 (Nicobar) |
| 29 | Chandigarh | 133.46 |
| 30 | Dadra & Nagar Haveli | 99.80 |
| 31 | Delhi | 140.00 |
| 32 | Pondicherry | |
| | (i) Pondicherry/Karaikal | 54.00 (For 5 Hours) |
| | (ii) Yanam Region | 55.00 (For 6 Hours) |

@: The Minimum Wages also include the variable dearness allowance, where provided.
*Source:* Ministry of Labour, New Delhi.

## Table 19.1

*Livestock Population in India*

*(Million Numbers)*

| *Species* | *1951* | *1956* | *1961* | *1966* | *1972* | *1977* | *1982* | *1987* | *1992* | *1997* | *2003* |
|---|---|---|---|---|---|---|---|---|---|---|---|
| *1* | 2 | 3 | 4 | 5 | 6 | 7 | 8 | 9 | *10* | *11* | *12* |
| Cattle | 155.3 | 158.7 | 175.6 | 176.2 | 178.3 | 180.0 | 192.5 | 199.7 | 204.6 | 198.8 | 185.2 |
| Adult Female Cattle | 54.4 | 47.3 | 51.0 | 51.8 | 53.4 | 54.6 | 59.2 | 62.1 | 64.4 | 63.6 | 64.5 |
| Buffalo | 43.4 | 44.9 | 51.2 | 53.0 | 57.4 | 62.0 | 69.8 | 76.0 | 84.2 | 89.9 | 97.9 |
| Adult Female Buffalo | 21.0 | 21.7 | 24.3 | 25.4 | 28.6 | 31.3 | 32.5 | 39.1 | 43.8 | 46.8 | 51.0 |
| Total Bovins | 198.7 | 203.6 | 226.8 | 229.2 | 235.7 | 242.0 | 262.4 | 275.8 | 289.0 | 289.0 | 283.4 |
| Sheep | 39.1 | 39.3 | 40.2 | 42.4 | 40.0 | 41.0 | 48.8 | 45.7 | 50.8 | 57.5 | 61.5 |
| Goats | 47.2 | 55.4 | 60.9 | 64.6 | 67.5 | 75.6 | 95.3 | 110.2 | 115.3 | 122.7 | 124.4 |
| Horses & Ponies | 1.5 | 1.5 | 1.3 | 1.1 | 0.9 | 0.9 | 0.9 | 0.8 | 0.8 | 0.8 | 0.8 |
| Camels | 0.6 | 0.8 | 0.9 | 1.0 | 1.1 | 1.1 | 1.1 | 1.0 | 1.0 | 0.9 | 0.6 |
| Pigs | 4.4 | 4.9 | 5.2 | 5.0 | 6.9 | 7.6 | 10.1 | 10.6 | 12.8 | 13.3 | 13.5 |
| Mules | 0.1 | 0.0 | 0.1 | 0.1 | 0.1 | 0.1 | 0.1 | 0.2 | 0.2 | 0.2 | 0.2 |
| Donkeys | 1.3 | 1.1 | 1.1 | 1.1 | 1.0 | 1.0 | 1.0 | 1.0 | 1.0 | 0.9 | 0.6 |
| Total Livestock | 292.8 | 306.6 | 335.4 | 344.1 | 353.6 | 369.0 | 419.6 | 445.3 | 470.9 | 485.4 | 485.0 |
| Poultry | 73.5 | 94.8 | 114.2 | 115.4 | 138.5 | 159.2 | 207.7 | 275.3 | 307.1 | 347.6 | 489.0 |
| Dogs | NC | NC | NC | NC | NC | NC | 18.5 | 18.0 | 21.8 | 25.5 | 29.0 |

*NC:* Not Collected

*Note:* Total Bovins include Yaks and Mithans

*Source:* Department of Animal Husbandry, Dairying & Fisheries, New Delhi.

## Table 19.2

*All India Production of Milk, Eggs and Wool from 1985-86 to 2006-07*

| Year | Milk (Million Tonnes) | Eggs (Billion Nos.) | Wool (Million Kgs.) |
|---|---|---|---|
| 1 | 2 | 3 | 4 |
| 1985-86 | 44.0 | 16.1 | 39.1 |
| 1986-87 | 46.1 | 17.3 | 40.0 |
| 1987-88 | 46.7 | 17.8 | 40.1 |
| 1988-89 | 48.4 | 19.0 | 40.8 |
| 1989-90 | 51.4 | 20.2 | 41.7 |
| 1990-91 | 53.9 | 21.1 | 41.2 |
| 1991-92 | 55.7 | 22.0 | 41.6 |
| 1992-93 | 58.0 | 22.9 | 38.8 |
| 1993-94 | 60.6 | 24.2 | 39.9 |
| 1994-95 | 64.0 | 26.0 | 40.6 |
| 1995-96 | 66.2 | 27.2 | 42.4 |
| 1996-97 | 69.1 | 27.5 | 44.4 |
| 1997-98 | 72.1 | 28.7 | 45.6 |
| 1998-99 | 75.4 | 29.5 | 46.9 |
| 1999-2000 | 78.3 | 30.4 | 47.9 |
| 2000-01 | 80.6 | 36.6 | 48.4 |
| 2001-02 | 84.4 | 38.7 | 49.5 |
| 2002-03 | 86.2 | 39.8 | 50.5 |
| 2003-04 | 88.1 | 40.4 | 48.5 |
| 2004-05 | 92.6 | 45.2 | 44.6 |
| 2005-06 | 97.1 | 46.2 | 44.9 |
| 2006-07 | 100.9 | 50.7 | 45.1 |

*Source:* Department of Animal Husbandry & Dairying. New Delhi.

## Table 19.3

*State-wise Production of Milk, Eggs and Wool from 2004-05 to 2006-07*

| States/UTs | Milk ('000 tonnes) | | | Eggs (Lakh nos) | | | Wool ('000 kgs) | | |
|---|---|---|---|---|---|---|---|---|---|
| | 2004-05 | 2005-06 | 2006-07 | 2004-05 | 2005-06 | 2006-07 | 2004-05 | 2005-06 | 2006-07 |
| 1 | 2 | 3 | 4 | 5 | 6 | 7 | 8 | 9 | 10 |
| Andhra Pradesh | 7257 | 7624 | 7939 | 158040 | 164534 | 159411 | 3678 | 3978 | 4230 |
| Arunachal Pradesh | 48 | 48 | 49 | 90 | 73 | 73 | 16 | 14 | 14 |
| Assam | 739 | 747 | 751 | 5216 | 5359 | 5350 | | | |
| Bihar | 4743 | 5060 | 5450 | 7894 | 10012 | 9454 | 378 | 220 | 231 |
| Chhattisgarh | 831 | 839 | 849 | 8795 | 8875 | 9216 | 255 | | 245 |
| Goa | 57 | 56 | 57 | 146 | 146 | 135 | | | |
| Gujarat | 6745 | 6980 | 7533 | 5031 | 5775 | 7757 | 2950 | 3123 | 2962 |
| Haryana | 5222 | 5299 | 5367 | 14816 | 15126 | 39596 | 1304 | 1136 | 1121 |
| Himachal Pradesh | 870 | 869 | 872 | 811 | 753 | 772 | 1600 | 1603 | 1605 |
| Jammu & Kashmir | 1422 | 1400 | 1400 | 6105 | 6320 | 6320 | 7120 | 7400 | 7400 |
| Jharkhand | 1330 | 1335 | 1401 | 6980 | 6970 | 7110 | 348 | 353 | 146 |
| Karnataka | 3917 | 4022 | 4124 | 17719 | 18348 | 19498 | 6448 | 150 | 5599 |
| Kerala | 2025 | 2063 | 2119 | 11974 | 11956 | 11987 | | | |
| Madhya Pradesh | 5506 | 6283 | 6375 | 9023 | 9414 | 9518 | 681 | 5598 | 428 |
| Maharashtra | 6567 | 6769 | 6978 | 34362 | 35227 | 33950 | 1620 | 1640 | 1667 |
| Manipur | 75 | 77 | 77 | 822 | 835 | 836 | | | |
| Meghalaya | 71 | 73 | 75 | 946 | 973 | 978 | | | |
| Mizoram | 16 | 15 | 16 | 317 | 326 | 348 | | | |
| Nagaland | 69 | 74 | 67 | 755 | 868 | 868 | | | |

...contd...

*...contd...*

| 1 | 2 | 3 | 4 | 5 | 6 | 7 | 8 | 9 | 10 |
|---|---|---|---|---|---|---|---|---|---|
| Orissa | 1283 | 1342 | 1431 | 12160 | 12787 | 14246 | | | |
| Punjab | 8554 | 8909 | 9168 | 36800 | 35200 | 37740 | 566 | 712 | 513 |
| Rajasthan | 8310 | 8713 | 9375 | 6932 | 7029 | 6631 | 15049 | 15405 | 15685 |
| Sikkim | 46 | 46 | 49 | 140 | 146 | 144 | 4 | 2 | 1 |
| Tamil Nadu | 4784 | 5474 | 5560 | 63948 | 62225 | 80435 | 750 | 750 | 750 |
| Tripura | 86 | 87 | 89 | 1071 | 1100 | 1193 | | | |
| Uttar Pradesh | 18512 | 17358 | 18095 | 9018 | 9228 | 9483 | 1965 | 1459 | 1461 |
| Uttarakhand | 1195 | 1206 | 1213 | 1866 | 1673 | 1889 | 148 | 150 | 355 |
| West Bengal | 3790 | 3891 | 3982 | 28877 | 29637 | 30386 | 659 | 866 | 673 |
| A & N Islands | 24 | 20 | 23 | 530 | 477 | 535 | | | |
| Chandigarh | 43 | 46 | 46 | 280 | 280 | 291 | | | |
| Daman & Diu | 1 | 1 | 1 | 9 | 11 | 11 | | | |
| D & N Haveli | 4 | 5 | 5 | 50 | 48 | 48 | | | |
| Delhi | 303 | 310 | 289 | 264 | 196 | 186 | | | |
| Lakshadweep | 1 | 2 | 2 | 120 | 124 | 127 | | | |
| Pondicherry | 41 | 43 | 45 | 102 | 105 | 107 | | | |
| All India | 92484 | 97066 | 100869 | 452009 | 462307 | 506630 | 44575 | 44884 | 45085 |

*Note:* Totals may not tally due to rounding off.

*Source:* Department of Animal Husbandry & Dairying. New Delhi.

# Table 20.1

*Fish Production in India in 1950-51, 1960-61, 1970-71 and 1980-81 to 2006-07*

*(Lakh Tonnes)*

| Year | Marine | Inland | Total |
|---|---|---|---|
| 1 | 2 | 3 | 4 |
| 1950-51 | 5.34 | 2.18 | 7.52 |
| 1960-61 | 8.80 | 2.80 | 11.60 |
| 1970-71 | 10.86 | 6.70 | 17.56 |
| 1980-81 | 15.55 | 8.87 | 24.42 |
| 1981-82 | 14.45 | 9.99 | 24.44 |
| 1982-83 | 14.27 | 9.40 | 23.67 |
| 1983-84 | 15.19 | 9.87 | 25.06 |
| 1984-85 | 16.98 | 11.03 | 28.01 |
| 1985-86 | 17.16 | 11.60 | 28.76 |
| 1986-87 | 17.13 | 12.29 | 29.42 |
| 1987-88 | 16.58 | 13.01 | 29.59 |
| 1988-89 | 18.17 | 13.35 | 31.52 |
| 1989-90 | 22.75 | 14.02 | 36.77 |
| 1990-91 | 23.00 | 15.36 | 38.36 |
| 1991-92 | 24.47 | 17.10 | 41.57 |
| 1992-93 | 25.76 | 17.89 | 43.65 |
| 1993-94 | 26.49 | 19.95 | 46.44 |
| 1994-95 | 26.92 | 20.97 | 47.89 |
| 1995-96 | 27.07 | 22.42 | 49.49 |
| 1996-97 | 29.67 | 23.81 | 53.48 |
| 1997-98 | 29.50 | 24.38 | 53.88 |
| 1998-99 | 26.96 | 26.02 | 52.98 |
| 1999-2000 | 28.52 | 28.23 | 56.75 |
| 2000-01 | 28.11 | 28.45 | 56.56 |
| 2001-02 | 28.30 | 31.26 | 59.56 |
| 2002-03 | 29.90 | 32.10 | 62.00 |
| 2003-04 | 29.41 | 34.58 | 63.99 |
| 2004-05 | 27.79 | 35.25 | 63.05 |
| 2005-06 | 28.16 | 37.56 | 65.72 |
| 2006-07(P) | 30.24 | 38.45 | 68.69 |

(P): Provisional

*Source:* Department of Animal Husbandry, Dairying & Fisheries, Ministry of Agriculture, New Delhi

## Table 20.2

*State-wise Fish Production in 2006-07 (Provisional)*

*(In '000 Tonnes)*

| *State/UTs* | *Marine* | *Inland* | *Total* |
|---|---|---|---|
| *1* | *2* | *3* | *4* |
| Andhra Pradesh | 240.20 | 616.73 | 856.93 |
| Arunachal Pradesh | 0.00 | 2.77 | 2.77 |
| Assam | 0.00 | 181.48 | 181.48 |
| Bihar | 0.00 | 267.04 | 267.04 |
| Goa | 98.96 | 3.43 | 102.39 |
| Gujarat | 670.51 | 76.82 | 747.33 |
| Haryana | 0.00 | 60.08 | 60.08 |
| Himachal Pradesh | 0.00 | 6.89 | 6.89 |
| Jammu & Kashmir | 0.00 | 19.20 | 19.20 |
| Karnataka | 168.54 | 123.92 | 292.46 |
| Kerala | 598.06 | 79.57 | 677.63 |
| Madhya Pradesh | 0.00 | 65.04 | 65.04 |
| Maharashtra | 464.09 | 131.85 | 595.94 |
| Manipur | 0.00 | 18.61 | 18.61 |
| Meghalaya | 0.00 | 5.49 | 5.49 |
| Mizoram | 0.00 | 3.76 | 3.76 |
| Nagaland | 0.00 | 5.80 | 5.80 |
| Orissa | 128.14 | 213.90 | 342.04 |
| Punjab | 0.00 | 86.70 | 86.70 |
| Rajasthan | 0.00 | 22.20 | 22.20 |
| Sikkim | 0.00 | 0.15 | 0.15 |
| Tamil Nadu | 387.25 | 155.03 | 542.28 |
| Tripura | 0.00 | 28.63 | 28.63 |
| Uttar Pradesh | 0.00 | 306.73 | 306.73 |
| West Bengal | 178.10 | 1181.00 | 1359.10 |
| A & N Islands | 28.60 | 0.08 | 28.68 |
| Chandigarh | 0.00 | 0.17 | 0.17 |
| D & N Haveli | 0.00 | 0.05 | 0.05 |
| Daman & Diu | 16.35 | 0.06 | 16.41 |
| Delhi | 0.00 | 0.61 | 0.61 |
| Lakshadweep | 11.75 | 0.00 | 11.75 |
| Pondicherry | 33.61 | 6.05 | 39.66 |
| Chattisgarh | 0.00 | 137.75 | 137.75 |
| Uttarakhand | 0.00 | 3.03 | 3.03 |
| Jharkhand | 0.00 | 34.27 | 34.27 |
| Total | 3024.16 | 3844.89 | 6869.05 |

*Source*: Department of Animal Husbandry, Dairying & Fisheries, Ministry of Agriculture, New Delhi.

## 21.1: ANNUAL CLIMATE SCENARIO DURING 2007

### Highlights

The year 2007 witnessed extreme weather events across the country, both in terms of temperature and precipitation. The year was characterised by above normal temperature over different parts of the country. 2007 was the fourth warmest year on the record since 1901 after, 2002, 2006 and 2003. During the Southwest monsoon season, rainfall for the country as a whole was above normal. However, it was characterised by remarkable spatial variability with Peninsular parts of the country receiving excess rainfall, while Northern parts of the country receiving deficient rainfall. Northeast monsoon activity over South Peninsula was also normal.

### Temperature

Maximum, minimum and mean temperatures were above normal over most parts of the country except some parts of Peninsula and Northeastern region. Minimum temperatures were above normal by more 2°C over parts of Himachal Pradesh, maximum temperatures were above normal by more the 2°C over Northern parts of Rajasthan.

Spatial pattern of trends of mean annual temperatures anomalies suggest significant positive (increasing) trend over most parts of the country except over some parts of Rajasthan, Gujarat and Bihar, where significant negative (decreasing) trends were observed.

During January, maximum temperatures were above normal throughout the country except over some parts of Northeast India. They were above normal by about 2°C over parts of Central India and by more than 4°C over parts of Jammu and Kashmir. Minimum temperatures were above normal over Central, extreme Northwestern and Northern parts (except over Punjab, Haryana and parts of West Uttar Pradesh) of the country. They were below normal elsewhere. They were above normal by more then 2°C over parts of central, Northern and Northwestern India.

During February, maximum and minimum temperatures were generally above normal over most parts of the country. Northern parts of Jammu & Kashmir were warmer by more than 3°C during the month. Over most of the Northern parts of the country, minimum temperatures were above normal by about 2°C to 3°C.

During March, maximum temperatures were above normal over the Peninsula, extreme Northern & Northeastern parts of the country and were below normal elsewhere. Over parts of East Uttar Pradesh and Bihar, maximum temperatures were below normal by more than 2°C. Minimum temperatures were above normal over most parts of the country except over Eastern region, parts of Jammu & Kashmir, Punjab and extreme South Peninsula where they were slightly below normal.

## Rainfall in 2007

In 2007, the annual rainfall over the country as a whole was 99 per cent of Long Period Average (LPA). Season wise rainfall distribution over the country as a whole was as follows:

| | |
|---|---|
| Winter (January to February): | 87 per cent of LPA |
| Pre-monsoon (March to May): | 84 per cent of LPA |
| Monsoon (June to September): | 106 per cent of LPA |
| Post-Monsoon (Oct. to Dec.): | 68 per cent of LPA |

## Annual Rainfall Activity 2007

Rainfall activity over the country as a whole was normal during the year. Gangetic West Bengal, Uttaranchal, Saurashtra & Kutch, Rayalseema, South Interior Karnataka and Lakshadweep received excess rainfall. West UP, Punjab, Himachal Pradesh, East Madhya Pradesh and Marathwada received deficient rainfall. Out of 36 meteorological subdivisions, 6 received excess rainfall, 25 received normal rainfall and 5 received deficient rainfall.

During the winter season (January-February), rainfall activity over the country as a whole was generally subdued. However, many subdivisions of North and East India received excess rainfall. Peninsular India either received scanty rainfall or no rain. Out of 36 meteorological subdivisions, 11 received excess rainfall, 6 received normal rainfall, 5 received deficient rainfall and 10 received scanty rainfall. Four (4) subdivisions did not received any rain.

Rainfall activity during the pre-monsoon season (March-May) over the country as a whole was generally subdued. However, East Uttar Pradesh, West UP, Haryana, Chandigarh & Delhi, Punjab and West Rajasthan received excess rainfall. Out of 36 meteorological subdivisions, 5 received excess rainfall, 11 received normal rainfall, 16 received deficient rainfall and 4 received scanty rainfall.

During the monsoon season (June-September) for the country as whole, seasonal rainfall at the end of season was 105 per cent of its LPA value. The

LPA value of southwest monsoon rainfall calculated with the data of the period, 1941-1990 is 89 cm. During the season, 13 subdivisions received excess rainfall, 18 received normal rainfall and the remaining 5 subdivisions received deficient rainfall. Moderate drought conditions (rainfall deficiency of 26% to 50%) prevailed over five subdivisions *viz.*: West UP (-39%), Haryana, Chandigarh and Delhi (-33%), Punjab (-32%), Himachal Pradesh (-36%) and East Madhya Pradesh (-28%). Out of 504 meteorological districts for which rainfall data were available, 73 per cent of the meteorological districts received excess/normal rainfall and the remaining 27 per cent districts received deficient rainfall (rainfall deficiency more than 19%). During the season, 74 districts (15%) experienced moderate drought conditions (rainfall deficiency of 26% to 50%) and 29 districts (6) experiences severe drought conditions (rainfall deficiency 51% and more). During the season, rainfall was excess (rainfall above its LPA by 20% or more) in 162 districts (32%).

Northeast monsoon (post-monsoon: October-December) over South Peninsula was normal during the season. Tamil Nadu & Pondicherry received excess rainfall. However, Rayalaseema, South Interior Karnataka and Kerala received normal rainfall. During the post-monsoon season: 2 subdivisions received excess rainfall, 7 received normal rainfall, 9 received deficient rainfall and 18 received scanty rainfall. Northeast monsoon seasonal rainfall over south peninsula (comprising of 5 subdivisions *viz.* Coastal Andhra Pradesh, Rayalaseema, Tamil Nadu & Pondicherry, South Interior Karnataka and Kerala) was 104 per cent of its LPA.

## Extreme weather events

Severe cold wave/cold wave conditions prevailed over Northern parts of the country during first fortnight of January. Heat wave conditions with maximum temperatures above normal by 5 to 7°C prevailed over northern and central parts of the country during the first 10 days of June. Heat wave conditions also prevailed over parts of Orissa, Coastal Andhra Pradesh and Telangana during second fortnight of May. Many stations of western Himalayas were abnormally warmer from last week of March to first week of May. Similarly, many stations of eastern Himalayas were abnormally warmer in the first week of May.

---

Source of the above material: National Climate Centre, India Meteorological Department.

## Table 21.2

*Brief History of Most Intense Cyclones from 1970 to 2007*

| S.No. | Name of Cyclone | Date and Year | Intensity (T-Number) | Observed/ Estimated Max. Wind | Damage |
|---|---|---|---|---|---|
| 1 | 2 | 3 | 4 | 5 | 6 |
| **I. Over the Arabian Sea** | | | | | |
| 1 | Severe Cyclonic Storm over the Arabian sea | October 19-24, 1975 | N/A | 97 Knots 180 Kmph | 85 people died in the districts of Junagarh. Jamnager and Rajkot of Gujarat state. This Cyclone caused considerable damage, estimated to be about Rs. 75 Crores. |
| 2 | Severe Cyclonic Storm over the Arabian sea | May 31-June 5, 1976 | N/A | 90 Knots 167 Kmph | This Cyclone caused damage to property which was estimated to be about Rs. 3 Crore. Four Burges each. Containing Rs. 5 Lakh and 6 fishing boats were swept away. Mehasana, Bhavnagar, Kaira, Panchmahal, Rajkot and Broach districts of Gujarat State were most affected areas. |
| 3 | Severe Cyclonic Storm over the Arabian sea | November 13-23, 1977 | T- 5.5 (as per US satellite) | 90 Knots 167 Kmph | Kerala and Lakshadweep were most affected areas due to this storm. People killed—72. Houses damaged—8400 and 620 fishing Vessels damaged in Kerala coast. Total loss was estimated to be about Rs. 10 crores. |

...contd...

*...contd...*

| *1* | *2* | *3* | *4* | *5* | *6* |
|---|---|---|---|---|---|
| 4 | Gujarat Cyclone | November 5-13,1978 | N/A | 278 Kmph | Damage to Property reported. |
| 5 | Severe Cyclonic Storm over the Arabian sea | October 28-November 3, 1981 | T-4.0 (as per US satellite) | 65 Knots 120 Kmph | Junagarh, Rajkot and Jamnagar of Gujarat state were most affected areas, Total loss of damage to property was estimated to be about Rs. 52 Crores. |
| 6 | Severe Cyclonic Storm over the Arabian sea | November 4-9, 1982 | N/A | N/A | Saurastra Coast of Gujarat about 45 km east of Veraval was affected very much by this storm. 507 people died and 1.5 lakh livestock perished. |
| 7 | Severe Cyclonic Storm over the Arabian sea | November 4-9, 1982 | T- 4.0 | 50 Knots 93 Kmph | 50 fishermen were reported missing in Gujarat Coast. |
| 8 | Severe Cyclonic Storm over the Arabian sea | June 17-20, 1996 | T-3.5 | 60 Knots 111 Kmph | 19 Districts of Gujarat State were affected. 33 people died. 27964 pucca houses were destroyed. Total estimated loss was Rs. 1803.52 lakh. |
| 9 | Severe Cyclonic Storm over the Arabian sea | October 23-28, 1996 | T-4.0 | 60 Knots 111 Kmph | As the system did not cross the coast no significant damage was reported. |
| 10 | Very severe Cyclonic Storm over the Arabian sea | June 4-10, 1998 | T-5.0 | 90 Knots 167 Kmph | Gujarat & Rajasthan states were affected. Porbander of Gujarat state was the most affected area. Loss incurred due to storm was estimated to be about Rs.1855.38 |

*...contd...*

*...contd...*

| *1* | *2* | *3* | *4* | *5* | *6* |
|---|---|---|---|---|---|
| | | | | | Crores in Kandla. Number of lines lost 1173 and number of persons missing 1774 in Gujarat. |
| 11 | Very severe Cyclonic Storm over the Arabian sea | May 16-22, 1999 | T- 5.5 | 55 Knots 102 Kmph | This system caused severe damage in Kutch and Jamnagar districts of Gujarat. 453 people died. Loss of property was estimated to be about Rs. 80 crores. In Rajasthan one person died and 5104 cattle heads perished. 5133 houses were partially damaged. |
| 12 | Severe Cyclonic Storm over Arabian sea | May 5-10, 2004 | 3.5 | Weakened over sea | Widespread damage caused at Lakshadweep. Several houses damaged due to wind, sea wave, falling of trees and floods. 45 houses, 35 boats and small mechanised cargo ships lost in sea near land. 16 boats reported sunk and cargo boat damaged. Sea erosion reported over Kerala and 9 people dead and 17 people missing. Total loss worth Rs.30 crores. |
| 13 | Severe Cyclonic "Storm MUKDA" over Arabian sea | September 21-24, 2006 | 3.5 | 55 Knots 102 Kmph | No damage as dissipated over the Arabian Sea. |
| 14 | Super cyclone | 01-07 June | T-6.5 | 127 knots | The system crossed Makaran |

*...contd...*

*...contd...*

| 1 | 2 | 3 | 4 | 5 | 6 |
|---|---|---|---|---|---|
| | "GONU" | 2007 | | | coast and hence there was no damage in India. |
| 15 | Cyclonic storm "YEMYIN" | 25-26 June, 2007 | T - 2.5 | 35 knots | The system crossed Pakistan coast and hence there was no damage in India |
| **II. Over the Bay of Bengal** | | | | | |
| 1 | Bangladesh (Chittagong) | November 12-13, 1970 | N/A | 224 Kmph | 3,00,000 People died. |
| 2 | Balasore (Orissa) | October 26-31, 1971 | N/A | 185 Kmph | 10,000 people died and 1 million people rendered homeless. |
| 3 | Andhra Cyclone | November 14-20, 1977 | N/A | 259 Kmph | 8547 People died. Strom surge height 5 m. |
| 4 | Sri Harikota (A.P.) | November 9-14, 1984 | T-6.0 | 213 Kmph | 604 people died. |
| 5 | Indo-Bangladesh Border | November 24-30, 1988 | T-6.0 | 213 Kmph | 2000 people died. 6000 people missing in Bangladesh. |
| 6 | Kavali (A.P.) | November 1-9, 1989 | T-6.5 | 235 Kmph | 69 people died. |
| 7 | Machilipatnam (A.P.) | May 4-9, 1990 | T-6.5 | 235 Kmph | 967 people died. |
| 8 | Bangladesh (Chittagong) | April 25-30, 1991 | T-6.5 | 235 Kmph | 1,38,882 people died. 1,39,054 people injured. |
| 9 | Bangladesh (Teknaf) | April 24-May 2, 1994 | T-6.0 | 215 Kmph | 184 people killed |
| 10 | Bangladesh | May 15-19, | T-6.0 | 230 Kmph as | 155 people died. |

*...contd...*

*...contd...*

| 1 | 2 | 3 | 4 | 5 | 6 |
|---|---|---|---|---|---|
| | (Teknaf) | 1997 | | reported by Bangladesh Met. Office | 9663 people injured. |
| 11 | Gopalpur Cyclone (Orissa) | October 15-19, 1999 | T-5.0 | 170 Kmph | 198 people died. 402 persons injured. |
| 12 | Super Cyclone (Orissa) | October 25-31, 1999 | T-7.0 | 260 Kmph | 9887 people died. 129.22 lakh people affected. |
| 13 | Very severe Cyclonic Storm over the Bay of Bengal. | November 26-30, 2000 | T-5.5 | 102 Knots 189 kmph | Two states *viz.*, Tamil Nadu and Pondicherry were mainly affected by this storm.The loss was mainly due to crop damage. Uprooting of big trees and partial damages to more than one thousand kuchha houses. 30,000, Plantain trees and 50,000 Plantain saplings got destroyed and 30,000 trees were uprooted in Tamil Nadu state. In Pondicherry two persons lost their lives. Damage to paddy crops, plantains, coconut plantation were the major losses in Pondicherry. |
| 14 | Very severe Cyclonic Storm over the Bay of Bengal | December 23-28, 2000 | T-5.0 | 90 Knots 167 Kmph | Three districts of Tamil Nadu state were affected by this storm in the Ramnathanpuram district, 350 houses were damaged in Thirunelveli. Houses damaged— |

*...contd...*

*...contd...*

| 1 | 2 | 3 | 4 | 5 | 6 |
|---|---|---|---|---|---|
| | | | | | 318. In Tutocorin houses damaged—318, Fishing boats lost —95, loss to crops—281 hectares of paddy, 650 hectares of plaintain and 80 hectares of betal destroyed. |
| 15 | Cyclonic Storm over Bay of Bengal | October 14-17, 2001 | 2.5 | 35 Knots 65 kmph | 108 people dead, 21 people missing, damage to crops about 125000 Hectares and 55747 house damaged in Andhra Pradesh. |
| 16 | Severe Cyclonic Storm over Bay of Bengal | November 10-12, 2002 | 3.5 | 55 Knots 102 kmph | Two travelers damaged and 18 people died in Orissa and another two trawlers missing. In West Bengal 2 people died. |
| 17 | Severe Cyclonic Storm over Bay of Bengal | December 11-16, 2003 | 3.5 | 55 Knots 102 kmph | 81 people dead in Andhra Pradesh, 1637 Building destroyed completely, 7453 Building partially damaged 61898.5 hectares of Agricultural land damaged. Loss of property to the tune of Rs. 23903.13 Lakh. No of villages electric failure are 2000 and telecommunication disruption are in 41 villages. |
| 18 | Cyclonic Storm "PYARR" over Bay of Bengal | September 17-21, 2005 | T-2.5 | 35 Knots 65 kmph | Damages reported in Andhra Pradesh are as follows Number of Human deaths—1, Number of livestock died—291, Buildings partly/completely destroyed— |

*...contd...*

*...contd...*

| 1 | 2 | 3 | 4 | 5 | 6 |
|---|---|---|---|---|---|
| | | | | | 12041, (estimated cost Rs.177.28 lakhs), Crop loss—4,82,188 hectors (estimated cost Rs.627.74 lakhs), Total loss estimated by Government—5029.82 lakhs. |
| 19 | Cyclonic storm "BAAZ" over Bay of Bengal | November 28-December-02, 2005 | T-3.0 | 45 Knots 83 Kmph | According to press reports, heavy rain caused floods in Nellore. Chittoor and Cuddapah districts of Andhra Pradesh. Number of deaths—11 (Nellore 7, Chittoor 3 Cuddapah 1), Number of tanks breached—27 (Nellore district), Many villages were reported to be marooned in the above districts. |
| 20 | The Very Severe Cyclonic Storm "Mala" over the Bay of Bengal | May 25- 30, 2006 | 5.5 | 100 Knots 185 Kmph | lt crossed the coast of Myanmar and hence there was no damage in India. |
| 21 | Cyclonic Storm "OGNI" over Bay of Bengal | October 29-30, 2006 | 2.5 | 35 Knots 65 Kmph | Andhra Pradesh Loss of life: 24 Livestock: 3,61,553 Loss of crops: 1,99,986 acres Villages submerged: 900 Damage to houses (fully): 26,853 Damage to houses (partly): 73,218. Total loss: Rs. 21,601 lakhs. |

*...contd...*

*...contd...*

| 1 | 2 | 3 | 4 | 5 | 6 |
|---|---|---|---|---|---|
| 22 | Cyclonic storm "AKASH" | 13-15 May, 2007 | T-6.0 | 115 knots | The system crossed Bangladesh coast and hence there was no damage in India |
| 23 | Very serere cyclonic storm "SIDR" | 11-16 November | T-6.0 | 115 knots | The system crossed Bangladesh coast. However, one person died. 46 Villages and thousands of people were affected in west Bengal. Crops were damaged in thousands of hectares of land. There was disruption of electricity supply in the coastal belt of West Bangal due to breaking of dam over Bidyadhari river. Extensive ares were flooded near Gajikhali and Kheaghat. Some houses were partially/totally damaged in Mizoram. Shillong road was also affected at several places. |

*Source:* India Meteorological Department, New Delhi.

# Table 22

*Conversion Factors between Important Primary and Secondary Agricultural Commodities*

| *Commodity* | *Conversion Factor* |
|---|---|
| 1 | 2 |
| Rice (Cleaned) Production | 2/3 of Paddy Production |
| Cotton | |
| Cotton Lint Production | 1/3 of Kapas Production |
| Cotton Seed Production | 2/3 of Kapas Production |
| | 2 Times of Cotton Lint Production |
| **Jute** | |
| 100 Yards of Hessian | 54 lbs. of Raw Jute |
| 4148 Yards of Hessian | 1 Tonne of Raw Jute |
| | 5.55 Bales of Raw Jute (of 180 Kgs. Each) |
| 1 Tonne of Sacking | 1.11 Tonne of Raw Jute |
| | 6.17 Bales of Raw Jute (of 180 Kgs. Each) |
| 1 Tonne of Hessian | 1.05 Tonnes of Raw Jute |
| Sacking etc. | 5.85 Bales of Raw Jute (of 180 Kgs. Each) |
| **Groundnut** | |
| Kernel to Nuts in Shell | 70 Per cent |
| Oil to Nuts in Shell | 28 Per cent |
| Oil to Kernels Crushed | 40 Per cent |
| Cake to Kernels Crushed | 60 Per cent |
| **Sesamum** | |
| Oil to Seeds Crushed | 40 Per cent |
| Cake to Seeds Crushed | 60 Per cent |
| **Rapseed and Mustard** | |
| Oil to Seeds Crushed | 33 Per cent |
| Cake to Seeds Crushed | 67 Per cent |
| **Linseed** | |
| Oil to seeds Crushed | 33 Per cent |
| Cake to Seeds Crushed | 67 Per cent |
| **Castorseed** | |
| Oil to Seeds Crushed | 37 Per cent |
| Cake to Seeds Crushed | 63 Per cent |
| **Cotton Seed** | |
| Oil to Seeds Crushed | 14 to 18 Per cent |
| Cake to Seeds Crushed | 82 to 86 Per cent |

...contd...

...contd...

| 1 | 2 |
|---|---|
| **Coconut** | |
| Copra to Nuts | One Tonne of Copra = 6773 Nuts |
| Oil to Copra Crushed | 62 Per cent |
| Cake to Copra Crushed | 38 Per cent |
| **Nigerseed** | |
| Oil to Seeds Crushed | 28 Per cent |
| Cake to Seeds Crushed | 72 Per cent |
| **Kardi Seed** | |
| Oil to Seeds Crushed | 40 Per cent |
| Cake to Seeds Crushed | 60 Per cent |
| **Mahua Seed** | |
| Oil to Seeds Crushed | 36 Per cent |
| Cake to Seeds Crushed | 64 Per cent |
| **Neem Seed** | |
| Oil to Kernels Crushed | 45 to 50 Per cent |
| Cake to Kernels Crushed | 50 to 55 Per cent |
| **Soyabean Seed** | |
| Oil to Soyabean Seed Crushed | 18 Per cent |
| Meal to Soyabean Seed Crushed | 73 Per cent |
| Hull from Soyabean Seed Crushed | 8 Per cent |
| Wastage from Soyabean Seed Crushed | 1 Per cent |
| **Sugar** | |
| *Gur* from Cane Crushed | 11.20 Per cent to 11.50 Per cent |
| Crystal Sugar from *Gur* Refined (*Gur* Refineries) | 62.5 Per cent |
| Crystal Sugar from Cane Crushed (Cane Factories) | 10.20 Per cent |
| *Khandasari* Sugar (Sulpher and Non-sulpher) from standard *Gur* Refined | 46 Per cent |
| Molasses from Cane Crushed | 4.0 Per cent to 4.5 Per cent |
| Cane–Trash* from Cane Harvested | 8.0 Per cent to 12.0 Per cent |
| **Lac** | |
| Seed Lac | 66.0 Per cent of Stick Lac |
| Shell Lac | 57.4 Per cent of Stick, or<br>87.0 Per cent of Seed Lac |
| **Cashewnut** | |
| Cashew Kernel | 25 Per cent of Cashewnuts |

* This consists of leaves and portion of the top of stalk which are removed from the canestalk, while harvesting and before sending the cane for milling.

# Appendix I

*Explanatory Notes*

## Poverty Line

1. The Task Force, constituted by the Planning Commission, defined the poverty line as per capita consumption expenditure level, which meets the average per capita daily calorie requirement of 2400 calorie in rural areas and 2100 calorie in urban areas alongwith a minimum of non-food expenditure.
2. The poverty line in 2004-05 at all-India level as estimated by the Planning Commission is Rs. 356.30 per capita per month for rural areas and Rs. 538.60 per capita per month for urban areas.

## Advance Estimates

3. The estimates of area, production and yield rates for 2007-08 are "advance estimates" as on 09.07.2008 and are based on deliberations held in weekly inter-disciplinary meetings of Crop Weather Watch Group, availability of water in important reservoirs in the country, availability/supply of important inputs including credit to farmers community, feed back received from states etc. These estimates are "advance" in the sense that these have been prepared in advance of availability of results of CCE (Crop Cutting Experiments) and are subject to revision as more precise information flows from states to DES.

## Yield

4. Yield rate is defined as production per unit of area. However, yield rates of various crops given in this publication may not exactly tally with the ratio of production to the corresponding area because figures of area and production of various crops have been generally rounded off in ten thousands.

## Growth Rates

5. The growth rates of area, production and yield of various crops given in the tables contained in Chapter 6 of this publication depict percentage growth based on index numbers of corresponding parameter. The methodology adopted for working out index numbers is given in Appendix II.

## Principal Crops

6. Principal Crops as referred to in this publication include only those crops which are covered for preparation of index numbers. The details of these crops are given in Appendix II.

## Estimation Procedure for Non-Availability of Data for 2007-08

7. The index numbers of area, production and yield rate for 2007-08 (from where growth rates have been derived) are based on advance estimates of principal crops as on 09.07.2008. However, there are a number of 'non-forecast' crops such as plantation crops, cardamom and spices for which no advance estimates for 2007-08 (or for that matter any year) are prepared. In such cases, the estimates of area and production for 2006-07 are repeated for 2007-08 for the purpose of index numbers.

## Nine Oilseeds

8. Wherever "Nine oilseeds" is mentioned in this publication, it includes Castorseed, Groundnut, Linseed, Nigerseed, Safflower, Sesamum, Soyabean, Sunflower and Rapeseed & Mustard.

## Total Oilseeds

9. Total oilseeds include cottonseed and coconut besides nine oilseeds mentioned in the above paragraph.

## Net Availability of Foodgrains

10. The Net availability of foodgrains is defined as

$$NAF = GP - SFW - e + i + s$$

Where NAF = Net availability of Foodgrains

GP = Gross Production of Foodgrains

SFW = Seed, feed and wastages of Foodgrains

e = Exports of Foodgrains

i = Imports of Foodgrains

s = Change in Stocks of Foodgrains

To work out per capita net availability of foodgrains in terms of kg. per year, NAF is divided by the estimated population for a particular year. When this is further divided by the number of days in a year i.e. 365 days, it gives net availability of foodgrains per day.

# Appendix II

*Methodology of Index Numbers of Area, Production and Yield*

## Base Year

The Directorate of Economics and Statistics (DES) had adopted Triennium Ending (TE) 1981-82 as base year for the purpose of Index Numbers of Area, Production and Yield in Agriculture until 1999-2000. In 2000-01, it decided to adopt TE 1993-1994 as a way of updating the base to a recent year and keep it in harmony with the other series of indices such as Index of Industrial Production, Whole Sale Price Index and the series of National Accounts Statistics. Since area and production in agriculture can fluctuate a great deal from year to year, the average of area/production over the TE 1993-94 is taken for determining the base level of area/production against which the area/production of the succeeding years is to be compared.

In a chapter of this publication, growth rates of various crops worked out from index numbers based on the above mentioned two base years have been separately given.

## Weighting Diagram

The weight of a commodity for the production index is taken as the average production of the commodity in the TE 1993-94 and the national average price of the commodity during 1993-94 as obtained from the National Accounts Statistics. This has been done to fix the base production at its average level by eliminating the cyclical variation and to evaluate the production with the same price for all states in view of wide variations observed in the state prices. A similar exercise was done for preparing weighting diagram for TE 1981-82.

## Coverage

The index numbers of agriculture commodities cover 46 crops under two main groups and eight sub-groups. The crops and groups are listed below:

*Foodgrains*

| | | |
|---|---|---|
| i. Cereals | - | Rice, wheat, *jowar, bajra*, maize, *ragi*, barley and small millets. (Crops except rice and wheat constitute the sub-group coarse cereals). |
| ii. Pulses | - | Gram, *tur* and other pulses. |

*Non-foodgrains*

| | | |
|---|---|---|
| i. Oilseeds | - | Groundnut, sesamum, rapeseed & mustard, linseed, castorseed, safflower, nigerseed, soyabean, sunflower, coconut and cottonseed. (The oilseed crops except coconut and cottonseed constitute 9 oilseeds) |
| ii. Fibres | - | Cotton, jute, mesta and sannhemp. (Jute and Mesta constitute a sub-group) |
| iii. Plantation crops | - | Tea, coffee and rubber. |
| iv. Condiments and spices | - | Pepper, ginger, garlic, chillies, turmeric, arecanut, coriander and cardamom. |
| v. Fruits & vegetables | - | Potato, onion, banana, cashewnut, tapioca and sweet potato. |
| vi. Other crops | - | Sugarcane, tobacco and guarseed |

## Methodology

Let aijk - the area under $i^{th}$ crop in the $j^{th}$ year in $k^{th}$ state.

aiok - the area under $i^{th}$ crop in base year period in $k^{th}$ state.

pijk - production of $i^{th}$ crop in the jth year in $k^{th}$ state.

piok - production of $i^{th}$ crop in base year period in $k^{th}$ state.

wik - weight of $i^{th}$ crop in $k^{th}$ state.

Pio - price per unit of the $i^{th}$ crop in the base period.

For the state k and for the year j, individual crop indices are calculated as below:

(a) Index number of area $= \dfrac{a_{ijk}}{a_{iok}} \times 100 = IA_{ijk}$

(b) Index number of production $= \dfrac{p_{ijk}}{p_{iok}} \times 100 = IP_{ijk}$

(c) Index number of yield $= \dfrac{IP_{ijk}}{IA_{ijk}} \times 100$

For any sub-group G of commodities, the indices for the year j and state k are calculated as below:

(a) Index number of area $= \dfrac{\Sigma a_{ijk}}{\Sigma a_{iok}} \times 100$

The state index is obtained by including all the items of the state in sub-group G where the summation is taken over items in G.

(b) Index number of production $= \dfrac{\Sigma p_{ijk} P_{io} \times 100}{\Sigma p_{iok} P_{io}}$

(c) Index number of yield $= \dfrac{\text{Index number of production} \times 100}{\text{Index number of area}}$

# Appendix III

*Glossary of English, Botanical and Hindi Names of Important Crops*

| Crop/Group of Crops | English | Botanical | Hindi |
|---|---|---|---|
| 1 | 2 | 3 | 4 |
| Cereals | Bara (Bulrush or spiked millet) | Pennisetum Typhoides | *Bajra* |
| | Barley | Hordeum vulgare | *Jau* |
| | Barnyard millet | Echinochloa frumentacea | *Kutki* |
| | Cholam (Great Millet) | Sorghum bicolor | *Jowar* |
| | Common Millet | Panicum milliaceum | *Cheena* |
| | Little Millet | Penicum milliare | *Sawan* |
| | Italian foxtail Millet | Setarisiaitalica | *Kangani* |
| | Kodo Millet | Paspalum scrobiculatum | *Koden* |
| | Maize or Indian corn | Zea mays | *Makka* |
| | Oat | Avena sativa | *Jaie* |
| | Ragi | Eleusine coracana | *Mundua* |
| | Paddy (Rice) | Oryza sativa | *Dhan (Chawal)* |
| | Wheat | Triticum acstivum | *Gehun* |
| Pulses and Beans | Black gram | Vignamungo | *Urad* |
| | Chickpea (Bengal gram) | Cicer arietinum | *Chana* |
| | Chicking vetch | Lathyrus sativus | *Khesari* |
| | Cluster Bean | Cyamopiss tetragonoloba | *Guar* |
| | Cowpea | Vigna unguiculate | *Lobia* |
| | Green Gram | Vigna radiata | *Mung* |
| | Horsegram | Dilichos biflorus | *Kulthi* |
| | Kidney bean | Vigna aconitifolia | *Moth* |
| | Lentil | Lens culimaris | *Masur* |
| | Peas | Pisum sativum vararvense | *Matar* |
| | Red gram (Pigeon pea) | Cajanus cajan | *Tur, Arhar* |
| | Soyabean | Glucine max | Soyabean |
| Sugar | Sugarcane | Saccharum Officinarum | *Ganna* |
| Fruits | Apple | Malus sylvespris | *Seb* |
| | Apricot | Prunus armeniaca | *Khoobani* |
| | Cashewnut | Anaardium occidentale | *Kaju* |
| | Fig | Ficus carica | *Anjeer* |
| | Grape | Vitis vinifera | *Angur* |
| | Guvava | Psidium guajava | *Amrood* |
| | Jackfruit | Artocarpur heterophyllus | *Katahal* |
| | Lemon | Citrus Lemon | *Nimbu* |
| | Lime | Citrus Urantifolia | *Bara Nimbu* |
| | Litchi | Litchi chinensis | *Litchi* |
| | Mango | Magnifera indica | *Aam* |
| | Orange Mandar | Citrus reticulata | *Santara, Narangi* |
| | Papaya | Carica papaya | *Papeeta* |
| | Pear | Pyrus communis | *Naspati* |
| | Pineapple | Ananas comosus | *Ananas* |
| | Banana | Musa paradisiaca | *Kela* |
| | Pomegranate | Punica granatum | *Anaar* |
| | Sweet Orange | Citrus sincensis | *Melta, Mosambi* |

*...contd...*

...contd...

| 1 | 2 | 3 | 4 |
|---|---|---|---|
| Vegetables | Ash gourd | Benincasa hispida | *Petha* |
| | Beet | Beta vulgaris | *Chukandar* |
| | Bitter gourd | Momordica charantia | *Karela* |
| | Bottle gourd | Lagenaria siceraria | *Lauki* |
| | Brinjal | Lolanum melongena | *Baingan* |
| | Cabbage | Brassica oleracca var, Capitata | *Band gobi* |
| | Carrot | Daucus carota | *Gajar* |
| | Cauliflower | Brassica oleracca var Botrytis | *Phul gobi* |
| | Cowpea | Vigna unguiculate | *Lobia* |
| | Cucumber | Cucumis satius | *Kheera* |
| | French bean | Phaseolus vulgaris | *Faras bean* |
| | Indian flat bean or sem | Dolichos lablab | *Sem* |
| | Kaol Khol | Brassica oleracea var, Gongylodes | *Ganth gobi* |
| | Lady's finger | Abelmoschus esculentus | *Bhindi* |
| | Little gourd | Cuccinia cordifolia | *Kundur* |
| | Musk melon | Cucumis melo | *Kharbooza* |
| | Onion | Allium cepa | *Piyaz* |
| | Pointed gourd | Trichosanthes dioica | *Parwal, Potal* |
| | Potato | Solanum tuberosum | *Aaloo* |
| | Pumpkin | Curcurbita moschata | *Sitaphal, Lal Kaddu, Kumbhra* |
| | Radish | Raphanus satius | *Muli* |
| | Round gourd of India | Citrullus vulgaris var, fistulosus | *Tinda* |
| | Snap melon | Cucumis melovvar, momordica | *Phoot* |
| | Snake gourd | Trichosanthes anguina | *Chachinda* |
| | Tomato | Lycopersicon escen lentum | *Tamatar* |
| | Turnip | Brassica rapa | *Shalgam* |
| | Water melon | Cirtrullus vulgaris | *Tarbooz* |
| Drugs and Narcotics | | | |
| | Betel Leave | | *Paan* |
| | Betelnut (arecanut) | Areca catechu | *Supari* |
| | Indian hemp | Cannabis sativa | *Bhang* |
| | Opium | Papaver somniferum | *Afeem* |
| | Tobacco | Nicotiana tabacum and Nicotiana rustica | *Tambaku* |
| Condiments and Spices | | | |
| | Black pepper | Piper nigrum | *Kalimirch* |
| | Cardamom, Cardamum (lesser) | Elettaria cardamomum | *Chhoti Ilaichi* |
| | Chillies | Capsicum annuum | *Lalmirch* |

# Appendix IV(i)

## *All-India Estimates of Yield of Major Crops during 1999-2000 to 2007-08*

| *State/UTs* | *Season* | Yield (Kg/Hectare) | | | | | | | | |
|---|---|---|---|---|---|---|---|---|---|---|
| | | *1999-2000* | *2000-2001* | *2001-2002* | *2002-2003* | *2003-2004* | *2004-2005* | *2005-2006* | *2006-2007* | *2007-2008** |
| *1* | *2* | *3* | *4* | *5* | *6* | *7* | *8* | *9* | *10* | *11* |
| Rice | Autumn | 1534 | 1404 | 1532 | 1202 | 1385 | 1327 | 1349 | 1526 | |
| | Winter | 1736 | 1726 | 1938 | 1595 | 1868 | 1756 | 1795 | 2000 | |
| | *Kharif* | 2049 | 1882 | 2074 | 1766 | 2174 | 2039 | 2199 | 2086 | |
| | Total *Kharif* | 1892 | 1788 | 1982 | 1658 | 2004 | 1883 | 1990 | 2024 | 2103 |
| | *Rabi*/Summer | 2897 | 3042 | 2992 | 2783 | 2947 | 3077 | 3127 | 3130 | 3103 |
| | Total | 1986 | 1901 | 2079 | 1744 | 2078 | 1984 | 2102 | 2131 | 2203 |
| Wheat | *Rabi* | 2778 | 2708 | 2762 | 2610 | 2713 | 2602 | 2619 | 2708 | 2785 |
| *Jowar* | *Kharif* | 998 | 938 | 945 | 995 | 1085 | 987 | 1082 | 992 | 1161 |
| | *Rabi* | 713 | 594 | 625 | 552 | 377 | 641 | 726 | 727 | 834 |
| | Total | 847 | 764 | 771 | 754 | 716 | 797 | 880 | 844 | 981 |
| *Bajra* | *Kharif* | 650 | 688 | 869 | 610 | 1141 | 859 | 802 | 886 | 1030 |
| Maize | *Kharif* | 1695 | 1707 | 1896 | 1552 | 1932 | 1740 | 1799 | 1660 | 2123 |
| | *Rabi*/Summer | 2594 | 2921 | 2952 | 2851 | 2987 | 3224 | 3076 | 3793 | 3692 |
| | Total | 1792 | 1822 | 2000 | 1681 | 2041 | 1907 | 1938 | 1912 | 2337 |
| Ragi | *Kharif* | 1401 | 1553 | 1442 | 930 | 1180 | 1567 | 1534 | 1226 | 1458 |
| Small Millets | *Kharif* | 438 | 412 | 440 | 383 | 473 | 434 | 443 | 475 | 523 |
| Barley | *Rabi* | 1997 | 1840 | 2160 | 2006 | 1975 | 1958 | 1938 | 2055 | 1997 |
| Coarse Cereals | *Kharif* | 1032 | 1042 | 1167 | 972 | 1314 | 1168 | 1178 | 1144 | 1402 |
| | *Rabi* | 1040 | 973 | 1005 | 947 | 858 | 1102 | 1152 | 1316 | 1479 |
| | Total | 1034 | 1027 | 1131 | 966 | 1221 | 1153 | 1172 | 1182 | 1418 |
| Cereals | *Kharif* | 1587 | 1512 | 1688 | 1417 | 1739 | 1618 | 1693 | 1706 | 5206 |
| | *Rabi* | 2483 | 2438 | 2476 | 2319 | 2413 | 2382 | 2427 | 2526 | |
| | Total | 1926 | 1844 | 1980 | 1753 | 1983 | 1903 | 1968 | 2022 | |
| *Tur (Arhar)* | *Kharif* | 786 | 618 | 679 | 651 | 670 | 667 | 765 | 650 | 824 |
| Pulses other than *Tur* | *Kharif* | 333 | 313 | 349 | 298 | 466 | 304 | 300 | 349 | |
| Gram | *Rabi* | 833 | 744 | 853 | 717 | 811 | 815 | 808 | 845 | 780 |
| Pulses other than gram | *Rabi* | 673 | 615 | 628 | 590 | 640 | 622 | 610 | 611 | |
| Pulses | *Kharif* | 492 | 417 | 451 | 417 | 528 | 417 | 456 | 449 | 557 |
| | *Rabi* | 760 | 684 | 756 | 661 | 742 | 735 | 727 | 751 | 709 |
| | Total | 635 | 544 | 607 | 543 | 635 | 577 | 598 | 612 | 635 |
| Foodgrains | *Kharif* | 1441 | 1357 | 1510 | 1272 | 1551 | 1430 | 1511 | 1522 | 1644 |
| | *Rabi* | 2091 | 2067 | 2076 | 1933 | 2004 | 1988 | 2020 | 2091 | 2157 |
| | Total | 1704 | 1626 | 1734 | 1535 | 1727 | 1652 | 1715 | 1756 | 1854 |

* Advance Estimates released as on 09.07.2008.

## Appendix IV(ii)

*State-wise Yield of Kharif Foodgrins during 1999-2000 to 2007-08*

*Yield (Kg/hect.)*

| *State/UT* | *1999-2000* | *2000-2001* | *2001-2002* | *2002-2003* | *2003-2004* | *2004-2005* | *2005-2006* | *2006-2007* | *2007-2008** |
|---|---|---|---|---|---|---|---|---|---|
| *1* | *2* | *3* | *4* | *5* | *6* | *7* | *8* | *9* | *10* |
| Andhra Pradesh | 1927 | 2030 | 1991 | 1702 | 2003 | 2096 | 2167 | 2073 | 2632 |
| Arunachal Pradesh | 1121 | 1140 | 1150 | 1221 | 1273 | 1158 | 1201 | 1204 | |
| Assam | 1352 | 1419 | 1435 | 1396 | 1483 | 1374 | 1409 | 1210 | 1343 |
| Bihar | 1410 | 1481 | 1457 | 1410 | 1516 | 835 | 1096 | 1469 | 1200 |
| Chhattisgarh | | 595 | 1196 | 656 | 1325 | 1072 | 1212 | 1236 | 1336 |
| Goa | 3589 | 2452 | 2538 | 2669 | 3195 | 2740 | 2793 | 2440 | |
| Gujarat | 1054 | 682 | 1273 | 997 | 1405 | 1146 | 1238 | 1037 | 1342 |
| Haryana | 1778 | 1888 | 2061 | 1867 | 2134 | 2186 | 2214 | 2466 | 2629 |
| Himachal Pradesh | 1962 | 1983 | 2195 | 1393 | 2078 | 1975 | 1652 | 2044 | 2005 |
| Jammu & Kashmir | 1430 | 1552 | 1583 | 1474 | 1688 | 1615 | 1656 | 1691 | 1657 |
| Jharkhand | | 1077 | 1181 | 1037 | 1516 | 1207 | 1077 | 1595 | 2642 |
| Karnataka | 1460 | 1603 | 1423 | 1116 | 1194 | 1539 | 1934 | 1470 | 1619 |
| Kerala | 2065 | 1993 | 2078 | 2103 | 1914 | 2261 | 2214 | 2329 | 2211 |
| Madhya Pradesh | 1024 | 703 | 992 | 783 | 1083 | 821 | 963 | 792 | 863 |
| Maharashtra | 1015 | 825 | 985 | 972 | 1090 | 920 | 1077 | 1093 | 1275 |
| Manipur | 2328 | 2359 | 2306 | 2217 | 2355 | 2417 | 2241 | 2241 | |
| Meghalaya | 1552 | 1559 | 1585 | 1604 | 1616 | 1487 | 1405 | 1632 | |
| Mizoram | 1757 | 2034 | 1922 | 1883 | 1888 | 1894 | 1745 | 767 | |
| Nagaland | 930 | 1427 | 1377 | 1584 | 1590 | 1634 | 1614 | 1503 | |
| Orissa | 940 | 914 | 1396 | 661 | 1326 | 1282 | 1346 | 1352 | 1489 |
| Punjab | 3248 | 3418 | 3442 | 3388 | 3609 | 3836 | 3765 | 3792 | 3951 |
| Rajasthan | 399 | 447 | 688 | 298 | 1040 | 591 | 505 | 637 | 820 |
| Sikkim | 1320 | 1403 | 1334 | 1373 | 1447 | 1455 | 1403 | 1403 | |
| Tamil Nadu | 2623 | 2678 | 2404 | 1755 | 1588 | 2073 | 2025 | 2805 | 2442 |
| Tripura | 2106 | 2071 | 2229 | 2287 | 2096 | 2156 | 2169 | 2433 | |
| Uttar Pradesh | 1858 | 1747 | 1811 | 1558 | 1312 | 1584 | 1697 | 1628 | 1787 |
| Uttarakhand | | 1659 | 1678 | 1326 | 1578 | 1502 | 1551 | 1606 | 1642 |
| West Bengal | 1964 | 1933 | 2326 | 2271 | 2295 | 2389 | 2350 | 2356 | 2278 |
| A & N Islands | 2679 | 2954 | 2786 | 2945 | 2880 | 2713 | 2879 | 2870 | |
| D & N Haveli | 1481 | 1369 | 1838 | 1495 | 1471 | 1524 | 1545 | 1545 | |
| Delhi | 615 | 580 | 1541 | 1534 | 1817 | 1791 | 2054 | 2492 | |
| Daman & Diu | 3733 | 1600 | 2222 | 2125 | 2167 | 2353 | 2056 | 2056 | |
| Pondicherry | 2333 | 2416 | 2541 | 2247 | 2644 | 2602 | 2372 | 2372 | |
| All India | 1441 | 1357 | 1510 | 1272 | 1551 | 1430 | 1511 | 1522 | 1644 |

* Advance Estimates released as on 09.07.2008.

# Appendix IV(iii)

*State-wise Yield of Rabi Foodgrins during 1999-2000 to 2007-08*

*Yield (Kg/hect.)*

| *State/UT* | *1999-2000* | *2000-2001* | *2001-2002* | *2002-2003* | *2003-2004* | *2004-2005* | *2005-2006* | *2006-2007* | *2007-2008** |
|---|---|---|---|---|---|---|---|---|---|
| *1* | *2* | *3* | *4* | *5* | *6* | *7* | *8* | *9* | *10* |
| Andhra Pradesh | 1903 | 2208 | 2265 | 1681 | 2029 | 2212 | 2669 | 2452 | 2584 |
| Arunachal Pradesh | 1269 | 1333 | 1246 | 1330 | 1368 | 1642 | 1390 | 1402 | |
| Assam | 1690 | 1635 | 1570 | 1510 | 1419 | 1543 | 1446 | 1594 | 1500 |
| Bihar | 1965 | 1958 | 1910 | 1756 | 1704 | 1585 | 1563 | 1876 | 1945 |
| Chhattisgarh | | 539 | 650 | 615 | 731 | 470 | 572 | 661 | 709 |
| Goa | 913 | 921 | 910 | 793 | 957 | 919 | 1067 | 1383 | |
| Gujarat | 1765 | 1979 | 2136 | 1680 | 2319 | 2204 | 2298 | 2123 | 2311 |
| Haryana | 3989 | 3899 | 3888 | 3950 | 3759 | 3744 | 3651 | 4066 | 4000 |
| Himachal Pradesh | 1539 | 709 | 1698 | 1369 | 1364 | 1863 | 1813 | 1373 | 1643 |
| Jammu & Kashmir | 1695 | 529 | 1320 | 1592 | 1731 | 1855 | 1740 | 1829 | 1801 |
| Jharkhand | | 1519 | 1515 | 1353 | 1173 | 1560 | 1016 | 1173 | 1116 |
| Karnataka | 1007 | 1039 | 895 | 674 | 558 | 1071 | 1464 | 980 | 1164 |
| Kerala | 2220 | 2598 | 2202 | 2428 | 2056 | 2391 | 2268 | 2338 | 2851 |
| Madhya Pradesh | 1402 | 1131 | 1253 | 1057 | 1356 | 1321 | 1234 | 1398 | 1204 |
| Maharashtra | 808 | 637 | 707 | 640 | 480 | 709 | 778 | 827 | 1024 |
| Manipur | | | | | | | | | |
| Meghalaya | 1205 | 2047 | 2577 | 2609 | 3055 | 3868 | 1886 | 3273 | |
| Mizoram | 2917 | 2154 | 1824 | 1500 | 1167 | 1720 | 1939 | 1587 | |
| Nagaland | 1767 | 1179 | 1404 | 1414 | 1342 | 1120 | 1622 | 1174 | |
| Orissa | 1735 | 1437 | 1431 | 1529 | 1351 | 1492 | 1372 | 1414 | 1582 |
| Punjab | 4670 | 4533 | 4507 | 4178 | 4188 | 4204 | 4165 | 4197 | 4491 |
| Rajasthan | 2035 | 1974 | 2197 | 2315 | 2045 | 2128 | 1998 | 2224 | 2023 |
| Sikkim | 1303 | 1150 | 1101 | 1176 | 1176 | 1185 | 1142 | 1142 | |
| Tamil Nadu | 1495 | 1545 | 1546 | 1099 | 1024 | 1213 | 1203 | 1923 | 1417 |
| Tripura | 2156 | 2025 | 2561 | 2295 | 2199 | 2247 | 2272 | 2304 | |
| Uttar Pradesh | 2448 | 2380 | 2442 | 2301 | 2498 | 2233 | 2327 | 2392 | 2518 |
| Uttarakhand | | 1785 | 1826 | 1737 | 1792 | 1955 | 1544 | 1946 | 1715 |
| West Bengal | 2676 | 2839 | 2646 | 2607 | 2706 | 2688 | 2593 | 2858 | 2889 |
| A & N Islands | 333 | 571 | 500 | 462 | 571 | 571 | 500 | 500 | |
| D & N Haveli | 909 | 939 | 942 | 1143 | 1000 | 974 | 975 | 975 | |
| Delhi | 909 | 3461 | | 3970 | 3514 | 3939 | 4290 | 4333 | |
| Daman & Diu | 1000 | 1000 | | 769 | 846 | 923 | 846 | 846 | |
| Pondicherry | 1526 | 1513 | 2186 | 2192 | 2022 | 1766 | 1785 | 1785 | |
| All India | 2091 | 2067 | 2076 | 1933 | 2004 | 1988 | 2020 | 2091 | 2157 |

* Advance Estimates released as on 09.07.2008.

# Appendix IV(iv)

*State-wise Yield of Total Foodgrins during 1999-2000 to 2007-08*

*Yield (Kg/hect.)*

| State/UT | 1999-2000 | 2000-2001 | 2001-2002 | 2002-2003 | 2003-2004 | 2004-2005 | 2005-2006 | 2006-2007 | 2007-2008* |
|---|---|---|---|---|---|---|---|---|---|
| Andhra Pradesh | 1919 | 2089 | 2102 | 1694 | 2012 | 2138 | 2365 | 2231 | 2612 |
| Arunachal Pradesh | 1125 | 1148 | 1154 | 1226 | 1277 | 1178 | 1212 | 1216 | |
| Assam | 1409 | 1457 | 1460 | 1417 | 1472 | 1405 | 1416 | 1286 | 1379 |
| Bihar | 1617 | 1694 | 1664 | 1568 | 1601 | 1192 | 1311 | 1656 | 1540 |
| Chhattisgarh | | 589 | 1118 | 651 | 1231 | 979 | 1111 | 1148 | 1239 |
| Goa | 3228 | 2227 | 2293 | 2382 | 2866 | 2456 | 2509 | 2254 | |
| Gujarat | 1186 | 827 | 1414 | 1114 | 1629 | 1412 | 1551 | 1423 | 1731 |
| Haryana | 3047 | 3060 | 3128 | 3103 | 3068 | 3092 | 3045 | 3393 | 3432 |
| Himachal Pradesh | 1754 | 1366 | 1954 | 1381 | 1733 | 1923 | 1731 | 1714 | 1825 |
| Jammu & Kashmir | 1508 | 1224 | 1505 | 1508 | 1701 | 1686 | 1680 | 1733 | 1699 |
| Jharkhand | | 1095 | 1199 | 1053 | 1490 | 1234 | 1073 | 1550 | 1588 |
| Karnataka | 1286 | 1412 | 1212 | 949 | 954 | 1388 | 1776 | 1289 | 1470 |
| Kerala | 2094 | 2094 | 2097 | 2151 | 1935 | 2278 | 2219 | 2331 | 2338 |
| Madhya Pradesh | 1208 | 945 | 1148 | 943 | 1248 | 1131 | 1130 | 1167 | 1072 |
| Maharashtra | 931 | 757 | 874 | 846 | 860 | 836 | 948 | 940 | 1166 |
| Manipur | 2328 | 2359 | 2306 | 2217 | 2355 | 2390 | 2241 | 2241 | |
| Meghalaya | 1531 | 1598 | 1667 | 1686 | 1733 | 1674 | 1455 | 1800 | |
| Mizoram | 1781 | 2036 | 1917 | 1866 | 1854 | 1888 | 1754 | 822 | |
| Nagaland | 977 | 1406 | 1380 | 1565 | 1561 | 1577 | 1615 | 1482 | |
| Orissa | 1025 | 950 | 1399 | 716 | 1328 | 1300 | 1349 | 1359 | 1500 |
| Punjab | 4028 | 4032 | 4040 | 3828 | 3929 | 4040 | 3986 | 4017 | 4252 |
| Rajasthan | 975 | 883 | 1099 | 873 | 1288 | 1008 | 919 | 991 | 1181 |
| Sikkim | 1317 | 1356 | 1289 | 1334 | 1395 | 1406 | 1354 | 1354 | |
| Tamil Nadu | 2343 | 2461 | 2209 | 1612 | 1539 | 1874 | 1847 | 2610 | 2178 |
| Tripura | 2119 | 2060 | 2311 | 2289 | 2121 | 2179 | 2194 | 2399 | |
| Uttar Pradesh | 2188 | 2105 | 2164 | 1995 | 2192 | 1961 | 2057 | 2057 | 2202 |
| Uttarakhand | | 1712 | 1742 | 1508 | 1672 | 1697 | 1548 | 1760 | 1675 |
| West Bengal | 2178 | 2231 | 2424 | 2374 | 2421 | 2479 | 2423 | 2511 | 2476 |
| A & N Islands | 2496 | 2810 | 2696 | 2680 | 2739 | 2583 | 2713 | 2746 | |
| D & N Haveli | 1394 | 1305 | 1646 | 1431 | 1401 | 1430 | 1445 | 1445 | |
| Delhi | 802 | 2340 | 2691 | 2840 | 3034 | 2892 | 3175 | 3405 | |
| Daman & Diu | 2464 | 1364 | 1645 | 1517 | 1613 | 1733 | 1548 | 1548 | |
| Pondicherry | 2041 | 2096 | 2452 | 2231 | 2445 | 2331 | 2180 | 2180 | |
| All India | 1704 | 1626 | 1734 | 1535 | 1727 | 1652 | 1715 | 1756 | 1854 |

* Advance Estimates released as on 09.07.2008.

# Appendix IV(v)

*State-wise Yield of Rice during 1999-2000 to 2007-08*

*Yield (Kg/hectare)*

| *State/ UT* | *Season* | *1999-2000* | *2000-2001* | *2001-2002* | *2002-2003* | *2003-2004* | *2004-2005* | *2005-2006* | *2006-2007* | *2007-2008** |
|---|---|---|---|---|---|---|---|---|---|---|
| *1* | *2* | *3* | *4* | *5* | *6* | *7* | *8* | *9* | *10* | *11* |
| Andhra Pradesh | *Kharif* | 2477 | 2741 | 2679 | 2396 | 2770 | 2886 | 2525 | 2631 | 3178 |
| | *Rabi* | 3073 | 3409 | 3496 | 3188 | 3592 | 3683 | 3659 | 3682 | 3679 |
| | Total | 2650 | 2936 | 2978 | 2596 | 3009 | 3111 | 2939 | 2984 | 3352 |
| Arunachal Pradesh | *Kharif* | 1099 | 1119 | 1126 | 1224 | 1297 | 1110 | 1195 | 1195 | |
| Assam | Autumn | 923 | 1034 | 984 | 957 | 975 | 667 | 999 | 887 | |
| | Winter | 1501 | 1553 | 1582 | 1527 | 1626 | 1573 | 1520 | 1302 | |
| | Kharif Total | 1364 | 1432 | 1448 | 1407 | 1496 | 1385 | 1421 | 1218 | 1358 |
| | Summer | 2217 | 2068 | 2003 | 1902 | 1796 | 1959 | 1779 | 2019 | 1920 |
| | Total | 1459 | 1511 | 1519 | 1471 | 1534 | 1460 | 1468 | 1332 | 1449 |
| Bihar | Autumn | 1159 | 1330 | 1240 | 552 | 1242 | 893 | 940 | 962 | |
| | Winter | 1486 | 1511 | 1493 | 1460 | 1579 | 740 | 1090 | 1584 | |
| | *Kharif* Total | 1436 | 1481 | 1449 | 1411 | 1525 | 767 | 1064 | 1486 | 1230 |
| | Summer | 1968 | 1702 | 1932 | 1643 | 1453 | 1430 | 1369 | 1499 | 1491 |
| | Total | 1450 | 1489 | 1465 | 1419 | 1523 | 792 | 1075 | 1486 | 1242 |
| Chattisgarh | *Kharif* | | 629 | 1332 | 697 | 1454 | 1170 | 1337 | 1354 | 1446 |
| Goa | Autumn | 3565 | 2356 | 2412 | 2674 | 3268 | 2850 | 2866 | 2406 | |
| | Winter | 3981 | 2809 | 2855 | 2709 | 3162 | 2608 | 2731 | 2566 | |
| | Total *Kharif* | 3684 | 2484 | 2561 | 2685 | 3233 | 2771 | 2822 | 2458 | |
| Gujarat | *Kharif* | 1482 | 810 | 1557 | 1155 | 1832 | 1745 | 1949 | 1799 | 1895 |
| | Summer | | | | | 3506 | 3506 | | | 3108 |
| | Total | | | | | 1891 | 1806 | 1949 | 1894 | 1954 |
| Haryana | *Kharif* | 2385 | 2557 | 2652 | 2724 | 2749 | 2941 | 3051 | 3238 | 3361 |
| Himachal Pradesh | *Kharif* | 1501 | 1533 | 1705 | 1029 | 1483 | 1506 | 1412 | 1559 | 1532 |
| Jammu & Kashmir | *Kharif* | 1561 | 1700 | 1689 | 1782 | 1941 | 1969 | 2150 | 2194 | 2144 |
| Jharkhand | Autumn | | 606 | 654 | 545 | 925 | 710 | 628 | | |
| | Winter | | 1181 | 1274 | 1062 | 1803 | 1382 | 1223 | | |
| | *Kharif* Total | | 1111 | 1198 | 998 | 1695 | 1300 | 1150 | 1832 | 1839 |
| | *Rabi* | | | | | | 2000 | | | 1000 |
| | Total | | 1111 | 1198 | 998 | 1695 | 1305 | 1150 | 1828 | 1834 |
| Karnataka | *Kharif* | 2444 | 2492 | 2200 | 2041 | 2224 | 2359 | 3657 | 2402 | 2423 |
| | *Rabi* | 2032 | 2044 | 1864 | 2026 | 2894 | 3182 | 3200 | 2227 | 1918 |
| | Summer | 3093 | 3048 | 2680 | 2495 | 3896 | 4404 | 4574 | 2761 | |
| | *Rabi*/Summer | 2935 | 2904 | 2547 | 2304 | 3474 | 4190 | 4423 | 2690 | |
| | Total | 2564 | 2593 | 2281 | 2070 | 2375 | 2712 | 3868 | 2470 | 2543 |
| Kerala | Autumn | 2081 | 2005 | 2024 | 2075 | 2141 | 2296 | 2162 | 2264 | |
| | Winter | 2193 | 2071 | 2238 | 2190 | 1809 | 2253 | 2299 | 2408 | |

...contd...

*...contd...*

| 1 | 2 | 3 | 4 | 5 | 6 | 7 | 8 | 9 | 10 | 11 |
|---|---|---|---|---|---|---|---|---|---|---|
| | *Kharif* Total | 2147 | 2042 | 2149 | 2142 | 1946 | 2271 | 2246 | 2352 | 2245 |
| | Summer | 2493 | 2796 | 2392 | 2723 | 2237 | 2515 | 2726 | 2622 | 2956 |
| | Total | 2204 | 2162 | 2182 | 2219 | 1983 | 2301 | 2284 | 2390 | 2384 |
| Madhya Pradesh | *Kharif* | 1191 | 575 | 953 | 614 | 1018 | 720 | 999 | 824 | 824 |
| Maharashtra | *Kharif* | 1674 | 1264 | 1742 | 1198 | 1846 | 1410 | 1770 | 1669 | 1889 |
| | *Rabi* | 2024 | 2000 | 2259 | 2346 | 2200 | 2129 | 2098 | 2105 | 2194 |
| | Total | 1684 | 1276 | 1751 | 1217 | 1853 | 1425 | 1779 | 1680 | 1895 |
| Manipur | *Kharif* | 2323 | 2431 | 2382 | 2302 | 2416 | 2472 | 2322 | 2322 | |
| Meghalaya | *Kharif* | 1604 | 1617 | 1638 | 1655 | 1668 | 1528 | 1432 | 1695 | |
| | *Rabi* | | 3275 | 3842 | 3877 | 4027 | 4065 | 2214 | 4000 | |
| | Total | 1604 | 1679 | 1755 | 1773 | 1830 | 1737 | 1508 | 1916 | |
| Mizoram | *Kharif* | 1777 | 1990 | 1888 | 1905 | 1922 | 1895 | 1770 | 510 | |
| | *Rabi* | 2000 | 2667 | 2500 | 2800 | 2667 | 2400 | 2600 | 1682 | |
| | Total | 1779 | 1998 | 1894 | 1912 | 1926 | 1900 | 1778 | 559 | |
| Nagaland | *Kharif* | 892 | 1533 | 1516 | 1490 | 1600 | 1698 | 1682 | 1600 | |
| Orissa | Autumn | 932 | 543 | 999 | 370 | 982 | 988 | 872 | 1011 | |
| | Winter | 1032 | 1100 | 1694 | 778 | 1581 | 1501 | 1623 | 1580 | |
| | *Kharif* Total | 1013 | 987 | 1554 | 697 | 1459 | 1401 | 1482 | 1476 | 1633 |
| | Summer | 2389 | 2135 | 2125 | 2371 | 2111 | 2132 | 2163 | 2293 | 2402 |
| | Total | 1127 | 1041 | 1588 | 767 | 1496 | 1446 | 1531 | 1534 | 1689 |
| Punjab | *Kharif* | 3347 | 3506 | 3545 | 3510 | 3694 | 3943 | 3858 | 3868 | 4019 |
| Rajasthan | *Kharif* | 1262 | 936 | 1247 | 809 | 1653 | 1485 | 1425 | 1577 | 2031 |
| Sikkim | *Kharif* | 1472 | 1408 | 1420 | 1432 | 1442 | 1469 | 1433 | 1433 | |
| Tamil Nadu | Autumn | 4101 | 4106 | 3916 | 3371 | 3312 | 3881 | 3655 | 4914 | |
| | Winter | 3335 | 3432 | 2999 | 2144 | 2119 | 2483 | 2338 | 3144 | |
| | *Kharif* Total | 3456 | 3545 | 3156 | 2307 | 2251 | 2637 | 2483 | 3339 | 2879 |
| | Summer | 3671 | 3511 | 3498 | 2925 | 2918 | 3416 | 3216 | 4325 | 3248 |
| | Total | 3481 | 3541 | 3196 | 2359 | 2308 | 2703 | 2546 | 3423 | 2907 |
| Tripura | Autumn | 1461 | 1745 | 1640 | 1874 | 1624 | 1669 | 1705 | 1915 | |
| | Winter | 2342 | 2274 | 2480 | 2449 | 2327 | 2377 | 2384 | 2678 | |
| | *Kharif* Total | 2149 | 2136 | 2281 | 2342 | 2147 | 2213 | 2225 | 2500 | |
| | Summer | 2265 | 2108 | 2698 | 2415 | 2261 | 2356 | 2372 | 2392 | |
| | Total | 2178 | 2129 | 2381 | 2359 | 2173 | 2247 | 2260 | 2472 | |
| Uttar Pradesh | *Kharif* | 2176 | 1977 | 2118 | 1840 | 2187 | 1789 | 1996 | 1878 | 2062 |
| | *Rabi*/Summer | | 2300 | 2333 | 2000 | 2226 | 2231 | 2179 | 2278 | 2250 |
| | Total | 2176 | 1977 | 2118 | 1841 | 2187 | 1790 | 1996 | 1879 | 2062 |
| Uttarakhand | *Kharif* | | 1988 | 2056 | 1707 | 1906 | 1839 | 1925 | 1952 | 1993 |
| | *Zaid/Rabi* | | | | | 4000 | 3143 | 2889 | 2875 | 2875 |
| | Total | | | | | 1942 | 1869 | 1954 | 1979 | 2018 |
| West Bengal | Autumn | 1938 | 1736 | 2091 | 2069 | 2117 | 2036 | 2102 | 2026 | |
| | Winter | 1992 | 1979 | 2374 | 2319 | 2339 | 2441 | 2397 | 2411 | |

*...contd...*

*...contd...*

| 1 | 2 | 3 | 4 | 5 | 6 | 7 | 8 | 9 | 10 | 11 |
|---|---|---|---|---|---|---|---|---|---|---|
| | *Kharif* Total | 1987 | 1955 | 2350 | 2297 | 2322 | 2411 | 2378 | 2386 | 2327 |
| | Summer | 3031 | 3240 | 3034 | 2986 | 3086 | 3093 | 2928 | 3226 | 3250 |
| | Total | 2237 | 2287 | 2514 | 2463 | 2504 | 2574 | 2509 | 2593 | 2570 |
| A & N Islands | *Kharif* | 2679 | 2954 | 2786 | 2945 | 2915 | 2729 | 2896 | 2896 | |
| D & N Haveli | *Kharif* | 1674 | 1471 | 2200 | 1672 | 1652 | 1733 | 1743 | 1743 | |
| Delhi | *Kharif* | 827 | 820 | 2813 | 3180 | 1708 | 2383 | 3200 | 4203 | |
| Daman & Diu | *Kharif* | 3733 | 1600 | 2222 | 2125 | 2167 | 2353 | 2056 | 2056 | |
| Pondicherry | Autumn | 2534 | 2677 | 2557 | 2818 | 2746 | 2774 | 2810 | 2810 | |
| | Winter | 2229 | 2299 | 2531 | 2008 | 2613 | 2558 | 2195 | 2195 | |
| | *Kharif* Total | 2318 | 2413 | 2539 | 2253 | 2658 | 2628 | 2387 | 2387 | |
| | Summer | 2411 | 2509 | 2561 | 2797 | 2703 | 2719 | 2719 | 2719 | |
| | Total | 2339 | 2433 | 2544 | 2386 | 2669 | 2649 | 2465 | 2465 | |
| All India | Autumn | 1534 | 1404 | 1532 | 1202 | 1385 | 1327 | 1349 | 1526 | |
| | Winter | 1736 | 1726 | 1938 | 1595 | 1868 | 1756 | 1795 | 2000 | |
| | *Kharif* | 2049 | 1882 | 2074 | 1766 | 2174 | 2039 | 2199 | 2086 | |
| | *Kharif* Total | 1892 | 1788 | 1982 | 1658 | 2004 | 1883 | 1990 | 2024 | 2103 |
| | Summer/*Rabi* | 2897 | 3042 | 2992 | 2783 | 2947 | 3077 | 3127 | 3130 | 3103 |
| | Total | 1986 | 1901 | 2079 | 1744 | 2078 | 1984 | 2102 | 2131 | 2203 |

* Advance Estimates released as on 09.07.2008.

# Appendix IV(vi)

## *State-wise Yield of Wheat during 1999-2000 to 2007-08*

*Yield (Kg/Hectare)*

| *State/UT* | *1999-2000* | *2000-2001* | *2001-2002* | *2002-2003* | *2003-2004* | *2004-2005* | *2005-2006* | *2006-2007* | *2007-2008** |
|---|---|---|---|---|---|---|---|---|---|
| *1* | *5* | *6* | *7* | *8* | *9* | *10* | *11* | *12* | *13* |
| Andhra Pradesh | 667 | 571 | 643 | 1273 | 583 | 556 | 818 | 900 | 700 |
| Arunachal Pradesh | 1308 | 1409 | 1395 | 1537 | 1500 | 2023 | 1525 | 1575 | |
| Assam | 1289 | 1219 | 1181 | 1130 | 1043 | 1066 | 1074 | 1117 | 1086 |
| Bihar | 2186 | 2146 | 2065 | 1896 | 1776 | 1609 | 1617 | 1908 | 2058 |
| Chhatisgarh | | 1022 | 1057 | 1063 | 1024 | 853 | 886 | 1002 | 1065 |
| Gujarat | 2116 | 2268 | 2435 | 1966 | 2681 | 2482 | 2700 | 2498 | 2700 |
| Haryana | 4165 | 4106 | 4103 | 4053 | 3937 | 3901 | 3844 | 4232 | 4160 |
| Himachal Pradesh | 1574 | 693 | 1738 | 1379 | 1380 | 1890 | 1894 | 1385 | 1648 |
| Jammu & Kashmir | 1768 | 529 | 1325 | 1646 | 1804 | 1910 | 1790 | 1893 | 1874 |
| Jharkhand | | 1626 | 1765 | 1625 | 552 | 2381 | 1340 | 1529 | 1512 |
| Karnataka | 833 | 917 | 763 | 598 | 415 | 740 | 858 | 762 | 883 |
| Madhya Pradesh | 1863 | 1471 | 1620 | 1456 | 1800 | 1735 | 1613 | 1835 | 1675 |
| Maharashtra | 1369 | 1257 | 1388 | 1295 | 1170 | 1344 | 1393 | 1325 | 1659 |
| Meghalaya | 1628 | 1643 | 1714 | 1741 | 1667 | 1778 | 1714 | 2000 | |
| Nagaland | 2500 | 2000 | 2500 | 2125 | 2000 | 1585 | 1583 | 867 | |
| Orissa | 1305 | 1466 | 1514 | 1261 | 1471 | 1250 | 1364 | 1487 | 1275 |
| Punjab | 4696 | 4563 | 4532 | 4200 | 4207 | 4221 | 4179 | 4210 | 4506 |
| Rajasthan | 2540 | 2402 | 2793 | 2709 | 2794 | 2839 | 2762 | 2751 | 2749 |
| Sikkim | 1580 | 1403 | 1273 | 1413 | 1421 | 1456 | 1385 | 1385 | |
| Tripura | 1846 | 2000 | 2083 | 2000 | 4250 | 2545 | 2636 | 1800 | |
| Uttar Pradesh | 2764 | 2724 | 2755 | 2596 | 2794 | 2502 | 2627 | 2721 | 2790 |
| Uttarakhand | | 1885 | 1933 | 1825 | 1877 | 2038 | 1633 | 2049 | 1781 |
| West Bengal | 2336 | 2485 | 2215 | 2189 | 2315 | 2103 | 2109 | 2282 | 2323 |
| D & N Haveli | 1500 | 1600 | 2000 | 2000 | 2000 | 1833 | 1833 | 1833 | |
| Delhi | 919 | 3522 | 3510 | 4005 | 3515 | 3944 | 4339 | 4341 | |
| All India | 2778 | 2708 | 2762 | 2610 | 2713 | 2602 | 2619 | 2708 | 2785 |

* Advance Estimates released as on 09.07.2008.

# Appendix IV(vii)

*State-wise Yield of Bajra during 1999-2000 to 2007-08*

*Yield (kgs/hect.)*

| *State/ UT* | *Season* | *1999-2000* | *2000-2001* | *2001-2002* | *2002-2003* | *2003-2004* | *2004-2005* | *2005-2006* | *2006-2007* | *2007-2008** |
|---|---|---|---|---|---|---|---|---|---|---|
| *1* | *2* | *3* | *4* | *5* | *6* | *7* | *8* | *9* | *12* | *13* |
| Andhra Pradesh | *Kharif* | 838 | 1042 | 776 | 655 | 1093 | 881 | 1043 | 816 | |
| | *Rabi* | | | | | 778 | 700 | 833 | 583 | |
| | Total | | | | | 1072 | 862 | 1012 | 770 | 1233 |
| Bihar | | 621 | 867 | 750 | 1000 | 1000 | 1083 | 1070 | 1073 | 1111 |
| Chattisgarh | | - | - | - | - | - | - | - | - | |
| Gujarat | *Kharif* | 641 | 650 | 1261 | 849 | 1370 | 1021 | 1010 | 940 | |
| | Summer | 2062 | 1915 | 1812 | 1614 | 2144 | 1926 | 2000 | 1861 | |
| | Total | 918 | 832 | 1343 | 965 | 1494 | 1172 | 1169 | 1088 | 1319 |
| Haryana | | 991 | 1079 | 1422 | 891 | 1607 | 1316 | 1147 | 1649 | 1843 |
| Himachal Pradesh | | | | 1000 | 552 | 667 | 667 | 333 | 500 | |
| Jammu & Kashmir | | 478 | 466 | 596 | 597 | 595 | 600 | 600 | 591 | 588 |
| Jharkhand | | | 619 | 558 | | 3000 | 833 | 500 | 500 | |
| Karnataka | *Kharif* | 596 | 728 | 534 | 393 | 600 | 602 | 977 | 482 | |
| | Summer | 1167 | 1000 | 923 | 857 | 1000 | 500 | 1000 | 1000 | |
| | Total | 597 | 729 | 536 | 394 | 601 | 601 | 977 | 483 | 818 |
| Madhya Pradesh | | 1008 | 1253 | 1271 | 754 | 1403 | 1368 | 1491 | 1365 | 1400 |
| Maharashtra | | 653 | 604 | 594 | 741 | 676 | 736 | 720 | 729 | 878 |
| Nagaland | | 867 | 867 | 600 | 2000 | 1667 | 1481 | 1000 | | |
| Orissa | | 500 | 400 | 561 | 474 | 515 | 500 | 552 | 583 | 667 |
| Punjab | | 800 | 1000 | 1000 | 857 | 1000 | 1000 | 1000 | 1000 | 1000 |
| Rajasthan | | 330 | 441 | 745 | 223 | 1136 | 658 | 556 | 701 | 832 |
| Tamil Nadu | | 1530 | 1318 | 1224 | 870 | 1085 | 1274 | 1158 | 1511 | 1370 |
| Uttar Pradesh | | 1395 | 1449 | 1137 | 1290 | 1276 | 1534 | 1434 | 1455 | 1520 |
| West Bengal | | | 1000 | 750 | 1000 | 667 | | | | |
| Pondicherry | | | 2500 | 2500 | 2000 | 2000 | 2000 | 2500 | 2500 | |
| Delhi | | 556 | 548 | 1962 | 1750 | 2000 | 2435 | 1824 | 1813 | |
| All India | | 650 | 688 | 869 | 610 | 1141 | 859 | 802 | 886 | 1030 |

* Advance Estimates released as on 09.07.2008.

# Appendix IV(viii)

*State-wise Yield of Maize during 1999-2000 to 2007-08*

*Yield (Kgs/hect.)*

| State/ UT | Season | 1999-2000 | 2000-2001 | 2001-2002 | 2002-2003 | 2003-2004 | 2004-2005 | 2005-2006 | 2006-2007 | 2007-2008* |
|---|---|---|---|---|---|---|---|---|---|---|
| 1 | 2 | 3 | 4 | 5 | 6 | 7 | 8 | 9 | 10 | 11 |
| Andhra Pradesh | *Kharif* | 3031 | 2568 | 2917 | 2203 | 2996 | 2453 | 3538 | 2402 | 4125 |
| | *Rabi* | 4141 | 4887 | 5233 | 5125 | 4939 | 5428 | 5994 | 6195 | 5542 |
| | Total | 3257 | 2994 | 3404 | 2825 | 3436 | 3142 | 4073 | 3396 | 4606 |
| Arunachal Pradesh | *Kharif* | 1357 | 1362 | 1359 | 1392 | 1411 | 1455 | 1373 | 1363 | |
| | *Rabi* | - | - | 1400 | 1400 | 1455 | 1455 | 1474 | 1452 | |
| | Total | 1357 | 1362 | 1360 | 1393 | 1412 | 1455 | 1382 | 1371 | |
| Assam | *Kharif* | 700 | 723 | 700 | 700 | 700 | 724 | 721 | 778 | 737 |
| Bihar | Autumn | 1554 | 1748 | 1882 | 1679 | 1691 | 1571 | 1518 | 1532 | 963 |
| | *Rabi* | 3009 | 2920 | 3195 | 552 | 3030 | 3094 | 2641 | 3597 | 3208 |
| | Garma | 2396 | 2955 | 2650 | 2682 | 2736 | 2919 | 2385 | 3249 | |
| | Total | 2125 | 2413 | 2504 | 2236 | 2390 | 2386 | 2098 | 2678 | 2274 |
| Chhattisgarh | *Kharif* | | 1346 | 743 | 1304 | 1369 | 1365 | 1076 | 1225 | 1566 |
| Goa | *Kharif* | 4000 | 4000 | 4000 | 4000 | 4000 | 4000 | 5000 | 5000 | |
| Gujarat | *Kharif* | 1289 | 753 | 1995 | 1706 | 1717 | 898 | 1124 | 698 | 1375 |
| Haryana | *Kharif* | 2400 | 2267 | 2611 | 1813 | 2412 | 2500 | 2125 | 2286 | 2643 |
| Himachal Pradesh | *Kharif* | 2272 | 2293 | 2550 | 1612 | 2444 | 2272 | 1839 | 2326 | 2277 |
| Jammu & Kashmir | *Kharif* | 1485 | 1592 | 1648 | 1412 | 1658 | 1526 | 1413 | 1505 | 1526 |
| Jharkhand | Autumn | - | 1255 | 1452 | 1797 | 1602 | 1500 | 1309 | 1203 | 1500 |
| | *Rabi* | - | 3000 | 2435 | 2000 | 1667 | 1400 | 1622 | 1806 | 1500 |
| | Total | - | 1267 | 1495 | 1799 | 1604 | 1497 | 1315 | 1230 | 1500 |
| Karnataka | *Kharif* | 2698 | 3270 | 2522 | 2013 | 1954 | 2966 | 2919 | 2839 | 2888 |
| | *Rabi* | 2386 | 2469 | 2344 | 2508 | 1924 | 2807 | 3000 | 2658 | 2967 |
| | Summer | 1811 | 2777 | 2581 | 2531 | 2417 | 3000 | 2000 | 3125 | |
| | Total | 2644 | 3193 | 2502 | 2068 | 1957 | 2955 | 2915 | 2829 | 2894 |
| Madhya Pradesh | *Kharif* | 1574 | 1449 | 1968 | 1738 | 2052 | 1398 | 1450 | 976 | 1400 |
| Maharashtra | Kharif | 1531 | 832 | 1804 | 2093 | 2151 | 1720 | 2134 | 1996 | 2690 |
| | *Rabi* | 1571 | 1239 | 1803 | 1629 | 1421 | 1947 | 1976 | 1924 | 2682 |
| | Total | 1541 | 920 | 1804 | 2005 | 2038 | 1759 | 2106 | 1983 | 2689 |
| Manipur | *Kharif* | 2488 | 2319 | 1980 | 1667 | 3261 | 2781 | 2724 | 2724 | |
| Meghalaya | *Kharif* | 1452 | 1438 | 1491 | 1533 | 1533 | 1420 | 1426 | 1471 | |
| Mizoram | *Kharif* | 1830 | 2516 | 2509 | 2000 | 1939 | 2132 | 1917 | 1942 | |
| | *Rabi* | 1000 | 1500 | 1526 | 1833 | 1857 | 1714 | 2222 | 2668 | |

...contd...

*...contd...*

| 1 | 2 | 3 | 4 | 5 | 6 | 7 | 8 | 9 | 10 | 11 |
|---|---|---|---|---|---|---|---|---|---|---|
| | Total | 1800 | 2485 | 2257 | 1987 | 1933 | 2093 | 1940 | 1963 | |
| Nagaland | *Kharif* | 1250 | 1429 | 1375 | 2000 | 1780 | 1800 | 1800 | 1674 | |
| Orissa | *Kharif* | 1499 | 1268 | 902 | 969 | 1394 | 1607 | 1587 | 1655 | 1986 |
| | *Rabi* | 1556 | 1200 | 1600 | 1545 | 1960 | 2000 | 2188 | 2368 | 2000 |
| | Total | 1500 | 1267 | 909 | 984 | 1420 | 1631 | 1602 | 1677 | 1986 |
| Punjab | *Kharif* | 2577 | 2794 | 2721 | 2039 | 2981 | 2740 | 2723 | 3123 | 3405 |
| Rajasthan | *Kharif* | 1038 | 1047 | 1454 | 885 | 1863 | 1211 | 1098 | 1085 | 1859 |
| | *Rabi* | | 714 | 2000 | 2667 | 1571 | 1571 | 1429 | 1429 | 1000 |
| | Total | 1038 | 1047 | 1454 | 885 | 1863 | 1211 | 1098 | 1086 | 1858 |
| Sikkim | *Kharif* | 1340 | 1494 | 1374 | 1450 | 1556 | 1586 | 1491 | 1491 | |
| Tamil Nadu | *Kharif* | 1583 | 1716 | 1521 | 1583 | 1567 | 1433 | 1170 | 3224 | 3826 |
| | *Rabi* | 1588 | 1718 | 1655 | 1581 | 1568 | 1733 | 1215 | 5047 | 3838 |
| | Total | 1587 | 1717 | 1616 | 1582 | 1568 | 1552 | 1189 | 3838 | 3831 |
| Tripura | *Kharif* | 769 | 1000 | 1000 | 957 | 1000 | 1071 | 1000 | 960 | |
| Uttar Pradesh | *Kharif* | 1445 | 1622 | 1628 | 1107 | 1392 | 1539 | 1284 | 1326 | 1437 |
| | *Rabi* | | | | | | 3000 | 3341 | 1567 | 2500 |
| | Total | 1445 | 1622 | 1628 | 1107 | 1392 | 1705 | 1295 | 1335 | 1442 |
| Uttarakhand | *Kharif* | | 1633 | 1474 | 1086 | 1478 | 1467 | 1333 | 1333 | 1500 |
| West Bengal | *Kharif* | 1986 | 2501 | 2595 | 1996 | 2105 | 2606 | 2076 | 2166 | 2057 |
| | *Rabi* | | | | | 3657 | 4327 | 3913 | 4265 | 4000 |
| | Summer | | | | | 2041 | | | | |
| | Total | 1986 | 2501 | 2595 | 1996 | 2275 | 2977 | 2533 | 2968 | 3200 |
| A & N Islands | *Kharif* | | | | | 1000 | 1000 | 1000 | 2000 | |
| Delhi | *Kharif* | | | | 2000 | 5000 | 2000 | 1000 | 1000 | |
| All India | *Kharif* | 1695 | 1707 | 1896 | 1552 | 1932 | 1740 | 1799 | 1660 | 2123 |
| | *Rabi* | 2594 | 2921 | 2952 | 2851 | 2987 | 3224 | 3076 | 3793 | 3692 |
| | Total | 1792 | 1822 | 2000 | 1681 | 2041 | 1907 | 1938 | 1912 | 2337 |

* Advance Estimates released as on 09.07.2008.

# Appendix IV(ix)

*State-wise Yield of Total Pulses during 1999-2000 to 2007-08*

*Yield (Kgs/hect.)*

| State/UT | 1999-2000 | 2000-2001 | 2001-2002 | 2002-2003 | 2003-2004 | 2004-2005 | 2005-2006 | 2006-2007 | 2007-2008* |
|---|---|---|---|---|---|---|---|---|---|
| 1 | 2 | 3 | 4 | 5 | 6 | 7 | 8 | 9 | 10 |
| Andhra Pradesh | 486 | 554 | 593 | 506 | 567 | 565 | 772 | 679 | 823 |
| Arunachal Pradesh | 1000 | 1115 | 1044 | 1055 | 1088 | 1046 | 1078 | 1078 | |
| Assam | 552 | 559 | 558 | 541 | 557 | 571 | 537 | 557 | 558 |
| Bihar | 863 | 865 | 788 | 804 | 814 | 710 | 749 | 722 | 788 |
| Chhattisgarh | | 400 | 517 | 468 | 608 | 395 | 477 | 543 | 588 |
| Goa | 882 | 889 | 894 | 798 | 939 | 904 | 1045 | 1358 | |
| Gujarat | 579 | 300 | 520 | 469 | 748 | 675 | 704 | 593 | 812 |
| Haryana | 571 | 622 | 794 | 655 | 719 | 793 | 622 | 824 | 961 |
| Himachal Pradesh | 470 | 629 | 360 | 497 | 552 | 560 | 713 | 932 | 1000 |
| Jammu & Kashmir | 490 | 471 | 472 | 486 | 482 | 492 | 504 | 505 | 467 |
| Jharkhand | | 825 | 733 | 806 | 508 | 586 | 567 | 686 | 753 |
| Karnataka | 442 | 467 | 397 | 337 | 304 | 376 | 487 | 377 | 513 |
| Kerala | 796 | 797 | 792 | 773 | 833 | 853 | 775 | 857 | 800 |
| Madhya Pradesh | 766 | 640 | 773 | 574 | 761 | 759 | 754 | 780 | 602 |
| Maharashtra | 613 | 460 | 555 | 577 | 569 | 492 | 584 | 602 | 776 |
| Manipur | | 525 | 517 | 481 | 508 | 380 | 523 | 523 | |
| Meghalaya | 742 | 745 | 745 | 717 | 745 | 1385 | 750 | 744 | |
| Mizoram | 1780 | 1625 | 1444 | 1087 | 865 | 1297 | 1215 | 1160 | |
| Nagaland | 794 | 827 | 849 | 933 | 1000 | 797 | 1281 | 1200 | |
| Orissa | 387 | 352 | 398 | 354 | 382 | 388 | 416 | 445 | 563 |
| Punjab | 688 | 740 | 670 | 783 | 823 | 801 | 804 | 850 | 781 |
| Rajasthan | 360 | 308 | 425 | 269 | 590 | 375 | 261 | 462 | 398 |
| Sikkim | 910 | 852 | 862 | 930 | 958 | 957 | 897 | 897 | |
| Tamil Nadu | 431 | 455 | 427 | 345 | 374 | 410 | 337 | 541 | 358 |
| Tripura | 587 | 584 | 647 | 625 | 619 | 618 | 629 | 654 | |
| Uttar Pradesh | 968 | 803 | 886 | 822 | 890 | 847 | 811 | 725 | 745 |
| Uttarakhand | | 765 | 635 | 624 | 690 | 622 | 590 | 642 | 797 |
| West Bengal | 768 | 800 | 704 | 695 | 840 | 740 | 785 | 703 | 747 |
| A & N Islands | 333 | 571 | 500 | 462 | 571 | 571 | 500 | 571 | |
| D & N Haveli | 763 | 877 | 825 | 923 | 850 | 831 | 862 | 862 | |
| Delhi | 250 | 571 | 667 | 750 | 8000 | 1000 | 1000 | 1750 | |
| Daman & Diu | 1000 | 1000 | 846 | 769 | 846 | 923 | 846 | 846 | |
| Pondicherry | 672 | 600 | 538 | 350 | 346 | 316 | 324 | 324 | |
| All India | 635 | 544 | 607 | 543 | 635 | 577 | 597 | 612 | 635 |

* Advance Estimates released as on 09.07.2008.

## Appendix IV(x)

*State-wise Yield of Gram during 1999-2000 to 2007-08*

*Yield/Kg(Hectare)*

| *State/UT* | *1999-2000* | *2000-2001* | *2001-2002* | *2002-2003* | *2003-2004* | *2004-2005* | *2005-2006* | *2006-2007* | *2007-2008** |
|---|---|---|---|---|---|---|---|---|---|
| *1* | *2* | *3* | *4* | *5* | *6* | *7* | *8* | *9* | *10* |
| Andhra Pradesh | 583 | 1139 | 1274 | 979 | 1083 | 1012 | 1591 | 1085 | 1469 |
| Assam | 500 | 481 | 500 | 500 | 500 | 524 | 500 | 600 | 500 |
| Bihar | 975 | 1033 | 957 | 1010 | 979 | 835 | 902 | 818 | 941 |
| Chhattisgarh | | 515 | 735 | 644 | 964 | 542 | 706 | 843 | 873 |
| Gujarat | 512 | 529 | 554 | 503 | 884 | 803 | 850 | 870 | 902 |
| Haryana | 580 | 640 | 853 | 745 | 813 | 850 | 554 | 843 | 1000 |
| Himachal Pradesh | 1500 | 1154 | 1000 | 714 | 1000 | 1333 | 538 | 769 | 1000 |
| Jammu & Kashmir | | 500 | 667 | | | | 500 | | |
| Karnataka | 549 | 648 | 587 | 525 | 336 | 536 | 548 | 473 | 589 |
| Madhya Pradesh | 986 | 819 | 943 | 694 | 552 | 928 | 926 | 980 | 698 |
| Maharashtra | 644 | 519 | 595 | 564 | 530 | 561 | 691 | 706 | |
| Meghalaya | 750 | 600 | 600 | 600 | 600 | 600 | 600 | 600 | 917 |
| Nagaland | 1200 | 800 | 1000 | 1000 | 1333 | 1000 | 1000 | 1250 | |
| Orissa | 588 | 476 | 651 | 619 | 619 | 607 | 646 | 652 | 667 |
| Punjab | 968 | 948 | 873 | 957 | 900 | 863 | 750 | 1000 | 1000 |
| Rajasthan | 695 | 590 | 759 | 757 | 633 | 747 | 443 | 863 | 466 |
| Tamil Nadu | 628 | 649 | 714 | 677 | 652 | 657 | 678 | 644 | 558 |
| Tripura | | 667 | 750 | 750 | 500 | 500 | 500 | | |
| Uttar Pradesh | 948 | 844 | 971 | 893 | 1035 | 910 | 893 | 1000 | 745 |
| Uttarakhand | | 909 | 833 | 1000 | 1000 | 667 | 1000 | 742 | 1000 |
| West Bengal | 827 | 918 | 850 | 781 | 1026 | 1024 | 913 | 1000 | 800 |
| D & N Haveli | | 1000 | 500 | 500 | 500 | 500 | 500 | 769 | |
| Delhi | | | | | | | 1000 | | |
| All India | 833 | 744 | 853 | 717 | 811 | 815 | 808 | 845 | 780 |

* Advance Estimates released as on 09.07.2008.

# Appendix IV(xi)

*State-wise Yield of Tur during 1999-2000 to 2007-08*

*Yield (Kgs/hect.)*

| *State/ UT* | *1999-2000* | *2000-2001* | *2001-2002* | *2002-2003* | *2003-2004* | *2004-2005* | *2005-2006* | *2006-2007* | *2007-2008** |
|---|---|---|---|---|---|---|---|---|---|
| *1* | *2* | *3* | *4* | *5* | *6* | *7* | *8* | *9* | *10* |
| Andhra Pradesh | 358 | 427 | 448 | 347 | 418 | 456 | 609 | 401 | 652 |
| Arunachal Pradesh | | 800 | 1000 | 800 | 800 | 800 | 1000 | 1000 | |
| Assam | 714 | 712 | 714 | 714 | 714 | 716 | 692 | 714 | 714 |
| Bihar | 1235 | 1348 | 1155 | 1141 | 1237 | 1235 | 1291 | 989 | 1098 |
| Chhattisgarh | | 429 | 402 | 433 | 602 | 511 | 445 | 426 | 519 |
| Gujarat | 812 | 337 | 563 | 630 | 870 | 928 | 1102 | 751 | 1109 |
| Haryana | 798 | 768 | 812 | 829 | 1202 | 1032 | 1067 | 1100 | 1167 |
| Himachal Pradesh | | 333 | 333 | 333 | | 333 | | | |
| Jharkhand | | 1238 | 914 | 1511 | 595 | 544 | 633 | 645 | 803 |
| Karnataka | 570 | 452 | 306 | 469 | 552 | 516 | 727 | 470 | 704 |
| Kerala | | 1000 | | | | | | | 1500 |
| Madhya Pradesh | 854 | 672 | 821 | 619 | 811 | 785 | 739 | 683 | 674 |
| Maharashtra | 834 | 602 | 756 | 733 | 658 | 613 | 720 | 726 | 928 |
| Meghalaya | 667 | 667 | 750 | 750 | 750 | 750 | 750 | 750 | |
| Mizoram | | 3000 | 1000 | 600 | 667 | 1000 | 1000 | 1000 | |
| Nagaland | 714 | 875 | 714 | 1000 | 1067 | 938 | 950 | 1200 | |
| Orissa | 625 | 503 | 646 | 643 | 700 | 683 | 742 | 803 | 825 |
| Punjab | 831 | 874 | 859 | 859 | 928 | 865 | 885 | 931 | 833 |
| Rajasthan | 609 | 378 | 576 | 236 | 815 | 746 | 647 | 503 | 889 |
| Tamil Nadu | 711 | 712 | 641 | 546 | 611 | 625 | 540 | 732 | 565 |
| Tripura | 625 | 727 | 750 | 714 | 643 | 667 | 692 | 750 | |
| Uttar Pradesh | 1312 | 1254 | 1157 | 955 | 1101 | 982 | 987 | 749 | 897 |
| Uttarakhand | | | | | | | 1000 | 500 | 667 |
| West Bengal | 742 | 663 | 846 | 903 | 1000 | 733 | 889 | 700 | |
| D & N Haveli | 882 | 786 | 813 | 933 | 875 | 813 | 867 | 867 | |
| Delhi | | 333 | 800 | 1000 | | 667 | 1000 | 1500 | |
| All India | 786 | 618 | 679 | 651 | 670 | 667 | 765 | 650 | 824 |

* Advance Estimates released as on 09.07.2008.

# Appendix IV (xii)

*State-wise Yield of Total Oilseeds during 1999-2000 to 2007-08*

*Yield (Kgs./Hect.)*

| *State* | *1999-2000* | *2000-2001* | *2001-2002* | *2002-2003* | *2003-2004* | *2004-2005* | *2005-2006* | *2006-2007* | *2007-2008** |
|---|---|---|---|---|---|---|---|---|---|
| *1* | *2* | *3* | *4* | *5* | *6* | *7* | *8* | *9* | *10* |
| Andhra Pradesh | 536 | 927 | 661 | 543 | 634 | 757 | 698 | 609 | 1259 |
| Arunachal Pradesh | 1008 | 1036 | 1076 | 1060 | 1030 | 1042 | 838 | 838 | |
| Assam | 459 | 515 | 506 | 502 | 525 | 529 | 465 | 495 | 500 |
| Bihar | 725 | 847 | 839 | 763 | 881 | 887 | 982 | 1031 | 949 |
| Chhattisgarh | | 329 | 404 | 381 | 432 | 426 | 419 | 503 | 574 |
| Goa | 1769 | 1778 | 1563 | 1792 | 1727 | 1588 | 2394 | 1769 | |
| Gujarat | 619 | 603 | 1287 | 603 | 1901 | 990 | 1544 | 908 | 1593 |
| Haryana | 1311 | 1363 | 1477 | 1139 | 1547 | 1176 | 1124 | 1344 | 1214 |
| Himachal Pradesh | 526 | 552 | 552 | 522 | 568 | 632 | 344 | 477 | 438 |
| Jammu & Kashmir | 704 | 382 | 536 | 402 | 668 | 1125 | 429 | 610 | 625 |
| Jharkhand | | 625 | 625 | 386 | 235 | 241 | 311 | 422 | 540 |
| Karnataka | 602 | 804 | 587 | 535 | 412 | 587 | 600 | 478 | 652 |
| Kerala | 611 | 614 | 636 | 483 | 590 | 792 | 667 | 889 | 750 |
| Madhya Pradesh | 982 | 741 | 818 | 611 | 1049 | 833 | 1009 | 955 | 856 |
| Maharashtra | 974 | 820 | 944 | 947 | 1057 | 817 | 925 | 963 | 1279 |
| Manipur | 462 | 464 | 400 | 500 | 500 | 444 | 7000 | | |
| Meghalaya | 667 | 670 | 674 | 663 | 653 | 663 | 684 | 673 | |
| Mizoram | 1000 | 750 | 654 | 739 | 720 | 767 | 1125 | 927 | |
| Nagaland | 1007 | 981 | 1075 | 1233 | 1148 | 1165 | 926 | 901 | |
| Orissa | 469 | 425 | 422 | 353 | 515 | 568 | 565 | 550 | 594 |
| Punjab | 1067 | 1010 | 1004 | 883 | 1189 | 1098 | 1097 | 1111 | 1237 |
| Rajasthan | 937 | 768 | 1008 | 716 | 1239 | 1079 | 1134 | 1146 | 1086 |
| Sikkim | 760 | 690 | 690 | 724 | 747 | 768 | 727 | 727 | |
| Tamil Nadu | 1483 | 1716 | 1681 | 1283 | 1386 | 1483 | 1624 | 1829 | 1657 |
| Tripura | 677 | 723 | 717 | 700 | 731 | 722 | 709 | 705 | |
| Uttar Pradesh | 870 | 803 | 841 | 749 | 852 | 861 | 993 | 837 | 943 |
| Uttarakhand | | 502 | 714 | 719 | 919 | 927 | 857 | 750 | 1000 |
| West Bengal | 808 | 964 | 819 | 837 | 952 | 964 | 952 | 918 | 954 |
| D & N Havelli | 1000 | 1000 | 3000 | 1000 | 1000 | 1000 | 1000 | 1000 | |
| Delhi | 2600 | 1769 | 1429 | 1353 | 405 | 1220 | 763 | 1500 | |
| Pondicherry | 222 | 171 | 1000 | 1029 | 1800 | 1941 | 1778 | 1778 | |
| All India | 853 | 810 | 913 | 691 | 1064 | 885 | 1004 | 916 | 1086 |

* Advance Estimates released as on 09.07.2008.

# Appendix IV(xiii)

*State-wise Yield of Kharif Oilseeds during 1999-2000 to 2007-08*

*Yield (Kgs./Hect.)*

| States | 1999-2000 | 2000-2001 | 2001-2002 | 2002-2003 | 2003-2004 | 2004-2005 | 2005-2006 | 2006-2007 | 2007-2008* |
|---|---|---|---|---|---|---|---|---|---|
| 1 | 2 | 3 | 4 | 5 | 6 | 7 | 8 | 9 | 10 |
| Andhra Pradesh | 404 | 851 | 510 | 428 | 509 | 688 | 570 | 397 | 1202 |
| Arunachal Pradesh | 1095 | 1163 | 1238 | 1188 | 1098 | 1156 | 878 | 878 | |
| Assam | 533 | 525 | 538 | 538 | 560 | 543 | 533 | 571 | 565 |
| Bihar | 697 | 550 | 647 | 711 | 934 | 899 | 951 | 932 | 1143 |
| Chhattisgarh | | 387 | 492 | 438 | 497 | 542 | 521 | 623 | 708 |
| Goa | 1667 | 1500 | 1000 | 2000 | 1333 | 750 | 1333 | 2000 | |
| Gujarat | 550 | 541 | 1294 | 551 | 1958 | 910 | 1543 | 779 | 1590 |
| Haryana | 333 | 370 | 375 | 504 | 412 | 518 | 804 | 844 | 900 |
| Himachal Pradesh | 526 | 623 | 615 | 574 | 615 | 500 | 511 | 489 | 400 |
| Jammu & Kashmir | 422 | 424 | 413 | 422 | 429 | 426 | 462 | 429 | 500 |
| Jharkhand | | 638 | 638 | 315 | 188 | 231 | 287 | 220 | 600 |
| Karnataka | 567 | 862 | 563 | 527 | 446 | 612 | 569 | 439 | 677 |
| Kerala | 611 | 614 | 636 | 483 | 590 | 792 | 667 | 889 | 750 |
| Madhya Pradesh | 1008 | 752 | 823 | 616 | 1064 | 824 | 1020 | 962 | 860 |
| Maharashtra | 1104 | 917 | 1074 | 1059 | 1210 | 859 | 981 | 1031 | 1368 |
| Manipur | 500 | 462 | 500 | 667 | 500 | 500 | 6000 | | |
| Meghalaya | 696 | 696 | 696 | 692 | 692 | 692 | 731 | 692 | |
| Mizoram | 1143 | 720 | 547 | 735 | 755 | 765 | 1222 | 972 | |
| Nagaland | 1039 | 1106 | 1261 | 1314 | 1330 | 1523 | 1106 | 1099 | |
| Orissa | 406 | 340 | 329 | 264 | 348 | 390 | 403 | 423 | 436 |
| Punjab | 507 | 500 | 480 | 385 | 507 | 535 | 463 | 504 | 545 |
| Rajasthan | 871 | 592 | 938 | 469 | 1160 | 1083 | 972 | 1050 | 1223 |
| Sikkim | 810 | 810 | 810 | 795 | 821 | 846 | 825 | 825 | |
| Tamil Nadu | 1537 | 1342 | 1315 | 1007 | 1126 | 1045 | 1340 | 1251 | 1432 |
| Tripura | 552 | 621 | 591 | 600 | 600 | 609 | 583 | 600 | |
| Uttar Pradesh | 472 | 417 | 491 | 315 | 399 | 506 | 555 | 349 | 339 |
| Uttarakhand | | 306 | 835 | 789 | 958 | 1125 | 1059 | 1000 | 1313 |
| West Bengal | 690 | 893 | 832 | 792 | 865 | 832 | 837 | 922 | 920 |
| D & N Haveli | 1000 | 1000 | 3000 | 1000 | 1000 | 1000 | 1000 | 1000 | |
| Pondicherry | 2600 | 1769 | 1429 | 1353 | 1800 | 1941 | 1778 | 1778 | |
| All India | 813 | 757 | 883 | 625 | 1096 | 820 | 965 | 836 | 1111 |

* Advance Estimates released as on 09.07.2008.

# Appendix IV(xiv)

*State-wise Yield of Rabi Oilseeds during 1999-2000 to 2007-08*

*Yield (Kgs./Hect.)*

| *States* | *1999-2000* | *2000-2001* | *2001-2002* | *2002-2003* | *2003-2004* | *2004-2005* | *2005-2006* | *2006-2007* | *2007-2008** |
|---|---|---|---|---|---|---|---|---|---|
| *1* | *2* | *3* | *4* | *5* | *6* | *7* | *8* | *9* | *10* |
| Andhra Pradesh | 1110 | 1354 | 1325 | 942 | 1035 | 1027 | 1193 | 1235 | 1450 |
| Arunachal Pradesh | 991 | 1010 | 1045 | 1032 | 1014 | 1019 | 829 | 829 | |
| Assam | 453 | 515 | 504 | 498 | 522 | 528 | 458 | 488 | 494 |
| Bihar | 733 | 855 | 844 | 765 | 878 | 886 | 984 | 1037 | 939 |
| Chhattisgarh | | 260 | 309 | 306 | 351 | 287 | 284 | 319 | 345 |
| Goa | 1800 | 1857 | 1600 | 1773 | 1767 | 1700 | 2500 | 1727 | |
| Gujarat | 1062 | 1319 | 1220 | 1220 | 1491 | 1582 | 1552 | 1544 | 1609 |
| Haryana | 1324 | 1374 | 1488 | 1152 | 1557 | 1181 | 1128 | 1351 | 1220 |
| Himachal Pradesh | 526 | 523 | 527 | 495 | 545 | 667 | 274 | 471 | 455 |
| Jammu & Kashmir | 729 | 378 | 547 | 400 | 695 | 1171 | 405 | 630 | 638 |
| Jharkhand | | 556 | 556 | 500 | 278 | 250 | 333 | 461 | 521 |
| Karnataka | 684 | 678 | 621 | 548 | 371 | 541 | 655 | 543 | 603 |
| Madhya Pradesh | 844 | 649 | 779 | 566 | 943 | 890 | 955 | 912 | 820 |
| Maharashtra | 668 | 540 | 591 | 581 | 455 | 632 | 677 | 662 | 742 |
| Manipur | 429 | 467 | 333 | 400 | | 400 | | | |
| Meghalaya | 657 | 662 | 667 | 653 | 639 | 653 | 667 | 667 | |
| Mizoram | 696 | 818 | 880 | 750 | 636 | 773 | 833 | 600 | |
| Nagaland | 978 | 855 | 892 | 1120 | 784 | 778 | 756 | 729 | |
| Orissa | 626 | 657 | 704 | 640 | 933 | 951 | 925 | 838 | 896 |
| Punjab | 1201 | 1197 | 1227 | 1024 | 1331 | 1204 | 1236 | 1260 | 1396 |
| Rajasthan | 964 | 916 | 1055 | 868 | 1279 | 1078 | 1205 | 1185 | 1004 |
| Sikkim | 724 | 603 | 603 | 678 | 700 | 717 | 661 | 661 | |
| Tamil Nadu | 1381 | 2808 | 2717 | 2059 | 2120 | 2606 | 2073 | 3466 | 1989 |
| Tripura | 788 | 806 | 806 | 800 | 813 | 806 | 806 | 842 | |
| Uttar Pradesh | 967 | 920 | 946 | 862 | 968 | 940 | 1105 | 1024 | 1131 |
| Uttarakhand | | 744 | 598 | 615 | 846 | 647 | 667 | 563 | 643 |
| West Bengal | 844 | 981 | 816 | 849 | 981 | 1004 | 988 | 916 | 969 |
| Delhi | 222 | 171 | 1000 | 1029 | 405 | 1220 | 763 | 1500 | |
| All India | 922 | 929 | 972 | 821 | 1007 | 993 | 1068 | 1055 | 1035 |

* Advance Estimates released as on 09.07.2008.

# Appendix IV(xv)

*State-wise Yield of Groundnut during 1999-2000 to 2007-08*

*Yield (Kgs/hect.)*

| *States* | *Season* | *1999-2000* | *2000-2001* | *2001-2002* | *2002-2003* | *2003-2004* | *2004-2005* | *2005-2006* | *2006-2007* | *2007-2008** |
|---|---|---|---|---|---|---|---|---|---|---|
| *1* | *2* | *3* | *4* | *5* | *6* | *7* | *8* | *9* | *10* | *11* |
| Andhra Pradesh | *Kharif* | 448 | 1060 | 568 | 426 | 482 | 778 | 565 | 301 | 1357 |
| | *Rabi* | 1489 | 1636 | 1714 | 1397 | 1613 | 1657 | 1739 | 1806 | 1856 |
| | Total | 607 | 1144 | 739 | 558 | 660 | 890 | 728 | 557 | 1434 |
| Bihar | *Kharif* | 1208 | 2000 | 500 | 500 | 400 | 600 | 556 | 556 | 1000 |
| Chhattisgarh | *Kharif* | | 936 | 1234 | 1111 | 1107 | 1110 | 1078 | 1142 | 1250 |
| Goa | *Kharif* | 1667 | 1500 | 1000 | 2000 | 1333 | 750 | 1333 | 2000 | |
| | *Rabi* | 1800 | 1857 | 1600 | 1773 | 1767 | 1700 | 2500 | 1727 | |
| | Total | 1769 | 1778 | 1563 | 1792 | 1727 | 1588 | 2394 | 1769 | |
| Gujarat | *Kharif* | 349 | 368 | 1401 | 508 | 2267 | 891 | 1707 | 725 | 1781 |
| | *Rabi* | 1734 | 1788 | 1473 | 1641 | 1556 | 2391 | 2225 | 2000 | 1901 |
| | Total | 393 | 395 | 1402 | 539 | 2235 | 943 | 1734 | 809 | 1791 |
| Haryana | *Kharif* | 500 | 1000 | 800 | 792 | 733 | 750 | 733 | 840 | 1000 |
| Himachal Pradesh | *Kharif* | 1333 | 2000 | 1000 | 500 | 1000 | | | - | |
| Karnataka | *Kharif* | 600 | 998 | 592 | 549 | 435 | 737 | 590 | 434 | 779 |
| | *Rabi* | 1119 | 1108 | 1142 | 1144 | 996 | 917 | 924 | 760 | 725 |
| | Total | 686 | 1017 | 685 | 639 | 530 | 766 | 645 | 497 | 770 |
| Kerala | *Kharif* | 739 | 730 | 750 | 524 | 741 | 944 | 727 | 1000 | 1000 |
| Madhya Pradesh | *Kharif* | 992 | 1059 | 1121 | 632 | 1159 | 1158 | 1126 | 948 | 940 |
| Maharashtra | *Kharif* | 951 | 894 | 1081 | 957 | 1096 | 1035 | 831 | 743 | 1112 |
| | *Rabi* | 1388 | 1267 | 1447 | 1676 | 1491 | 1430 | 1429 | 1355 | 1487 |
| | Total | 1059 | 959 | 1146 | 1074 | 1153 | 1123 | 958 | 889 | 1184 |
| Nagaland | *Kharif* | 1200 | 1143 | 1200 | 1250 | 1500 | 1273 | 1000 | 1000 | |
| Orissa | *Kharif* | 900 | 650 | 765 | 688 | 865 | 879 | 835 | 904 | 971 |
| | *Rabi* | 926 | 946 | 1207 | 1042 | 1438 | 1453 | 1414 | 1292 | 1302 |
| | Total | 914 | 794 | 985 | 870 | 1207 | 1233 | 1171 | 1113 | 1172 |
| Punjab | *Kharif* | 1000 | 1000 | 1000 | 717 | 909 | 837 | 882 | 864 | 1000 |
| Rajasthan | *Kharif* | 965 | 924 | 1227 | 687 | 1566 | 1552 | 1549 | 1310 | 1728 |
| Tamil Nadu | *Kharif* | 1829 | 1557 | 1511 | 1146 | 1298 | 1159 | 1526 | 1329 | 1794 |
| | *Rabi* | 1402 | 2836 | 2753 | 2085 | 2145 | 2678 | 2103 | 3729 | 2025 |
| | Total | 1659 | 1942 | 1885 | 1429 | 1552 | 1632 | 1775 | 1981 | 1905 |
| Tripura | *Kharif* | 889 | 1125 | 1000 | 1000 | 1000 | 1000 | 1000 | 1000 | |
| | *Rabi* | 1000 | 1000 | 500 | 1000 | 1000 | 1000 | 667 | 1000 | |
| | Total | 917 | 1083 | 833 | 1000 | 1000 | 1000 | 889 | 1000 | |
| Uttar Pradesh | *Kharif* | 825 | 835 | 839 | 618 | 637 | 816 | 851 | 730 | 596 |
| Uttarakhand | *Kharif* | | | | | 667 | 1000 | 1000 | 2000 | 1000 |
| West Bengal | *Kharif* | 688 | 800 | 1000 | 929 | 1000 | 938 | 933 | 1000 | 1000 |
| | *Rabi* | 1451 | 1510 | 1542 | 1554 | 1678 | 1644 | 1727 | 1744 | 1750 |
| | Total | 1343 | 1471 | 1509 | 1526 | 1651 | 1620 | 1703 | 1713 | 1714 |
| Pondicherry | *Kharif* | 2778 | 1833 | 1583 | 1538 | 1923 | 2000 | 1938 | 1938 | |
| All India | *Kharif* | 655 | 861 | 1030 | 587 | 1320 | 909 | 1097 | 689 | 1412 |
| | *Rabi* | 1373 | 1756 | 1808 | 1548 | 1602 | 1771 | 1702 | 1880 | 1691 |
| | Total | 766 | 977 | 1127 | 694 | 1357 | 1020 | 1187 | 866 | 1460 |

* Advance Estimates released as on 09.07.2008.

# Appendix IV(xvi)

*State-wise Yield of Rapseed & Mustard during 1999-2000 to 2007-08*

*Yield (kgs/hect.)*

| *States* | *1999-2000* | *2000-2001* | *2001-2002* | *2002-2003* | *2003-2004* | *2004-2005* | *2005-2006* | *2006-2007* | *2007-2008** |
|---|---|---|---|---|---|---|---|---|---|
| *1* | *2* | *3* | *4* | *5* | *6* | *7* | *8* | *9* | *10* |
| Andhra Pradesh | 333 | 250 | 333 | 333 | | 250 | 333 | 600 | 333 |
| Arunachal Pradesh | 991 | 1010 | 1045 | 1032 | 1014 | 1019 | 829 | 829 | |
| Assam | 451 | 514 | 504 | 498 | 523 | 528 | 456 | 487 | 491 |
| Bihar | 788 | 863 | 839 | 688 | 824 | 805 | 926 | 1029 | 869 |
| Chhattisgarh | | 332 | 372 | 328 | 412 | 375 | 330 | 399 | 396 |
| Gujarat | 941 | 1236 | 1182 | 1072 | 1469 | 1390 | 1349 | 1396 | 1501 |
| Haryana | 1322 | 1371 | 1488 | 1148 | 1559 | 1177 | 1117 | 1343 | 1203 |
| Himachal Pradesh | 570 | 575 | 575 | 540 | 556 | 667 | 292 | 506 | 500 |
| Jammu & Kashmir | 733 | 378 | 546 | 399 | 696 | 1175 | 406 | 649 | 638 |
| Karnataka | 281 | 267 | 268 | 256 | 283 | 333 | 400 | 250 | 333 |
| Madhya Pradesh | 999 | 773 | 908 | 649 | 1081 | 988 | 1047 | 999 | 914 |
| Maharashtra | 275 | 232 | 250 | 250 | 300 | 333 | 308 | 357 | 444 |
| Manipur | 429 | 467 | 333 | 400 | | 400 | | | |
| Meghalaya | 657 | 662 | 667 | 653 | 639 | 653 | 667 | 667 | |
| Mizoram | 696 | 818 | 880 | 750 | 636 | 773 | 833 | 600 | |
| Nagaland | 1000 | 1000 | 1000 | 1063 | 800 | 779 | 800 | 753 | |
| Orissa | 176 | 137 | 155 | 131 | 195 | 188 | 190 | 171 | 188 |
| Punjab | 1125 | 1208 | 1200 | 909 | 1192 | 1033 | 1102 | 1122 | 1214 |
| Rajasthan | 965 | 918 | 1056 | 868 | 1279 | 1078 | 1205 | 1185 | 1004 |
| Sikkim | 724 | 603 | 603 | 678 | 700 | 717 | 661 | 661 | |
| Tamil Nadu | 200 | 200 | | 200 | 250 | 167 | 200 | 333 | |
| Tripura | 767 | 781 | 828 | 773 | 800 | 793 | 821 | 824 | |
| Uttar Pradesh | 1023 | 1000 | 998 | 895 | 1008 | 979 | 1149 | 1057 | 1159 |
| Uttarakhand | | 744 | 598 | 615 | 846 | 647 | 667 | 563 | 643 |
| West Bengal | 805 | 956 | 766 | 805 | 928 | 934 | 909 | 803 | 850 |
| Delhi | 222 | 171 | 1000 | 1029 | 405 | 1220 | 763 | 1500 | |
| All India | 960 | 935 | 1002 | 854 | 1159 | 1038 | 1117 | 1095 | 1009 |

* Advance Estimates released as on 09.07.2008.

# Appendix IV(xvii)

*State-wise Yield of Linseed during 1999-2000 to 2007-08*

*Yield (Kgs/Hect)*

| States | 1999-2000 | 2000-2001 | 2001-2002 | 2002-2003 | 2003-2004 | 2004-2005 | 2005-2006 | 2006-2007 | 2007-2008* |
|---|---|---|---|---|---|---|---|---|---|
| 1 | 2 | 3 | 4 | 5 | 6 | 7 | 8 | 9 | 10 |
| Andhra Pradesh | 56 | 200 | 167 | 250 | 200 | 250 | 250 | 333 | 500 |
| Assam | 500 | 520 | 500 | 500 | 500 | 516 | 500 | 500 | 556 |
| Bihar | 561 | 682 | 715 | 760 | 784 | 841 | 895 | 850 | 857 |
| Chhattisgarh | | 214 | 270 | 291 | 308 | 221 | 248 | 250 | 305 |
| Himachal Pradesh | 367 | 292 | 304 | 250 | 500 | 667 | 100 | 267 | |
| Jammu & Kashmir | 429 | 375 | 750 | 667 | 500 | 500 | 400 | 426 | |
| Jharkhand | | 556 | 556 | 500 | 278 | 250 | 333 | 389 | 375 |
| Karnataka | 338 | 368 | 337 | 351 | 78 | 214 | 308 | 333 | 385 |
| Madhya Pradesh | 402 | 333 | 383 | 353 | 430 | 393 | 412 | 411 | 298 |
| Maharashtra | 289 | 233 | 304 | 222 | 238 | 220 | 277 | 235 | 284 |
| Nagaland | 750 | 462 | 625 | 1125 | 746 | 767 | 600 | 648 | |
| Orissa | 384 | 381 | 368 | 327 | 396 | 379 | 420 | 422 | 444 |
| Punjab | 1000 | 1000 | 1000 | 1000 | 667 | 667 | 500 | 500 | - |
| Rajasthan | 642 | 598 | 795 | 667 | 1000 | 767 | 833 | 833 | 1000 |
| Uttar Pradesh | 433 | 360 | 414 | 456 | 506 | 420 | 414 | 373 | 267 |
| West Bengal | 408 | 336 | 416 | 416 | 317 | 400 | 284 | 240 | 333 |
| All India | 406 | 351 | 390 | 393 | 412 | 378 | 395 | 385 | 372 |

* Advance Estimates released as on 09.07.2008.

# Appendix IV(xviii)

*State-wise Yield of Safflower during 1999-2000 to 2007-08*

*Yield (Kgs./Hect.)*

| States | 1999-2000 | 2000-2001 | 2001-2002 | 2002-2003 | 2003-2004 | 2004-2005 | 2005-2006 | 2006-2007 | 2007-2008* |
|---|---|---|---|---|---|---|---|---|---|
| 1 | 2 | 3 | 4 | 5 | 6 | 7 | 8 | 9 | 10 |
| Andhra Pradesh | 471 | 375 | 300 | 337 | 294 | 421 | 471 | 643 | 467 |
| Bihar | 571 | 667 | 571 | 500 | 1000 | 1000 | 1000 | 500 | |
| Chhattisgarh | | 222 | 333 | 200 | 222 | 231 | 273 | 286 | |
| Karnataka | 751 | 654 | 791 | 650 | 422 | 468 | 741 | 753 | 838 |
| Madhya Pradesh | 333 | | 333 | 333 | 333 | 167 | 300 | 333 | |
| Maharashtra | 552 | 412 | 484 | 441 | 355 | 476 | 605 | 606 | 643 |
| Orissa | 500 | 1000 | 550 | 385 | 438 | 500 | 615 | 583 | 1000 |
| All India | 594 | 476 | 546 | 483 | 371 | 470 | 627 | 637 | 694 |

* Advance Estimates released as on 09.07.2008.

# Appendix IV(xix)

*State-wise Yield of Sunflower during 1999-2000 to 2007-08*

*Yield (Kgs./Hect.)*

| *States* | *Season* | *1999-2000* | *2000-2001* | *2001-2002* | *2002-2003* | *2003-2004* | *2004-2005* | *2005-2006* | *2006-2007* | *2007-2008** |
|---|---|---|---|---|---|---|---|---|---|---|
| *1* | *2* | *3* | *4* | *5* | *6* | *7* | *8* | *9* | *10* | *11* |
| Andhra Pradesh | *Kharif* | 449 | 771 | 621 | 614 | 632 | 552 | 351 | 406 | 629 |
| | *Rabi*/Summer | 639 | 912 | 930 | 685 | 697 | 634 | 805 | 871 | 1164 |
| | Total | 572 | 853 | 820 | 663 | 678 | 609 | 671 | 738 | 1017 |
| Bihar | *Kharif* | | | | | | | | 1233 | 1333 |
| | *Rabi*/Summer | | 1463 | 1363 | 1299 | 1365 | 1457 | 1374 | 1347 | 1368 |
| | Total | 921 | 1463 | 1363 | 1299 | 1383 | 1401 | 1345 | 1332 | 1364 |
| Chhattisgarh | *Kharif* | | 417 | 414 | 385 | 344 | 545 | 366 | 600 | 1000 |
| Haryana | Summer | 1429 | 1500 | 1544 | 1536 | 1483 | 1657 | 1667 | 1667 | 1583 |
| Karnataka | *Kharif* | 409 | 535 | 479 | 458 | 459 | 470 | 504 | 365 | 541 |
| | *Rabi*/Summer | 404 | 450 | 441 | 411 | 257 | 472 | 593 | 472 | 541 |
| | Total | 406 | 483 | 449 | 427 | 321 | 471 | 552 | 420 | 541 |
| Madhya Pradesh | *Kharif* | | 250 | 500 | 286 | 400 | 500 | 333 | 333 | |
| | Summer | 446 | 375 | | 412 | 533 | 571 | 583 | 571 | 1000 |
| | Total | 352 | 344 | 500 | 375 | 500 | 550 | 533 | 500 | 500 |
| Maharashtra | *Kharif* | 394 | 493 | 357 | 485 | 360 | 478 | 532 | 529 | 580 |
| | *Rabi* | 587 | 536 | 524 | 497 | 355 | 537 | 593 | 545 | 719 |
| | Summer | 733 | 556 | 833 | | 750 | 810 | 733 | | |
| | Total | 523 | 519 | 482 | 495 | 362 | 525 | 580 | 548 | 664 |
| Nagaland | *Kharif* | 1300 | 750 | 1400 | 2000 | 2000 | 600 | 800 | 750 | |
| | Summer | 1400 | 1100 | 1300 | 2000 | | 1000 | 1000 | 800 | |
| | Total | 1371 | 820 | 1333 | 2000 | 2000 | 714 | 875 | 786 | |
| Orissa | *Kharif* | 500 | 1000 | 400 | | 625 | | | 667 | 1000 |
| | *Rabi*/Summer | 807 | 1000 | 775 | 667 | 750 | 857 | 830 | 949 | 1000 |
| | Total | 787 | 1000 | 755 | 650 | 733 | 857 | 821 | 932 | 1000 |
| Punjab | Summer | 1357 | 1155 | 1419 | 1560 | 1700 | 1826 | 1612 | 1641 | 1650 |
| Rajasthan | *Rabi*/Summer | 667 | | | | | | 1000 | 1000 | |
| Tamil Nadu | *Kharif* | 1083 | 1250 | 1263 | 1143 | 1288 | 1069 | 1304 | 2710 | 1393 |
| | *Rabi* | 843 | 1300 | 1313 | 1171 | 1289 | 1050 | 1165 | 1284 | 1545 |
| | Total | 957 | 1273 | 1286 | 1156 | 1288 | 1060 | 1240 | 1828 | 1460 |
| Uttar Pradesh | *Rabi*/Summer | 1311 | 1111 | 1927 | 1667 | 1708 | 2049 | 1278 | 1500 | 1625 |
| West Bengal | Summer | 500 | 1000 | 1375 | 1333 | 659 | 1500 | 1256 | 1292 | 1333 |
| All India | *Kharif* | 429 | 576 | 501 | 510 | 501 | 494 | 496 | 425 | 596 |
| | *Rabi* | 602 | 618 | 604 | 541 | 448 | 587 | 692 | 661 | 841 |
| | Total | 538 | 602 | 577 | 531 | 464 | 549 | 615 | 567 | 741 |

* Advance Estimates released as on 09.07.2008

# Appendix IV(xx)

*State-wise Yield of Soyabean during 1999-2000 to 2007-08*

*Yield (Kgs./Hect.)*

| States | 1999-2000 | 2000-2001 | 2001-2002 | 2002-2003 | 2003-2004 | 2004-2005 | 2005-2006 | 2006-2007 | 2007-2008* |
|---|---|---|---|---|---|---|---|---|---|
| 1 | 2 | 3 | 4 | 5 | 6 | 7 | 8 | 9 | 10 |
| Andhra Pradesh | 939 | 1009 | 1015 | 1015 | 1569 | 1567 | 1949 | 1515 | 1977 |
| Arunachal Pradesh | 1321 | 1448 | 1448 | 1343 | 1237 | 1344 | 947 | 947 | |
| Chhattisgarh | | 543 | 821 | 546 | 885 | 1018 | 900 | 997 | 1159 |
| Gujarat | 810 | 810 | 810 | 667 | 714 | 556 | 853 | 510 | 382 |
| Himachal Pradesh | 1833 | 1667 | 1667 | 1571 | 1000 | 1000 | 1333 | 667 | 1000 |
| Karnataka | 915 | 894 | 831 | 831 | 482 | 604 | 534 | 718 | 832 |
| Madhya Pradesh | 1068 | 767 | 839 | 638 | 1105 | 835 | 1058 | 1006 | 894 |
| Maharashtra | 1392 | 1109 | 1254 | 1255 | 1396 | 900 | 1077 | 1147 | 1493 |
| Meghalaya | 900 | 900 | 900 | 900 | 1000 | 900 | 1000 | 1000 | |
| Mizoram | 1273 | 1333 | 923 | 900 | 1000 | 1067 | 1929 | 1000 | |
| Nagaland | 1120 | 1269 | 1389 | 1346 | 1385 | 1600 | 1199 | 1200 | |
| Rajasthan | 1221 | 691 | 1091 | 501 | 1400 | 1425 | 1150 | 1203 | 1342 |
| Sikkim | 810 | 810 | 810 | 795 | 821 | 846 | 825 | 825 | |
| Uttar Pradesh | 778 | 586 | 725 | 170 | 1071 | 441 | 769 | 875 | 800 |
| Uttarakhand | | 638 | 1024 | 813 | 1105 | 1278 | 1154 | 1111 | 1462 |
| West Bengal | 571 | 600 | 600 | 600 | | 600 | 500 | 600 | 1000 |
| All India | 1138 | 822 | 940 | 762 | 1193 | 908 | 1073 | 1063 | 1124 |

* Advance Estimates released as on 09.07.2008.

# Appendix V

## *Crop Calender of Major Crops: State-wise and All India*

| *States/UTs* | *Period* | *Kharif* | *Rabi* | *Kharif* | *Summer* | *Rabi* |
|---|---|---|---|---|---|---|
| | | *Paddy* | *Paddy* | *Bajra* | *Bajra* | *Wheat* |
| Andhra Pradesh | Sowing | Jun.( B)-Aug.(E) | Nov.(E)-Feb(B) | Jun(B)-Jul(M) | | Nov(B)-Dec(M) |
| | Harvesting | Oct.(E)-Feb.(B) | Mar.(E)-May(M) | Aug(B)-Oct(B) | | Mar(B)-Apr(E) |
| Assam | Sowing | Jun(B)-Sep(B) | | | | |
| | Harvesting | Oct.(B)-Feb.(E) | | | | |
| Bihar | Sowing | Jun(M)-Jul(B) | | | | Nov(M)-Dec(E) |
| | Harvesting | Nov(E)-Dec(E) | | | | Mar(M)-Apr(E) |
| Goa | Sowing | May(M)-Jul(M) | Nov(B)-Jan(E) | | | |
| | Harvesting | Sep(B)-Dec(E) | Mar(B)-Apr(E) | | | |
| Gujarat | Sowing | Jun(B)-Aug(E) | | Jun(B)-Jul(E) | Feb(B)-Feb(E) | Oct(B)-Nov(E) |
| | Harvesting | Oct(B)-Dec(E) | | Sep(B)-Nov(E) | May(B)-May(E) | Feb(B)-Mar(E) |
| Haryana | Sowing | Jun(B)-Jul(B) | | Jun(M)-Jul(B) | | Oct(E)-Dec(B) |
| | Harvesting | Oct(M)-Nov(E) | | Oct(M)-Nov(M) | | Apr(M)-Apr(E) |
| Himachal Pradesh | Sowing | May(B)-Jul(E) | | | | Oct(B)-Nov(E) |
| | Harvesting | Sep(M)-Oct(M) | | | | Apr(M)-Jun(E) |
| Jammu & Kashmir | Sowing | May(B)-Jun(E) | | | | Oct(B)-Dec(E) |
| | Harvesting | Oct(B)-Nov(E) | | | | May(B)-May(E) |
| Karnataka | Sowing | May(B)-Jul(E) | Aug(B)-Sep(E) | Jul(B)-Sep(E) | Jan(B)-Feb(E) | Oct(B)-Dec(E) |
| | Harvesting | Oct(B)-Dec(E) | Jan(B)-Feb(E) | Oct(B)-Nov(E) | Apr(B)-May(E) | Jan(B)-Feb(E) |
| Kerala | Sowing | Aug((B)-Nov(E) | | | | |
| | Harvesting | Nov(B)-Jan(E) | | | | |
| Madhya Pradesh | Sowing | Jun(M)_Jul(E) | | Jun(M)-Jul(M) | | Oct(M)-Dec(E) |
| | Harvesting | Sep(E)-Dec(E) | | Sep(E)-Dec(M) | | Feb(M)Apr(E) |
| Maharashtra | Sowing | Jul(B)-Agu(E) | | Jun(B)-Jul(E) | | Oct(B)-Dec(E) |
| | Harvesting | Oct(B)-Dec(E) | | Sep(B)-Oct(E) | | Feb(B)-Ar(E) |
| Orissa | Sowing | Jun-Aug | | Jun-Jul | | Oct-Nov |
| | Harvesting | Dec-Jan | | Sep-Oct | | Mar-Apr |
| Punjab | Sowing | May(B)-Jul(E) | | | | Oct(B)-Dec(E) |
| | Harvesting | Sep(B)-Nov(E) | | | | Apr(B)-May(E) |
| Rajasthan | Sowing | Jun(B)-Jul(E) | | Jun(B)-Jul(E) | | Nov(B)-Dec(E) |
| | Harvesting | Oct(B)-Nov(E) | | Sep(B)-Oct(E) | | Mar(B)-May(E) |
| U.P. | Sowing | Jun(B)-Sep(M) | | Jun(B)-Jul(E) | | Oct(B)-Dec(E) |
| | Harvesting | Oct(B)-Nov(E) | | Oct(B)-Nov(E) | | Apr(B)-Jun(E) |
| West Bengal | Sowing | Jul(B)-Aug(E) | | | | Nov(B)-Dec(E) |
| | Harvesting | Nov(B)-Jan(E) | | | | Mar(B)-Apr(E) |
| All India | Sowing | May-Aug | Aug-Sep | Jun-Jul | Jan-Feb | Oct-Dec |
| | Harvesting | Sep-Jan | Jan-May | Sep-Nov | Apr-May | Feb-Jun |

| *States/Uts* | *Period* | *Early Kharif* | *Kharif* | *Rabi* | *Summer* |
|---|---|---|---|---|---|
| | | *Arhar/tur* | *Arhar/tur* | *Redgram* | *Redgram* |
| Andhra Pradesh | Sowing | | Jun(M)-Jul(M) | | |
| | Harvesting | | Jan(M)-Feb(M) | | |
| Karnataka | Sowing | | Jun(B)-Jul(E) | | |
| | Harvesting | | Nov(B)-Feb(E) | | |
| Kerala | Sowing | May(B)-Aug(B) | Aug(B)-Nov(E) | | |
| | Harvesting | Aug(B)-Oct(E) | Oct(B)-Jan(E) | | |
| Maharashtra | Sowing | | Jul(B)-Jul(E) | | |
| | Harvesting | | Dec(B)-Jan(E) | | |
| Manipur | Sowing | | Jun-Jul | | |
| | Harvesting | | Oct-Nov | | |
| Meghalaya | Sowing | | Apr(B)-May(E) | | |
| | Harvesting | | Dec(B)-Jan(E) | | |
| Rajasthan | Sowing | | Jul(B)-Jul(E) | | |
| | Harvesting | | Dec(B)-Jan(E) | | |
| Uttar Pradesh | Sowing | | | | |
| | Harvesting | | | | |
| West Bengal | Sowing | | Jun(B)-Jun(E) | | |
| | Harvesting | | Feb(B)-Mar(E) | | |
| All India | Sowing | | Jun(B)-Jul | Jun(B)-Jul(E) | Feb(B)-Mar(E) |
| | Harvesting | | Dec-Feb | Mar(B)-Apr(E) | May(B)-May(E) |

| *States/Uts* | *Period* | *Kharif* | *Rabi* | *Summer* | *Early Kharif* |
|---|---|---|---|---|---|
| | | *Blackgram/Urad* | *Blackgram/Urad* | *Blackgram/Urad* | *Greengram* |
| Andhra Pradesh | Sowing | Jun(M)-Jul(M) | Nov(M)-Jan(E) | | |
| | Harvesting | Sep(B)-Sep(E) | Feb(M)-Apr(E) | | |
| Assam | Sowing | | Aug(B)-Sep(E) | | |
| | Harvesting | | Nov(B)-Dec(E) | | |
| Karnataka | Sowing | Jun(B)-Jul(E) | Sep(B)-Oct(E) | | |
| | Harvesting | Sep(B)-Oct(E) | Dec(B)-Jan(E) | | |
| Kerala | Sowing | Mar(B)-Jun(B) | | Sep(B)-Oct(E) | |
| | Harvesting | Jun(B)-Sep(E) | | Nov(B)-Dec(E) | |
| Madhya Pradesh | Sowing | Jun(M)-Aug(M) | Oct(M)-Nov(E) | | |
| | Harvesting | Sep(M)-Nov(E) | Feb(M)-May(M) | | |
| Maharashtra | Sowing | Jul(B)-Aug(E) | | | |
| | Harvesting | Oct(B)-Oct(E) | | | |
| Orissa | Sowing | | Nov-Dec | | |
| | Harvesting | | Mar-Apr | | |
| Rajasthan | Sowing | Jul(B)-Aug(E) | | | |
| | Harvesting | Oct(B)-Nov(E) | | | |
| Tripura | Sowing | Jun-Jul | Oct-Dec | | |
| | Harvesting | Sep-Oct | Feb-Mar | | |
| Uttar Pradesh | Sowing | | Apr(B)-Apr(E) | | |
| | Harvesting | | Jun(B)-Jun(E) | | |
| All India | Sowing | Jun(M)-Jul(B) | Apr(B)-Dec(B) | | Jun(B)-Aug(E) |
| | Harvesting | Sep(B)-Nov(E) | Sep(M)-Nov(E) | | Aug(B)-Sep(E) |

| *States/Uts* | *Period* | *Kharif* | *Rabi* | *Summer* | *Early Kharif* |
|---|---|---|---|---|---|
| | | *Greengram* | *Greengram* | *Greengram* | *Horsegram* |
| Andhra Pradesh | Sowing | Jun(M)-Jul(M) | Nov(M)-Dec(E) | | |
| | Harvesting | Aug(M)-Spe(B) | Feb(E)-May(M) | | |
| Assam | Sowing | | Aug(B)-Sep(E) | | |
| | Harvesting | | Nov(B)-Dec(E) | | |
| Jammu & Kashmir | Sowing | Jun(B)-Jun(E) | | | |
| | Harvesting | Aug(M)-Sep(M) | | | |
| Karnataka | Sowing | Jun(B)-Jul(E) | Sep(B)-Oct(E) | | |
| | Harvesting | Sep(B)-Oct(E) | Dec(B)-Jan(E) | | |
| Kerala | Sowing | Oct(B)-Nov(E) | | Sep(B)-Oct(E) | |
| | Harvesting | Nov(B)-Nov(E) | | Nov(B)-Dec(E) | |
| Madhya Pradesh | Sowing | Jun(M)-Aug(M) | Oct(M)-Dec(M) | | |
| | Harvesting | Sep(M)-Dec(M) | Feb(B)-May(B) | | |
| Maharashtra | Sowing | Jul(B)-Jul(E) | | | |
| | Harvesting | Aug(B)-Sep(E) | | | |
| Rajasthan | Sowing | Jul(B)-Aug(E) | | | |
| | Harvesting | Oct(B)-Oct(E) | | | |
| Tripura | Sowing | Jun-Jul | | | |
| | Harvesting | Sep-Oct | | | |
| Uttar Pradesh | Sowing | | | Apr(B)-Apr(E) | |
| West Bengal | Sowing | Jul(B)-Aug(E) | | Feb(B)-Mar(E) | |
| | Harvesting | Oct(B)-Nov(E) | | Apr(B)-May(E) | |
| All India | Sowing | Jun(M)-Aug(B) | Aug-Oct | Sep-Oct | Feb(B)-Apr(E) |
| | Harvesting | Sep(M)-Dec(M) | Sep-Oct | Feb-May | Apr(B)-Jun(B) |

| *States/UTs* | *Period* | *Rabi* | *Rabi* | *Kharif* | *Rabi* |
|---|---|---|---|---|---|
| | | *Gram* | *Masur* | *Pulses/Lentil* | *Pulses/Lentil* |
| Assam | Sowing | | | Oct(B)-Nov(E) | |
| | Harvesting | | | Mar(B)-Apr(E) | |
| Bihar | Sowing | | | Jun(M)-Ju;l(B) | Oct(M)-Nov(M) |
| | Harvesting | | | Nov(B)-Dec(E) | Mar(B)-Mar(E) |
| Chhattisgarh | | | | | |
| Goa | Sowing | | | Aug(B)-Sep(E) | Dec(B)-Jan(E) |
| | H .rvesting | | | Nov(B)-Dec(E) | Mar(B)-Mayu(E) |
| Gujarat | Sowing | Oct(B)-Nov(E) | | | |
| | Harvesting | Feb(B)-Mar(E) | | | |
| Haryana | Sowing | Oct(B)-Oct(M) | | | |
| | Harvesting | Mar(M)-Mar(E) | | | |
| Karnataka | Sowing | Oct(B)-Nov(E) | | | |
| | Harvesting | Jan(B)-Mar(E) | | | |
| Madhya Pradesh | Sowing | Oct(B)-Dec(B) | Sep(M)-Nov(E) | | |
| | Harvesting | Feb(B)-Apr(B) | Feb(M)-Apr(E) | | |
| Maharashtra | Sowing | Sep(B)-Oct)E) | | | |
| | Harvesting | Feb(B)-Mar(E) | | | |
| Manipur | Sowing | Sep-Oct | | | |
| | Harvesting | Mar-Apr | | | |
| Meghalaya | Sowing | Sep(B)-Oct(E) | | Sep(B)-Oct(E) | |
| | Harvesting | Dec(B)-Jan(E) | | Feb(B)-Mar(E) | |
| Punjab | Sowing | | | Jun(B)-Jul(E) | |
| | Harvesting | | | Oct(B)-Nov(E) | |
| Uttar Pradesh | Sowing | Oct(B)-Nov(E) | Oct(B)-Nov(E) | | |
| | Harvesting | Mar(B)-Apr(E) | Mar(B)-Mar(E) | | |
| West Bengal | Sowing | Nov(M)-Dec(B) | | Nov(B)-Nov(E) | |
| | Harvesting | Mar(B)-Mar(E) | | Feb(B)-Mar(E) | |
| All India | Sowing | Oct-Dec | Sep-Nov | Jun-Oct | Oct-Nov |
| | Harvesting | Mar-Apr | Feb-Apr | Nov-Dec | Mar-May |

| *States/Uts* | *Period* | *Kharif* | *Rabi* | *Summer* |
|---|---|---|---|---|
| | | *Groundnut* | *Goundnut* | *Goundnut* |
| Andhra Pradesh | Sowing | Jun(B)-Jul(E) | Nov(M)-Jan(E) | |
| | Harvesting | Sep(M)-Nov(M) | Feb(E)-May(B) | |
| Assam | Sowing | | Jul(B)-Aug(E) | |
| | Harvesting | | Nov(B)-Dec(E) | |
| Gujarat | Sowing | Jun(B)-Jul(E) | | Jan(B)-Jan(E) |
| | Harvesting | Sep(B)-Nov(E) | | May(B)-Jun(E) |
| Karnataka | Sowing | May(B)-Jul(E) | | Dec(B)-Jan(E) |
| | Harvesting | Sep(B)-Dec(E) | | Mar(B)-Apr(E) |
| Madhya Pradesh | Sowing | Jun(M)-Jul(E) | | |
| | Harvesting | Sep(M)-Dec(E) | | |
| Maharashtra | Sowing | Jul(B)-Aug(E) | | Jan(B)-Feb(E) |
| | Harvesting | Oct(B)-Nov(E) | | Apr(B)-May(E) |
| Manipur | Sowing | May-Jun | | |
| | Harvesting | Oct-Nov | | |
| Rajasthan | Sowing | Jun(B)-Jul(E) | | |
| | Harvesting | Oct(B)-Nov(E) | | |
| Uttar Pradesh | Sowing | Jul(B)-Jul(E) | | |
| | Harvesting | Oct(B)-Nov(E) | | |
| West Bengal | Sowing | Nov(B)-Dec(E) | | Feb(B)-Mar(E) |
| | Harvesting | Feb(B)-Mar(E) | | May(B)-Jun(E) |
| Pondicherry | Sowing | | Dec(B)-Jan(E) | |
| | Harvesting | | Mar(B)-Apr(E) | |
| All India | Sowing | Jun-Jul | Nov-Jan | Dec-Jan |
| | Harvesting | Oct-Nov | Feb-May | Mar-Jun |

| States | Period | Kharif | Rabi | Kharif | Early Kharif |
|---|---|---|---|---|---|
| | | Linseed | Linseed | Castor | Sesamum |
| Andhra Pradesh | Sowing | | | Jun(B)-Sep(M) | |
| | Harvesting | | | Oct(M)-Mar(M) | |
| Karnataka | Sowing | | Sep(B)-Oct(E) | Jun(B)-Jul(E) | |
| | Harvesting | | Jan(B)-Mar(E) | Oct(B)-Feb(E) | |
| Kerala | Sowing | | | | Apr(B)-Aug(E) |
| | Harvesting | | | | Aug(B)-Oct(E) |
| Madhya Pradesh | Sowing | | Oct(B)-Nov(E) | | |
| | Harvesting | | Feb(M)-Apr(M) | | |
| Maharashtra | Sowing | | Oct(B)-Oct(E) | | |
| | Harvesting | | Jan(B)-Feb(E) | | |
| Manipur | Sowing | | Sep-Oct | May-Jun | |
| | Harvesting | | Feb-Mar | Feb-Mar | |
| Meghalaya | Sowing | | | Apr(B)-May(E) | |
| | Harvesting | | | Jan(B)-Feb(E) | |
| Orissa | Sowing | | | Jun-Jul | |
| | Harvesting | | | Dec-Jan | |
| Rajasthan | Sowing | | Oct(B)-Oct(E) | | |
| | Harvesting | | Mar(B)-Mar(E) | | |
| Uttar Pradesh | Sowing | | Oct(B)-Nov(E) | | |
| | Harvesting | | Mar(B)-Mar(E) | | |
| West Bengal | Sowing | Nov(B)-Dec(E) | | Jun(B)-Jul(E) | Jun(B)-Jul(E) |
| | Harvesting | Feb(B)-Mar(E) | | Dec(B)-Jan(E) | Oct(B)-Nov(E) |
| All India | Sowing | Nov(B)-Dec(E) | Sep-Oct | Apr-Jun | Apr-Aug |
| | Harvesting | Feb(B)-Mar(E) | Jan-Mar | Oct-Feb | Oct-Nov |

| *States/UTs* | *Period* | *Kharif* | *Rabi* | *Summer* | *Kharif Soyabean* |
|---|---|---|---|---|---|
| | | *Sesamum/TII* | *Sesamum/TII* | *Sesamum* | |
| Andhra Pradesh | Sowing | May(E)-Jun(E) | Oct(B)-Feb(B) | | |
| | Harvesting | Aug(M)-Sep(E) | Feb(M)-Mar(M) | | |
| Assam | Sowing | Jul(B)-Aug(M) | | | |
| | Harvesting | Oct(M)-Nov(E) | | | |
| Karnataka | Sowing | Apr(B)-Jul(E) | | | Jun(B)-Aug(E) |
| | Harvesting | Jul(B)-Oct(E) | | | Sep(B)-Nov(E) |
| Kerala | Sowing | Aug(B)-Oct(E) | | Dec(B)-Feb(E) | |
| | Harvesting | Dec(B)-Apr(E) | | Mar(B)-May(E) | |
| Madhya Pradesh | Sowing | Jun(M)-Jul(E) | | | Jun(M)-Jul(E) |
| | Harvesting | Sep(M)-Dec(B) | | | Oct(M)-Nov(E) |
| Maharashtra | Sowing | Jul(B)-Jul(E) | | | Jul(B)-Aug(E) |
| | Harvesting | Sep(B)-Oct(E) | | | Oct(B)-Nov(E) |
| Manipur | Sowing | Jun-Jul | | | |
| | Harvesting | Oct-Nov | | | |
| Meghalaya | Sowing | May(B)-Jun(E) | | | |
| | Harvesting | Nov(B)-Dec(E) | | | |
| Rajasthan | Sowing | | | | Jun(B)-Jul(E) |
| | Harvesting | | | | Oct(B)-Nov(E) |
| All India | Sowing | May-Jul | Oct(B)-Oct(E) | Dec(B)-Feb(E) | Jun-Aug |
| | Harvesting | Aug-Oct | Jan(B)-Feb(E) | Mar(B)-May(E) | Sep-Nov |

| *States/UTs* | *Period* | *Kharif* | *Rabi* | *Rabi* |
|---|---|---|---|---|
| | | *Sunflower* | *Sunflower* | *Rape & Mustard* |
| Andhra Pradesh | Sowing | Jun(B)-Aug(B) | Dec(B)-Jan(E) | |
| | Harvesting | Sep(M)-Nov(M) | Mar(E)-Apr(M) | |
| Assam | Sowing | | | Oct(M)-Nov(M) |
| | Harvesting | | | Feb(B)-Mar(E) |
| Karnataka | Sowing | Jun(B)-Jul(E) | Sep(B)-Oct(E) | |
| | Harvesting | Sep(B)-Oct(E) | Dec(B)-Jan(E0 | |
| Madhya Pradesh | Sowing | | | Sep(E)-Nov(E) |
| | Harvesting | | | Dec(M)-Mar(E) |
| Maharashtra | Sowing | Jul(B)-Aug(E) | Oct(B)-Nov(E) | |
| | Harvesting | Oct(B)-Nov(E) | Jan(B)-Feb(E) | |
| Manipur | Sowing | | | Sep-Oct |
| | Harvesting | | | Feb-Mar |
| Meghalaya | Sowing | | | Oct(B)-Nov(E) |
| | Harvesting | | | Jan(B)-Feb(E) |
| Orissa | Sowing | | | Nov-Dec |
| | Harvesting | | | Feb-Mar |
| Punjab | Sowing | | | Oct(B)-Nov(E) |
| | Harvesting | | | Mar(B)-Apr(E) |
| Rajasthan | Sowing | | | Oct(B)-Nov(E) |
| | Harvesting | | | Feb(B)-Mar(E) |
| Uttar Pradesh | Sowing | | | Oct(B)-Nov(E) |
| | Harvesting | | | Feb(B)-Mar(E) |
| West Bengal | Sowing | | Dec(B)-Jan(E) | Oct(B)-Nov(E) |
| | Harvesting | | Apr(B)-Apr(E) | Jan(B)-Mar(E) |
| All India | Sowing | Jun-Aug | Sep-Jun | Sep-Nov |
| | Harvesting | Sep-Oct | Jan-Mar | Feb-Mar |

| *States/UTs* | *Period* | *Kharif* | *Rabi* |
|---|---|---|---|
| | *Oilseeds* | *Oilseeds* | |
| Bihar | Sowing | | Sep(B)-Nov(E) |
| | Harvesting | | Feb(B)-Mar(E) |
| Haryana | Sowing | | Sep(B)-Oct(B) |
| | Harvesting | | Mar(B)-Mar(M) |
| Jammu & Kashmir | Sowing | | Oct(B)-Nov(E) |
| | Harvesting | | Apr(B)-Apr(E) |
| All India | Sowing | Jun(M)-Jul(E) | Sep-Nov |
| | Harvesting | Nov(B)-Dec(E) | Feb-Mar |

| *States/UTs* | *Period* | *Rabi/Kharif* | *Rabi/Kharif* | *Rabi/Summer* |
|---|---|---|---|---|
| | | *Onion* | *Potato* | *Potato* |
| Andhra Pradesh | Sowing | Jul.-Nov | Oct.- Nov | |
| | Harvesting | Mar(B)-May(E) | Jan - Mar | |
| Bihar | Sowing | Oct-Jan | Oct(B)-Feb(E) | |
| | Harvesting | Feb-Apr | Dec-Mar | |
| Karnataka | Sowing | | March | |
| | Harvesting | Aug-Dec | Aug-Nov | |
| Madhya pradesh | Sowing | Oct-Dec | Oct-Dec | |
| | Harvesting | Apr-July | Feb-Jun | |
| Maharashtra | Sowing | Nov-Dec | Nov-Dec | |
| | Harvesting | Mar-May | Feb-May | |
| Rajasthan | Sowing | - | Oct-Dec | |
| | Harvesting | Mar-May | Jan-Mar | |
| Uttar Pradesh | Sowing | Apr(B)-Apr(E) | | Oct(B)-Nov(E) |
| | Harvesting | Jun(B)-Jun(E) | | Jan(B)-Mar(E) |
| West Bengal | Sowing | | Oct(B)-Dec(E) | |
| | Harvesting | | Jan(B)-Mar(E) | |
| All India | Sowing | Apr(B)-Apr(E) | Jul-Dec | Jan-Mar |
| | Harvesting | Jun(B)-Jun(E) | Dec-Feb | Jun-Nov |

* =Sweet Potato.

| *States/UTs* | *Period* | *Kharif* | *Kharif* | *Rabi* |
|---|---|---|---|---|
| | | *Cotton* | *Maize* | *Maize* |
| Andhra Pradesh | Sowing | Jun(E)-Jul(E) | Jun(B)-Jul(M) | Oct(E)-Jan(M) |
| | Harvesting | Dec(E)-Mar(M) | Sep(E)-Oct(E) | Feb(E)-May(B) |
| Bihar | Sowing | | Jun(M)-Jul(B) | Oct(M)-Nov(M) |
| | Harvesting | | Nov(E)-Dec(E) | Feb(B)-Mar(B) |
| Gujarat | Sowing | May(B)-May(E) | Jun(B)-Jul(E) | |
| | Harvesting | Oct(B)-Apr(E) | Sep(B)-Nov(E) | |
| Haryana | Sowing | Apr(B)-Apr(M) | Jul(M)-Aug(B) | |
| | Harvesting | Oct(M)-Nov(M) | Oct(M)-OctE) | |
| Himachal Pradesh | Sowing | | May(M)-Jun(E) | |
| | Harvesting | | Sep(M)-Oct(M) | |
| Karnataka | Sowing | | May(B)-Jun(E) | Sep(B)-Oct(E) |
| | Harvesting | | Sep(B)-Oct(E) | Jan(B)-Mar(E) |
| Kerala | Sowing | Jun(B)-Jul(E) | | |
| | Harvesting | Nov(B)-Dec(E) | | |
| Madhya Pradesh | Sowing | | Jun(M)-Jul(E) | |
| | Harvesting | | Aug(M)-Dec(E) | |
| Maharashtra | Sowing | | Jul(B)-Aug(E) | |
| | Harvesting | | Oct(B)-Nov(E) | |
| Orissa | Sowing | Jun-Jul | Jun-Jul | |
| | Harvesting | Nov-Dec | Sep-Oct | |
| Punjab | Sowing | Apr(B)-Jun(E) | Jun(B)-Jul(E) | |
| | Harvesting | Sep(B)-Dec(E) | Sep(B)-Oct(E) | |
| Rajasthan | Sowing | Apr(B)-May(E) | Jun(B)-Jul(E) | |
| | Harvesting | Nov(B)-Dec(E) | Oct(B)-Nov(E) | |
| Uttar Pradesh | Sowing | Apr(B)-Jun(E) | Jun(B)-Jul(E) | |
| | Harvesting | Sep(B)-Nov(E) | Sep(B)-Sep(E) | |
| West Bengal | Sowing | Oct(B)-Nov(E) | Mar(B)-May(E) | Nov(B)-Nov(E) |
| | Harvesting | Sep(B)-Sep(E) | Jun(B)-Aug(E) | Mar(B)-Mar(E) |
| All India | Sowing | Apr-Jul | Mar-Jul | Sep-Jan |
| | Harvesting | Sep-Dec | Sep-Dec | Jan-May |

| *States/UTs* | *Period* | *Kharif* | *Rabi* | *Kharif* |
|---|---|---|---|---|
| | | *Sugarcane* | *Sugarcane* | *Jute* |
| Andhra Pradesh | Sowing | Dec(E)-Jun(M) | | |
| | Harvesting | Dec(E)_May(M) | | |
| Assam | Sowing | Mar(B)-Apr(E) | | |
| | Harvesting | Dec(B)-Jan(E) | | |
| Haryana | Sowing | Feb(M)-Mar(M) | | |
| | Harvesting | Dec(M)-Apr(E) | | |
| Karnataka | Sowing | Dec(B)-Mar(E)* | | |
| | Harvesting | Aug(B)-May(E) | | |
| Kerala | Sowing | Oct(B)-Feb(E)** | | |
| | Harvesting | Oct(B)-Dec(E) | | |
| Madhya Pradesh | Sowing | Oct(B)-Apr(E) | | |
| | Harvesting | Oct(E)-Mar(E) | | |
| Maharashtra | Sowing | Jul(B)-Aug(E) | | |
| | Harvesting | Oct(B)-Nov(E) | | |
| Manipur | Sowing | | | Feb-Mar |
| | Harvesting | | | Aug-Sep |
| Orissa | Sowing | Feb-May | | May-Jun |
| | Harvesting | Nov-Dec | | Aug-Sep |
| Punjab | Sowing | Feb(B)-Mar(E) | | |
| | Harvesting | Dec(B)-Mar(E) | | |
| Rajasthan | Sowing | Mar(B)-Apr(E) | | |
| | Harvesting | Dec(B)-Mar(E) | | |
| Tamil Nadu | Sowing | Dec(B)-Jan(E)*** | | |
| | Harvesting | Aug(B)-Dec(E) | | |
| Tripura | Sowing | Feb-May | | Mar-May |
| | Harvesting | Dec-Mar | | Aug-Sep |
| West Bengal | Sowing | | | Mar(B)-May(E) |
| | Harvesting | | | Jul(B)-Aug(E) |
| All India | Sowing | Feb(B)-Aug(E) | Jun(B)-Oct(E) | Feb-Jun |
| | Harvesting | Aug(B)-nov(E) | Oct(B)-Jan(E) | Aug-Oct |

* =Annual; **Early *Kharif*; ***Early sugarcane.

# Appendix VI

## *General Seed Rate of Sowing for Important Field Crops**

*Kg/ha.*

| *Crop* | *Variety* | *Hybrid* |
|---|---|---|
| **Cereal Crops** | | |
| Rice | | |
| Direct sowing | 75-100 | - |
| Transplanting | 50-75 | 15 |
| Wheat | 100-125 | - |
| Maize | 20 | 15 |
| Barley | 100 | - |
| *Jowar* | 12-15 | 7.5 |
| *Bajra* | 5 | 5 |
| Ragi | 5 | - |
| **Pulses** | | |
| Gram | 75-100 | - |
| *Tur (Arhar)* | 10 | 10 |
| Green Gram | 20 | - |
| Black Gram | 20 | - |
| Lentil | 25-40 | - |
| Peas | 80-100 | - |
| Cowpea | 20-25 | - |
| Rajmash | 75-100 | - |
| *Moth* | 15-18 | - |
| *Kulthi* | 22-30 | - |
| **Oilseed** | | |
| Groundnut | 150 (With pods)<br>100 (With kernel) | -<br>- |
| Rape/Mustard | 5 | - |
| Soyabean | 75 | - |
| *Til* (Sesamum) | 5 | - |
| Castor | 12.5 | 10 |
| Sunflower | 20 | 7.5 |
| Safflower | 12 | 7.5 |
| Linseed | 25 | - |
| Niger | 25 | - |
| Cotton | 7.5 | 2.5 |
| Jute | 8-10 | - |
| Mesta | 12-5 | - |
| Sunhemp | 25 | - |
| **Foder Crops** | | |
| *Jowar* | 25 | 12 |
| Barseem | 20-25 | - |
| Lucern | 7.5 | - |
| Oats | 62.5 | - |
| Maize | 25 | - |
| **Others** | | |
| Sugarcane | 5500-6000<br>(35000-40000<br>sets with 3-buds) | - |

* The quantity of seed may vary according to variety (duration), seed size, recommended spacing, method of sowing, agro-climatic conditions and cropping system.